# Modern Robotics
## Mechanics, Planning, and Control

This introduction to robotics offers a distinct and unified perspective of the mechanics, planning, and control of robots. It is ideal for self-learning, or for courses, as it assumes only freshman-level physics, ordinary differential equations, linear algebra, and a little bit of computing background. *Modern Robotics*, and its accompanying website `http://modernrobotics.org`:

- Presents state-of-the-art screw-theoretic techniques capturing the most salient physical features of a robot in an intuitive geometrical way;
- Includes numerous exercises at the end of each chapter;
- Has accompanying, freely-downloadable software written to reinforce book concepts;
- Provides virtual hands-on experience with algorithms and controllers using a state-of-the-art robot simulator;
- Offers numerous practice exercises with solutions;
- Provides freely-downloadable video lectures aimed at changing the classroom experience, which students can watch in their own time, whilst class time is focused on collaborative problem-solving;
- Offers instructors the opportunity to design both one- and two-semester courses tailored to emphasize a range of topics, such as kinematics of robots and wheeled vehicles, kinematics and motion planning, mechanics of manipulation, and robot control;
- Supports self-learning and augmented traditional classroom learning through online courses.

KEVIN M. LYNCH received his B.S.E. in electrical engineering from Princeton, New Jersey in 1989, and Ph.D. in robotics from Carnegie Mellon University, Pennsylvania in 1996. He has been a professor of mechanical engineering at Northwestern University, Illinois since 1997 and has held visiting positions at California Institute of Technology, Carnegie Mellon University, Tsukuba University, Japan, and Northeastern University in Shenyang, China. His research focuses on dynamics, motion planning and control for robot manipulation and locomotion; self-organizing multi-agent systems; and physically interacting human–robot systems. A Fellow of the Institute of Electrical and Electronics Engineers (IEEE), he also was the recipient of the IEEE Early Career Award in Robotics and Automation, Northwestern's Professorship of Teaching Excellence, and the Northwestern Teacher of the Year award in engineering. He is Editor-in-Chief of the *IEEE Transactions on Robotics* and former Editor-in-Chief of the *IEEE International Conference on Robotics and Automation*. This is his third book.

FRANK C. PARK received his BS in electrical engineering from MIT and his PhD in applied mathematics from Harvard University. From 1991 to 1995 he was assistant professor of mechanical and aerospace engineering at the University of California, Irvine. Since 1995 he has been professor of mechanical and aerospace engineering at Seoul National University. His research interests are in robot mechanics, planning and control, vision and image processing, and related areas of applied mathematics. He has been an IEEE Robotics and Automation Society Distinguished Lecturer, and received best paper awards for his work on visual tracking and parallel robot design. He has served on the editorial boards of the Springer *Handbook of Robotics*, Springer Advanced Tracts in Robotics (STAR), *Robotica*, and the *ASME Journal of Mechanisms and Robotics*. He has held adjunct faculty positions at the HKUST Robotics Institute, NYU Courant Institute, and the Interactive Computing Department at Georgia Tech. In 2014 he received the Seoul National University Teaching Excellence Award. He is a Fellow of the IEEE, former Editor-in-Chief of the *IEEE Transactions on Robotics*, and developer of the edX courses Robot Mechanics and Control I, II.

# Modern Robotics: Mechanics, Planning, and Control

Kevin M. Lynch
*Northwestern University*

and

Frank C. Park
*Seoul National University*

# CAMBRIDGE
## UNIVERSITY PRESS

University Printing House, Cambridge CB2 8BS, United Kingdom

One Liberty Plaza, 20th Floor, New York, NY 10006, USA

477 Williamstown Road, Port Melbourne, VIC 3207, Australia

314-321, 3rd Floor, Plot 3, Splendor Forum, Jasola District Centre, New Delhi – 110025, India

103 Penang Road, #05-06/07, Visioncrest Commercial, Singapore 238467

Cambridge University Press is part of the University of Cambridge.

It furthers the University's mission by disseminating knowledge in the pursuit of education, learning, and research at the highest international levels of excellence.

www.cambridge.org
Information on this title: www.cambridge.org/9781107156302
DOI: 10.1017/9781316661239

© Kevin M. Lynch and Frank C. Park 2017

First published 2017
Reprinted with additions and corrections, 2019
7th printing 2022

Printed in the United Kingdom by TJ Books Limited, Padstow Cornwall

*A catalogue record for this publication is available from the British Library.*

*Library of Congress Cataloging-in-Publication Data*

ISBN 978-1-107-15630-2 Hardback
ISBN 978-1-316-60984-2 Paperback

# Contents

# Foreword by Roger Brockett

In the 1870s, Felix Klein was developing his far-reaching Erlangen Program, which cemented the relationship between geometry and group theoretic ideas. With Sophus Lie's nearly simultaneous development of a theory of continuous (Lie) groups, important new tools involving infinitesimal analysis based on Lie algebraic ideas became available for the study of a very wide range of geometric problems. Even today, the thinking behind these ideas continues to guide developments in important areas of mathematics. Kinematic mechanisms are, of course, more than just geometry; they need to accelerate, avoid collisions, etc., but first of all they are geometrical objects and the ideas of Klein and Lie apply. The groups of rigid motions in two or three dimensions, as they appear in robotics, are important examples in the work of Klein and Lie.

In the mathematics literature the representation of elements of a Lie group in terms of exponentials usually takes one of two different forms. These are known as exponential coordinates of the first kind and exponential coordinates of the second kind. For the first kind one has $X = e^{(A_1 x_1 + A_2 x_2 \cdots)}$. For the second kind this is replaced by $X = e^{A_1 x_1} e^{A_2 x_2} \cdots$. Up until now, the first choice has found little utility in the study of kinematics whereas the second choice, a special case having already shown up in Euler parametrizations of the orthogonal group, turns out to be remarkably well-suited for the description of open kinematic chains consisting of the concatenation of single degree of freedom links. This is all nicely explained in Chapter 4 of this book. Together with the fact that $Pe^A P^{-1} = e^{PAP^{-1}}$, the second form allows one to express a wide variety of kinematic problems very succinctly. From a historical perspective, the use of the product of exponentials to represent robotic movement, as the authors have done here, can be seen as illustrating the practical utility of the 150-year-old ideas of the geometers Klein and Lie.

In 1983 I was invited to speak at the triennial Mathematical Theory of Networks and Systems Conference in Beer Sheva, Israel, and after a little thought I decided to try to explain something about what my recent experiences had taught me. By then I had some experience in teaching a robotics course that discussed kinematics, including the use of the product of exponentials representation of kinematic chains. From the 1960s onward $e^{At}$ had played a central role in system theory and signal processing, so at this conference a familiarity, even an affection, for the matrix exponential could be counted on. Given this, it was natural for me to pick something $e^{Ax}$-related for the talk. Although I had no reason to think that there would be many in the audience with an interest in kinematics, I still hoped that I could say something interesting and maybe even inspire further developments. The result was the paper referred to in the preface that follows.

In this book, Frank and Kevin have provided a wonderfully clear and patient explanation of their subject. They translate the foundation laid out by Klein and Lie 150 years ago to the modern practice of robotics, at a level appropriate for undergraduate engineers. After an elegant discussion of the fundamental properties of configuration spaces, they introduce the Lie group representations of rigid-body configurations, and the corresponding representations of velocities and forces, used throughout the book. This consistent perspective is carried through foundational robotics topics including the forward, inverse, and differential kinematics of open chains, robot dynamics, trajectory generation, robot control, and more specialized topics such as the kinematics of closed chains, motion planning, robot manipulation, planning and control for wheeled mobile robots, and the control of mobile manipulators.

I am confident that this book will be a valuable resource for a generation of students and practitioners of robotics.

Roger Brockett
Cambridge, Massachusetts, USA
November 2016

# Foreword by Matthew Mason

Robotics is about turning ideas into action. Somehow, robots turn abstract goals into physical action: sending power to motors, monitoring motions, and guiding things towards the goal. Every human can perform this trick, but it is nonetheless so intriguing that it has captivated philosophers and scientists, including Descartes and many others.

What is the secret? Did some roboticist have a eureka moment? Did some pair of teenage entrepreneurs hit on the key idea in their garage? To the contrary, it is not a single idea. It is a substantial body of scientific and engineering results, accumulated over centuries. It draws primarily from mathematics, physics, mechanical engineering, electrical engineering and computer science, but also from philosophy, psychology, biology and other fields.

Robotics is the gathering place of these ideas. Robotics provides motivation. Robotics tests ideas and steers continuing research. Finally, robotics is the proof. Observing a robot's behavior is the nearly compelling proof that machines can be aware of their surroundings, can develop meaningful goals, and can act effectively to accomplish those goals. The same principles apply to a thermostat or a flyball governor, but few are persuaded by watching a thermostat. Nearly all are persuaded by watching a robot soccer team.

The heart of robotics is motion – controlled programmable motion – which brings us to the present text. *Modern Robotics* imparts the most important insights of robotics: the nature of motion, the motions available to rigid bodies, the use of kinematic constraint to organize motions, the mechanisms that enable general programmable motion, the static and dynamic character of mechanisms, and the challenges and approaches to control, programming, and planning motions. *Modern Robotics* presents this material with a clarity that makes it accessible to undergraduate students. It is distinguished from other undergraduate texts in two important ways.

First, in addressing rigid-body motion, *Modern Robotics* presents not only the classical geometrical underpinnings and representations, but also their expression using modern matrix exponentials, and the connection to Lie algebras. The rewards to the students are two-fold: a deeper understanding of motion, and better practical tools.

Second, *Modern Robotics* goes beyond a focus on robot mechanisms to address the interaction with objects in the surrounding world. When robots make contact with the real world, the result is an *ad hoc* kinematic mechanism, with associated statics and dynamics. The mechanism includes kinematic loops, unactuated joints, and nonholonomic constraints, all of which will be familiar concepts to students of *Modern Robotics*.

Even if this is the only robotics course students take, it will enable them to analyze, control, and program a wide range of physical systems. With its introduction to the mechanics of physical interaction, *Modern Robotics* is also an excellent beginning for the student who intends to continue with advanced courses or with original research in robotics.

Matthew T. Mason
Pittsburgh, Pennsylvania, USA
November 2016

# Preface

It was at the IEEE International Conference on Robotics and Automation in Pasadena in 2008 when, over a beer, we decided to write an undergraduate textbook on robotics. Since 1996, Frank had been teaching robot kinematics to Seoul National University undergraduates using his own lecture notes; by 2008 these notes had evolved to the kernel around which this book was written. Kevin had been teaching his introductory robotics class at Northwestern University from his own set of notes, with content drawn from an eclectic collection of papers and books.

We believe that there is a distinct and unifying perspective to the mechanics, planning, and control for robots that is lost if these subjects are studied independently or as part of other more traditional subjects. At the 2008 meeting, we noted the lack of a textbook that (a) treated these topics in a unified way, with plenty of exercises and figures, and (b), most importantly, was written at a level appropriate for a first robotics course for undergraduates with only freshman-level physics, ordinary differential equations, linear algebra, and a little bit of computing background. We decided that the only sensible recourse was to write such a book ourselves. (We didn't know then that it would take us more than eight years to finish the project!)

A second motivation for this book, and one that we believe sets it apart from other introductory treatments on robotics, is its emphasis on modern geometric techniques. Often the most salient physical features of a robot are best captured by a geometric description. The advantages of the geometric approach have been recognized for quite some time by practitioners of classical screw theory. What has made these tools largely inaccessible to undergraduates – the primary target audience for this book – is that they require an entirely new language of concepts and constructs (screws, twists, wrenches, reciprocity, transversality, conjugacy, etc.), and their often obscure rules for manipulation and transformation. However, the mostly algebraic alternatives to screw theory often mean that students end up buried in the details of calculation, losing the simple and elegant geometric interpretation that lies at the heart of what they are calculating.

The breakthrough that made the techniques of classical screw theory accessible to a more general audience arrived in the early 1980s, when Roger Brockett showed how to describe kinematic chains mathematically in terms of the Lie group structure of rigid-body motions (Brockett, 1983b). This discovery allowed one, among other things, to re-invent screw theory simply by appealing to basic linear algebra and linear differential equations. With this "modern screw theory" the powerful tools of modern differential geometry can be brought to bear on a wide-ranging collection of robotics problems, some of which we explore here, and others of which are covered in the excellent but more advanced graduate textbook by Murray et al. (1994).

As the title indicates, this book covers what we feel to be the fundamentals of robot mechanics, together with the basics of planning and control. A thorough treatment of all the chapters would likely take two semesters, particularly when coupled with programming assignments or experiments with robots. The contents of Chapters 2–6 constitute the minimum essentials, and these topics should probably be covered in sequence.

The instructor can then selectively choose content from the remaining chapters. At Seoul National University, the undergraduate course M2794.0027 Introduction to Robotics covers, in one semester, Chapters 2–7 and parts of Chapters 10–12. At Northwestern, ME 449 Robotic Manipulation covers, in an 11-week quarter, Chapters 2–6 and 8, and then touches on different topics in Chapters 9–13 depending on the interests of the students and instructor. A course focusing on the kinematics of robot arms and wheeled vehicles could cover Chapters 2–7 and 13, while one on kinematics and motion planning could additionally include Chapters 9 and 10. A course on the mechanics of manipulation would cover Chapters 2–6, 8, and 12, while another on robot control would cover Chapters 2–6, 8, 9, and 11. If the instructor prefers to avoid dynamics (Chapter 8), the basics of robot control (Chapters 11 and 13) can be covered by assuming that the velocity at each actuator is controlled, not the forces and torques. A course focusing only on motion planning could cover Chapters 2 and 3, Chapter 10 in depth (possibly supplemented by research papers or other references cited in that chapter), and Chapter 13.

To help the instructor choose which topics to teach and to help the student keep track of what she has learned, we have included summaries at the ends of chapters and a summary of important notation and formulas used throughout the book in Appendix A. For those whose primary interest in this text is as an introductory reference, we have attempted to provide a reasonably comprehensive, though by no means exhaustive, set of references and bibliographic notes at the end of each chapter. Some of the exercises provided at the end of each chapter extend the basic results covered in the book and, for those who wish to probe further, these should be of interest in their own right. Some of the more advanced material in the book can be used to support independent study projects.

Another important component of the book is the software, which is written to reinforce the concepts in the book and to make the formulas operational. The software was developed primarily by Kevin's ME 449 students at Northwestern and is freely downloadable from http://modernrobotics.org. Video lectures that accompany the textbook are also available at the website. The intention of the video content is to help the instructor to "flip" the classroom: students watch brief lectures on their own time, rewinding as needed, and class time is focused more on the collaborative problem-solving that has traditionally occurred between classes. This way, the professor is present when the students are applying the material and discovering the gaps in their understanding; this creates the opportunity for interactive mini-lectures addressing the concepts that need most reinforcing. We believe that the added value of the professor is greatest in this interactive role, not in delivering a lecture in the same way that it was delivered the previous year. This approach has worked well for Kevin's introduction to mechatronics course, `http://nu32.org`.

The video content is generated using the Lightboard, `http://lightboard.info`, created by Michael Peshkin at Northwestern University. We thank him for sharing this convenient and effective tool for creating instructional videos.

We have also found the V-REP robot simulation software to be a valuable supplement to the book and its software. This simulation software allows students to explore interactively the kinematics of robot arms and mobile manipulators and to animate trajectories that are the result of exercises on kinematics, dynamics, and control.

While this book presents our own perspective on how to introduce the fundamental topics in first courses on robot mechanics, planning, and control, we acknowledge the excellent textbooks that already exist and that have served our field well. Among these, we would like to mention as particularly influential the books by Murray et al. (1994), Craig (2004), Spong et al. (2005), Siciliano et al. (2009), Mason (2001), and Corke (2017), and the motion planning books by Latombe (1991), LaValle (2006), and Choset et al. (2005). In addition, the *Handbook of Robotics* (Siciliano and Khatib, 2016), with a multimedia extension edited by Kröger (`http://handbookofrobotics.org`), is a landmark in our field, collecting the perspectives of hundreds of leading researchers on a huge variety of topics relevant to modern robotics.

It is our pleasure to acknowledge the many people who have been the sources of help and inspiration in writing this book. In particular, we would like to thank our Ph.D. advisors, Roger Brockett and Matt Mason. Brockett laid down much of the foundation for the geometric approach to robotics employed in this book. Mason's pioneering contributions to analysis and planning for manipulation form a cornerstone of modern robotics. We also thank the many students who provided feedback on various versions of this material, in M2794.0027 at Seoul National University and in ME 449 at Northwestern University. In particular, Frank would like to thank Seunghyeon Kim, Keunjun Choi, Jisoo Hong, Jinkyu Kim, Youngsuk Hong, Wooyoung Kim, Cheongjae Jang, Taeyoon Lee, Soocheol Noh, Kyumin Park, Seongjae Jeong, Sukho Yoon, Jaewoon Kwen, Jinhyuk Park, and Jihoon Song, as well as Jim Bobrow and Scott Ploen from his time at UC Irvine. Kevin would like to thank Matt Elwin, Sherif Mostafa, Nelson Rosa, Jarvis Schultz, Jian Shi, Mikhail Todes, Huan Weng, and Zack Woodruff.

We would also like to thank Susan Parkinson and David Tranah at Cambridge

University Press for their diligence and expertise in matters to do with copy-editing, proof-reading, and layout.

Finally, and most importantly, we thank our wives and families, for putting up with our late nights and our general unavailability, and for supporting us as we made the final push to finish the book. Without the love and support of Hyunmee, Shiyeon, and Soonkyu (Frank's family) and Yuko, Erin, and Patrick (Kevin's family), this book would not exist. We dedicate this book to them.

Kevin M. Lynch
Evanston, Illinois, USA

Frank C. Park
Seoul, Korea

November 2016

**Publication note.**

The authors consider themselves to be equal contributors to this book. The author order is alphabetical.

**Notes on the second printing.**

Readers are encouraged to consult the companion website

http://modernrobotics.org

for more information on the Modern Robotics software library, videos, online courses, robot simulations, practice problems with solutions, errata, a linear algebra refresher chapter, and more.

Various minor corrections have been made in this second printing. Thanks to the following people who provided corrections, starting from the preliminary version of the book posted online in October, 2016:

H. Andy Nam, Eric Lee, Yuchen Rao, Chainatee Tanakulrongson, Mengjiao Hong, Kevin Cheng, Jens Lundell, Elton Cheng, Michael Young, Jarvis Schultz, Logan Springgate, Sofya Akhmametyeva, Aykut Onol, Josh Holcomb, Yue Chen, Mark Shi, AJ Ibraheem, Yalun Wen, Seongjae Jeong, Josh Mehling, Felix Wang, Drew Warren, Chris Miller, Clemens Eppner, Zack Woodruff, Jian Shi, Jixiang Zhang, Shachar Liberman, Will Wu, Dirk Boysen, Awe Wang, Ville Kyrki, John Troll, Andrew Taylor, and Nikhil Bakshi.

# 1 Preview

As an academic discipline, robotics is a relatively young field with highly ambitious goals, the ultimate one being the creation of machines that can behave and think like humans. This attempt to create intelligent machines naturally leads us first to examine ourselves – to ask, for example, why our bodies are designed the way they are, how our limbs are coordinated, and how we learn and perform complex tasks. The sense that the fundamental questions in robotics are ultimately questions about ourselves is part of what makes robotics such a fascinating and engaging endeavor.

Our focus in this book is on mechanics, planning, and control for **robot mechanisms**. Robot arms are one familiar example. So are wheeled vehicles, as are robot arms mounted on wheeled vehicles. Basically, a mechanism is constructed by connecting rigid bodies, called **links**, together by means of **joints**, so that relative motion between adjacent links becomes possible. **Actuation** of the joints, typically by electric motors, then causes the robot to move and exert forces in desired ways.

The links of a robot mechanism can be arranged in serial fashion, like the familiar open-chain arm shown in Figure 1.1(a). Robot mechanisms can also have links that form closed loops, such as the Stewart–Gough platform shown in Figure 1.1(b). In the case of an open chain, all the joints are actuated, while in the case of mechanisms with closed loops, only a subset of the joints may be actuated.

Let us examine more closely the current technology behind robot mechanisms. The links are moved by actuators, which typically are electrically driven (e.g., by DC or AC motors, stepper motors, or shape memory alloys) but can also be driven by pneumatic or hydraulic cylinders. In the case of rotating electric motors, these would ideally be lightweight, operate at relatively low rotational speeds (e.g., in the range of hundreds of RPM), and be able to generate large forces and torques. Since most currently available motors operate at low torques and at up to thousands of RPM, speed reduction and torque amplification are required. Examples of such transmissions or transformers include gears, cable drives, belts and pulleys, and chains and sprockets. These speed-reduction devices should have zero or low slippage and **backlash** (defined as the amount of rotation available at the output of the speed-reduction device without motion at

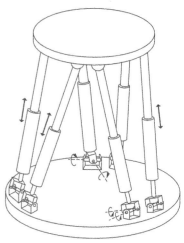

(a) An open-chain industrial manipulator, visualized in V-REP (Rohmer et al., 2013).

(b) Stewart–Gough platform. Closed loops are formed from the base platform, through the legs, through the top platform, and through the legs back to the base platform.

**Figure 1.1** Open-chain and closed-chain robot mechanisms.

the input). Brakes may also be attached to stop the robot quickly or to maintain a stationary posture.

Robots are also equipped with sensors to measure the motion at the joints. For both revolute and prismatic joints, encoders, potentiometers, or resolvers measure the displacement and sometimes tachometers are used to measure velocity. Forces and torques at the joints or at the end-effector of the robot can be measured using various types of force–torque sensors. Additional sensors may be used to help localize objects or the robot itself, such as vision-only cameras, RGB-D cameras which measure the color (RGB) and depth (D) to each pixel, laser range finders, and various types of acoustic sensor.

The study of robotics often includes artificial intelligence and computer perception, but an essential feature of any robot is that it moves in the physical world. Therefore, this book, which is intended to support a first course in robotics for undergraduates and graduate students, focuses on mechanics, motion planning, and control of robot mechanisms.

In the rest of this chapter we provide a preview of the rest of the book.

## Chapter 2: Configuration Space

As mentioned above, at its most basic level a robot consists of rigid bodies connected by joints, with the joints driven by actuators. In practice the links may not be completely rigid, and the joints may be affected by factors such as elasticity, backlash, friction, and hysteresis. In this book we ignore these effects for the most part and assume that all links are rigid.

With this assumption, Chapter 2 focuses on representing the **configuration** of a robot system, which is a specification of the position of every point of the robot. Since the robot consists of a collection of rigid bodies connected by joints, our study begins with understanding the configuration of a rigid body. We see that the configuration of a rigid body in the plane can be described using three variables (two for the position and one for the orientation) and the configuration of a rigid body in space can be described using six variables (three for the position and three for the orientation). The number of variables is the number of **degrees of freedom** (dof) of the rigid body. It is also the dimension of the **configuration space**, the space of all configurations of the body.

The dof of a robot, and hence the dimension of its configuration space, is the sum of the dof of its rigid bodies minus the number of constraints on the motion of those rigid bodies provided by the joints. For example, the two most popular joints, revolute (rotational) and prismatic (translational) joints, allow only one motion freedom between the two bodies they connect. Therefore a revolute or prismatic joint can be thought of as providing five constraints on the motion of one spatial rigid body relative to another. Knowing the dof of a rigid body and the number of constraints provided by joints, we can derive **Grübler's formula** for calculating the dof of general robot mechanisms. For **open-chain** robots such as the industrial manipulator of Figure 1.1(a), each joint is independently actuated and the dof is simply the sum of the freedoms provided by each joint. For **closed chains** like the Stewart–Gough platform in Figure 1.1(b), Grübler's formula is a convenient way to calculate the dof. Unlike open-chain robots, some joints of closed chains are not actuated.

Apart from calculating the dof, other configuration space concepts of interest include the **topology** (or "shape") of the configuration space and its **representation**. Two configuration spaces of the same dimension may have different shapes, just like a two-dimensional plane has a different shape from the two-dimensional surface of a sphere. These differences become important when determining how to represent the space. The surface of a unit sphere, for example, could be represented using a minimal number of coordinates, such as latitude and longitude, or it could be represented by three numbers $(x, y, z)$ subject to the constraint $x^2 + y^2 + z^2 = 1$. The former is an **explicit parametrization** of the space and the latter is an **implicit parametrization** of the space. Each type of representation has its advantages, but in this book we will use implicit representations of configurations of rigid bodies.

A robot arm is typically equipped with a hand or gripper, more generally called an **end-effector**, which interacts with objects in the surrounding world. To accomplish a task such as picking up an object, we are concerned with the configuration of a reference frame rigidly attached to the end-effector, and not necessarily the configuration of the entire arm. We call the space of positions and orientations of the end-effector frame the **task space** and note that there is not a one-to-one mapping between the robot's configuration space and the task space. The **workspace** is defined to be the subset of the task space that the end-effector frame can reach.

## Chapter 3: Rigid-Body Motions

This chapter addresses the problem of how to describe mathematically the motion of a rigid body moving in three-dimensional physical space. One convenient way is to attach a reference frame to the rigid body and to develop a way to quantitatively describe the frame's position and orientation as it moves. As a first step, we introduce a $3 \times 3$ matrix representation for describing a frame's orientation; such a matrix is referred to as a **rotation matrix**.

A rotation matrix is parametrized by three independent coordinates. The most natural and intuitive way to visualize a rotation matrix is in terms of its **exponential coordinate** representation. That is, given a rotation matrix $R$, there exists some unit vector $\hat{\omega} \in \mathbb{R}^3$ and angle $\theta \in [0, \pi]$ such that the rotation matrix can be obtained by rotating the identity frame (that is, the frame corresponding to the identity matrix) about $\hat{\omega}$ by $\theta$. The exponential coordinates are defined as $\omega = \hat{\omega}\theta \in \mathbb{R}^3$, which is a three-parameter representation. There are several other well-known coordinate representations, e.g., Euler angles, Cayley–Rodrigues parameters, and unit quaternions, which are discussed in Appendix B.

Another reason for focusing on the exponential description of rotations is that they lead directly to the exponential description of rigid-body motions. The latter can be viewed as a modern geometric interpretation of classical screw theory. Keeping the classical terminology as much as possible, we cover in detail the linear algebraic constructs of screw theory, including the unified description of linear and angular velocities as six-dimensional **twists** (also known as **spatial velocities**), and an analogous description of three-dimensional forces and moments as six-dimensional **wrenches** (also known as **spatial forces**).

## Chapter 4: Forward Kinematics

For an open chain, the position and orientation of the end-effector are uniquely determined from the joint positions. The **forward kinematics** problem is to find the position and orientation of the reference frame attached to the end-effector given the set of joint positions. In this chapter we present the **product of exponentials (PoE)** formula describing the forward kinematics of open chains. As the name implies, the PoE formula is directly derived from the exponential coordinate representation for rigid-body motions. Aside from providing an intuitive and easily visualizable interpretation of the exponential coordinates as the twists of the joint axes, the PoE formula offers other advantages, like eliminating the need for link frames (only the base frame and end-effector frame are required, and these can be chosen arbitrarily).

In Appendix C we also present the Denavit–Hartenberg (D–H) representation for forward kinematics. The D–H representation uses fewer parameters but requires that reference frames be attached to each link following special rules of assignment, which can be cumbersome. Details of the transformation from the D–H to the PoE representation are also provided in Appendix C.

## Chapter 5: Velocity Kinematics and Statics

Velocity kinematics refers to the relationship between the joint linear and angular velocities and those of the end-effector frame. Central to velocity kinematics is the **Jacobian** of the forward kinematics. By multiplying the vector of joint-velocity rates by this configuration-dependent matrix, the twist of the end-effector frame can be obtained for any given robot configuration. **Kinematic singularities**, which are configurations in which the end-effector frame loses the ability to move or rotate in one or more directions, correspond to those configurations at which the Jacobian matrix fails to have maximal rank. The **manipulability ellipsoid**, whose shape indicates the ease with which the robot can move in various directions, is also derived from the Jacobian.

Finally, the Jacobian is also central to static force analysis. In static equilibrium settings, the Jacobian is used to determine what forces and torques need to be exerted at the joints in order for the end-effector to apply a desired wrench.

The definition of the Jacobian depends on the representation of the end-effector velocity, and our preferred representation of the end-effector velocity is as a six-dimensional twist. We touch briefly on other representations of the end-effector velocity and their corresponding Jacobians.

## Chapter 6: Inverse Kinematics

The **inverse kinematics** problem is to determine the set of joint positions that achieves a desired end-effector configuration. For open-chain robots, the inverse kinematics is in general more involved than the forward kinematics: for a given set of joint positions there usually exists a unique end-effector position and orientation but, for a particular end-effector position and orientation, there may exist multiple solutions to the joint positions, or no solution at all.

In this chapter we first examine a popular class of six-dof open-chain structures whose inverse kinematics admits a closed-form analytic solution. Iterative numerical algorithms are then derived for solving the inverse kinematics of general open chains by taking advantage of the inverse of the Jacobian. If the open-chain robot is **kinematically redundant**, meaning that it has more joints than the dimension of the task space, then we use the pseudoinverse of the Jacobian.

## Chapter 7: Kinematics of Closed Chains

While open chains have unique forward kinematics solutions, closed chains often have multiple forward kinematics solutions, and sometimes even multiple solutions for the inverse kinematics as well. Also, because closed chains possess both actuated and passive joints, the kinematic singularity analysis of closed chains presents subtleties not encountered in open chains. In this chapter we study the basic concepts and tools for the kinematic analysis of closed chains. We begin with a detailed case study of mechanisms such as the planar five-bar linkage and the Stewart–Gough platform. These results are then generalized into a systematic methodology for the kinematic analysis of more general closed chains.

## Chapter 8: Dynamics of Open Chains

Dynamics is the study of motion taking into account the forces and torques that cause it. In this chapter we study the dynamics of open-chain robots. In analogy to the notions of a robot's forward and inverse kinematics, the **forward dynamics** problem is to determine the resulting joint accelerations for a given set of joint forces and torques. The **inverse dynamics** problem is to determine the input joint torques and forces needed for desired joint accelerations. The dynamic equations relating the forces and torques to the motion of the robot's links are given by a set of second-order ordinary differential equations.

The dynamics for an open-chain robot can be derived using one of two approaches. In the Lagrangian approach, first a set of coordinates – referred to as generalized coordinates in the classical dynamics literature – is chosen to parametrize the configuration space. The sum of the potential and kinetic energies of the robot's links are then expressed in terms of the generalized coordinates and their time derivatives. These are then substituted into the **Euler–Lagrange equations**, which then lead to a set of second-order differential equations for the dynamics, expressed in the chosen coordinates for the configuration space.

The **Newton–Euler** approach builds on the generalization of $f = \mathfrak{m}a$, i.e., the equations governing the acceleration of a rigid body given the wrench acting on it. Given the joint variables and their time derivatives, the Newton–Euler approach to inverse dynamics is: to propagate the link velocities and accelerations outward from the proximal link to the distal link, in order to determine the velocity and acceleration of each link; to use the equations of motion for a rigid body to calculate the wrench (and therefore the joint force or torque) that must be acting on the outermost link; and to proceed along the links back toward the base of the robot, calculating the joint forces or torques needed to create the motion of each link and to support the wrench transmitted to the distal links. Because of the open-chain structure, the dynamics can be formulated recursively.

In this chapter we examine both approaches to deriving a robot's dynamic equations. Recursive algorithms for both the forward and inverse dynamics, as well as analytical formulations of the dynamic equations, are presented.

## Chapter 9: Trajectory Generation

What sets a robot apart from an automated machine is that it should be easily reprogrammable for different tasks. Different tasks require different motions, and it would be unreasonable to expect the user to specify the entire time-history of each joint for every task; clearly it would be desirable for the robot's control computer to "fill in the details" from a small set of task input data.

This chapter is concerned with the automatic generation of joint trajectories from this set of task input data. Formally, a trajectory consists of a **path**, which is a purely geometric description of the sequence of configurations achieved by a robot, and a **time scaling**, which specifies the times at which those configurations are reached.

Often the input task data is given in the form of an ordered set of joint values, called control points, together with a corresponding set of control times. On the basis of this data the trajectory generation algorithm produces a trajectory for each joint which satisfies various user-supplied conditions. In this chapter we focus on three cases: (i) point-to-point straight-line trajectories in both joint space and task space; (ii) smooth trajectories passing through a sequence of timed "via points"; and (iii) time-optimal trajectories along specified paths, subject to the robot's dynamics and actuator limits. Finding paths that avoid collisions is the subject of the next chapter on motion planning.

## Chapter 10: Motion Planning

This chapter addresses the problem of finding a collision-free motion for a robot through a cluttered workspace, while avoiding joint limits, actuator limits, and other physical constraints imposed on the robot. The **path planning** problem is a subproblem of the general motion planning problem that is concerned with finding a collision-free path between a start and goal configuration, usually without regard to the dynamics, the duration of the motion, or other constraints on the motion or control inputs.

There is no single planner applicable to all motion planning problems. In this chapter we consider three basic approaches: grid-based methods, sampling methods, and methods based on virtual potential fields.

## Chapter 11: Robot Control

A robot arm can exhibit a number of different behaviors depending on the task and its environment. It can act as a source of programmed motions for tasks such as moving an object from one place to another, or tracing a trajectory for manufacturing applications. It can act as a source of forces, for example when grinding or polishing a workpiece. In tasks such as writing on a chalkboard, it must control forces in some directions (the force pressing the chalk against the board) and motions in other directions (the motion in the plane of the board).

In certain applications, e.g., haptic displays, we may want the robot to act like a programmable spring, damper, or mass, by controlling its position, velocity, or acceleration in response to forces applied to it.

In each of these cases, it is the job of the robot controller to convert the task specification to forces and torques at the actuators. Control strategies to achieve the behaviors described above are known as **motion** (or **position**) **control**, **force control**, **hybrid motion–force control**, and **impedance control**. Which of these behaviors is appropriate depends on both the task and the environment. For example, a force-control goal makes sense when the end-effector is in contact with something, but not when it is moving in free space. We also have a fundamental constraint imposed by the mechanics, irrespective of the environment: the robot cannot independently control both motions and forces in the same direction. If the robot imposes a motion then the environment determines the force, and vice versa.

Most robots are driven by actuators that apply a force or torque to each joint. Hence, precisely controlling a robot requires an understanding of the relationship between the joint forces and torques and the motion of the robot; this is the domain of dynamics. Even for simple robots, however, the dynamic equations are complex and dependent on a precise knowledge of the mass and inertia of each link, which may not be readily available. Even if it were, the dynamic equations would still not reflect physical phenomena such as friction, elasticity, backlash, and hysteresis.

Most practical control schemes compensate for these uncertainties by using **feedback control**. After examining the performance limits of feedback control without a dynamic model of the robot, we study motion control algorithms, such as **computed torque control**, that combine approximate dynamic modeling with feedback control. The basic lessons learned for robot motion control are then applied to force control, hybrid motion–force control, and impedance control.

## Chapter 12: Grasping and Manipulation

The focus of earlier chapters is on characterizing, planning, and controlling the motion of the robot itself. To do useful work, the robot must be capable of manipulating objects in its environment. In this chapter we model the contact between the robot and an object, specifically the constraints on the object motion imposed by a contact and the forces that can be transmitted through a frictional contact. With these models we study the problem of choosing contacts to immobilize an object by **form closure** and **force closure** grasping. We also apply contact modeling to manipulation problems other than grasping, such as pushing an object, carrying an object dynamically, and testing the stability of a structure.

## Chapter 13: Wheeled Mobile Robots

The final chapter addresses the kinematics, motion planning, and control of wheeled mobile robots and of wheeled mobile robots equipped with robot arms. A mobile robot can use specially designed **omniwheels** or **mecanum wheels** to achieve omnidirectional motion, including spinning in place or translating in any direction. Many mobile bases, however, such as cars and differential-drive robots, use more typical wheels, which do not slip sideways. These no-slip constraints are fundamentally different from the loop-closure constraints found in closed chains; the latter are **holonomic**, meaning that they are configuration constraints, while the former are **nonholonomic**, meaning that the velocity constraints cannot be integrated to become equivalent configuration constraints.

Because of the different properties of omnidirectional mobile robots versus nonholonomic mobile robots, we consider their kinematic modeling, motion planning, and control separately. In particular, the motion planning and control of nonholonomic mobile robots is more challenging than for omnidirectional mobile robots.

Once we have derived their kinematic models, we show that the **odometry** problem – the estimation of the chassis configuration based on wheel encoder data – can be solved in the same way for both types of mobile robots. Similarly, for mobile manipulators consisting of a wheeled base and a robot arm, we show that feedback control for **mobile manipulation** (controlling the motion of the end-effector using the arm joints and wheels) is the same for both types of mobile robots. The fundamental object in mobile manipulation is the Jacobian mapping joint rates and wheel velocities to end-effector twists.

Each chapter concludes with a summary of important concepts from the chapter, and Appendix A compiles some of the most used equations into a handy reference. Videos supporting the book can be found at the book's website, http://modernrobotics.org. Some chapters have associated software, downloadable from the website. The software is meant to be neither maximally robust nor efficient but to be readable and to reinforce the concepts in the book. You are encouraged to read the software, not just use it, to cement your understanding of the material. Each function contains a sample usage in the comments. The software package may grow over time, but the core functions are documented in the chapters themselves.

# 2 Configuration Space

A robot is mechanically constructed by connecting a set of bodies, called **links**, to each other using various types of **joints**. **Actuators**, such as electric motors, deliver forces or torques that cause the robot's links to move. Usually an **end-effector**, such as a gripper or hand for grasping and manipulating objects, is attached to a specific link. All the robots considered in this book have links that can be modeled as rigid bodies.

Perhaps the most fundamental question one can ask about a robot is, where is it? The answer is given by the robot's **configuration**: a specification of the positions of all points of the robot. Since the robot's links are rigid and of a known shape,[1] only a few numbers are needed to represent its configuration. For example, the configuration of a door can be represented by a single number, the angle $\theta$ about its hinge. The configuration of a point on a plane can be described by two coordinates, $(x, y)$. The configuration of a coin lying heads up on a flat table can be described by three coordinates: two coordinates $(x, y)$ that specify the location of a particular point on the coin, and one coordinate $(\theta)$ that specifies the coin's orientation. (See Figure 2.1).

The above coordinates all take values over a continuous range of real numbers. The number of **degrees of freedom (dof)** of a robot is the smallest number of real-valued coordinates needed to represent its configuration. In the example above, the door has one degree of freedom. The coin lying heads up on a table has three degrees of freedom. Even if the coin could lie either heads up or tails up, its configuration space still would have only three degrees of freedom; a fourth variable, representing which side of the coin faces up, takes values in the discrete set {heads, tails}, and not over a continuous range of real values like the other three coordinates.

**Definition 2.1** The **configuration** of a robot is a complete specification of the position of every point of the robot. The minimum number $n$ of real-valued coordinates needed to represent the configuration is the number of **degrees of freedom (dof)** of the robot. The $n$-dimensional space containing all possible configurations of the robot is called the **configuration space (C-space)**. The configuration of a robot is represented by a point in its C-space.

In this chapter we study the C-space and degrees of freedom of general robots.

---

[1] Compare with trying to represent the configuration of a soft object like a pillow.

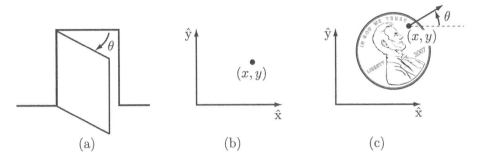

**Figure 2.1** (a) The configuration of a door is described by the angle $\theta$. (b) The configuration of a point in a plane is described by coordinates $(x, y)$. (c) The configuration of a coin on a table is described by $(x, y, \theta)$, where $\theta$ defines the direction in which Abraham Lincoln is looking.

Since our robots are constructed from rigid links, we examine first the degrees of freedom of a single rigid body, and then the degrees of freedom of general multi-link robots. Next we study the shape (or topology) and geometry of C-spaces and their mathematical representation. The chapter concludes with a discussion of the C-space of a robot's end-effector, its **task space**. In the following chapter we study in more detail the mathematical representation of the C-space of a single rigid body.

## Degrees of Freedom of a Rigid Body

Continuing with the example of the coin lying on the table, choose three points $A$, $B$, and $C$ on the coin (Figure 2.2(a)). Once a coordinate frame $\hat{x}$–$\hat{y}$ is attached to the plane,[2] the positions of these points in the plane are written $(x_A, y_A)$, $(x_B, y_B)$, and $(x_C, y_C)$. If the points could be placed independently anywhere in the plane, the coin would have six degrees of freedom – two for each of the three points. But, according to the definition of a rigid body, the distance between point $A$ and point $B$, denoted $d(A, B)$, is always constant regardless of where the coin is. Similarly, the distances $d(B, C)$ and $d(A, C)$ must be constant. The following equality constraints on the coordinates $(x_A, y_A)$, $(x_B, y_B)$, and $(x_C, y_C)$ must therefore always be satisfied:

$$d(A, B) = \sqrt{(x_A - x_B)^2 + (y_A - y_B)^2} = d_{AB},$$
$$d(B, C) = \sqrt{(x_B - x_C)^2 + (y_B - y_C)^2} = d_{BC},$$
$$d(A, C) = \sqrt{(x_A - x_C)^2 + (y_A - y_C)^2} = d_{AC}.$$

To determine the number of degrees of freedom of the coin on the table, first

---

[2] The unit axes of coordinate frames are written with a hat, indicating they are unit vectors, and in a non-italic font, e.g., $\hat{x}$, $\hat{y}$, and $\hat{z}$.

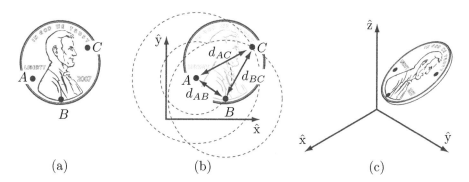

(a)                    (b)                    (c)

**Figure 2.2** (a) Choosing three points fixed to the coin. (b) Once the location of $A$ is chosen, $B$ must lie on a circle of radius $d_{AB}$ centered at $A$. Once the location of $B$ is chosen, $C$ must lie at the intersection of circles centered at $A$ and $B$. Only one of these two intersections corresponds to the "heads up" configuration. (c) The configuration of a coin in three-dimensional space is given by the three coordinates of $A$, two angles to the point $B$ on the sphere of radius $d_{AB}$ centered at $A$, and one angle to the point $C$ on the circle defined by the intersection of the a sphere centered at $A$ and a sphere centered at $B$.

choose the position of point $A$ in the plane (Figure 2.2(b)). We may choose it to be anything we want, so we have two degrees of freedom to specify, namely $(x_A, y_A)$. Once $(x_A, y_A)$ is specified, the constraint $d(A, B) = d_{AB}$ restricts the choice of $(x_B, y_B)$ to those points on the circle of radius $d_{AB}$ centered at $A$. A point on this circle can be specified by a single parameter, e.g., the angle specifying the location of $B$ on the circle centered at $A$. Let's call this angle $\phi_{AB}$ and define it to be the angle that the vector $\overrightarrow{AB}$ makes with the $\hat{x}$-axis.

Once we have chosen the location of point $B$, there are only two possible locations for $C$: at the intersections of the circle of radius $d_{AC}$ centered at $A$ and the circle of radius $d_{BC}$ centered at $B$ (Figure 2.2(b)). These two solutions correspond to heads or tails. In other words, once we have placed $A$ and $B$ and chosen heads or tails, the two constraints $d(A, C) = d_{AC}$ and $d(B, C) = d_{BC}$ eliminate the two apparent freedoms provided by $(x_C, y_C)$, and the location of $C$ is fixed. The coin has exactly three degrees of freedom in the plane, which can be specified by $(x_A, y_A, \phi_{AB})$.

Suppose that we choose to specify the position of an additional point $D$ on the coin. This introduces three additional constraints: $d(A, D) = d_{AD}$, $d(B, D) = d_{BD}$, and $d(C, D) = d_{CD}$. One of these constraints is **redundant**, i.e., it provides no new information; only two of the three constraints are independent. The two freedoms apparently introduced by the coordinates $(x_D, y_D)$ are then immediately eliminated by these two independent constraints. The same would hold for any other newly chosen point on the coin, so that there is no need to consider additional points.

We have been applying the following general rule for determining the number of degrees of freedom of a system:

$$\text{degrees of freedom} = \text{(sum of freedoms of the points)} -$$

$$\text{(number of independent constraints)}. \quad (2.1)$$

This rule can also be expressed in terms of the number of variables and independent equations that describe the system:

$$\text{degrees of freedom} = \text{(number of variables)} -$$

$$\text{(number of independent equations)}. \quad (2.2)$$

This general rule can also be used to determine the number of freedoms of a rigid body in three dimensions. For example, assume our coin is no longer confined to the table (Figure 2.2(c)). The coordinates for the three points $A$, $B$, and $C$ are now given by $(x_A, y_A, z_A)$, $(x_B, y_B, z_B)$, and $(x_C, y_C, z_C)$, respectively. Point $A$ can be placed freely (three degrees of freedom). The location of point $B$ is subject to the constraint $d(A, B) = d_{AB}$, meaning it must lie on the sphere of radius $d_{AB}$ centered at $A$. Thus we have $3 - 1 = 2$ freedoms to specify, which can be expressed as the latitude and longitude for the point on the sphere. Finally, the location of point $C$ must lie at the intersection of spheres centered at $A$ and $B$ of radius $d_{AC}$ and $d_{BC}$, respectively. In the general case the intersection of two spheres is a circle, and the location of point $C$ can be described by an angle that parametrizes this circle. Point $C$ therefore adds $3 - 2 = 1$ freedom. Once the position of point $C$ is chosen, the coin is fixed in space.

In summary, a rigid body in three-dimensional space has six freedoms, which can be described by the three coordinates parametrizing point $A$, the two angles parametrizing point $B$, and one angle parametrizing point $C$, provided $A$, $B$, and $C$ are noncollinear. Other representations for the configuration of a rigid body are discussed in Chapter 3.

We have just established that a rigid body moving in three-dimensional space, which we call a **spatial rigid body**, has six degrees of freedom. Similarly, a rigid body moving in a two-dimensional plane, which we henceforth call a **planar rigid body**, has three degrees of freedom. This latter result can also be obtained by considering the planar rigid body to be a spatial rigid body with six degrees of freedom but with the three independent constraints $z_A = z_B = z_C = 0$.

Since our robots consist of rigid bodies, Equation (2.1) can be expressed as follows:

$$\text{degrees of freedom} = \text{(sum of freedoms of the bodies)} -$$

$$\text{(number of independent constraints)}. \quad (2.3)$$

Equation (2.3) forms the basis for determining the degrees of freedom of general robots, which is the topic of the next section.

## 2.2     Degrees of Freedom of a Robot

Consider once again the door example of Figure 2.1(a), consisting of a single rigid body connected to a wall by a hinge joint. From the previous section we know that the door has only one degree of freedom, conveniently represented by the hinge joint angle $\theta$. Without the hinge joint, the door would be free to move in three-dimensional space and would have six degrees of freedom. By connecting the door to the wall via the hinge joint, five independent constraints are imposed on the motion of the door, leaving only one independent coordinate ($\theta$). Alternatively, the door can be viewed from above and regarded as a planar body, which has three degrees of freedom. The hinge joint then imposes two independent constraints, again leaving only one independent coordinate ($\theta$). The door's C-space is represented by some range in the interval $[0, 2\pi)$ over which $\theta$ is allowed to vary.

In both cases the joints constrain the motion of the rigid body, thus reducing the overall degrees of freedom. This observation suggests a formula for determining the number of degrees of freedom of a robot, simply by counting the number of rigid bodies and joints. In this section we derive precisely such a formula, called Grübler's formula, for determining the number of degrees of freedom of planar and spatial robots.

### 2.2.1     Robot Joints

Figure 2.3 illustrates the basic joints found in typical robots. Every joint connects exactly two links; joints that simultaneously connect three or more links are not allowed. The **revolute joint** (R), also called a hinge joint, allows rotational motion about the joint axis. The **prismatic joint** (P), also called a sliding or linear joint, allows translational (or rectilinear) motion along the direction of the joint axis. The **helical joint** (H), also called a screw joint, allows simultaneous rotation and translation about a screw axis. Revolute, prismatic, and helical joints all have one degree of freedom.

Joints can also have multiple degrees of freedom. The **cylindrical joint** (C) has two degrees of freedom and allows independent translations and rotations about a single fixed joint axis. The **universal joint** (U) is another two-degree-of-freedom joint that consists of a pair of revolute joints arranged so that their joint axes are orthogonal. The **spherical joint** (S), also called a ball-and-socket joint, has three degrees of freedom and functions much like our shoulder joint.

A joint can be viewed as providing freedoms to allow one rigid body to move relative to another. It can also be viewed as providing constraints on the possible motions of the two rigid bodies it connects. For example, a revolute joint can be viewed as allowing one freedom of motion between two rigid bodies in space, or it can be viewed as providing five constraints on the motion of one rigid body relative to the other. Generalizing, the number of degrees of freedom of a rigid body (three for planar bodies and six for spatial bodies) minus the number of

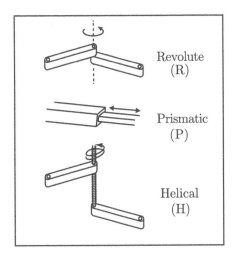
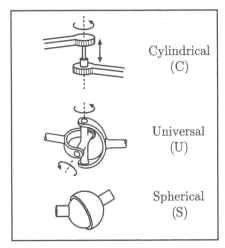

**Figure 2.3** Typical robot joints.

| Joint type | dof $f$ | Constraints $c$ between two planar rigid bodies | Constraints $c$ between two spatial rigid bodies |
|---|---|---|---|
| Revolute (R) | 1 | 2 | 5 |
| Prismatic (P) | 1 | 2 | 5 |
| Helical (H) | 1 | N/A | 5 |
| Cylindrical (C) | 2 | N/A | 4 |
| Universal (U) | 2 | N/A | 4 |
| Spherical (S) | 3 | N/A | 3 |

**Table 2.1** The number of degrees of freedom $f$ and constraints $c$ provided by common joints.

constraints provided by a joint must equal the number of freedoms provided by that joint.

The freedoms and constraints provided by the various joint types are summarized in Table 2.1.

### 2.2.2    Grübler's Formula

The number of degrees of freedom of a mechanism with links and joints can be calculated using **Grübler's formula**, which is an expression of Equation (2.3).

**Proposition 2.2**    *Consider a mechanism consisting of $N$ links, where ground is also regarded as a link. Let $J$ be the number of joints, $m$ be the number of degrees of freedom of a rigid body ($m = 3$ for planar mechanisms and $m = 6$ for spatial mechanisms), $f_i$ be the number of freedoms provided by joint $i$, and $c_i$ be the number of constraints provided by joint $i$, where $f_i + c_i = m$ for all $i$. Then*

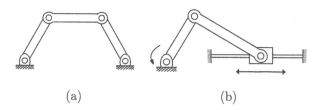

**Figure 2.4** (a) Four-bar linkage. (b) Slider–crank mechanism.

*Grübler's formula for the number of degrees of freedom of the robot is*

$$\text{dof} = \underbrace{m(N-1)}_{\text{rigid body freedoms}} - \underbrace{\sum_{i=1}^{J} c_i}_{\text{joint constraints}}$$

$$= m(N-1) - \sum_{i=1}^{J}(m - f_i)$$

$$= m(N - 1 - J) + \sum_{i=1}^{J} f_i. \tag{2.4}$$

*This formula holds in "generic" cases, but fails under certain configurations of the links and joints, such as when the joint constraints are not independent.*

Below we apply Grübler's formula to several planar and spatial mechanisms. We distinguish between two types of mechanism: **open-chain mechanisms** (also known as **serial mechanisms**) and **closed-chain mechanisms**. A closed-chain mechanism is any mechanism that has a closed loop. A person standing with both feet on the ground is an example of a closed-chain mechanism, since a closed loop can be traced from the ground, through the right leg, through the waist, through the left leg, and back to ground (recall that the ground itself is a link). An open-chain mechanism is any mechanism without a closed loop; an example is your arm when your hand is allowed to move freely in space.

**Example 2.3** (Four-bar linkage and slider–crank mechanism)   The planar four-bar linkage shown in Figure 2.4(a) consists of four links (one of them ground) arranged in a single closed loop and connected by four revolute joints. Since all the links are confined to move in the same plane, we have $m = 3$. Substituting $N = 4$, $J = 4$, and $f_i = 1$, $i = 1, \ldots, 4$, into Grübler's formula, we see that the four-bar linkage has one degree of freedom.

The slider–crank closed-chain mechanism of Figure 2.4(b) can be analyzed in two ways: (i) the mechanism consists of three revolute joints and one prismatic joint ($J = 4$ and each $f_i = 1$) and four links ($N = 4$, including the ground link), or (ii) the mechanism consists of two revolute joints ($f_i = 1$) and one RP joint (the RP joint is a concatenation of a revolute and prismatic joint, so that $f_i = 2$)

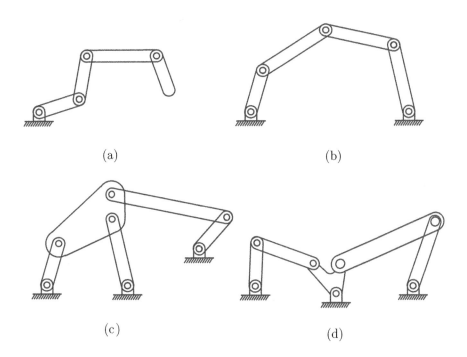

(a)                                          (b)

(c)                                          (d)

**Figure 2.5** (a) A $k$-link planar serial chain. (b) Five-bar planar linkage. (c) Stephenson six-bar linkage. (d) Watt six-bar linkage.

and three links ($N = 3$; remember that each joint connects precisely two bodies). In both cases the mechanism has one degree of freedom.

**Example 2.4** (Some classical planar mechanisms)    Let us now apply Grübler's formula to several classical planar mechanisms. The $k$-link planar serial chain of revolute joints in Figure 2.5(a) (called a $k$R robot for its $k$ revolute joints) has $N = k + 1$ links ($k$ links plus ground), and $J = k$ joints, and, since all the joints are revolute, $f_i = 1$ for all $i$. Therefore,

$$\text{dof} = 3((k+1) - 1 - k) + k = k$$

as expected. For the planar five-bar linkage of Figure 2.5(b), $N = 5$ (four links plus ground), $J = 5$, and since all joints are revolute, each $f_i = 1$. Therefore,

$$\text{dof} = 3(5 - 1 - 5) + 5 = 2.$$

For the Stephenson six-bar linkage of Figure 2.5(c), we have $N = 6$, $J = 7$, and $f_i = 1$ for all $i$, so that

$$\text{dof} = 3(6 - 1 - 7) + 7 = 1.$$

Finally, for the Watt six-bar linkage of Figure 2.5(d), we have $N = 6$, $J = 7$, and $f_i = 1$ for all $i$, so that, like the Stephenson six-bar linkage,

$$\text{dof} = 3(6 - 1 - 7) + 7 = 1.$$

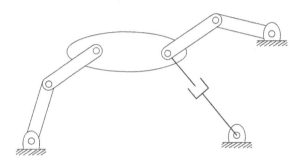

**Figure 2.6** A planar mechanism with two overlapping joints.

**Example 2.5** (A planar mechanism with overlapping joints)   The planar mechanism illustrated in Figure 2.6 has three links that meet at a single point on the right of the large link. Recalling that a joint by definition connects exactly two links, the joint at this point of intersection should not be regarded as a single revolute joint. Rather, it is correctly interpreted as two revolute joints overlapping each other. Again, there is more than one way to derive the number of degrees of freedom of this mechanism using Grübler's formula: (i) The mechanism consists of eight links ($N = 8$), eight revolute joints, and one prismatic joint. Substituting into Grübler's formula yields

$$\text{dof} = 3(8 - 1 - 9) + 9(1) = 3.$$

(ii) Alternatively, the lower-right revolute–prismatic joint pair can be regarded as a single two-dof joint. In this case the number of links is $N = 7$, with seven revolute joints, and a single two-dof revolute–prismatic pair. Substituting into Grübler's formula yields

$$\text{dof} = 3(7 - 1 - 8) + 7(1) + 1(2) = 3.$$

**Example 2.6** (Redundant constraints and singularities)   For the parallelogram linkage of Figure 2.7(a), $N = 5$, $J = 6$, and $f_i = 1$ for each joint. From Grübler's

(a)                                             (b)

**Figure 2.7** (a) A parallelogram linkage. (b) The five-bar linkage in a regular and singular configuration.

formula, the number of degrees of freedom is $3(5 - 1 - 6) + 6 = 0$. A mechanism with zero degrees of freedom is by definition a rigid structure. It is clear from examining the figure, though, that the mechanism can in fact move with one degree of freedom. Indeed, any one of the three parallel links, with its two joints, has no effect on the motion of the mechanism, so we should have calculated dof $= 3(4 - 1 - 4) + 4 = 1$. In other words, the constraints provided by the joints are not independent, as required by Grübler's formula.

A similar situation arises for the two-dof planar five-bar linkage of Figure 2.7(b). If the two joints connected to ground are locked at some fixed angle, the five-bar linkage should then become a rigid structure. However, if the two middle links are the same length and overlap each other, as illustrated in Figure 2.7(b), these overlapping links can rotate freely about the two overlapping joints. Of course, the link lengths of the five-bar linkage must meet certain specifications in order for such a configuration to even be possible. Also note that if a different pair of joints is locked in place, the mechanism does become a rigid structure as expected.

**Example 2.7** (Delta robot)   The Delta robot of Figure 2.8 consists of two platforms – the lower one mobile, the upper one stationary – connected by three legs. Each leg contains a parallelogram closed chain and consists of three revolute joints, four spherical joints, and five links. Adding the two platforms, there are $N = 17$ links and $J = 21$ joints (nine revolute and 12 spherical). By Grübler's formula,

$$\mathrm{dof} = 6(17 - 1 - 21) + 9(1) + 12(3) = 15.$$

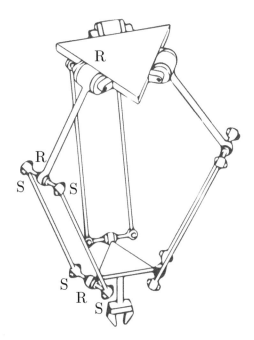

**Figure 2.8** The Delta robot.

Of these 15 degrees of freedom, however, only three are visible at the end-effector on the moving platform. In fact, the parallelogram leg design ensures that the moving platform always remains parallel to the fixed platform, so that the Delta robot acts as an $x$–$y$–$z$ Cartesian positioning device. The other 12 internal degrees of freedom are accounted for by torsion of the 12 links in the parallelograms (each of the three legs has four links in its parallelogram) about their long axes.

**Example 2.8** (Stewart–Gough platform)   The Stewart–Gough platform of Figure 1.1(b) consists of two platforms – the lower one stationary and regarded as ground, the upper one mobile – connected by six universal–prismatic–spherical (UPS) legs. The total number of links is 14 ($N = 14$). There are six universal joints (each with two degrees of freedom, $f_i = 2$), six prismatic joints (each with a single degree of freedom, $f_i = 1$), and six spherical joints (each with three degrees of freedom, $f_i = 3$). The total number of joints is 18. Substituting these values into Grübler's formula with $m = 6$ yields

$$\text{dof} = 6(14 - 1 - 18) + 6(1) + 6(2) + 6(3) = 6.$$

In some versions of the Stewart–Gough platform the six universal joints are replaced by spherical joints. By Grübler's formula this mechanism has 12 degrees of freedom; replacing each universal joint by a spherical joint introduces an extra degree of freedom in each leg, allowing torsional rotations about the leg axis. Note, however, that this torsional rotation has no effect on the motion of the mobile platform.

The Stewart–Gough platform is a popular choice for car and airplane cockpit simulators, as the platform moves with the full six degrees of freedom of motion of a rigid body. On the one hand, the parallel structure means that each leg needs to support only a fraction of the weight of the payload. On the other hand, this structure also limits the range of translational and rotational motion of the platform relative to the range of motion of the end-effector of a six-dof open chain.

## 2.3      Configuration Space: Topology and Representation

### 2.3.1      Configuration Space Topology

Until now we have been focusing on one important aspect of a robot's C-space – its dimension, or the number of degrees of freedom. However, the *shape* of the space is also important.

Consider a point moving on the surface of a sphere. The point's C-space is two dimensional, as the configuration can be described by two coordinates, latitude and longitude. As another example, a point moving on a plane also has a two-dimensional C-space, with coordinates $(x, y)$. While both a plane and the surface of a sphere are two dimensional, clearly they do not have the same shape – the plane extends infinitely while the sphere wraps around.

**Figure 2.9** An open interval of the real line, denoted $(a, b)$, can be deformed to an open semicircle. This open semicircle can then be deformed to the real line by the mapping illustrated: beginning from a point at the center of the semicircle, draw a ray that intersects the semicircle and then a line above the semicircle. These rays show that every point of the semicircle can be stretched to exactly one point on the line, and vice versa. Thus an open interval can be continuously deformed to a line, so an open interval and a line are topologically equivalent.

Unlike the plane, a larger sphere has the same shape as the original sphere, in that it wraps around in the same way. Only its size is different. For that matter, an oval-shaped American football also wraps around similarly to a sphere. The only difference between a football and a sphere is that the football has been stretched in one direction.

The idea that the two-dimensional surfaces of a small sphere, a large sphere, and a football all have the same kind of shape, which is different from the shape of a plane, is expressed by the **topology** of the surfaces. We do not attempt a rigorous treatment in this book,[3] but we say that two spaces are **topologically equivalent** if one can be continuously deformed into the other without cutting or gluing. A sphere can be deformed into a football simply by stretching, without cutting or gluing, so those two spaces are topologically equivalent. You cannot turn a sphere into a plane without cutting it, however, so a sphere and a plane are not topologically equivalent.

Topologically distinct one-dimensional spaces include the circle, the line, and a closed interval of the line. The circle is written mathematically as $S$ or $S^1$, a one-dimensional "sphere." The line can be written as $\mathbb{E}$ or $\mathbb{E}^1$, indicating a one-dimensional Euclidean (or "flat") space. Since a point in $\mathbb{E}^1$ is usually represented by a real number (after choosing an origin and a length scale), it is often written as $\mathbb{R}$ or $\mathbb{R}^1$ instead. A closed interval of the line, which contains its endpoints, can be written $[a, b] \subset \mathbb{R}^1$. (An open interval $(a, b)$ does not include the endpoints $a$ and $b$ and is topologically equivalent to a line, since the open interval can be stretched to a line, as shown in Figure 2.9. A closed interval is not topologically equivalent to a line, since a line does not contain endpoints.)

In higher dimensions, $\mathbb{R}^n$ is the $n$-dimensional Euclidean space and $S^n$ is the $n$-dimensional surface of a sphere in $(n+1)$-dimensional space. For example, $S^2$ is the two-dimensional surface of a sphere in three-dimensional space.

Note that the topology of a space is a fundamental property of the space itself and *is independent of how we choose coordinates to represent points in the space.*

---

[3] For those familiar with concepts in topology, all the spaces we consider can be viewed as embedded in a higher-dimensional Euclidean space, inheriting the Euclidean topology of that space.

For example, to represent a point on a circle, we could refer to the point by the angle $\theta$ from the center of the circle to the point, relative to a chosen zero angle. Or, we could choose a reference frame with its origin at the center of the circle and represent the point by the two coordinates $(x, y)$ subject to the constraint $x^2 + y^2 = 1$. No matter what our choice of coordinates is, the space itself does not change.

Some C-spaces can be expressed as the **Cartesian product** of two or more spaces of lower dimension; that is, points in such a C-space can be represented as the union of the representations of points in the lower-dimensional spaces. For example:

- The C-space of a rigid body in the plane can be written as $\mathbb{R}^2 \times S^1$, since the configuration can be represented as the concatenation of the coordinates $(x, y)$ representing $\mathbb{R}^2$ and an angle $\theta$ representing $S^1$.
- The C-space of a PR robot arm can be written $\mathbb{R}^1 \times S^1$. (We will occasionally ignore joint limits, i.e., bounds on the travel of the joints, when expressing the topology of the C-space; with joint limits, the C-space is the Cartesian product of two closed intervals of the line.)
- The C-space of a 2R robot arm can be written $S^1 \times S^1 = T^2$, where $T^n$ is the $n$-dimensional surface of a torus in an $(n + 1)$-dimensional space. (See Table 2.2.) Note that $S^1 \times S^1 \times \cdots \times S^1$ ($n$ copies of $S^1$) is equal to $T^n$, not $S^n$; for example, a sphere $S^2$ is not topologically equivalent to a torus $T^2$.
- The C-space of a planar rigid body (e.g., the chassis of a mobile robot) with a 2R robot arm can be written as $\mathbb{R}^2 \times S^1 \times T^2 = \mathbb{R}^2 \times T^3$.
- As we saw in Section 2.1 when we counted the degrees of freedom of a rigid body in three dimensions, the configuration of a rigid body can be described by a point in $\mathbb{R}^3$, plus a point on a two-dimensional sphere $S^2$, plus a point on a one-dimensional circle $S^1$, giving a total C-space of $\mathbb{R}^3 \times S^2 \times S^1$.

### 2.3.2     Configuration Space Representation

To perform computations, we must have a numerical **representation** of the space, consisting of a set of real numbers. We are familiar with this idea from linear algebra – a vector is a natural way to represent a point in a Euclidean space. It is important to keep in mind that the representation of a space involves a choice, and therefore it is not as fundamental as the topology of the space, which is independent of the representation. For example, the same point in a three-dimensional space can have different coordinate representations depending on the choice of reference frame (the origin and the direction of the coordinate axes) and the choice of length scale, but the topology of the underlying space is the same regardless of theses choices.

While it is natural to choose a reference frame and length scale and to use a vector to represent points in a Euclidean space, representing a point on a

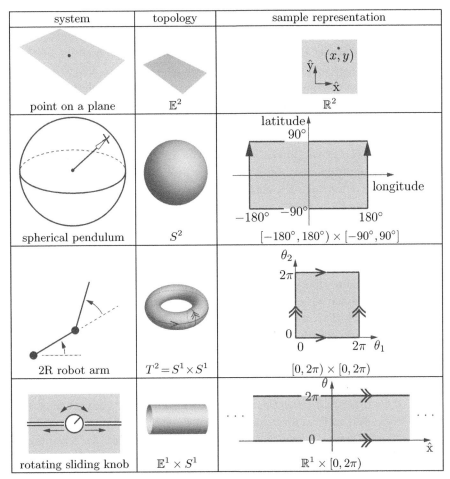

| system | topology | sample representation |
|--------|----------|----------------------|
| point on a plane | $\mathbb{E}^2$ | $\mathbb{R}^2$ |
| spherical pendulum | $S^2$ | $[-180°, 180°) \times [-90°, 90°]$ |
| 2R robot arm | $T^2 = S^1 \times S^1$ | $[0, 2\pi) \times [0, 2\pi)$ |
| rotating sliding knob | $\mathbb{E}^1 \times S^1$ | $\mathbb{R}^1 \times [0, 2\pi)$ |

**Table 2.2** Four topologically different two-dimensional C-spaces and example coordinate representations. In the latitude–longitude representation of the sphere, the latitudes $-90°$ and $90°$ each correspond to a single point (the south pole and the north pole, respectively), and the longitude parameter wraps around at $180°$ and $-180°$; the edges with arrows are glued together. Similarly, the coordinate representations of the torus and cylinder wrap around at the edges marked with corresponding arrows.

curved space, such as a sphere, is less obvious. One solution for a sphere is to use latitude and longitude coordinates. A choice of $n$ coordinates, or parameters, to represent an $n$-dimensional space is called an **explicit parametrization** of the space. Such an explicit parametrization is valid for a particular range of the parameters (e.g., $[-90°, 90°]$ for latitude and $[-180°, 180°)$ for longitude for a sphere, where, on Earth, negative values correspond to "south" and "west," respectively).

The latitude–longitude representation of a sphere is unsatisfactory if you are walking near the North Pole (where the latitude equals $90°$) or South Pole (where the latitude equals $-90°$), where taking a very small step can result in a large change in the coordinates. The North and South Poles are **singularities** of the representation, and the existence of singularities is a result of the fact that a sphere does not have the same topology as a plane, i.e., the space of the two real numbers that we have chosen to represent the sphere (latitude and longitude). The location of these singularities has nothing to do with the sphere itself, which looks the same everywhere, and everything to do with the chosen representation of it. Singularities of the parametrization are particularly problematic when representing velocities as the time rate of change of coordinates, since these representations may tend to infinity near singularities even if the point on the sphere is moving at a constant speed $\sqrt{\dot{x}^2 + \dot{y}^2 + \dot{z}^2}$ (which is what the speed would be had you represented the point as $(x, y, z)$ instead).

If you can assume that the configuration never approaches a singularity of the representation, you can ignore this issue. If you cannot make this assumption, there are two ways to overcome the problem.

- Use more than one **coordinate chart** on the space, where each coordinate chart is an explicit parametrization covering only a portion of the space such that, within each chart, there is no singularity. As the configuration representation approaches a singularity in one chart, e.g., the North or South Pole, you simply switch to another chart where the North and South Poles are far from singularities.

  If we define a set of singularity-free coordinate charts that overlap each other and cover the entire space, like the two charts above, the charts are said to form an **atlas** of the space, much as an atlas of the Earth consists of several maps that together cover the Earth. An advantage of using an atlas of coordinate charts is that the representation always uses the minimum number of numbers. A disadvantage is the extra bookkeeping required to switch representations between coordinate charts to avoid singularities. (Note that Euclidean spaces can be covered by a single coordinate chart without singularities.)

- Use an **implicit representation** of the space instead of an explicit parametrization. An implicit representation views the $n$-dimensional space as embedded in a Euclidean space of more than $n$ dimensions, just as a two-dimensional unit sphere can be viewed as a surface embedded in a three-dimensional Euclidean space. An implicit representation uses the coordinates of the higher-dimensional space (e.g., $(x, y, z)$ in the three-dimensional space), but subjects these coordinates to constraints that reduce the number of degrees of freedom (e.g., $x^2 + y^2 + z^2 = 1$ for the unit sphere).

  A disadvantage of this approach is that the representation has more numbers than the number of degrees of freedom. An advantage is that there are no singularities in the representation – a point moving smoothly around the sphere is represented by a smoothly changing $(x, y, z)$, even at the North and South poles. A single representation is used for the whole sphere; multiple coordinate charts are not needed.

Another advantage is that while it may be very difficult to construct an explicit parametrization, or atlas, for a closed-chain mechanism, it is easy to find an implicit representation: the set of all joint coordinates subject to the **loop-closure equations** that define the closed loops (Section 2.4).

We will use implicit representations throughout the book, beginning in the next chapter. In particular, we use nine numbers, subject to six constraints, to represent the three orientation freedoms of a rigid body in space. This is called a **rotation matrix**. In addition to being singularity-free (unlike three-parameter representations such as roll–pitch–yaw angles[4]), the rotation matrix representation allows us to use linear algebra to perform computations such as rotating a rigid body or changing the reference frame in which the orientation of a rigid body is expressed.[5]

In summary, the non-Euclidean shape of many C-spaces motivates our use of implicit representations of C-space throughout this book. We return to this topic in the next chapter.

## .4 Configuration and Velocity Constraints

For robots containing one or more closed loops, usually an implicit representation is more easily obtained than an explicit parametrization. For example, consider the planar four-bar linkage of Figure 2.10, which has one degree of freedom. The fact that the four links always form a closed loop can be expressed by the following three equations:

$$L_1 \cos\theta_1 + L_2 \cos(\theta_1 + \theta_2) + \cdots + L_4 \cos(\theta_1 + \cdots + \theta_4) = 0,$$
$$L_1 \sin\theta_1 + L_2 \sin(\theta_1 + \theta_2) + \cdots + L_4 \sin(\theta_1 + \cdots + \theta_4) = 0,$$
$$\theta_1 + \theta_2 + \theta_3 + \theta_4 - 2\pi = 0.$$

These equations are obtained by viewing the four-bar linkage as a serial chain with four revolute joints in which (i) the tip of link $L_4$ always coincides with the origin and (ii) the orientation of link $L_4$ is always horizontal.

These equations are sometimes referred to as **loop-closure equations**. For the four-bar linkage they are given by a set of three equations in four unknowns. The set of all solutions forms a one-dimensional curve in the four-dimensional joint space and constitutes the C-space.

---

[4] Roll–pitch–yaw angles and *Euler* angles use three parameters for the space of rotations $S^2 \times S^1$ (two for $S^2$ and one for $S^1$), and therefore are subject to singularities as discussed above.

[5] Another singularity-free implicit representation of orientations, the unit quaternion, uses only four numbers subject to the constraint that the 4-vector be of unit length. In fact, this representation is a double covering of the set of orientations: for every orientation, there are two unit quaternions.

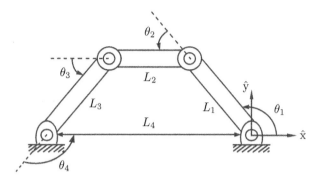

**Figure 2.10** The four-bar linkage.

In this book, when vectors are used in a linear algebra computation, they are generally treated as column vectors, e.g., $p = [1\ 2\ 3]^{\mathrm{T}}$. When a computation is not imminent, however, we often think of a vector simply as an ordered list of variables, e.g., $p = (1, 2, 3)$.

Thus, for general robots containing one or more closed loops, the configuration space can be implicitly represented by the column vector $\theta = [\theta_1\ \cdots\ \theta_n]^{\mathrm{T}} \in \mathbb{R}^n$ and loop-closure equations of the form

$$g(\theta) = \begin{bmatrix} g_1(\theta_1, \ldots, \theta_n) \\ \vdots \\ g_k(\theta_1, \ldots, \theta_n) \end{bmatrix} = 0, \tag{2.5}$$

a set of $k$ independent equations, with $k \leq n$. Such constraints are known as **holonomic constraints**, ones that reduce the dimension of the C-space.[6] The C-space can be viewed as a surface of dimension $n - k$ (assuming that all constraints are independent) embedded in $\mathbb{R}^n$.

Suppose that a closed-chain robot with loop-closure equations $g(\theta) = 0$, $g : \mathbb{R}^n \to \mathbb{R}^k$, is in motion, following the time trajectory $\theta(t)$. Differentiating both sides of $g(\theta(t)) = 0$ with respect to $t$, we obtain

$$\frac{d}{dt} g(\theta(t)) = 0;$$

$$\tag{2.6}$$

thus

$$\begin{bmatrix} \dfrac{\partial g_1}{\partial \theta_1}(\theta)\dot{\theta}_1 + \cdots + \dfrac{\partial g_1}{\partial \theta_n}(\theta)\dot{\theta}_n \\ \vdots \\ \dfrac{\partial g_k}{\partial \theta_1}(\theta)\dot{\theta}_1 + \cdots + \dfrac{\partial g_k}{\partial \theta_n}(\theta)\dot{\theta}_n \end{bmatrix} = 0.$$

[6] Viewing a rigid body as a collection of points, the distance constraints between the points, as we saw earlier, can be viewed as holonomic constraints.

This can be expressed as a matrix multiplying a column vector $[\dot{\theta}_1 \;\cdots\; \dot{\theta}_n]^{\mathrm{T}}$:

$$
\begin{bmatrix}
\dfrac{\partial g_1}{\partial \theta_1}(\theta) & \cdots & \dfrac{\partial g_1}{\partial \theta_n}(\theta) \\
\vdots & \ddots & \vdots \\
\dfrac{\partial g_k}{\partial \theta_1}(\theta) & \cdots & \dfrac{\partial g_k}{\partial \theta_n}(\theta)
\end{bmatrix}
\begin{bmatrix}
\dot{\theta}_1 \\
\vdots \\
\dot{\theta}_n
\end{bmatrix}
= 0,
$$

which we can write as

$$
\frac{\partial g}{\partial \theta}(\theta)\dot{\theta} = 0. \tag{2.7}
$$

Here, the joint-velocity vector $\dot{\theta}_i$ denotes the derivative of $\theta_i$ with respect to time $t$, $\partial g(\theta)/\partial\theta \in \mathbb{R}^{k\times n}$, and $\theta, \dot{\theta} \in \mathbb{R}^n$. The constraints (2.7) can be written

$$
A(\theta)\dot{\theta} = 0, \tag{2.8}
$$

where $A(\theta) \in \mathbb{R}^{k\times n}$. Velocity constraints of this form are called **Pfaffian constraints**. For the case of $A(\theta) = \partial g(\theta)/\partial\theta$, one could regard $g(\theta)$ as being the "integral" of $A(\theta)$; for this reason, holonomic constraints of the form $g(\theta) = 0$ are also called **integrable constraints** – the velocity constraints that they imply can be integrated to give equivalent configuration constraints.

We now consider another class of Pfaffian constraints that are fundamentally different from the holonomic type. To illustrate this with a concrete example, consider an upright coin of radius $r$ rolling on a plane as shown in Figure 2.11. The configuration of the coin is given by the contact point $(x, y)$ on the plane, the steering angle $\phi$, and the angle of rotation $\theta$. The C-space of the coin is therefore $\mathbb{R}^2 \times T^2$, where $T^2$ is the two-dimensional torus parametrized by the angles $\phi$ and $\theta$. This C-space is four dimensional.

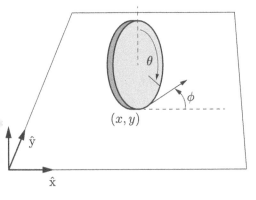

**Figure 2.11** A coin rolling on a plane without slipping.

We now express, in mathematical form, the fact that the coin rolls without slipping. The coin must always roll in the direction indicated by $(\cos\phi, \sin\phi)$, with forward speed $r\dot{\theta}$:

$$
\begin{bmatrix}
\dot{x} \\
\dot{y}
\end{bmatrix}
= r\dot{\theta}
\begin{bmatrix}
\cos\phi \\
\sin\phi
\end{bmatrix}. \tag{2.9}
$$

Collecting the four C-space coordinates into a single vector $q = [q_1 \; q_2 \; q_3 \; q_4]^{\mathrm{T}} = [x \; y \; \phi \; \theta]^{\mathrm{T}} \in \mathbb{R}^2 \times T^2$, the above no-slip rolling constraint can then be expressed in the form

$$\begin{bmatrix} 1 & 0 & 0 & -r\cos q_3 \\ 0 & 1 & 0 & -r\sin q_3 \end{bmatrix} \dot{q} = 0. \tag{2.10}$$

These are Pfaffian constraints of the form $A(q)\dot{q} = 0$, $A(q) \in \mathbb{R}^{2\times 4}$.

These constraints are not integrable; that is, for the $A(q)$ given in (2.10), there does not exist a differentiable function $g : \mathbb{R}^4 \to \mathbb{R}^2$ such that $\partial g(q)/\partial q = A(q)$. If this were not the case then there would have to exist a differentiable $g_1(q)$ that satisfied the following four equalities:

$$
\begin{array}{llll}
\partial g_1(q)/\partial q_1 & = & 1 & \longrightarrow & g_1(q) = q_1 + h_1(q_2, q_3, q_4) \\
\partial g_1(q)/\partial q_2 & = & 0 & \longrightarrow & g_1(q) = h_2(q_1, q_3, q_4) \\
\partial g_1(q)/\partial q_3 & = & 0 & \longrightarrow & g_1(q) = h_3(q_1, q_2, q_4) \\
\partial g_1(q)/\partial q_4 & = & -r\cos q_3 & \longrightarrow & g_1(q) = -rq_4\cos q_3 + h_4(q_1, q_2, q_3),
\end{array}
$$

for some $h_i$, $i = 1, \ldots, 4$, differentiable in each of its variables. By inspection it should be clear that no such $g_1(q)$ exists. Similarly, it can be shown that $g_2(q)$ does not exist, so that the constraint (2.10) is nonintegrable. A Pfaffian constraint that is nonintegrable is called a **nonholonomic constraint.** Such constraints reduce the dimension of the feasible velocities of the system but do not reduce the dimension of the reachable C-space. The rolling coin can reach any point in its four-dimensional C-space despite the two constraints on its velocity.[7] See Exercise 2.30.

In a number of robotics contexts nonholonomic constraints arise that involve the conservation of momentum and rolling without slipping, e.g., wheeled vehicle kinematics and grasp contact kinematics. We examine nonholonomic constraints in greater detail in our study of wheeled mobile robots in Chapter 13.

## 2.5     Task Space and Workspace

We now introduce two more concepts relating to the configuration of a robot: the task space and the workspace. Both relate to the configuration of the end-effector of a robot, not to the configuration of the entire robot.

The **task space** is a space in which the robot's task can be naturally expressed. For example, if the robot's task is to plot with a pen on a piece of paper, the task space would be $\mathbb{R}^2$. If the task is to manipulate a rigid body, a natural representation of the task space is the C-space of a rigid body, representing the position and orientation of a frame attached to the robot's end-effector. This is the default representation of task space. The decision of how to define the task space is driven by the task, independently of the robot.

---

[7] Some texts define the number of degrees of freedom of a system to be the dimension of the feasible velocities, e.g., two for the rolling coin. We always refer to the dimension of the C-space as the number of degrees of freedom.

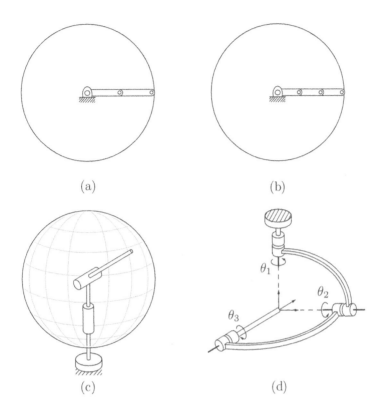

**Figure 2.12** Examples of workspaces for various robots: (a) a planar 2R open chain; (b) a planar 3R open chain; (c) a spherical 2R open chain; (d) a 3R orienting mechanism.

The **workspace** is a specification of the configurations that the end-effector of the robot can reach. The definition of the workspace is primarily driven by the robot's structure, independently of the task.

Both the task space and the workspace involve a choice by the user; in particular, the user may decide that some freedoms of the end-effector (e.g., its orientation) do not need to be represented.

The task space and the workspace are distinct from the robot's C-space. A point in the task space or the workspace may be achievable by more than one robot configuration, meaning that the point is not a full specification of the robot's configuration. For example, for an open-chain robot with seven joints, the six-dof position and orientation of its end-effector does not fully specify the robot's configuration.

Some points in the task space may not be reachable at all by the robot, such as some points on a chalkboard. By definition, however, all points in the workspace are reachable by at least one configuration of the robot.

Two mechanisms with different C-spaces may have the same workspace. For example, considering the end-effector to be the Cartesian tip of the robot (e.g., the location of a plotting pen) and ignoring orientations, the planar 2R open chain with links of equal length three (Figure 2.12(a)) and the planar 3R open chain with links of equal length two (Figure 2.12(b)) have the same workspace despite having different C-spaces.

Two mechanisms with the same C-space may also have different workspaces. For example, taking the end-effector to be the Cartesian tip of the robot and ignoring orientations, the 2R open chain of Figure 2.12(a) has a planar disk as its workspace, while the 2R open chain of Figure 2.12(c) has the surface of a sphere as its workspace.

Attaching a coordinate frame to the tip of the tool of the 3R open chain "wrist" mechanism of Figure 2.12(d), we see that the frame can achieve any orientation by rotating the joints but the Cartesian position of the tip is always fixed. This can be seen by noting that the three joint axes always intersect at the tip. For this mechanism, we would probably define the workspace to be the three-dof space of orientations of the frame, $S^2 \times S^1$, which is different from the C-space $T^3$. The task space depends on the task; if the job is to point a laser pointer, then rotations about the axis of the laser beam are immaterial and the task space would be $S^2$, the set of directions in which the laser can point.

**Example 2.9**  The SCARA robot of Figure 2.13 is an RRRP open chain that is widely used for tabletop pick-and-place tasks. The end-effector configuration is completely described by the four parameters $(x, y, z, \phi)$, where $(x, y, z)$ denotes the Cartesian position of the end-effector center point and $\phi$ denotes the orientation of the end-effector in the $x$–$y$-plane. Its task space would typically be defined as $\mathbb{R}^3 \times S^1$, and its workspace would typically be defined as the reachable points in $(x, y, z)$ Cartesian space, since all orientations $\phi \in S^1$ can be achieved at all reachable points.

**Example 2.10**  A standard 6R industrial manipulator can be adapted to spray-painting applications as shown in Figure 2.14. The paint spray nozzle attached to the tip can be regarded as the end-effector. What is important to the task is the Cartesian position of the spray nozzle, together with the direction in which the spray nozzle is pointing; rotations about the nozzle axis (which points in the direction in which paint is being sprayed) do not matter. The nozzle configuration can therefore be described by five coordinates: $(x, y, z)$ for the Cartesian position of the nozzle and spherical coordinates $(\theta, \phi)$ to describe the direction in which the nozzle is pointing. The task space can be written as $\mathbb{R}^3 \times S^2$. The workspace could be the reachable points in $\mathbb{R}^3 \times S^2$, or, to simplify visualization, the user could define the workspace to be the subset of $\mathbb{R}^3$ corresponding to the reachable Cartesian positions of the nozzle.

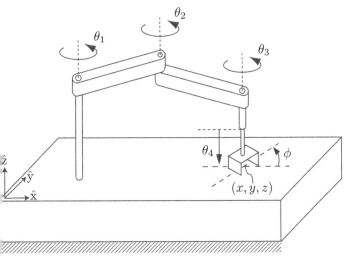

**Figure 2.13** SCARA robot.

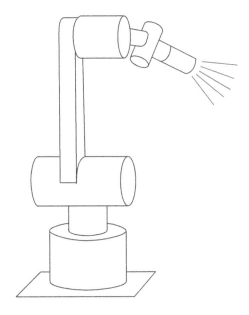

**Figure 2.14** A spray-painting robot.

## 2.6     Summary

- A robot is mechanically constructed from links that are connected by various types of joint. The links are usually modeled as rigid bodies. An end-effector such as a gripper may be attached to some link of the robot. Actuators deliver forces and torques to the joints, thereby causing motion of the robot.

- The most widely used one-dof joints are the revolute joint, which allows rotation about the joint axis, and the prismatic joint, which allows translation in the direction of the joint axis. Some common two-dof joints include the cylindrical joint, which is constructed by serially connecting a revolute and prismatic joint, and the universal joint, which is constructed by orthogonally connecting two revolute joints. The spherical joint, also known as the ball-and-socket joint, is a three-dof joint whose function is similar to the human shoulder joint.

- The configuration of a rigid body is a specification of the location of all its points. For a rigid body moving in the plane, three independent parameters are needed to specify the configuration. For a rigid body moving in three-dimensional space, six independent parameters are needed to specify the configuration.

- The configuration of a robot is a specification of the configuration of all its links. The robot's configuration space is the set of all possible robot configurations. The dimension of the C-space is the number of degrees of freedom of a robot.

- The number of degrees of freedom of a robot can be calculated using Grübler's formula,

$$\text{dof} = m(N - 1 - J) + \sum_{i=1}^{J} f_i,$$

where $m = 3$ for planar mechanisms and $m = 6$ for spatial mechanisms, $N$ is the number of links (including the ground link), $J$ is the number of joints, and $f_i$ is the number of degrees of freedom of joint $i$.

- A robot's C-space can be parametrized explicitly or represented implicitly. For a robot with $n$ degrees of freedom, an explicit parametrization uses $n$ coordinates, the minimum necessary. An implicit representation involves $m$ coordinates with $m \geq n$, with the $m$ coordinates subject to $m - n$ constraint equations. With an implicit parametrization, a robot's C-space can be viewed as a surface of dimension $n$ embedded in a space of higher dimension $m$.

- The C-space of an $n$-dof robot whose structure contains one or more closed loops can be implicitly represented using $k$ loop-closure equations of the form $g(\theta) = 0$, where $\theta \in \mathbb{R}^m$ and $g : \mathbb{R}^m \to \mathbb{R}^k$. Such constraint equations are called holonomic constraints. Assuming that $\theta$ varies with time $t$, the holonomic constraints $g(\theta(t)) = 0$ can be differentiated with respect to $t$ to yield

$$\frac{\partial g}{\partial \theta}(\theta)\dot{\theta} = 0,$$

where $\partial g(\theta)/\partial \theta$ is a $k \times m$ matrix.

- A robot's motion can also be subject to velocity constraints of the form

$$A(\theta)\dot{\theta} = 0,$$

where $A(\theta)$ is a $k \times m$ matrix that cannot be expressed as the differential of some function $g(\theta)$. In other words, there does not exist any $g(\theta), g : \mathbb{R}^m \to \mathbb{R}^k$, such that

$$A(\theta) = \frac{\partial g}{\partial \theta}(\theta).$$

Such constraints are said to be nonholonomic constraints, or nonintegrable constraints. These constraints reduce the dimension of feasible velocities of the system but do not reduce the dimension of the reachable C-space. Nonholonomic constraints arise in robot systems subject to conservation of momentum or rolling without slipping.

- A robot's task space is a space in which the robot's task can be naturally expressed. A robot's workspace is a specification of the configurations that the end-effector of the robot can reach.

## .7 Notes and References

In the kinematics literature, structures that consist of links connected by joints are also called mechanisms or linkages. The number of degrees of freedom of a mechanism, also referred to as its mobility, is treated in most texts on mechanism analysis and design, e.g., Erdman and Sandor (1996) or McCarthy and Soh (2011). The notion of a robot's configuration space was first formulated in the context of motion planning by Lozano-Perez (1980); more recent and advanced treatments can be found in Latombe (1991), LaValle (2006), and Choset et al. (2005). As apparent from examples in this chapter, a robot's configuration space can be nonlinear and curved, as can its task space. Such spaces often have the mathematical structure of a differentiable manifold, which are the central objects of study in differential geometry. Some accessible introductions to differential geometry are Millman and Parker (1977), do Carmo (1976) and Boothby (2002).

## .8 Exercises

In the exercises below, if you are asked to "describe" a C-space, you should indicate its dimension and whatever you know about its topology (e.g., using $\mathbb{R}$, $S$, and $T$, as with the examples in Sections 2.3.1 and 2.3.2).

**Exercise 2.1** Using the methods of Section 2.1 derive a formula, in terms of $n$, for the number of degrees of freedom of a rigid body in $n$-dimensional space. Indicate how many of these dof are translational and how many are rotational. Describe the topology of the C-space (e.g., for $n = 2$, the topology is $\mathbb{R}^2 \times S^1$).

**Exercise 2.2**   Find the number of degrees of freedom of your arm, from your torso to your palm (just past the wrist, not including finger degrees of freedom). Keep the center of the ball-and-socket joint of your shoulder stationary (do not "hunch" your shoulders). Find the number of degrees of freedom in two ways:

(a) add up the degrees of freedom at the shoulder, elbow, and wrist joints;
(b) fix your palm flat on a table with your elbow bent and, without moving the center of your shoulder joint, investigate with how many degrees of freedom you can still move your arm.

Do your answers agree? How many constraints were placed on your arm when you placed your palm at a fixed configuration on the table?

**Exercise 2.3**   In the previous exercise, we assumed that your arm is a serial chain. In fact, between your upper arm bone (the humerus) and the bone complex just past your wrist (the carpal bones), your forearm has two bones, the radius and the ulna, which are part of a closed chain. Model your arm, from your shoulder to your palm, as a mechanism with joints and calculate the number of degrees of freedom using Grübler's formula. Be clear on the number of freedoms of each joint you use in your model. Your joints may or may not be of the standard types studied in this chapter (R, P, H, C, U, and S).

**Exercise 2.4**   Assume each of your arms has $n$ degrees of freedom. You are driving a car, your torso is stationary relative to the car (owing to a tight seatbelt!), and both hands are firmly grasping the wheel, so that your hands do not move relative to the wheel. How many degrees of freedom does your arms-plus-steering wheel system have? Explain your answer.

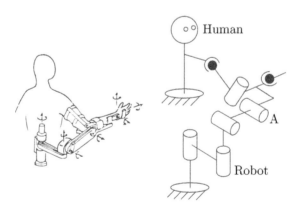

**Figure 2.15** Robot used for human arm rehabilitation.

**Exercise 2.5**   Figure 2.15 shows a robot used for human arm rehabilitation. Determine the number of degrees of freedom of the chain formed by the human arm and the robot.

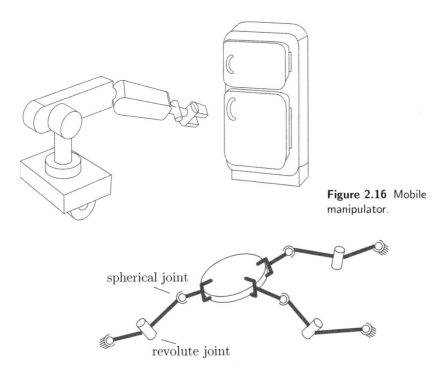

**Figure 2.16** Mobile manipulator.

**Figure 2.17** Three cooperating SRS arms grasping a common object.

**Exercise 2.6**   The mobile manipulator of Figure 2.16 consists of a 6R arm and multi-fingered hand mounted on a mobile base with a single wheel. You can think of the wheeled base as the same as the rolling coin in Figure 2.11 – the wheel (and base) can spin together about an axis perpendicular to the ground, and the wheel rolls without slipping. The base always remains horizontal. (Left unstated are the means to keep the base horizontal and to spin the wheel and base about an axis perpendicular to the ground.)

(a) Ignoring the multi-fingered hand, describe the configuration space of the mobile manipulator.

(b) Now suppose that the robot hand rigidly grasps a refrigerator door handle and, with its wheel and base completely stationary, opens the door using only its arm. With the door open, how many degrees of freedom does the mechanism formed by the arm and open door have?

(c) A second identical mobile manipulator comes along, and after parking its mobile base, also rigidly grasps the refrigerator door handle. How many degrees of freedom does the mechanism formed by the two arms and the open refrigerator door have?

**Exercise 2.7**   Three identical SRS open-chain arms are grasping a common object, as shown in Figure 2.17.

(a)  Find the number of degrees of freedom of this system.
(b)  Suppose there are now a total of $n$ such arms grasping the object. How many degrees of freedom does this system have?
(c)  Suppose the spherical wrist joint in each of the $n$ arms is now replaced by a universal joint. How many degrees of freedom does this system have?

**Exercise 2.8**    Consider a spatial parallel mechanism consisting of a moving plate connected to a fixed plate by $n$ identical legs. For the moving plate to have six degrees of freedom, how many degrees of freedom should each leg have, as a function of $n$? For example, if $n = 3$ then the moving plate and fixed plate are connected by three legs; how many degrees of freedom should each leg have for the moving plate to move with six degrees of freedom? Solve for arbitrary $n$.

**Exercise 2.9**    Use the planar version of Grübler's formula to determine the number of degrees of freedom of the mechanisms shown in Figure 2.18. Comment on whether your results agree with your intuition about the possible motions of these mechanisms.

**Exercise 2.10**    Use the planar version of Grübler's formula to determine the number of degrees of freedom of the mechanisms shown in Figure 2.19. Comment on whether your results agree with your intuition about the possible motions of these mechanisms.

**Exercise 2.11**    Use the spatial version of Grübler's formula to determine the number of degrees of freedom of the mechanisms shown in Figure 2.20. Comment on whether your results agree with your intuition about the possible motions of these mechanisms.

**Exercise 2.12**    Use the spatial version of Grübler's formula to determine the number of degrees of freedom of the mechanisms shown in Figure 2.21. Comment on whether your results agree with your intuition about the possible motions of these mechanisms.

**Exercise 2.13**    In the parallel mechanism shown in Figure 2.22, six legs of identical length are connected to a fixed and moving platform via spherical joints. Determine the number of degrees of freedom of this mechanism using Grübler's formula. Illustrate all possible motions of the upper platform.

**Exercise 2.14**    The $3 \times$ UPU platform of Figure 2.23 consists of two platforms – the lower one stationary, the upper one mobile – connected by three UPU legs.

(a)  Using the spatial version of Grübler's formula, verify that it has three degrees of freedom.

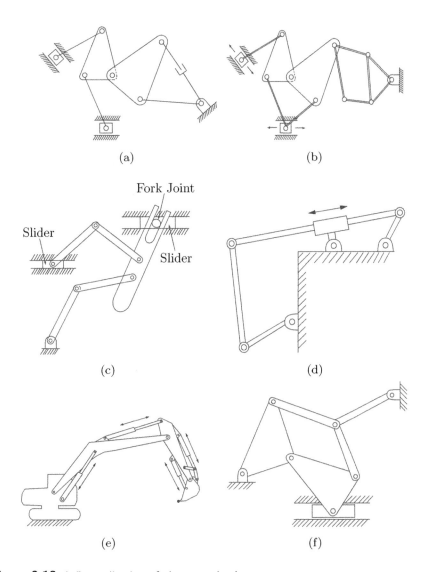

(a)                                    (b)

Fork Joint

Slider

Slider

(c)                                    (d)

(e)                                    (f)

**Figure 2.18** A first collection of planar mechanisms.

(b) Construct a physical model of the $3 \times$ UPU platform to see if it does indeed
have three degrees of freedom. In particular, lock the three P joints in place;
does the robot become a structure as predicted by Grübler's formula, or does
it move?

**Exercise 2.15**   Consider the mechanisms of Figures 2.24(a) and 2.24(b).

(a) The mechanism of Figure 2.24(a) consists of six identical squares arranged
in a single closed loop, connected by revolute joints. The bottom square is

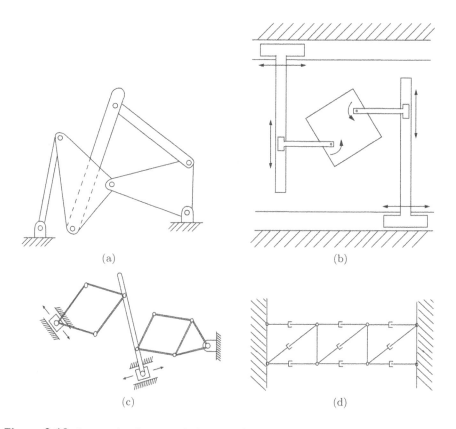

**Figure 2.19** A second collection of planar mechanisms.

fixed to ground. Determine the number of degrees of freedom using Grübler's formula.

(b) The mechanism of Figure 2.24(b) also consists of six identical squares connected by revolute joints, but arranged differently (as shown). Determine the number of degrees of freedom using Grübler's formula. Does your result agree with your intuition about the possible motions of this mechanism?

**Exercise 2.16**    Figure 2.25 shows a spherical four-bar linkage, in which four links (one of the links is the ground link) are connected by four revolute joints to form a single-loop closed chain. The four revolute joints are arranged so that they lie on a sphere such that their joint axes intersect at a common point.

(a) Use Grübler's formula to find the number of degrees of freedom. Justify your choice of formula.

(b) Describe the configuration space.

(c) Assuming that a reference frame is attached to the center link, describe its workspace.

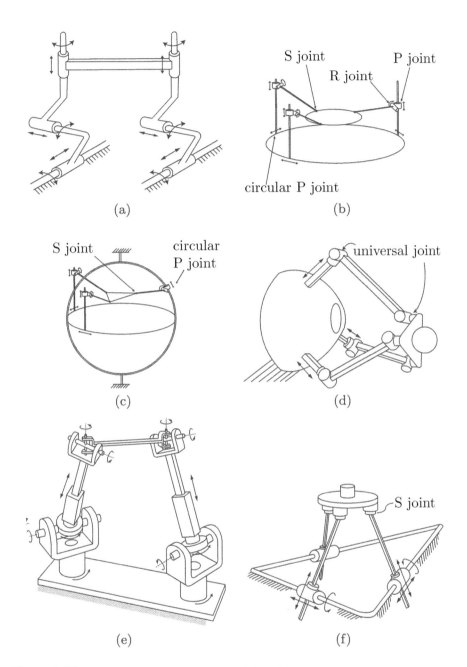

**Figure 2.20** A first collection of spatial parallel mechanisms.

**Exercise 2.17**  Figure 2.26 shows a parallel robot used for surgical applications. As shown in Figure 2.26(a), leg A is an RRRP chain, while legs B and C are RRRUR chains. A surgical tool is rigidly attached to the end-effector.

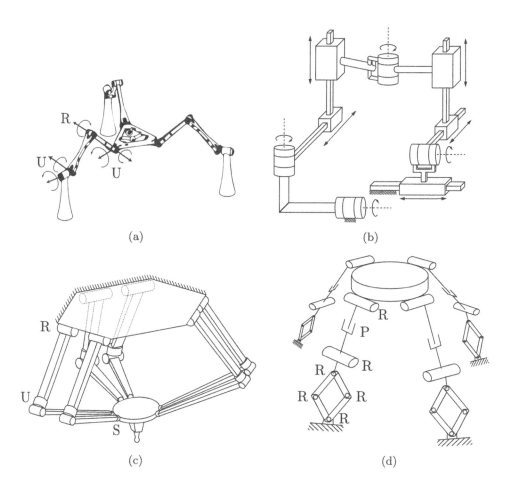

(a)

(b)

(c)

(d)

**Figure 2.21** A second collection of spatial parallel mechanisms.

(a) Use Grübler's formula to find the number of degrees of freedom of the mechanism in Figure 2.26(a).

(b) Now suppose that the surgical tool must always pass through point $A$ in Figure 2.26(a). How many degrees of freedom does the manipulator have?

(c) Legs A, B, and C are now replaced by three identical RRRR legs as shown in Figure 2.26(b). Furthermore, the axes of all R joints pass through point $A$. Use Grübler's formula to derive the number of degrees of freedom of this mechanism.

**Exercise 2.18**   Figure 2.27 shows a $3 \times$ PUP platform, in which three identical PUP legs connect a fixed base to a moving platform. The P joints on both the fixed base and moving platform are arranged symmetrically. Use Grübler's formula to find the number of degrees of freedom. Does your answer agree with your intuition about this mechanism? If not, try to explain any discrepancies without resorting to a detailed kinematic analysis.

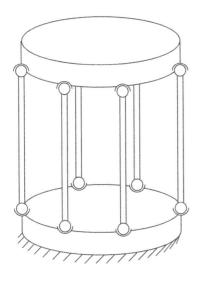

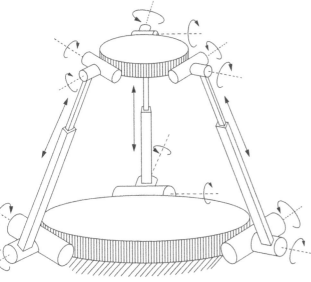

Figure 2.22 A 6×SS platform.

Figure 2.23 The 3×UPU platform.

**Exercise 2.19** The dual-arm robot of Figure 2.28 is rigidly grasping a box. The box can only slide on the table; the bottom face of the box must always be in contact with the table. How many degrees of freedom does this system have?

**Exercise 2.20** The dragonfly robot of Figure 2.29 has a body, four legs, and four wings as shown. Each leg is connected to each adjacent leg by a USP linkage. Use Grübler's formula to answer the following questions.

(a) Suppose the body is fixed and only the legs and wings can move. How many degrees of freedom does the robot have?

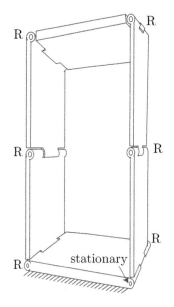
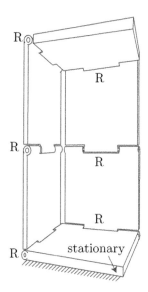

**Figure 2.24** Two mechanisms.

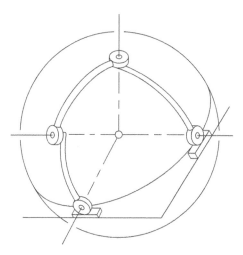

**Figure 2.25** The spherical four-bar linkage.

(b) Now suppose the robot is flying in the air. How many degrees of freedom does the robot have?

(c) Now suppose the robot is standing with all four feet in contact with the ground. Assume that the ground is uneven and that each foot–ground contact can be modeled as a point contact with no slip. How many degrees of freedom does the robot have? Explain your answer.

**Exercise 2.21**   A caterpillar robot.

(a) A caterpillar robot is hanging by its tail end as shown in Figure 2.30(a).

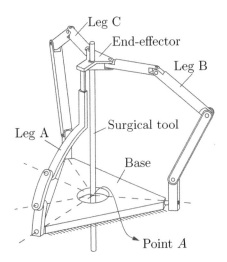
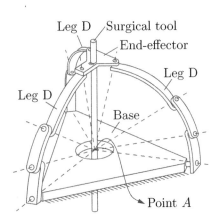

**Figure 2.26** Surgical manipulator.

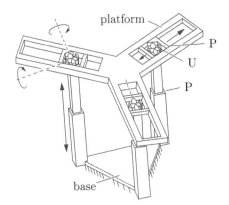

**Figure 2.27** The 3×PUP platform.

The robot consists of eight serially connected rigid links (one head, one tail, and six body links). The six body links are connected by revolute–prismatic–revolute joints, while the head and tail are connected to the body by revolute joints. Find the number of degrees of freedom of this robot.

(b) The caterpillar robot is now crawling on a leaf as shown in Figure 2.30(b). Suppose that all six body links must make contact with the leaf at all times but the links can slide and rotate on the leaf. Find the number of degrees of freedom of this robot during crawling.

(c) Now suppose the caterpillar robot crawls on the leaf as shown in Figure 2.30(c), with only the first and last body links in contact with the leaf. Find the number of degrees of freedom of this robot during crawling.

**Exercise 2.22** The four-fingered hand of Figure 2.31(a) consists of a palm and four URR fingers (the U joints connect the fingers to the palm).

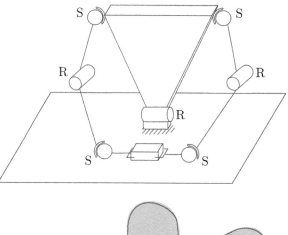

**Figure 2.28** Dual arm robot.

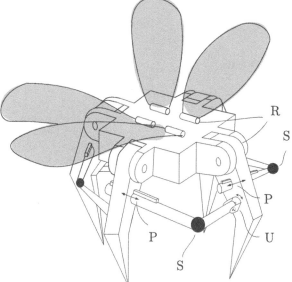

**Figure 2.29** Dragonfly robot.

(a) Assume that the fingertips are points and that one fingertip is in contact with the table surface (sliding of the fingertip point-contact is allowed). How many degrees of freedom does the hand have? What if two fingertips are in sliding point contact with the table? Three? All four?

(b) Repeat part (a) but with each URR finger replaced by an SRR finger (each universal joint is replaced by a spherical joint).

(c) The hand (with URR fingers) now grasps an ellipsoidal object, as shown in Figure 2.31(b). Assume that the palm is fixed in space and that no slip occurs between the fingertips and object. How many degrees of freedom does the system have?

(d) Now assume that the fingertips are hemispheres as shown in Figure 2.31(c). Each fingertip can roll on the object but cannot slip or break contact with the object. How many degrees of freedom does the system have? For a single

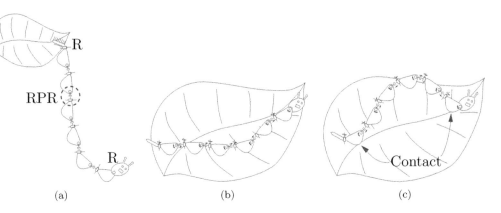

Figure 2.30 A caterpillar robot.

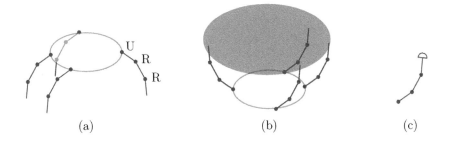

**Figure 2.31** (a) A four-fingered hand with palm. (b) The hand grasping an ellipsoidal object. (c) A rounded fingertip that can roll on the object without sliding.

fingertip in rolling contact with the object, comment on the dimension of the space of feasible fingertip velocities relative to the object versus the number of parameters needed to represent the fingertip configuration relative to the object (the number of degrees of freedom). (Hint: You may want to experiment by rolling a ball around on a tabletop to get some intuition.)

**Exercise 2.23** Consider the slider–crank mechanism of Figure 2.4(b). A rotational motion at the revolute joint fixed to ground (the "crank") causes a translational motion at the prismatic joint (the "slider"). Suppose that the two links connected to the crank and slider are of equal length. Determine the configuration space of this mechanism, and draw its projected version on the space defined by the crank and slider joint variables.

**Exercise 2.24** The planar four-bar linkage.

(a) Use Grübler's formula to determine the number of degrees of freedom of a planar four-bar linkage floating in space.

(b) Derive an implicit parametrization of the four-bar's configuration space as follows. First, label the four links 1, 2, 3, and 4, and choose three points

$A, B, C$ on link 1, $D, E, F$ on link 2, $G, H, I$ on link 3, and $J, K, L$ on link 4. The four-bar linkage is constructed in such a way that the following four pairs of points are each connected by a revolute joint: $C$ with $D$, $F$ with $G$, $I$ with $J$, and $L$ with $A$. Write down explicit constraints on the coordinates for the eight points $A, \dots, H$ (assume that a fixed reference frame has been chosen, and denote the coordinates for point $A$ by $p_A = (x_A, y_A, z_A)$, and similarly for the other points). Based on counting the number of variables and constraints, how many degrees of freedom does the configuration space have? If it differs from the result you obtained in (a), try to explain why.

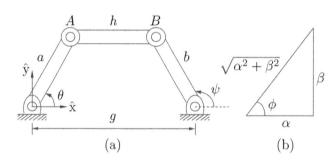

(a)          (b)

**Figure 2.32** Planar four-bar linkage.

**Exercise 2.25**   In this exercise we examine in more detail the representation of the C-space for the planar four-bar linkage of Figure 2.32. Attach a fixed reference frame and label the joints and link lengths as shown in the figure. The $(x, y)$ coordinates for joints $A$ and $B$ are given by

$$A(\theta) = (a \cos \theta, \ a \sin \theta),$$
$$B(\psi) = (g + b \cos \psi, \ b \sin \psi).$$

Using the fact that the link connecting $A$ and $B$ is of fixed length $h$, i.e., $\|A(\theta) - B(\psi)\|^2 = h^2$, we have the constraint

$$b^2 + g^2 + 2gb \cos \psi + a^2 - 2(a \cos \theta(g + b \cos \psi) + ab \sin \theta \sin \psi) = h^2.$$

Grouping the coefficients of $\cos \psi$ and $\sin \psi$, the above equation can be expressed in the form

$$\alpha(\theta) \cos \psi + \beta(\theta) \sin \psi = \gamma(\theta), \tag{2.11}$$

where

$$\alpha(\theta) = 2gb - 2ab \cos \theta,$$
$$\beta(\theta) = -2ab \sin \theta,$$
$$\gamma(\theta) = h^2 - g^2 - b^2 - a^2 + 2ag \cos \theta.$$

We now express $\psi$ as a function of $\theta$, by first dividing both sides of Equation (2.11) by $\sqrt{\alpha^2 + \beta^2}$ to obtain

$$\frac{\alpha}{\sqrt{\alpha^2 + \beta^2}} \cos \psi + \frac{\beta}{\sqrt{\alpha^2 + \beta^2}} \sin \psi = \frac{\gamma}{\sqrt{\alpha^2 + \beta^2}}. \tag{2.12}$$

Referring to Figure 2.32(b), the angle $\phi$ is given by $\phi = \tan^{-1}(\beta/\alpha)$, so that Equation (2.12) becomes

$$\cos(\psi - \phi) = \frac{\gamma}{\sqrt{\alpha^2 + \beta^2}}.$$

Therefore

$$\psi = \tan^{-1}\left(\frac{\beta}{\alpha}\right) \pm \cos^{-1}\left(\frac{\gamma}{\sqrt{\alpha^2 + \beta^2}}\right).$$

(a) Note that a solution exists only if $\gamma^2 \le \alpha^2 + \beta^2$. What are the physical implications if this constraint is not satisfied?
(b) Note that, for each value of the input angle $\theta$, there exist two possible values of the output angle $\psi$. What do these two solutions look like?
(c) Draw the configuration space of the mechanism in $\theta$–$\psi$ space for the following link length values: $a = b = g = h = 1$.
(d) Repeat (c) for the following link length values: $a = 1$, $b = 2$, $h = \sqrt{5}$, $g = 2$.
(e) Repeat (c) for the following link length values: $a = 1$, $b = 1$, $h = 1$, $g = \sqrt{3}$.

**Exercise 2.26**    The tip coordinates for the two-link planar 2R robot of Figure 2.33 are given by

$$x = 2\cos\theta_1 + \cos(\theta_1 + \theta_2)$$
$$y = 2\sin\theta_1 + \sin(\theta_1 + \theta_2).$$

(a) What is the robot's configuration space?
(b) What is the robot's workspace (i.e., the set of all points reachable by the tip)?
(c) Suppose infinitely long vertical barriers are placed at $x = 1$ and $x = -1$. What is the free C-space of the robot (i.e., the portion of the C-space that does not result in any collisions with the vertical barriers)?

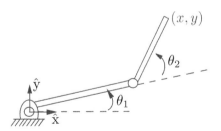

Figure 2.33 Two-link planar 2R open chain.

**Exercise 2.27**    The workspace of a planar 3R open chain.

(a) Consider a planar 3R open chain with link lengths (starting from the fixed base joint) 5, 2, and 1, respectively. Considering only the Cartesian point of the tip, draw its workspace.

(b) Now consider a planar 3R open chain with link lengths (starting from the fixed base joint) 1, 2, and 5, respectively. Considering only the Cartesian point of the tip, draw its workspace. Which of these two chains has a larger workspace?

(c) A not-so-clever designer claims that he can make the workspace of any planar open chain larger simply by increasing the length of the last link. Explain the fallacy behind this claim.

**Exercise 2.28**   Task space.

(a) Describe the task space for a robot arm writing on a blackboard.
(b) Describe the task space for a robot arm twirling a baton.

**Exercise 2.29**   Give a mathematical description of the topologies of the C-spaces of the following systems. Use cross products, as appropriate, of spaces such as a closed interval $[a, b]$ of a line and $\mathbb{R}^k$, $S^m$, and $T^n$, where $k$, $m$, and $n$ are chosen appropriately.

(a) The chassis of a car-like mobile robot rolling on an infinite plane.
(b) The car-like mobile robot (chassis only) driving around on a spherical asteroid.
(c) The car-like mobile robot (chassis only) on an infinite plane with an RRPR robot arm mounted on it. The prismatic joint has joint limits, but the revolute joints do not.
(d) A free-flying spacecraft with a 6R arm mounted on it and no joint limits.

**Exercise 2.30**   Describe an algorithm that drives the rolling coin of Figure 2.11 from any arbitrary initial configuration in its four-dimensional C-space to any arbitrary goal configuration, despite the two nonholonomic constraints. The control inputs are the rolling speed $\dot{\theta}$ and the turning speed $\dot{\phi}$. You should explain clearly in words or pseudocode how the algorithm would work. It is not necessary to give actual code or formulas.

**Exercise 2.31**   A differential-drive mobile robot has two wheels that do not steer but whose speeds can be controlled independently. The robot goes forward and backward by spinning the wheels in the same direction at the same speed, and it turns by spinning the wheels at different speeds. The configuration of the robot is given by five variables: the $(x, y)$ location of the point halfway between the wheels, the heading direction $\theta$ of the robot's chassis relative to the $x$-axis of the world frame, and the rotation angles $\phi_1$ and $\phi_2$ of the two wheels about the axis through the centers of the wheels (Figure 2.34). Assume that the radius of each wheel is $r$ and the distance between the wheels is $2d$.

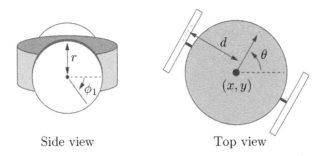

Side view            Top view

**Figure 2.34** A side view and a top view of a differential-drive robot.

(a) Let $q = (x, y, \theta, \phi_1, \phi_2)$ be the configuration of the robot. If the two control inputs are the angular velocities of the wheels $\omega_1 = \dot{\phi}_1$ and $\omega_2 = \dot{\phi}_2$, write down the vector differential equation $\dot{q} = g_1(q)\omega_1 + g_2(q)\omega_2$. The vector fields $g_1(q)$ and $g_2(q)$ are called *control vector fields* (see Section 13.3) and express how the system moves when the respective unit control signal is applied.

(b) Write the corresponding Pfaffian constraints $A(q)\dot{q} = 0$ for this system. How many Pfaffian constraints are there?

(c) Are the constraints holonomic or nonholonomic? Or how many are holonomic and how many nonholonomic?

**Exercise 2.32** Determine whether the following differential constraints are holonomic or nonholonomic:

(a) $(1 + \cos q_1)\dot{q}_1 + (1 + \cos q_2)\dot{q}_2 + (\cos q_1 + \cos q_2 + 4)\dot{q}_3 = 0.$

(b) 
$$-\dot{q}_1 \cos q_2 + \dot{q}_3 \sin(q_1 + q_2) - \dot{q}_4 \cos(q_1 + q_2) = 0$$
$$\dot{q}_3 \sin q_1 - \dot{q}_4 \cos q_1 = 0.$$

# 3    Rigid-Body Motions

In the previous chapter, we saw that a minimum of six numbers is needed to specify the position and orientation of a rigid body in three-dimensional physical space. In this chapter we develop a systematic way to describe a rigid body's position and orientation which relies on attaching a reference frame to the body. The configuration of this frame with respect to a fixed reference frame is then represented as a $4 \times 4$ matrix. This matrix is an example of an implicit representation of the C-space, as discussed in the previous chapter: the actual six-dimensional space of rigid-body configurations is obtained by applying ten constraints to the 16-dimensional space of $4 \times 4$ real matrices.

Such a matrix not only represents the configuration of a frame, but can also be used to (1) translate and rotate a vector or a frame, and (2) change the representation of a vector or a frame from coordinates in one frame to coordinates in another frame. These operations can be performed by simple linear algebra, which is a major reason why we choose to represent a configuration as a $4 \times 4$ matrix.

The non-Euclidean (i.e., non-"flat") nature of the C-space of positions and orientations leads us to use a matrix representation. A rigid body's velocity, however, can be represented simply as a point in $\mathbb{R}^6$, defined by three angular velocities and three linear velocities, which together we call a **spatial velocity** or **twist**. More generally, even though a robot's C-space may not be a vector space, the set of feasible velocities at any point in the C-space always forms a vector space. For example, consider a robot whose C-space is the sphere $S^2$: although the C-space is not flat, at any point on the sphere the space of velocities can be thought of as the plane (a vector space) tangent to that point on the sphere.

Any rigid-body configuration can be achieved by starting from the fixed (home) reference frame and integrating a constant twist for a specified time. Such a motion resembles the motion of a screw, rotating about and translating along the same fixed axis. The observation that all configurations can be achieved by a screw motion motivates a six-parameter representation of the configuration called the **exponential coordinates**. The six parameters can be divided into the parameters describing the direction of the screw axis and a scalar to indicate how far the screw motion must be followed to achieve the desired configuration.

This chapter concludes with a discussion of forces. Just as angular and linear velocities are packaged together into a single vector in $\mathbb{R}^6$, moments (torques) and forces are packaged together into a six-vector called a **spatial force** or **wrench**.

To illustrate the concepts and to provide a synopsis of the chapter, we begin with a motivating planar example. Before doing so, we make some remarks about vector notation.

### A Word about Vectors and Reference Frames

A **free vector** is a geometric quantity with a length and a direction. Think of it as an arrow in $\mathbb{R}^n$. It is called "free" because it is not necessarily rooted anywhere; only its length and direction matter. A linear velocity can be viewed as a free vector: the length of the arrow is the speed and the direction of the arrow is the direction of the velocity. A free vector is denoted by an upright text symbol, e.g., v.

If a reference frame and length scale have been chosen for the underlying space in which v lies then this free vector can be moved to a position such that the base of the arrow is at the origin without changing the orientation. The free vector v can then be represented by its coordinates in the reference frame. We write the vector in italics, $v \in \mathbb{R}^n$, where $v$ is at the "head" of the arrow in the frame's coordinates. If a different reference frame and length scale are chosen then the representation $v$ will change but the underlying free vector v is unchanged.

In other words, we say that v is **coordinate free**; it refers to a physical quantity in the underlying space, and it does not care how we represent it. However, $v$ is a representation of v that depends on the choice of coordinate frame.

A point p in physical space can also be represented as a vector. Given a choice of reference frame and length scale for physical space, the point p can be represented as a vector from the reference frame origin to p; its vector representation is denoted in italics by $p \in \mathbb{R}^n$. Here, as before, a different choice of reference frame and length scale for physical space leads to a different representation $p \in \mathbb{R}^n$ for the same point p in physical space. See Figure 3.1.

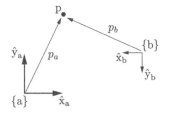

**Figure 3.1** The point p exists in physical space, and it does not care how we represent it. If we fix a reference frame {a}, with unit coordinate axes $\hat{x}_a$ and $\hat{y}_a$, we can represent p as $p_a = (1, 2)$. If we fix a reference frame {b} at a different location, a different orientation, and a different length scale, we can represent p as $p_b = (4, -2)$.

In the rest of this book, a choice of length scale will always be assumed, but we will be dealing with reference frames at different positions and orientations. A reference frame can be placed anywhere in space, and any reference frame leads to an equally valid representation of the underlying space and the objects in it. We always assume that exactly one stationary **fixed frame**, or **space frame**, denoted {s}, has been defined. This might be attached to a corner of a room, for

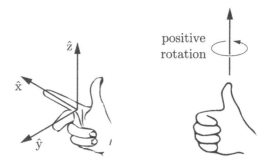

**Figure 3.2** (Left) The $\hat{x}$-, $\hat{y}$-, and $\hat{z}$-axes of a right-handed reference frame are aligned with the index finger, middle finger, and thumb of the right hand, respectively. (Right) A positive rotation about an axis is in the direction in which the fingers of the right hand curl when the thumb is pointed along the axis.

example. Similarly, we often assume that at least one frame has been attached to some moving rigid body, such as the body of a quadrotor flying in the room. This **body frame**, denoted {b}, is the stationary frame that is coincident with the body-attached frame at any instant.

While it is common to attach the origin of the {b} frame to some important point on the body, such as its center of mass, this is not necessary. The origin of the {b} frame does not even need to be on the physical body itself, as long as its configuration relative to the body, viewed from an observer stationary relative to the body, is constant.

**Important!** All frames in this book are stationary, inertial, frames. When we refer to a body frame {b}, we mean a motionless frame that is instantaneously coincident with a frame that is fixed to a (possibly moving) body. This is important to keep in mind, since you may have had a dynamics course that used non-inertial moving frames attached to rotating bodies. Do not confuse these with the stationary, inertial, body frames of this book.

For simplicity, we will usually refer to a body frame as a frame attached to a moving rigid body. Despite this, at any instant, by "body frame" we actually mean the stationary frame that is instantaneously coincident with the frame moving along with the body.

It is worth repeating one more time: **all frames are stationary**.

All reference frames are **right-handed**, as illustrated in Figure 3.2. A positive rotation about an axis is defined as the direction in which the fingers of the right hand curl when the thumb is pointed along the axis (Figure 3.2).

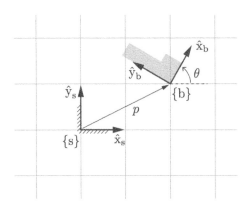

**Figure 3.3** The body frame $\{b\}$ is expressed in the fixed-frame coordinates $\{s\}$ by the vector $p$ and the directions of the unit axes $\hat{x}_b$ and $\hat{y}_b$. In this example, $p = (2, 1)$ and $\theta = 60°$, so $\hat{x}_b = (\cos\theta, \sin\theta) = (1/2, \sqrt{3}/2)$ and $\hat{y}_b = (-\sin\theta, \cos\theta) = (-\sqrt{3}/2, 1/2)$.

## .1    Rigid-Body Motions in the Plane

Consider the planar body (the gray shape) in Figure 3.3; its motion is confined to the plane. Suppose that a length scale and a fixed reference frame $\{s\}$ have been chosen as shown, with unit axes $\hat{x}_s$ and $\hat{y}_s$. (Throughout this book, the hat notation indicates a unit vector.) Similarly, we attach a reference frame with unit axes $\hat{x}_b$ and $\hat{y}_b$ to the planar body. Because this frame moves with the body, it is called the body frame and is denoted $\{b\}$.

To describe the configuration of the planar body, only the position and orientation of the body frame with respect to the fixed frame need to be specified. The body-frame origin p can be expressed in terms of the coordinate axes of $\{s\}$ as

$$p = p_x\hat{x}_s + p_y\hat{y}_s. \tag{3.1}$$

You are probably more accustomed to writing this vector as simply $p = (p_x, p_y)$; this is fine when there is no possibility of ambiguity about reference frames, but writing $p$ as in Equation (3.1) clearly indicates the reference frame with respect to which $(p_x, p_y)$ is defined.

The simplest way to describe the orientation of the body frame $\{b\}$ relative to the fixed frame $\{s\}$ is by specifying the angle $\theta$, as shown in Figure 3.3. Another (admittedly less simple) way is to specify the directions of the unit axes $\hat{x}_b$ and $\hat{y}_b$ of $\{b\}$ relative to $\{s\}$, in the form

$$\hat{x}_b = \cos\theta\,\hat{x}_s + \sin\theta\,\hat{y}_s, \tag{3.2}$$

$$\hat{y}_b = -\sin\theta\,\hat{x}_s + \cos\theta\,\hat{y}_s. \tag{3.3}$$

At first sight this seems to be a rather inefficient way of representing the body-frame orientation. However, imagine if the body were to move arbitrarily in three-dimensional space; a single angle $\theta$ would not suffice to describe the orientation of

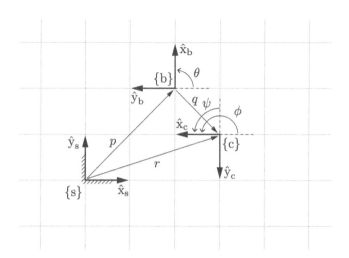

**Figure 3.4** The frame {b} in {s} is given by $(P, p)$, and the frame {c} in {b} is given by $(Q, q)$. From these we can derive the frame {c} in {s}, described by $(R, r)$. The numerical values of the vectors $p$, $q$, and $r$ and the coordinate-axis directions of the three frames are evident from the grid of unit squares.

the displaced reference frame. We would actually need three angles, but it is not yet clear how to define an appropriate set of three angles. However, expressing the directions of the coordinate axes of {b} in terms of coefficients of the coordinate axes of {s}, as we have done above for the planar case, is straightforward.

Assuming we agree to express everything in terms of {s} then, just as the point p can be represented as a column vector $p \in \mathbb{R}^2$ of the form

$$p = \begin{bmatrix} p_x \\ p_y \end{bmatrix}, \tag{3.4}$$

the two vectors $\hat{x}_b$ and $\hat{y}_b$ can also be written as column vectors and packaged into the following $2 \times 2$ matrix $P$:

$$P = [\hat{x}_b \ \hat{y}_b] = \begin{bmatrix} \cos\theta & -\sin\theta \\ \sin\theta & \cos\theta \end{bmatrix}. \tag{3.5}$$

The matrix $P$ is an example of a **rotation matrix**. Although $P$ consists of four numbers, they are subject to three constraints (each column of $P$ must be a unit vector, and the two columns must be orthogonal to each other), and the one remaining degree of freedom is parametrized by $\theta$. Together, the pair $(P, p)$ provides a description of the orientation and position of {b} relative to {s}.

Now refer to the three frames in Figure 3.4. Repeating the approach above, and expressing {c} in {s} as the pair $(R, r)$, we can write

$$r = \begin{bmatrix} r_x \\ r_y \end{bmatrix}, \qquad R = \begin{bmatrix} \cos\phi & -\sin\phi \\ \sin\phi & \cos\phi \end{bmatrix}. \tag{3.6}$$

We could also describe the frame {c} relative to {b}. Letting $q$ denote the vector from the origin of {b} to the origin of {c} expressed in {b} coordinates, and letting $Q$ denote the orientation of {c} relative to {b}, we can write {c} relative to {b} as the pair $(Q, q)$, where

$$q = \begin{bmatrix} q_x \\ q_y \end{bmatrix}, \qquad Q = \begin{bmatrix} \cos\psi & -\sin\psi \\ \sin\psi & \cos\psi \end{bmatrix}. \qquad (3.7)$$

If we know $(Q, q)$ (the configuration of {c} relative to {b}) and $(P, p)$ (the configuration of {b} relative to {s}), we can compute the configuration of {c} relative to {s} as follows:

$$R = PQ \quad \text{(convert } Q \text{ to the \{s\} frame)} \qquad (3.8)$$

$$r = Pq + p \quad \text{(convert } q \text{ to the \{s\} frame and vector-sum with } p\text{).} \qquad (3.9)$$

Thus $(P, p)$ not only represents a configuration of {b} in {s}; it can also be used to convert the representation of a point or frame from {b} coordinates to {s} coordinates.

Now consider a rigid body with two frames attached to it, {d} and {c}. The frame {d} is initially coincident with {s}, and {c} is initially described by $(R, r)$ in {s} (Figure 3.5(a)). Then the body is moved in such a way that {d} moves to {d′}, becoming coincident with a frame {b} described by $(P, p)$ in {s}. Where does {c} end up after this motion? Denoting the configuration of the new frame {c′} as $(R', r')$, you can verify that

$$R' = PR, \qquad (3.10)$$

$$r' = Pr + p, \qquad (3.11)$$

which is similar to Equations (3.8) and (3.9). The difference is that $(P, p)$ is expressed in the same frame as $(R, r)$, so the equations are not viewed as a change of coordinates, but instead as a **rigid-body displacement** (also known as a **rigid-body motion**): in Figure 3.5(a) transformation ① rotates {c} according to $P$ and transformation ② translates it by $p$ in {s}.

Thus we see that a rotation matrix–vector pair such as $(P, p)$ can be used for three purposes:

(a) to represent a configuration of a rigid body in {s} (Figure 3.3);
(b) to change the reference frame in which a vector or frame is represented (Figure 3.4);
(c) to displace a vector or a frame (Figure 3.5(a)).

Referring to Figure 3.5(b), note that the rigid-body motion illustrated in Figure 3.5(a), expressed as a rotation followed by a translation, can be obtained by simply rotating the body about a fixed point s by an angle $\beta$. This is a planar example of a **screw motion**.[1] The displacement can therefore be parametrized

---

[1] If the displacement is a pure translation without rotation, then s lies at infinity.

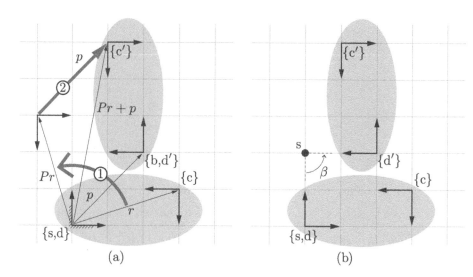

**Figure 3.5** (a) The frame {d}, fixed to an elliptical rigid body and initially coincident with {s}, is displaced to {d′} (which is coincident with the stationary frame {b}), by first rotating according to $P$ then translating according to $p$, where $(P, p)$ is the representation of {b} in {s}. The same transformation takes the frame {c}, also attached to the rigid body, to {c′}. The transformation marked ① rigidly rotates {c} about the origin of {s}, and then transformation ② translates the frame by $p$ expressed in {s}. (b) Instead of viewing this displacement as a rotation followed by a translation, both rotation and translation can be performed simultaneously. The displacement can be viewed as a rotation of $\beta = 90°$ about a fixed point s.

by the three screw coordinates $(\beta, s_x, s_y)$, where $(s_x, s_y) = (0, 2)$ denotes the coordinates for the point s (i.e., the screw axis out of the page) in the fixed frame {s}.

Another way to represent the screw motion is to consider it as the displacement obtained by following simultaneous angular and linear velocities for a given distance. Inspecting Figure 3.5(b), we see that rotating about s with a unit angular velocity ($\omega = 1$ rad/s) means that a point at the origin of the {s} frame moves at two units per second initially in the $+\hat{x}$-direction of the {s} frame, i.e., $v = (v_x, v_y) = (2, 0)$. We can package these together in the three-vector $\mathcal{S} = (\omega, v_x, v_y) = (1, 2, 0)$, a representation of the **screw axis**. Following this screw axis for an angle $\theta = \pi/2$ yields the final displacement. Thus we can represent the displacement using the three coordinates $\mathcal{S}\theta = (\pi/2, \pi, 0)$. These coordinates have some advantages, and we call these the **exponential coordinates** for the planar rigid-body displacement.

To represent the combination of an angular and a linear velocity, called a **twist**, we take a screw axis $\mathcal{S} = (\omega, v_x, v_y)$, where $\omega = 1$, and scale it by multiplying by some rotation speed, $\dot{\theta}$. The twist is $\mathcal{V} = \mathcal{S}\dot{\theta}$. The net displacement obtained by rotating about the screw axis $\mathcal{S}$ by an angle $\theta$ is equivalent to the displacement

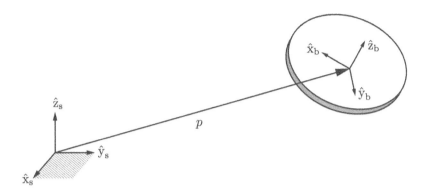

**Figure 3.6** Mathematical description of position and orientation.

obtained by rotating about $\mathcal{S}$ at a speed $\dot{\theta} = \theta$ for unit time, so $\mathcal{V} = \mathcal{S}\dot{\theta}$ can also be considered a set of exponential coordinates.

**Preview of the remainder of this chapter.**
In the rest of this chapter we generalize the concepts above to three-dimensional rigid-body motions. For this purpose consider a rigid body occupying three-dimensional physical space, as shown in Figure 3.6. Assume that a length scale for physical space has been chosen, and that both the fixed frame {s} and body frame {b} have been chosen as shown. Throughout this book all reference frames are right-handed – the unit axes $\{\hat{x}, \hat{y}, \hat{z}\}$ always satisfy $\hat{x} \times \hat{y} = \hat{z}$. Denote the unit axes of the fixed frame by $\{\hat{x}_s, \hat{y}_s, \hat{z}_s\}$ and the unit axes of the body frame by $\{\hat{x}_b, \hat{y}_b, \hat{z}_b\}$. Let p denote the vector from the fixed-frame origin to the body-frame origin. In terms of the fixed-frame coordinates, p can be expressed as

$$p = p_1\hat{x}_s + p_2\hat{y}_s + p_3\hat{z}_s. \tag{3.12}$$

The axes of the body frame can also be expressed as

$$\hat{x}_b = r_{11}\hat{x}_s + r_{21}\hat{y}_s + r_{31}\hat{z}_s, \tag{3.13}$$

$$\hat{y}_b = r_{12}\hat{x}_s + r_{22}\hat{y}_s + r_{32}\hat{z}_s, \tag{3.14}$$

$$\hat{z}_b = r_{13}\hat{x}_s + r_{23}\hat{y}_s + r_{33}\hat{z}_s. \tag{3.15}$$

Defining $p \in \mathbb{R}^3$ and $R \in \mathbb{R}^{3\times3}$ as

$$p = \begin{bmatrix} p_1 \\ p_2 \\ p_3 \end{bmatrix}, \qquad R = [\hat{x}_b \ \hat{y}_b \ \hat{z}_b] = \begin{bmatrix} r_{11} & r_{12} & r_{13} \\ r_{21} & r_{22} & r_{23} \\ r_{31} & r_{32} & r_{33} \end{bmatrix}, \tag{3.16}$$

the 12 parameters given by $(R, p)$ then provide a description of the position and orientation of the rigid body relative to the fixed frame.

Since the orientation of a rigid body has three degrees of freedom, only three of the nine entries in $R$ can be chosen independently. One three-parameter representation of rotations is provided by the exponential coordinates, which define an axis of rotation and the angle rotated about that axis. We leave other popular representations of orientations (the three-parameter **Euler angles** and the **roll–pitch–yaw angles**, the **Cayley–Rodrigues parameters**, and the **unit quaternions**, which use four variables subject to one constraint) to Appendix B.

We then examine the six-parameter exponential coordinates for the configuration of a rigid body that arise from integrating a six-dimensional twist consisting of the body's angular and linear velocities. This representation follows from the Chasles–Mozzi theorem which states that every rigid-body displacement can be obtained by a finite rotation and translation about a fixed screw axis.

We conclude with a discussion of forces and moments. Rather than treat these as separate three-dimensional quantities, we merge the moment and force vectors into a six-dimensional **wrench**. The twist and wrench, and rules for manipulating them, form the basis for the kinematic and dynamic analyses in subsequent chapters.

## 3.2    Rotations and Angular Velocities

### 3.2.1    Rotation Matrices

We argued earlier that, of the nine entries in the rotation matrix $R$, only three can be chosen independently. We begin by expressing a set of six explicit constraints on the entries of $R$. Recall that the three columns of $R$ correspond to the body-frame unit axes $\{\hat{x}_b, \hat{y}_b, \hat{z}_b\}$. The following conditions must therefore be satisfied.

(a) The unit norm condition: $\hat{x}_b$, $\hat{y}_b$, and $\hat{z}_b$ are all unit vectors, i.e.,

$$
\begin{aligned}
r_{11}^2 + r_{21}^2 + r_{31}^2 &= 1, \\
r_{12}^2 + r_{22}^2 + r_{32}^2 &= 1, \\
r_{13}^2 + r_{23}^2 + r_{33}^2 &= 1.
\end{aligned}
\tag{3.17}
$$

(b) The orthogonality condition: $\hat{x}_b \cdot \hat{y}_b = \hat{x}_b \cdot \hat{z}_b = \hat{y}_b \cdot \hat{z}_b = 0$ (here $\cdot$ denotes the inner product), or

$$
\begin{aligned}
r_{11}r_{12} + r_{21}r_{22} + r_{31}r_{32} &= 0, \\
r_{12}r_{13} + r_{22}r_{23} + r_{32}r_{33} &= 0, \\
r_{11}r_{13} + r_{21}r_{23} + r_{31}r_{33} &= 0.
\end{aligned}
\tag{3.18}
$$

These six constraints can be expressed more compactly as a single set of constraints on the matrix $R$,

$$
R^{\mathrm{T}} R = I,
\tag{3.19}
$$

where $R^{\mathrm{T}}$ denotes the transpose of $R$ and $I$ denotes the identity matrix.

There is still the matter of accounting for the fact that the frame is right-handed (i.e., $\hat{x}_b \times \hat{y}_b = \hat{z}_b$, where $\times$ denotes the cross-product) rather than left-handed (i.e., $\hat{x}_b \times \hat{y}_b = -\hat{z}_b$); our six equality constraints above do not distinguish between right- and left-handed frames. We recall the following formula for evaluating the determinant of a $3 \times 3$ matrix $M$: denoting the three columns of $M$ by $a$, $b$, and $c$, respectively, its determinant is given by

$$\det M = a^T(b \times c) = c^T(a \times b) = b^T(c \times a). \tag{3.20}$$

Substituting the columns for $R$ into this formula then leads to the constraint

$$\det R = 1. \tag{3.21}$$

Note that, had the frame been left-handed, we would have $\det R = -1$. In summary, the six equality constraints represented by Equation (3.19) imply that $\det R = \pm 1$; imposing the additional constraint $\det R = 1$ means that only right-handed frames are allowed. The constraint $\det R = 1$ does not change the number of independent continuous variables needed to parametrize $R$.

The set of $3 \times 3$ rotation matrices forms the **special orthogonal group** $SO(3)$, which we now formally define.

**Definition 3.1** The **special orthogonal group** $SO(3)$, also known as the group of rotation matrices, is the set of all $3 \times 3$ real matrices $R$ that satisfy (i) $R^T R = I$ and (ii) $\det R = 1$.

The set of $2 \times 2$ rotation matrices is a subgroup of $SO(3)$ and is denoted $SO(2)$.

**Definition 3.2** The **special orthogonal group** $SO(2)$ is the set of all $2 \times 2$ real matrices $R$ that satisfy (i) $R^T R = I$ and (ii) $\det R = 1$.

From the definition it follows that every $R \in SO(2)$ can be written

$$R = \begin{bmatrix} r_{11} & r_{12} \\ r_{21} & r_{22} \end{bmatrix} = \begin{bmatrix} \cos\theta & -\sin\theta \\ \sin\theta & \cos\theta \end{bmatrix},$$

where $\theta \in [0, 2\pi)$. The elements of $SO(2)$ represent planar orientations and the elements of $SO(3)$ represent spatial orientations.

### 3.2.1.1 Properties of Rotation Matrices

The sets of rotation matrices $SO(2)$ and $SO(3)$ are called groups because they satisfy the properties required of a mathematical group.[2] Specifically, a group consists of a set of elements and an operation on two elements (matrix multiplication for $SO(n)$) such that, for all $A$, $B$ in the group, the following properties are satisfied:

- **closure:** $AB$ is also in the group.
- **associativity:** $(AB)C = A(BC)$.
- **identity element existence:** There exists an element $I$ in the group (the identity matrix for $SO(n)$) such that $AI = IA = A$.

---

[2] More specifically, the $SO(n)$ groups are also called *matrix Lie groups* (where "Lie" is pronounced "Lee") because the elements of the group form a differentiable manifold.

- **inverse element existence:** There exists an element $A^{-1}$ in the group such that $AA^{-1} = A^{-1}A = I$.

Proofs of these properties are given below, using the fact that the identity matrix $I$ is a trivial example of a rotation matrix.

**Proposition 3.3**   *The inverse of a rotation matrix $R \in SO(3)$ is also a rotation matrix, and it is equal to the transpose of $R$, i.e., $R^{-1} = R^T$.*

*Proof*   The condition $R^T R = I$ implies that $R^{-1} = R^T$ and $RR^T = I$. Since $\det R^T = \det R = 1$, $R^T$ is also a rotation matrix.    □

**Proposition 3.4**   *The product of two rotation matrices is a rotation matrix.*

*Proof*   Given $R_1, R_2 \in SO(3)$, their product $R_1 R_2$ satisfies $(R_1 R_2)^T (R_1 R_2) = R_2^T R_1^T R_1 R_2 = R_2^T R_2 = I$. Further, $\det R_1 R_2 = \det R_1 \cdot \det R_2 = 1$. Thus $R_1 R_2$ satisfies the conditions for a rotation matrix.    □

**Proposition 3.5**   *Multiplication of rotation matrices is associative, $(R_1 R_2)R_3 = R_1(R_2 R_3)$, but generally not commutative, $R_1 R_2 \neq R_2 R_1$. For the special case of rotation matrices in $SO(2)$, rotations commute.*

*Proof*   Associativity and nocommutativity follows from the properties of matrix multiplication in linear algebra. Commutativity for planar rotations follows from a direct calculation.    □

Another important property is that the action of a rotation matrix on a vector (e.g., rotating the vector) does not change the length of the vector.

**Proposition 3.6**   *For any vector $x \in \mathbb{R}^3$ and $R \in SO(3)$, the vector $y = Rx$ has the same length as $x$.*

*Proof*   This follows from $\|y\|^2 = y^T y = (Rx)^T Rx = x^T R^T Rx = x^T x = \|x\|^2$.    □

### 3.2.1.2   Uses of Rotation Matrices

Analogously to the discussion after Equations 3.10 and (3.11) in Section 3.1, there are three major uses for a rotation matrix $R$:

(a) to represent an orientation;
(b) to change the reference frame in which a vector or a frame is represented;
(c) to rotate a vector or a frame.

In the first use, $R$ is thought of as representing a frame; in the second and third uses, $R$ is thought of as an operator that acts on a vector or frame (changing its reference frame or rotating it).

To illustrate these uses, refer to Figure 3.7, which shows three different coordinate frames – {a}, {b}, and {c} – representing the same space. These frames are chosen to have the same origin, since we are only representing orientations,

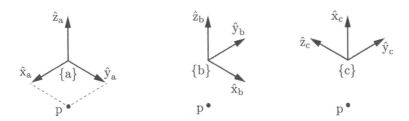

**Figure 3.7** The same space and the same point p represented in three different frames with different orientations.

but, to make the axes clear, the figure shows the same space drawn three times. A point p in the space is also shown. Not shown is a fixed space frame $\{s\}$, which is aligned with $\{a\}$. The orientations of the three frames relative to $\{s\}$ can be written

$$R_a = \begin{bmatrix} 1 & 0 & 0 \\ 0 & 1 & 0 \\ 0 & 0 & 1 \end{bmatrix}, \quad R_b = \begin{bmatrix} 0 & -1 & 0 \\ 1 & 0 & 0 \\ 0 & 0 & 1 \end{bmatrix}, \quad R_c = \begin{bmatrix} 0 & -1 & 0 \\ 0 & 0 & -1 \\ 1 & 0 & 0 \end{bmatrix},$$

and the location of the point p in these frames can be written

$$p_a = \begin{bmatrix} 1 \\ 1 \\ 0 \end{bmatrix}, \quad p_b = \begin{bmatrix} 1 \\ -1 \\ 0 \end{bmatrix}, \quad p_c = \begin{bmatrix} 0 \\ -1 \\ -1 \end{bmatrix}.$$

Note that $\{b\}$ is obtained by rotating $\{a\}$ about $\hat{z}_a$ by $90°$, and $\{c\}$ is obtained by rotating $\{b\}$ about $\hat{y}_b$ by $-90°$.

**Representing an orientation**
When we write $R_c$, we are implicitly referring to the orientation of frame $\{c\}$ relative to the fixed frame $\{s\}$. We can be more explicit about this by writing it as $R_{sc}$: we are representing the frame $\{c\}$ of the second subscript relative to the frame $\{s\}$ of the first subscript. This notation allows us to express one frame relative to another that is not $\{s\}$; for example, $R_{bc}$ is the orientation of $\{c\}$ relative to $\{b\}$.

If there is no possibility of confusion regarding the frames involved, we may simply write $R$.

Inspecting Figure 3.7, we see that

$$R_{ac} = \begin{bmatrix} 0 & -1 & 0 \\ 0 & 0 & -1 \\ 1 & 0 & 0 \end{bmatrix}, \quad R_{ca} = \begin{bmatrix} 0 & 0 & 1 \\ -1 & 0 & 0 \\ 0 & -1 & 0 \end{bmatrix}.$$

A simple calculation shows that $R_{ac}R_{ca} = I$; that is, $R_{ac} = R_{ca}^{-1}$ or, equivalently, from Proposition 3.3, $R_{ac} = R_{ca}^{\mathrm{T}}$. In fact, for any two frames $\{d\}$ and $\{e\}$,

$$R_{de} = R_{ed}^{-1} = R_{ed}^{\mathrm{T}}.$$

You can verify this fact using any two frames in Figure 3.7.

**Changing the reference frame**

The rotation matrix $R_{ab}$ represents the orientation of $\{b\}$ in $\{a\}$, and $R_{bc}$ represents the orientation of $\{c\}$ in $\{b\}$. A straightforward calculation shows that the orientation of $\{c\}$ in $\{a\}$ can be computed as

$$R_{ac} = R_{ab}R_{bc}. \tag{3.22}$$

In the previous equation, $R_{bc}$ can be viewed as a representation of the orientation of $\{c\}$, while $R_{ab}$ can be viewed as a mathematical operator that changes the reference frame from $\{b\}$ to $\{a\}$, i.e.,

$$R_{ac} = R_{ab}R_{bc} = \text{change\_reference\_frame\_from\_}\{b\}\text{\_to\_}\{a\}\,(R_{bc}).$$

A subscript cancellation rule helps us to remember this property. When multiplying two rotation matrices, if the second subscript of the first matrix matches the first subscript of the second matrix, the two subscripts "cancel" and a change of reference frame is achieved:

$$R_{ab}R_{bc} = R_{a\!\!\!/b}R_{\!\!\!/bc} = R_{ac}.$$

A rotation matrix is just a collection of three unit vectors, so the reference frame of a vector can also be changed by a rotation matrix using a modified version of the subscript cancellation rule:

$$R_{ab}p_b = R_{a\!\!\!/b}p_{\!\!\!/b} = p_a.$$

You can verify these properties using the frames and points in Figure 3.7.

**Rotating a vector or a frame**

The final use of a rotation matrix is to rotate a vector or a frame. Figure 3.8 shows a frame $\{c\}$ initially aligned with $\{s\}$ with axes $\{\hat{x}, \hat{y}, \hat{z}\}$. If we rotate the frame $\{c\}$ about a unit axis $\hat{\omega}$ by an amount $\theta$, the new frame, $\{c'\}$ (light gray), has coordinate axes $\{\hat{x}', \hat{y}', \hat{z}'\}$. The rotation matrix $R = R_{sc'}$ represents the orientation of $\{c'\}$ relative to $\{s\}$, but instead we can think of it as representing the rotation operation that takes $\{s\}$ to $\{c'\}$. Emphasizing our view of $R$ as a rotation operator, instead of as an orientation, we can write

$$R = \text{Rot}(\hat{\omega}, \theta),$$

meaning the operation that rotates the orientation represented by the identity matrix to the orientation represented by $R$. Examples of rotation operations about coordinate frame axes are

$$\text{Rot}(\hat{x}, \theta) = \begin{bmatrix} 1 & 0 & 0 \\ 0 & \cos\theta & -\sin\theta \\ 0 & \sin\theta & \cos\theta \end{bmatrix}, \quad \text{Rot}(\hat{y}, \theta) = \begin{bmatrix} \cos\theta & 0 & \sin\theta \\ 0 & 1 & 0 \\ -\sin\theta & 0 & \cos\theta \end{bmatrix},$$

$$\text{Rot}(\hat{z}, \theta) = \begin{bmatrix} \cos\theta & -\sin\theta & 0 \\ \sin\theta & \cos\theta & 0 \\ 0 & 0 & 1 \end{bmatrix}.$$

More generally, as we will see in Section 3.2.3.3, for $\hat{\omega} = (\hat{\omega}_1, \hat{\omega}_2, \hat{\omega}_3)$,

$$\text{Rot}(\hat{\omega}, \theta) =$$

$$\begin{bmatrix} c_\theta + \hat{\omega}_1^2(1 - c_\theta) & \hat{\omega}_1\hat{\omega}_2(1 - c_\theta) - \hat{\omega}_3 s_\theta & \hat{\omega}_1\hat{\omega}_3(1 - c_\theta) + \hat{\omega}_2 s_\theta \\ \hat{\omega}_1\hat{\omega}_2(1 - c_\theta) + \hat{\omega}_3 s_\theta & c_\theta + \hat{\omega}_2^2(1 - c_\theta) & \hat{\omega}_2\hat{\omega}_3(1 - c_\theta) - \hat{\omega}_1 s_\theta \\ \hat{\omega}_1\hat{\omega}_3(1 - c_\theta) - \hat{\omega}_2 s_\theta & \hat{\omega}_2\hat{\omega}_3(1 - c_\theta) + \hat{\omega}_1 s_\theta & c_\theta + \hat{\omega}_3^2(1 - c_\theta) \end{bmatrix},$$

where $s_\theta = \sin\theta$ and $c_\theta = \cos\theta$. Any $R \in SO(3)$ can be obtained by rotating from the identity matrix by some $\theta$ about some $\hat{\omega}$. Note also that $\text{Rot}(\hat{\omega}, \theta) = \text{Rot}(-\hat{\omega}, -\theta)$.

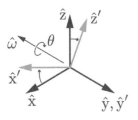

**Figure 3.8** A coordinate frame with axes $\{\hat{x}, \hat{y}, \hat{z}\}$ is rotated by $\theta$ about a unit axis $\hat{\omega}$ (which is aligned with $-\hat{y}$ in this figure). The orientation of the final frame, with axes $\{\hat{x}', \hat{y}', \hat{z}'\}$, is written as $R$ relative to the original frame.

Now, say that $R_{sb}$ represents some $\{b\}$ relative to $\{s\}$ and that we want to rotate $\{b\}$ by $\theta$ about a unit axis $\hat{\omega}$, i.e., by a rotation $R = \text{Rot}(\hat{\omega}, \theta)$. To be clear about what we mean, we have to specify whether the axis of rotation $\hat{\omega}$ is expressed in $\{s\}$ coordinates or $\{b\}$ coordinates. Depending on our choice, the same numerical $\hat{\omega}$ (and therefore the same numerical $R$) corresponds to different rotation axes in the underlying space, unless the $\{s\}$ and $\{b\}$ frames are aligned. Letting $\{b'\}$ be the new frame after a rotation by $\theta$ about $\hat{\omega}_s = \hat{\omega}$ (the rotation axis $\hat{\omega}$ is considered to be in the fixed frame, $\{s\}$), and letting $\{b''\}$ be the new frame after a rotation by $\theta$ about $\hat{\omega}_b = \hat{\omega}$ (the rotation axis $\hat{\omega}$ is considered to be in the body frame $\{b\}$), representations of these new frames can be calculated as

$$R_{sb'} = \text{rotate\_by\_}R\text{\_in\_}\{s\}\text{\_frame}\,(R_{sb}) = RR_{sb} \qquad (3.23)$$

$$R_{sb''} = \text{rotate\_by\_}R\text{\_in\_}\{b\}\text{\_frame}\,(R_{sb}) = R_{sb}R. \qquad (3.24)$$

In other words, premultiplying by $R = \text{Rot}(\hat{\omega}, \theta)$ yields a rotation about an axis $\hat{\omega}$ considered to be in the fixed frame, and postmultiplying by $R$ yields a rotation about $\hat{\omega}$ considered as being in the body frame.

Rotation by $R$ in the $\{s\}$ frame and in the $\{b\}$ frame is illustrated in Figure 3.9.

To rotate a vector $v$, note that there is only one frame involved, the frame in which $v$ is represented, and therefore $\hat{\omega}$ must be interpreted as being in this frame. The rotated vector $v'$, in that same frame, is

$$v' = Rv.$$

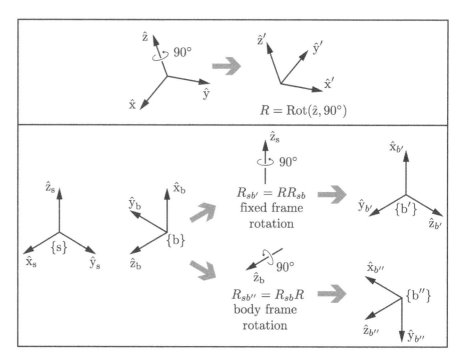

**Figure 3.9** (Top) The rotation operator $R = \mathrm{Rot}(\hat{z}, 90°)$ gives the orientation of the right-hand frame in the left-hand frame. (Bottom) On the left are shown a fixed frame $\{s\}$ and a body frame $\{b\}$, which can be expressed as $R_{sb}$. The quantity $RR_{sb}$ rotates $\{b\}$ by 90° about the fixed-frame axis $\hat{z}_s$ to $\{b'\}$. The quantity $R_{sb}R$ rotates $\{b\}$ by 90° about the body-frame axis $\hat{z}_b$ to $\{b''\}$.

### 3.2.2　Angular Velocities

Referring to Figure 3.10(a), suppose that a frame with unit axes $\{\hat{x}, \hat{y}, \hat{z}\}$ is attached to a rotating body. Let us determine the time derivatives of these unit axes. Beginning with $\dot{\hat{x}}$, first note that $\hat{x}$ is of unit length; only the direction of $\hat{x}$ can vary with time (the same goes for $\hat{y}$ and $\hat{z}$). If we examine the body frame at times $t$ and $t + \Delta t$, the change in frame orientation can be described as a rotation of angle $\Delta\theta$ about some unit axis $\hat{w}$ passing through the origin. The axis $\hat{w}$ is coordinate-free; it is not yet represented in any particular reference frame.

In the limit as $\Delta t$ approaches zero, the ratio $\Delta\theta / \Delta t$ becomes the rate of rotation $\dot{\theta}$, and $\hat{w}$ can similarly be regarded as the instantaneous axis of rotation. In fact, $\hat{w}$ and $\dot{\theta}$ can be combined to define the **angular velocity** $w$ as follows:

$$w = \hat{w}\dot{\theta}. \tag{3.25}$$

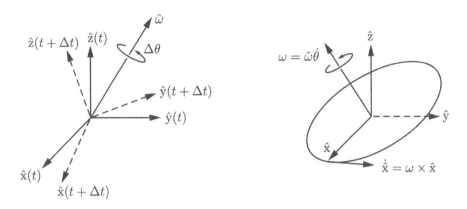

**Figure 3.10** (Left) The instantaneous angular velocity vector. (Right) Calculating $\dot{\hat{x}}$.

Referring to Figure 3.10(b), it should be evident that

$$\dot{\hat{x}} = w \times \hat{x}, \tag{3.26}$$

$$\dot{\hat{y}} = w \times \hat{y}, \tag{3.27}$$

$$\dot{\hat{z}} = w \times \hat{z}. \tag{3.28}$$

To express these equations in coordinates, we have to choose a reference frame in which to represent w. We can choose any reference frame, but two natural choices are the fixed frame {s} and the body frame {b}. Let us start with fixed-frame {s} coordinates. Let $R(t)$ be the rotation matrix describing the orientation of the body frame with respect to the fixed frame at time $t$; $\dot{R}(t)$ is its time rate of change. The first column of $R(t)$, denoted $r_1(t)$, describes $\hat{x}$ in fixed-frame coordinates; similarly, $r_2(t)$ and $r_3(t)$ respectively describe $\hat{y}$ and $\hat{z}$ in fixed-frame coordinates. At a specific time $t$, let $\omega_s \in \mathbb{R}^3$ be the angular velocity w expressed in fixed-frame coordinates. Then Equations (3.26)–(3.28) can be expressed in fixed-frame coordinates as

$$\dot{r}_i = \omega_s \times r_i, \qquad i = 1, 2, 3.$$

These three equations can be rearranged into the following single $3 \times 3$ matrix equation:

$$\dot{R} = [\omega_s \times r_1 \ \ \omega_s \times r_2 \ \ \omega_s \times r_3] = \omega_s \times R. \tag{3.29}$$

To eliminate the cross product on the right in Equation (3.29), we introduce some new notation, rewriting $\omega_s \times R$ as $[\omega_s]R$, where $[\omega_s]$ is a $3 \times 3$ **skew-symmetric** matrix representation of $\omega_s \in \mathbb{R}^3$:

**Definition 3.7** Given a vector $x = [x_1 \ x_2 \ x_3]^{\mathrm{T}} \in \mathbb{R}^3$, define

$$[x] = \begin{bmatrix} 0 & -x_3 & x_2 \\ x_3 & 0 & -x_1 \\ -x_2 & x_1 & 0 \end{bmatrix}. \tag{3.30}$$

The matrix $[x]$ is a $3 \times 3$ **skew-symmetric** matrix representation of $x$; that is,

$$[x] = -[x]^{\mathrm{T}}.$$

The set of all $3 \times 3$ real skew-symmetric matrices is called $so(3)$.[3]

A useful property involving rotations and skew-symmetric matrices is the following.

**Proposition 3.8**  *Given any* $\omega \in \mathbb{R}^3$ *and* $R \in SO(3)$, *the following always holds:*

$$R[\omega]R^{\mathrm{T}} = [R\omega]. \tag{3.31}$$

*Proof*  Letting $r_i^{\mathrm{T}}$ be the $i$th row of $R$, we have

$$R[\omega]R^{\mathrm{T}} = \begin{bmatrix} r_1^{\mathrm{T}}(\omega \times r_1) & r_1^{\mathrm{T}}(\omega \times r_2) & r_1^{\mathrm{T}}(\omega \times r_3) \\ r_2^{\mathrm{T}}(\omega \times r_1) & r_2^{\mathrm{T}}(\omega \times r_2) & r_2^{\mathrm{T}}(\omega \times r_3) \\ r_3^{\mathrm{T}}(\omega \times r_1) & r_3^{\mathrm{T}}(\omega \times r_2) & r_3^{\mathrm{T}}(\omega \times r_3) \end{bmatrix}$$

$$= \begin{bmatrix} 0 & -r_3^{\mathrm{T}}\omega & r_2^{\mathrm{T}}\omega \\ r_3^{\mathrm{T}}\omega & 0 & -r_1^{\mathrm{T}}\omega \\ -r_2^{\mathrm{T}}\omega & r_1^{\mathrm{T}}\omega & 0 \end{bmatrix}$$

$$= [R\omega], \tag{3.32}$$

where the second line makes use of the determinant formula for $3 \times 3$ matrices, i.e., if $M$ is a $3 \times 3$ matrix with columns $\{a, b, c\}$, then $\det M = a^{\mathrm{T}}(b \times c) = c^{\mathrm{T}}(a \times b) = b^{\mathrm{T}}(c \times a)$.  □

With the skew-symmetric notation, we can rewrite Equation (3.29) as

$$[\omega_s]R = \dot{R}. \tag{3.33}$$

We can post-multiply both sides of Equation (3.33) by $R^{-1}$ to get

$$[\omega_s] = \dot{R}R^{-1}. \tag{3.34}$$

Now let $\omega_b$ be w expressed in body-frame coordinates. To see how to obtain $\omega_b$ from $\omega_s$ and vice versa, we write $R$ explicitly as $R_{sb}$. Then $\omega_s$ and $\omega_b$ are two different vector representations of the same angular velocity w and, by our subscript cancellation rule, $\omega_s = R_{sb}\omega_b$. Therefore

$$\omega_b = R_{sb}^{-1}\omega_s = R^{-1}\omega_s = R^{\mathrm{T}}\omega_s. \tag{3.35}$$

---

[3] The set of skew-symmetric matrices $so(3)$ is called the *Lie algebra* of the Lie group $SO(3)$. It consists of all possible $\dot{R}$ when $R = I$.

Let us now express this relation in skew-symmetric matrix form:

$$\begin{aligned}
[\omega_b] &= [R^{\mathrm{T}}\omega_s] \\
&= R^{\mathrm{T}}[\omega_s]R \quad \text{(by Proposition 3.8)} \\
&= R^{\mathrm{T}}(\dot{R}R^{\mathrm{T}})R \\
&= R^{\mathrm{T}}\dot{R} = R^{-1}\dot{R}.
\end{aligned} \tag{3.36}$$

In summary, two equations relate $R$ and $\dot{R}$ to the angular velocity w:

**Proposition 3.9** *Let $R(t)$ denote the orientation of the rotating frame as seen from the fixed frame. Denote by w the angular velocity of the rotating frame. Then*

$$\dot{R}R^{-1} = [\omega_s], \tag{3.37}$$

$$R^{-1}\dot{R} = [\omega_b], \tag{3.38}$$

*where $\omega_s \in \mathbb{R}^3$ is the fixed-frame vector representation of w and $[\omega_s] \in so(3)$ is its $3 \times 3$ matrix representation, and where $\omega_b \in \mathbb{R}^3$ is the body-frame vector representation of w and $[\omega_b] \in so(3)$ is its $3 \times 3$ matrix representation.*

It is important to note that $\omega_b$ is *not* the angular velocity relative to a moving frame. Rather, $\omega_b$ is the angular velocity relative to the *stationary* frame {b} that is instantaneously coincident with a frame attached to the moving body.

It is also important to note that the fixed-frame angular velocity $\omega_s$ *does not depend on the choice of body frame.* Similarly, the body-frame angular velocity $\omega_b$ *does not depend on the choice of fixed frame.* While Equations (3.37) and (3.38) may appear to depend on both frames (since $R$ and $\dot{R}$ individually depend on both {s} and {b}), the product $\dot{R}R^{-1}$ is independent of {b} and the product $R^{-1}\dot{R}$ is independent of {s}.

Finally, an angular velocity expressed in an arbitrary frame {d} can be represented in another frame {c} if we know the rotation that takes {c} to {d}, using our now-familiar subscript cancellation rule:

$$\omega_c = R_{cd}\omega_d.$$

## .2.3 Exponential Coordinate Representation of Rotation

We now introduce a three-parameter representation for rotations, the **exponential coordinates for rotation**. The exponential coordinates parametrize a rotation matrix in terms of a rotation axis (represented by a unit vector $\hat{\omega}$) and an angle of rotation $\theta$ about that axis; the vector $\hat{\omega}\theta \in \mathbb{R}^3$ then serves as the three-parameter exponential coordinate representation of the rotation. Writing $\hat{\omega}$ and $\theta$ individually is the **axis-angle** representation of a rotation.

The exponential coordinate representation $\hat{\omega}\theta$ for a rotation matrix $R$ can be interpreted equivalently as:

- the axis $\hat{\omega}$ and rotation angle $\theta$ such that, if a frame initially coincident with {s} were rotated by $\theta$ about $\hat{\omega}$, its final orientation relative to {s} would be expressed by $R$; or
- the angular velocity $\hat{\omega}\theta$ expressed in {s} such that, if a frame initially coincident with {s} followed $\hat{\omega}\theta$ for one unit of time (i.e., $\hat{\omega}\theta$ is integrated over this time interval), its final orientation would be expressed by $R$; or
- the angular velocity $\hat{\omega}$ expressed in {s} such that, if a frame initially coincident with {s} followed by $\hat{\omega}$ for $\theta$ units of time (i.e., $\hat{\omega}$ is integrated over this time interval) its final orientation would be expressed by $R$.

The latter two views suggest that we consider exponential coordinates in the setting of linear differential equations. Below we briefly review some key results from linear differential equations theory.

### 3.2.3.1    Essential Results from Linear Differential Equations Theory

Let us begin with the simple scalar linear differential equation

$$\dot{x}(t) = ax(t), \tag{3.39}$$

where $x(t) \in \mathbb{R}$, $a \in \mathbb{R}$ is constant, and the initial condition $x(0) = x_0$ is given. Equation (3.39) has solution

$$x(t) = e^{at}x_0.$$

It is also useful to remember the series expansion of the exponential function:

$$e^{at} = 1 + at + \frac{(at)^2}{2!} + \frac{(at)^3}{3!} + \cdots.$$

Now consider the vector linear differential equation

$$\dot{x}(t) = Ax(t), \tag{3.40}$$

where $x(t) \in \mathbb{R}^n$, $A \in \mathbb{R}^{n \times n}$ is constant, and the initial condition $x(0) = x_0$ is given. From the above scalar result one can conjecture a solution of the form

$$x(t) = e^{At}x_0 \tag{3.41}$$

where the **matrix exponential** $e^{At}$ now needs to be defined in a meaningful way. Again mimicking the scalar case, we define the matrix exponential to be

$$e^{At} = I + At + \frac{(At)^2}{2!} + \frac{(At)^3}{3!} + \cdots \tag{3.42}$$

The first question to be addressed is under what conditions this series converges, so that the matrix exponential is well defined. It can be shown that if $A$ is constant and finite then this series is always guaranteed to converge to a finite limit; the proof can be found in most texts on ordinary linear differential equations and is not covered here.

The second question is whether Equation (3.41), using Equation (3.42), is indeed a solution to Equation (3.40). Taking the time derivative of $x(t) = e^{At}x_0$,

$$
\begin{aligned}
\dot{x}(t) &= \left( \frac{d}{dt} e^{At} \right) x_0 \\
&= \frac{d}{dt} \left( I + At + \frac{A^2 t^2}{2!} + \frac{A^3 t^3}{3!} + \cdots \right) x_0 \\
&= \left( A + A^2 t + \frac{A^3 t^2}{2!} + \cdots \right) x_0 \\
&= A e^{At} x_0 \\
&= A x(t),
\end{aligned}
\tag{3.43}
$$

which proves that $x(t) = e^{At}x_0$ is indeed a solution. That this is a unique solution follows from the basic existence and uniqueness result for linear ordinary differential equations, which we invoke here without proof.

While $AB \neq BA$ for arbitrary square matrices $A$ and $B$, it is always true that

$$
A e^{At} = e^{At} A
\tag{3.44}
$$

for any square $A$ and scalar $t$. You can verify this directly using the series expansion for the matrix exponential. Therefore, in line four of Equation (3.43), $A$ could also have been factored to the right, i.e.,

$$
\dot{x}(t) = e^{At} A x_0.
$$

While the matrix exponential $e^{At}$ is defined as an infinite series, closed-form expressions are often available. For example, if $A$ can be expressed as $A = PDP^{-1}$ for some $D \in \mathbb{R}^{n \times n}$ and invertible $P \in \mathbb{R}^{n \times n}$ then

$$
\begin{aligned}
e^{At} &= I + At + \frac{(At)^2}{2!} + \cdots \\
&= I + (PDP^{-1})t + (PDP^{-1})(PDP^{-1})\frac{t^2}{2!} + \cdots \\
&= P(I + Dt + \frac{(Dt)^2}{2!} + \cdots)P^{-1} \\
&= P e^{Dt} P^{-1}.
\end{aligned}
\tag{3.45}
$$

If moreover $D$ is diagonal, i.e., $D = \operatorname{diag}\{d_1, d_2, \ldots, d_n\}$, then its matrix exponential is particularly simple to evaluate:

$$
e^{Dt} =
\begin{bmatrix}
e^{d_1 t} & 0 & \cdots & 0 \\
0 & e^{d_2 t} & \cdots & 0 \\
\vdots & \vdots & \ddots & \vdots \\
0 & 0 & \cdots & e^{d_n t}
\end{bmatrix}.
\tag{3.46}
$$

We summarize the results above in the following proposition.

**Proposition 3.10** *The linear differential equation $\dot{x}(t) = Ax(t)$ with initial condition $x(0) = x_0$, where $A \in \mathbb{R}^{n \times n}$ is constant and $x(t) \in \mathbb{R}^n$, has solution*

$$x(t) = e^{At}x_0 \tag{3.47}$$

*where*

$$e^{At} = I + tA + \frac{t^2}{2!}A^2 + \frac{t^3}{3!}A^3 + \cdots . \tag{3.48}$$

*The matrix exponential $e^{At}$ further satisifies the following properties:*

(a) $d(e^{At})/dt = Ae^{At} = e^{At}A$.
(b) *If $A = PDP^{-1}$ for some $D \in \mathbb{R}^{n \times n}$ and invertible $P \in \mathbb{R}^{n \times n}$ then $e^{At} = Pe^{Dt}P^{-1}$.*
(c) *If $AB = BA$ then $e^A e^B = e^{A+B}$.*
(d) $(e^A)^{-1} = e^{-A}$.

The third property can be established by expanding the exponentials and comparing terms. The fourth property follows by setting $B = -A$ in the third property.

### 3.2.3.2 Exponential Coordinates of Rotations

The exponential coordinates of a rotation can be viewed equivalently as (1) a unit axis of rotation $\hat{\omega}$ ($\hat{\omega} \in \mathbb{R}^3$, $\|\hat{\omega}\| = 1$) together with a rotation angle about the axis $\theta \in \mathbb{R}$, or (2) as the 3-vector obtained by multiplying the two together, $\hat{\omega}\theta \in \mathbb{R}^3$. When we represent the motion of a robot joint in the next chapter, the first view has the advantage of separating the description of the joint axis from the motion $\theta$ about the axis.

Referring to Figure 3.11, suppose that a three-dimensional vector $p(0)$ is rotated by $\theta$ about $\hat{\omega}$ to $p(\theta)$; here we assume that all quantities are expressed in fixed-frame coordinates. This rotation can be achieved by imagining that $p(0)$ rotates at a constant rate of 1 rad/s (since $\hat{\omega}$ has unit magnitude) from time $t = 0$ to $t = \theta$. Let $p(t)$ denote the path traced by the tip of the vector. The velocity of $p(t)$, denoted $\dot{p}$, is then given by

$$\dot{p} = \hat{\omega} \times p. \tag{3.49}$$

To see why this is true, let $\phi$ be the constant angle between $p(t)$ and $\hat{\omega}$. Observe that $p$ traces a circle of radius $\|p\| \sin\phi$ about the $\hat{\omega}$-axis. Then $\dot{p}$ is tangent to the path with magnitude $\|p\| \sin\phi$, which is equivalent to Equation (3.49).

The differential equation (3.49) can be expressed as (see Equation (3.30))

$$\dot{p} = [\hat{\omega}]p \tag{3.50}$$

with initial condition $p(0)$. This is a linear differential equation of the form $\dot{x} = Ax$, which we studied earlier; its solution is given by

$$p(t) = e^{[\hat{\omega}]t}p(0).$$

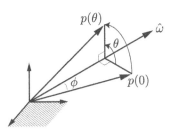

**Figure 3.11** The vector $p(0)$ is rotated by an angle $\theta$ about the axis $\hat{\omega}$ to $p(\theta)$.

Since $t$ and $\theta$ are interchangeable, the equation above can also be written

$$p(\theta) = e^{[\hat{\omega}]\theta} p(0).$$

Let us now expand the matrix exponential $e^{[\hat{\omega}]\theta}$ in series form. A straightforward calculation shows that $[\hat{\omega}]^3 = -[\hat{\omega}]$, and therefore we can replace $[\hat{\omega}]^3$ by $-[\hat{\omega}]$, $[\hat{\omega}]^4$ by $-[\hat{\omega}]^2$, $[\hat{\omega}]^5$ by $-[\hat{\omega}]^3 = [\hat{\omega}]$, and so on, obtaining

$$e^{[\hat{\omega}]\theta} = I + [\hat{\omega}]\theta + [\hat{\omega}]^2 \frac{\theta^2}{2!} + [\hat{\omega}]^3 \frac{\theta^3}{3!} + \cdots$$

$$= I + \left(\theta - \frac{\theta^3}{3!} + \frac{\theta^5}{5!} - \cdots\right)[\hat{\omega}] + \left(\frac{\theta^2}{2!} - \frac{\theta^4}{4!} + \frac{\theta^6}{6!} - \cdots\right)[\hat{\omega}]^2.$$

Now recall the series expansions for $\sin\theta$ and $\cos\theta$:

$$\sin\theta = \theta - \frac{\theta^3}{3!} + \frac{\theta^5}{5!} - \cdots$$

$$\cos\theta = 1 - \frac{\theta^2}{2!} + \frac{\theta^4}{4!} - \cdots$$

The exponential $e^{[\hat{\omega}]\theta}$ therefore simplifies to the following:

**Proposition 3.11**     *Given a vector $\hat{\omega}\theta \in \mathbb{R}^3$, such that $\theta$ is any scalar and $\hat{\omega} \in \mathbb{R}^3$ is a unit vector, the matrix exponential of $[\hat{\omega}]\theta = [\hat{\omega}\theta] \in so(3)$ is*

$$\text{Rot}(\hat{\omega}, \theta) = e^{[\hat{\omega}]\theta} = I + \sin\theta\,[\hat{\omega}] + (1 - \cos\theta)[\hat{\omega}]^2 \in SO(3). \qquad (3.51)$$

Equation (3.51) is also known as **Rodrigues' formula** for rotations.

We have shown how to use the matrix exponential to construct a rotation matrix from a rotation axis $\hat{\omega}$ and an angle $\theta$. Further, the quantity $e^{[\hat{\omega}]\theta}p$ has the effect of rotating $p \in \mathbb{R}^3$ about the fixed-frame axis $\hat{\omega}$ by an angle $\theta$. Similarly, considering that a rotation matrix $R$ consists of three column vectors, the rotation matrix $R' = e^{[\hat{\omega}]\theta} R = \text{Rot}(\hat{\omega}, \theta)R$ is the orientation achieved by rotating $R$ by $\theta$ about the axis $\hat{\omega}$ in the fixed frame. Reversing the order of matrix multiplication, $R'' = Re^{[\hat{\omega}]\theta} = R\,\text{Rot}(\hat{\omega}, \theta)$ is the orientation achieved by rotating $R$ by $\theta$ about $\hat{\omega}$ in the body frame.

**Example 3.12**     The frame {b} in Figure 3.12 is obtained by rotation from an initial orientation aligned with the fixed frame {s} about a unit axis $\hat{\omega}_1 =$

$(0, 0.866, 0.5)$ by an angle $\theta_1 = 30° = 0.524$ rad. The rotation matrix representation of $\{b\}$ can be calculated as

$$R = e^{[\hat{\omega}_1]\theta_1}$$

$$= I + \sin\theta_1[\hat{\omega}_1] + (1 - \cos\theta_1)[\hat{\omega}_1]^2$$

$$= I + 0.5\begin{bmatrix} 0 & -0.5 & 0.866 \\ 0.5 & 0 & 0 \\ -0.866 & 0 & 0 \end{bmatrix} + 0.134\begin{bmatrix} 0 & -0.5 & 0.866 \\ 0.5 & 0 & 0 \\ -0.866 & 0 & 0 \end{bmatrix}^2$$

$$= \begin{bmatrix} 0.866 & -0.250 & 0.433 \\ 0.250 & 0.967 & 0.058 \\ -0.433 & 0.058 & 0.899 \end{bmatrix}.$$

The frame $\{b\}$ can be represented by $R$ or by the unit axis $\hat{\omega}_1 = (0, 0.866, 0.5)$ and the angle $\theta_1 = 0.524$ rad, i.e., the exponential coordinates $\hat{\omega}_1\theta_1 = (0, 0.453, 0.262)$.
If $\{b\}$ is then rotated by $\theta_2$ about a fixed-frame axis $\hat{\omega}_2 \neq \hat{\omega}_1$, i.e.,

$$R' = e^{[\hat{\omega}_2]\theta_2}R,$$

then the frame ends up at a different location than that reached were $\{b\}$ to be rotated by $\theta_2$ about an axis expressed as $\hat{\omega}_2$ in the body frame, i.e.,

$$R'' = Re^{[\hat{\omega}_2]\theta_2} \neq R' = e^{[\hat{\omega}_2]\theta_2}R.$$

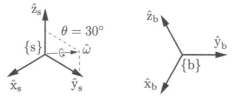

**Figure 3.12** The frame $\{b\}$ is obtained by a rotation from $\{s\}$ by $\theta_1 = 30°$ about $\hat{\omega}_1 = (0, 0.866, 0.5)$.

Our next task is to show that for any rotation matrix $R \in SO(3)$, one can always find a unit vector $\hat{\omega}$ and scalar $\theta$ such that $R = e^{[\hat{\omega}]\theta}$.

### 3.2.3.3   Matrix Logarithm of Rotations

If $\hat{\omega}\theta \in \mathbb{R}^3$ represents the exponential coordinates of a rotation matrix $R$, then the skew-symmetric matrix $[\hat{\omega}\theta] = [\hat{\omega}]\theta$ is the **matrix logarithm** of the rotation $R$.[4] The matrix logarithm is the inverse of the matrix exponential. Just as the matrix exponential "integrates" the matrix representation of an angular velocity $[\hat{\omega}]\theta \in so(3)$ for one second to give an orientation $R \in SO(3)$, the matrix logarithm "differentiates" an $R \in SO(3)$ to find the matrix representation of a constant

---

[4] We use the term "*the* matrix logarithm" to refer both to a specific matrix which is a logarithm of $R$ as well as to the algorithm that calculates this specific matrix. Also, while a matrix $R$ can have more than one matrix logarithm (just as $\sin^{-1}(0)$ has solutions $0, \pi, 2\pi$, etc.), we commonly refer to "the" matrix logarithm, i.e., the unique solution returned by the matrix logarithm algorithm.

angular velocity $[\hat{\omega}]\theta \in so(3)$ which, if integrated for one second, rotates a frame from $I$ to $R$. In other words,

$$\exp: \ [\hat{\omega}]\theta \in so(3) \ \rightarrow \ R \in SO(3),$$
$$\log: \ R \in SO(3) \ \rightarrow \ [\hat{\omega}]\theta \in so(3).$$

To derive the matrix logarithm, let us expand each entry for $e^{[\hat{\omega}]\theta}$ in Equation (3.51),

$$\begin{bmatrix} c_\theta + \hat{\omega}_1^2(1 - c_\theta) & \hat{\omega}_1\hat{\omega}_2(1 - c_\theta) - \hat{\omega}_3 s_\theta & \hat{\omega}_1\hat{\omega}_3(1 - c_\theta) + \hat{\omega}_2 s_\theta \\ \hat{\omega}_1\hat{\omega}_2(1 - c_\theta) + \hat{\omega}_3 s_\theta & c_\theta + \hat{\omega}_2^2(1 - c_\theta) & \hat{\omega}_2\hat{\omega}_3(1 - c_\theta) - \hat{\omega}_1 s_\theta \\ \hat{\omega}_1\hat{\omega}_3(1 - c_\theta) - \hat{\omega}_2 s_\theta & \hat{\omega}_2\hat{\omega}_3(1 - c_\theta) + \hat{\omega}_1 s_\theta & c_\theta + \hat{\omega}_3^2(1 - c_\theta) \end{bmatrix}, \quad (3.52)$$

where $\hat{\omega} = (\hat{\omega}_1, \hat{\omega}_2, \hat{\omega}_3)$, and we use again the shorthand notation $s_\theta = \sin\theta$ and $c_\theta = \cos\theta$. Setting the above matrix equal to the given $R \in SO(3)$ and subtracting the transpose from both sides leads to the following:

$$r_{32} - r_{23} = 2\hat{\omega}_1 \sin\theta,$$
$$r_{13} - r_{31} = 2\hat{\omega}_2 \sin\theta,$$
$$r_{21} - r_{12} = 2\hat{\omega}_3 \sin\theta.$$

Therefore, as long as $\sin\theta \neq 0$ (or, equivalently, $\theta$ is not an integer multiple of $\pi$), we can write

$$\hat{\omega}_1 = \frac{1}{2\sin\theta}(r_{32} - r_{23}),$$
$$\hat{\omega}_2 = \frac{1}{2\sin\theta}(r_{13} - r_{31}),$$
$$\hat{\omega}_3 = \frac{1}{2\sin\theta}(r_{21} - r_{12}).$$

The above equations can also be expressed in skew-symmetric matrix form as

$$[\hat{\omega}] = \begin{bmatrix} 0 & -\hat{\omega}_3 & \hat{\omega}_2 \\ \hat{\omega}_3 & 0 & -\hat{\omega}_1 \\ -\hat{\omega}_2 & \hat{\omega}_1 & 0 \end{bmatrix} = \frac{1}{2\sin\theta}\left(R - R^T\right). \quad (3.53)$$

Recall that $\hat{\omega}$ represents the axis of rotation for the given $R$. Because of the $\sin\theta$ term in the denominator, $[\hat{\omega}]$ is not well defined if $\theta$ is an integer multiple of $\pi$.[5] We address this situation next, but for now let us assume that $\sin\theta \neq 0$ and find an expression for $\theta$. Setting $R$ equal to (3.52) and taking the trace of both sides (recall that the trace of a matrix is the sum of its diagonal entries),

$$\operatorname{tr} R = r_{11} + r_{22} + r_{33} = 1 + 2\cos\theta. \quad (3.54)$$

The above follows since $\hat{\omega}_1^2 + \hat{\omega}_2^2 + \hat{\omega}_3^2 = 1$. For any $\theta$ satisfying $1 + 2\cos\theta = \operatorname{tr} R$ such that $\theta$ is not an integer multiple of $\pi$, $R$ can be expressed as the exponential $e^{[\hat{\omega}]\theta}$ with $[\hat{\omega}]$ as given in Equation (3.53).

---

[5] Singularities such as this are unavoidable for any three-parameter representation of rotation. Euler angles and roll–pitch–yaw angles suffer from similar singularities.

Let us now return to the case $\theta = k\pi$, where $k$ is some integer. When $k$ is an even integer, regardless of $\hat{\omega}$ we have rotated back to $R = I$ so the vector $\hat{\omega}$ is undefined. When $k$ is an odd integer (corresponding to $\theta = \pm\pi, \pm3\pi, \ldots$, which in turn implies $\operatorname{tr} R = -1$), the exponential formula (3.51) simplifies to

$$R = e^{[\hat{\omega}]\pi} = I + 2[\hat{\omega}]^2. \tag{3.55}$$

The three diagonal terms of Equation (3.55) can be manipulated to give

$$\hat{\omega}_i = \pm\sqrt{\frac{r_{ii}+1}{2}}, \qquad i = 1, 2, 3. \tag{3.56}$$

The off-diagonal terms lead to the following three equations:

$$\begin{aligned} 2\hat{\omega}_1\hat{\omega}_2 &= r_{12}, \\ 2\hat{\omega}_2\hat{\omega}_3 &= r_{23}, \\ 2\hat{\omega}_1\hat{\omega}_3 &= r_{13}, \end{aligned} \tag{3.57}$$

From Equation (3.55) we also know that $R$ must be symmetric: $r_{12} = r_{21}$, $r_{23} = r_{32}$, $r_{13} = r_{31}$. Equations (3.56) and (3.57) may both be necessary to obtain a solution for $\hat{\omega}$. Once such a solution has been found then $R = e^{[\hat{\omega}]\theta}$, where $\theta = \pm\pi, \pm3\pi, \ldots$

From the above it can be seen that solutions for $\theta$ exist at $2\pi$ intervals. If we restrict $\theta$ to the interval $[0, \pi]$ then the following algorithm can be used to compute the matrix logarithm of the rotation matrix $R \in SO(3)$.

**Algorithm:**
Given $R \in SO(3)$, find a $\theta \in [0, \pi]$ and a unit rotation axis $\hat{\omega} \in \mathbb{R}^3, \|\hat{\omega}\| = 1$, such that $e^{[\hat{\omega}]\theta} = R$. The vector $\hat{\omega}\theta \in \mathbb{R}^3$ comprises the exponential coordinates for $R$ and the skew-symmetric matrix $[\hat{\omega}]\theta \in so(3)$ is the matrix logarithm of $R$.

(a) If $R = I$ then $\theta = 0$ and $\hat{\omega}$ is undefined.
(b) If $\operatorname{tr} R = -1$ then $\theta = \pi$. Set $\hat{\omega}$ equal to any of the following three vectors that is a feasible solution:

$$\hat{\omega} = \frac{1}{\sqrt{2(1+r_{33})}} \begin{bmatrix} r_{13} \\ r_{23} \\ 1+r_{33} \end{bmatrix} \tag{3.58}$$

or

$$\hat{\omega} = \frac{1}{\sqrt{2(1+r_{22})}} \begin{bmatrix} r_{12} \\ 1+r_{22} \\ r_{32} \end{bmatrix} \tag{3.59}$$

or

$$\hat{\omega} = \frac{1}{\sqrt{2(1+r_{11})}} \begin{bmatrix} 1+r_{11} \\ r_{21} \\ r_{31} \end{bmatrix}. \tag{3.60}$$

(Note that if $\hat{\omega}$ is a solution, then so is $-\hat{\omega}$.)

(c) Otherwise $\theta = \cos^{-1}\left(\frac{1}{2}(\operatorname{tr} R - 1)\right) \in [0, \pi)$ and

$$[\hat{\omega}] = \frac{1}{2\sin\theta}(R - R^{\mathrm{T}}). \qquad (3.61)$$

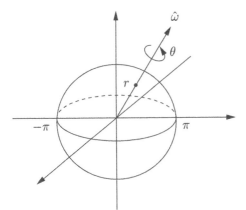

**Figure 3.13** $SO(3)$ as a solid ball of radius $\pi$. The exponential coordinates $r = \hat{\omega}\theta$ may lie anywhere within the solid ball

Since every $R \in SO(3)$ satisfies one of the three cases in the algorithm, for every $R$ there exists a matrix logarithm $[\hat{\omega}]\theta$ and therefore a set of exponential coordinates $\hat{\omega}\theta$.

Because the matrix logarithm calculates exponential coordinates $\hat{\omega}\theta$ satisfying $\|\hat{\omega}\theta\| \leq \pi$, we can picture the rotation group $SO(3)$ as a solid ball of radius $\pi$ (see Figure 3.13): given a point $r \in \mathbb{R}^3$ in this solid ball, let $\hat{\omega} = r/\|r\|$ be the unit axis in the direction from the origin to the point $r$ and let $\theta = \|r\|$ be the distance from the origin to $r$, so that $r = \hat{\omega}\theta$. The rotation matrix corresponding to $r$ can then be regarded as a rotation about the axis $\hat{\omega}$ by an angle $\theta$. For any $R \in SO(3)$ such that $\operatorname{tr} R \neq -1$, there exists a unique $r$ in the interior of the solid ball such that $e^{[r]} = R$. In the event that $\operatorname{tr} R = -1$, $\log R$ is given by two antipodal points on the surface of this solid ball. That is, if there exists some $r$ such that $R = e^{[r]}$ with $\|r\| = \pi$ then $R = e^{[-r]}$ also holds; both $r$ and $-r$ correspond to the same rotation $R$.

## 3.3    Rigid-Body Motions and Twists

In this section we derive representations for rigid-body configurations and velocities that extend, but otherwise are analogous to, those in Section 3.2 for rotations and angular velocities. In particular, the homogeneous transformation matrix $T$ is analogous to the rotation matrix $R$; a screw axis $\mathcal{S}$ is analogous to a rotation axis $\hat{\omega}$; a twist $\mathcal{V}$ can be expressed as $\mathcal{S}\dot{\theta}$ and is analogous to an angular velocity $\omega = \hat{\omega}\dot{\theta}$; and exponential coordinates $\mathcal{S}\theta \in \mathbb{R}^6$ for rigid-body motions are analogous to exponential coordinates $\hat{\omega}\theta \in \mathbb{R}^3$ for rotations.

### 3.3.1 Homogeneous Transformation Matrices

We now consider representations for the combined orientation and position of a rigid body. A natural choice would be to use a rotation matrix $R \in SO(3)$ to represent the orientation of the body frame {b} in the fixed frame {s} and a vector $p \in \mathbb{R}^3$ to represent the origin of {b} in {s}. Rather than identifying $R$ and $p$ separately, we package them into a single matrix as follows.

**Definition 3.13**   The **special Euclidean group** $SE(3)$, also known as the group of **rigid-body motions** or **homogeneous transformation matrices** in $\mathbb{R}^3$, is the set of all $4 \times 4$ real matrices $T$ of the form

$$
T = \begin{bmatrix} R & p \\ 0 & 1 \end{bmatrix} = \begin{bmatrix} r_{11} & r_{12} & r_{13} & p_1 \\ r_{21} & r_{22} & r_{23} & p_2 \\ r_{31} & r_{32} & r_{33} & p_3 \\ 0 & 0 & 0 & 1 \end{bmatrix}, \tag{3.62}
$$

where $R \in SO(3)$ and $p \in \mathbb{R}^3$ is a column vector.

An element $T \in SE(3)$ will sometimes be denoted $(R, p)$. In this section we will establish some basic properties of $SE(3)$ and explain why we package $R$ and $p$ into this matrix form.

Many robotic mechanisms we have encountered thus far are planar. With planar rigid-body motions in mind, we make the following definition:

**Definition 3.14**   The special Euclidean group $SE(2)$ is the set of all $3 \times 3$ real matrices $T$ of the form

$$
T = \begin{bmatrix} R & p \\ 0 & 1 \end{bmatrix}, \tag{3.63}
$$

where $R \in SO(2)$, $p \in \mathbb{R}^2$, and 0 denotes a row vector of two zeros.

A matrix $T \in SE(2)$ is always of the form

$$
T = \begin{bmatrix} r_{11} & r_{12} & p_1 \\ r_{21} & r_{22} & p_2 \\ 0 & 0 & 1 \end{bmatrix} = \begin{bmatrix} \cos\theta & -\sin\theta & p_1 \\ \sin\theta & \cos\theta & p_2 \\ 0 & 0 & 1 \end{bmatrix},
$$

where $\theta \in [0, 2\pi)$.

#### 3.3.1.1 Properties of Transformation Matrices

We now list some basic properties of transformation matrices, which can be proven by calculation. First, the identity $I$ is a trivial example of a transformation matrix. The first three properties confirm that $SE(3)$ is a group.

**Proposition 3.15**   *The inverse of a transformation matrix $T \in SE(3)$ is also a transformation matrix, and it has the following form:*

$$
T^{-1} = \begin{bmatrix} R & p \\ 0 & 1 \end{bmatrix}^{-1} = \begin{bmatrix} R^{\mathrm{T}} & -R^{\mathrm{T}}p \\ 0 & 1 \end{bmatrix}. \tag{3.64}
$$

**Proposition 3.16**   *The product of two transformation matrices is also a transformation matrix.*

**Proposition 3.17**   *The multiplication of transformation matrices is associative, so that $(T_1 T_2)T_3 = T_1(T_2 T_3)$, but generally not commutative: $T_1 T_2 \neq T_2 T_1$.*

Before stating the next proposition, we note that, just as in Section 3.1, it is often useful to calculate the quantity $Rx + p$, where $x \in \mathbb{R}^3$ and $(R, p)$ represents $T$. If we append a '1' to $x$, making it a four-dimensional vector, this computation can be performed as a single matrix multiplication:

$$T \begin{bmatrix} x \\ 1 \end{bmatrix} = \begin{bmatrix} R & p \\ 0 & 1 \end{bmatrix} \begin{bmatrix} x \\ 1 \end{bmatrix} = \begin{bmatrix} Rx + p \\ 1 \end{bmatrix}. \tag{3.65}$$

The vector $[x^T \ 1]^T$ is the representation of $x$ in **homogeneous coordinates**, and accordingly $T \in SE(3)$ is called a homogenous transformation. When, by an abuse of notation, we write $Tx$, we mean $Rx + p$.

**Proposition 3.18**   *Given $T = (R, p) \in SE(3)$ and $x, y \in \mathbb{R}^3$, the following hold:*

(a) $\|Tx - Ty\| = \|x - y\|$, *where $\| \cdot \|$ denotes the standard Euclidean norm in $\mathbb{R}^3$, i.e., $\|x\| = \sqrt{x^T x}$.*

(b) $\langle Tx - Tz, Ty - Tz \rangle = \langle x - z, y - z \rangle$ *for all $z \in \mathbb{R}^3$, where $\langle \cdot, \cdot \rangle$ denotes the standard Euclidean inner product in $\mathbb{R}^3$, $\langle x, y \rangle = x^T y$.*

In Proposition 3.18, $T$ is regarded as a transformation on points in $\mathbb{R}^3$; $T$ transforms a point $x$ to $Tx$. Property (a) then asserts that $T$ preserves distances, while property (b) asserts that $T$ preserves angles. Specifically, if $x, y, z \in \mathbb{R}^3$ represent the three vertices of a triangle then the triangle formed by the transformed vertices $\{Tx, Ty, Tz\}$ has the same set of lengths and angles as those of the triangle $\{x, y, z\}$ (the two triangles are said to be *isometric*). One can easily imagine taking $\{x, y, z\}$ to be the points on a rigid body, in which case $\{Tx, Ty, Tz\}$ represents a displaced version of the rigid body. It is in this sense that $SE(3)$ can be identified with rigid-body motions.

### .3.1.2   Uses of Transformation Matrices

As was the case for rotation matrices, there are three major uses for a transformation matrix $T$:

(a) to represent the configuration (position and orientation) of a rigid body;
(b) to change the reference frame in which a vector or frame is represented;
(c) to displace a vector or frame.

In the first use, $T$ is thought of as representing the configuration of a frame; in the second and third uses, $T$ is thought of as an operator that acts to change the reference frame or to move a vector or a frame.

To illustrate these uses, we refer to the three reference frames {a}, {b}, and {c}, and the point v, in Figure 3.14. The frames were chosen in such a way that the alignment of their axes is clear, allowing the visual confirmation of calculations.

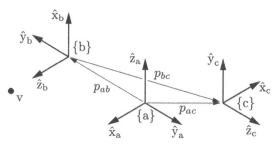

**Figure 3.14** Three reference frames in space, and a point v that can be represented in {b} as $v_b = (0, 0, 1.5)$.

### Representing a configuration

The fixed frame {s} is coincident with {a} and the frames {a}, {b}, and {c}, represented by $T_{sa} = (R_{sa}, p_{sa})$, $T_{sb} = (R_{sb}, p_{sb})$, and $T_{sc} = (R_{sc}, p_{sc})$, respectively, can be expressed relative to {s} by the rotations

$$R_{sa} = \begin{bmatrix} 1 & 0 & 0 \\ 0 & 1 & 0 \\ 0 & 0 & 1 \end{bmatrix}, \quad R_{sb} = \begin{bmatrix} 0 & 0 & 1 \\ 0 & -1 & 0 \\ 1 & 0 & 0 \end{bmatrix}, \quad R_{sc} = \begin{bmatrix} -1 & 0 & 0 \\ 0 & 0 & 1 \\ 0 & 1 & 0 \end{bmatrix}.$$

The location of the origin of each frame relative to {s} can be written

$$p_{sa} = \begin{bmatrix} 0 \\ 0 \\ 0 \end{bmatrix}, \quad p_{sb} = \begin{bmatrix} 0 \\ -2 \\ 0 \end{bmatrix}, \quad p_{sc} = \begin{bmatrix} -1 \\ 1 \\ 0 \end{bmatrix}.$$

Since {a} is collocated with {s}, the transformation matrix $T_{sa}$ constructed from $(R_{sa}, p_{sa})$ is the identity matrix.

Any frame can be expressed relative to any other frame, not just to {s}; for example, $T_{bc} = (R_{bc}, p_{bc})$ represents {c} relative to {b}:

$$R_{bc} = \begin{bmatrix} 0 & 1 & 0 \\ 0 & 0 & -1 \\ -1 & 0 & 0 \end{bmatrix}, \quad p_{bc} = \begin{bmatrix} 0 \\ -3 \\ -1 \end{bmatrix}.$$

It can also be shown, using Proposition 3.15, that

$$T_{de} = T_{ed}^{-1}$$

for any two frames {d} and {e}.

### Changing the reference frame of a vector or a frame

By a subscript cancellation rule analogous to that for rotations, for any three reference frames {a}, {b}, and {c}, and any vector v expressed in {b} as $v_b$,

$$T_{ab}T_{bc} = T_{a\cancel{b}}T_{\cancel{b}c} = T_{ac}$$
$$T_{ab}v_b = T_{a\cancel{b}}v_{\cancel{b}} = v_a,$$

where $v_a$ is the vector v expressed in $\{a\}$.

**Displacing (rotating and translating) a vector or a frame**

A transformation matrix $T$, viewed as the pair $(R, p) = (\text{Rot}(\hat{\omega}, \theta), p)$, can act on a frame $T_{sb}$ by rotating it by $\theta$ about an axis $\hat{\omega}$ and translating it by $p$. By a minor abuse of notation, we can extend the $3 \times 3$ rotation operator $R = \text{Rot}(\hat{\omega}, \theta)$ to a $4 \times 4$ transformation matrix that rotates without translating,

$$\text{Rot}(\hat{\omega}, \theta) = \begin{bmatrix} R & 0 \\ 0 & 1 \end{bmatrix},$$

and we can similarly define a translation operator that translates without rotating,

$$\text{Trans}(p) = \begin{bmatrix} 1 & 0 & 0 & p_x \\ 0 & 1 & 0 & p_y \\ 0 & 0 & 1 & p_z \\ 0 & 0 & 0 & 1 \end{bmatrix}.$$

(To parallel the rotation operator more directly, we could write $\text{Trans}(\hat{p}, \|p\|)$, a translation along the unit direction $\hat{p}$ by a distance $\|p\|$, but we will use the simpler notation with $p = \hat{p}\|p\|$.)

Whether we pre-multiply or post-multiply $T_{sb}$ by $T = (R, p)$ determines whether the $\hat{\omega}$-axis and $p$ are interpreted as in the fixed frame $\{s\}$ or in the body frame $\{b\}$:

$$T_{sb'} = T T_{sb} = \text{Trans}(p) \, \text{Rot}(\hat{\omega}, \theta) T_{sb} \qquad \text{(fixed frame)}$$

$$= \begin{bmatrix} R & p \\ 0 & 1 \end{bmatrix} \begin{bmatrix} R_{sb} & p_{sb} \\ 0 & 1 \end{bmatrix} = \begin{bmatrix} R R_{sb} & R p_{sb} + p \\ 0 & 1 \end{bmatrix} \qquad (3.66)$$

$$T_{sb''} = T_{sb} T = T_{sb} \, \text{Trans}(p) \, \text{Rot}(\hat{\omega}, \theta) \qquad \text{(body frame)}$$

$$= \begin{bmatrix} R_{sb} & p_{sb} \\ 0 & 1 \end{bmatrix} \begin{bmatrix} R & p \\ 0 & 1 \end{bmatrix} = \begin{bmatrix} R_{sb} R & R_{sb} p + p_{sb} \\ 0 & 1 \end{bmatrix}. \qquad (3.67)$$

The fixed-frame transformation (corresponding to pre-multiplication by $T$) can be interpreted as first rotating the $\{b\}$ frame by $\theta$ about an axis $\hat{\omega}$ in the $\{s\}$ frame (this rotation will cause the origin of $\{b\}$ to move if it is not coincident with the origin of $\{s\}$), then translating it by $p$ in the $\{s\}$ frame to get a frame $\{b'\}$. The body-frame transformation (corresponding to post-multiplication by $T$) can be interpreted as first translating $\{b\}$ by $p$ considered to be in the $\{b\}$ frame, then rotating about $\hat{\omega}$ in this new body frame (this does not move the origin of the frame) to get $\{b''\}$.

Fixed-frame and body-frame transformations are illustrated in Figure 3.15 for

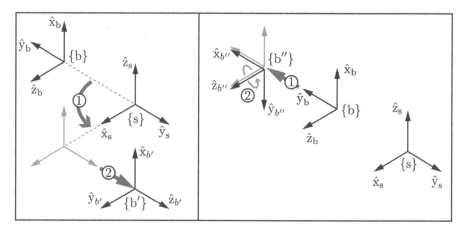

**Figure 3.15** Fixed-frame and body-frame transformations corresponding to $\hat{\omega} = (0, 0, 1)$, $\theta = 90°$, and $p = (0, 2, 0)$. (Left) The frame $\{b\}$ is rotated by $90°$ about $\hat{z}_s$ and then translated by two units in $\hat{y}_s$, resulting in the new frame $\{b'\}$. (Right) The frame $\{b\}$ is translated by two units in $\hat{y}_b$ and then rotated by $90°$ about its $\hat{z}$-axis, resulting in the new frame $\{b''\}$.

a transformation $T$ with $\hat{\omega} = (0, 0, 1)$, $\theta = 90°$, and $p = (0, 2, 0)$, yielding

$$T = (\text{Rot}(\hat{\omega}, \theta), p) = \begin{bmatrix} 0 & -1 & 0 & 0 \\ 1 & 0 & 0 & 2 \\ 0 & 0 & 1 & 0 \\ 0 & 0 & 0 & 1 \end{bmatrix}.$$

Beginning with the frame $\{b\}$ represented by

$$T_{sb} = \begin{bmatrix} 0 & 0 & 1 & 0 \\ 0 & -1 & 0 & -2 \\ 1 & 0 & 0 & 0 \\ 0 & 0 & 0 & 1 \end{bmatrix},$$

the new frame $\{b'\}$ achieved by a fixed-frame transformation $TT_{sb}$ and the new frame $\{b''\}$ achieved by a body-frame transformation $T_{sb}T$ are given by

$$TT_{sb} = T_{sb'} = \begin{bmatrix} 0 & 1 & 0 & 2 \\ 0 & 0 & 1 & 2 \\ 1 & 0 & 0 & 0 \\ 0 & 0 & 0 & 1 \end{bmatrix}, \quad T_{sb}T = T_{sb''} = \begin{bmatrix} 0 & 0 & 1 & 0 \\ -1 & 0 & 0 & -4 \\ 0 & -1 & 0 & 0 \\ 0 & 0 & 0 & 1 \end{bmatrix}.$$

**Example 3.19**    Figure 3.16 shows a robot arm mounted on a wheeled mobile platform moving in a room, and a camera fixed to the ceiling. Frames $\{b\}$ and $\{c\}$ are respectively attached to the wheeled platform and the end-effector of the robot arm, and frame $\{d\}$ is attached to the camera. A fixed frame $\{a\}$ has been established, and the robot must pick up an object with body frame

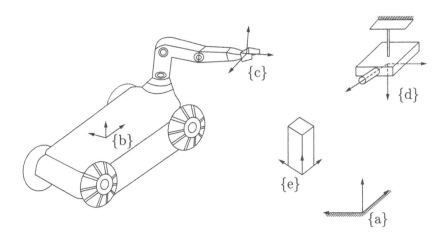

**Figure 3.16** Assignment of reference frames.

{e}. Suppose that the transformations $T_{db}$ and $T_{de}$ can be calculated from measurements obtained with the camera. The transformation $T_{bc}$ can be calculated using the arm's joint-angle measurements. The transformation $T_{ad}$ is assumed to be known in advance. Suppose these calculated and known transformations are given as follows:

$$
T_{db} = \begin{bmatrix} 0 & 0 & -1 & 250 \\ 0 & -1 & 0 & -150 \\ -1 & 0 & 0 & 200 \\ 0 & 0 & 0 & 1 \end{bmatrix},
$$

$$
T_{de} = \begin{bmatrix} 0 & 0 & -1 & 300 \\ 0 & -1 & 0 & 100 \\ -1 & 0 & 0 & 120 \\ 0 & 0 & 0 & 1 \end{bmatrix},
$$

$$
T_{ad} = \begin{bmatrix} 0 & 0 & -1 & 400 \\ 0 & -1 & 0 & 50 \\ -1 & 0 & 0 & 300 \\ 0 & 0 & 0 & 1 \end{bmatrix},
$$

$$
T_{bc} = \begin{bmatrix} 0 & -1/\sqrt{2} & -1/\sqrt{2} & 30 \\ 0 & 1/\sqrt{2} & -1/\sqrt{2} & -40 \\ 1 & 0 & 0 & 25 \\ 0 & 0 & 0 & 1 \end{bmatrix}.
$$

In order to calculate how to move the robot arm so as to pick up the object, the configuration of the object relative to the robot hand, $T_{ce}$, must be determined. We know that

$$
T_{ab}T_{bc}T_{ce} = T_{ad}T_{de},
$$

where the only quantity besides $T_{ce}$ not given to us directly is $T_{ab}$. However, since $T_{ab} = T_{ad}T_{db}$, we can determine $T_{ce}$ as follows:

$$T_{ce} = (T_{ad}T_{db}T_{bc})^{-1} T_{ad}T_{de}.$$

From the given transformations we obtain

$$T_{ad}T_{de} = \begin{bmatrix} 1 & 0 & 0 & 280 \\ 0 & 1 & 0 & -50 \\ 0 & 0 & 1 & 0 \\ 0 & 0 & 0 & 1 \end{bmatrix},$$

$$T_{ad}T_{db}T_{bc} = \begin{bmatrix} 0 & -1/\sqrt{2} & -1/\sqrt{2} & 230 \\ 0 & 1/\sqrt{2} & -1/\sqrt{2} & 160 \\ 1 & 0 & 0 & 75 \\ 0 & 0 & 0 & 1 \end{bmatrix},$$

$$(T_{ad}T_{db}T_{bc})^{-1} = \begin{bmatrix} 0 & 0 & 1 & -75 \\ -1/\sqrt{2} & 1/\sqrt{2} & 0 & 70/\sqrt{2} \\ -1/\sqrt{2} & -1/\sqrt{2} & 0 & 390/\sqrt{2} \\ 0 & 0 & 0 & 1 \end{bmatrix},$$

from which $T_{ce}$ is evaluated to be

$$T_{ce} = \begin{bmatrix} 0 & 0 & 1 & -75 \\ -1/\sqrt{2} & 1/\sqrt{2} & 0 & -260/\sqrt{2} \\ -1/\sqrt{2} & -1/\sqrt{2} & 0 & 160/\sqrt{2} \\ 0 & 0 & 0 & 1 \end{bmatrix}.$$

### 3.3.2    Twists

We now consider both the linear and angular velocities of a moving frame. As before, {s} and {b} denote the fixed (space) and moving (body) frames, respectively. Let

$$T_{sb}(t) = T(t) = \begin{bmatrix} R(t) & p(t) \\ 0 & 1 \end{bmatrix} \tag{3.68}$$

denote the configuration of {b} as seen from {s}. To keep the notation uncluttered, for the time being we write $T$ instead of the usual $T_{sb}$.

In Section 3.2.2 we discovered that pre- or post-multiplying $\dot{R}$ by $R^{-1}$ results in a skew-symmetric representation of the angular velocity vector, either in fixed- or body-frame coordinates. One might reasonably ask whether a similar property carries over to $\dot{T}$, i.e., whether $T^{-1}\dot{T}$ and $\dot{T}T^{-1}$ carry similar physical interpretations.

Let us first see what happens when we pre-multiply $\dot{T}$ by $T^{-1}$:

$$
\begin{aligned}
T^{-1}\dot{T} &= \begin{bmatrix} R^{\mathrm{T}} & -R^{\mathrm{T}}p \\ 0 & 1 \end{bmatrix} \begin{bmatrix} \dot{R} & \dot{p} \\ 0 & 0 \end{bmatrix} \\[6pt]
&= \begin{bmatrix} R^{\mathrm{T}}\dot{R} & R^{\mathrm{T}}\dot{p} \\ 0 & 0 \end{bmatrix} \\[6pt]
&= \begin{bmatrix} [\omega_b] & v_b \\ 0 & 0 \end{bmatrix}.
\end{aligned}
\tag{3.69}
$$

Recall that $R^{\mathrm{T}}\dot{R} = [\omega_b]$ is just the skew-symmetric matrix representation of the angular velocity expressed in {b} coordinates. Also, $\dot{p}$ is the linear velocity of the origin of {b} expressed in the fixed frame {s}, and $R^{\mathrm{T}}\dot{p} = v_b$ is this linear velocity expressed in the frame {b}. Putting these two observations together, we can conclude that $T^{-1}\dot{T}$ represents the linear and angular velocities of the moving frame relative to the stationary frame {b} currently aligned with the moving frame.

The above calculation of $T^{-1}\dot{T}$ suggests that it is reasonable to merge $\omega_b$ and $v_b$ into a single six-dimensional velocity vector. We define the **spatial velocity in the body frame**, or simply the **body twist**,[6] to be

$$
\mathcal{V}_b = \begin{bmatrix} \omega_b \\ v_b \end{bmatrix} \in \mathbb{R}^6.
\tag{3.70}
$$

Just as it is convenient to have a skew-symmetric matrix representation of an angular velocity vector, it is convenient to have a matrix representation of a twist, as shown in Equation (3.69). We will stretch the $[\cdot]$ notation, writing

$$
T^{-1}\dot{T} = [\mathcal{V}_b] = \begin{bmatrix} [\omega_b] & v_b \\ 0 & 0 \end{bmatrix} \in se(3),
\tag{3.71}
$$

where $[\omega_b] \in so(3)$ and $v_b \in \mathbb{R}^3$. The set of all $4 \times 4$ matrices of this form is called $se(3)$ and comprises the matrix representations of the twists associated with the rigid-body configurations $SE(3)$.[7]

Now that we have a physical interpretation for $T^{-1}\dot{T}$, let us evaluate $\dot{T}T^{-1}$:

$$
\begin{aligned}
\dot{T}T^{-1} &= \begin{bmatrix} \dot{R} & \dot{p} \\ 0 & 0 \end{bmatrix} \begin{bmatrix} R^{\mathrm{T}} & -R^{\mathrm{T}}p \\ 0 & 1 \end{bmatrix} \\[6pt]
&= \begin{bmatrix} \dot{R}R^{\mathrm{T}} & \dot{p} - \dot{R}R^{\mathrm{T}}p \\ 0 & 0 \end{bmatrix} \\[6pt]
&= \begin{bmatrix} [\omega_s] & v_s \\ 0 & 0 \end{bmatrix}.
\end{aligned}
\tag{3.72}
$$

---

[6] The term "twist" has been used in different ways in the mechanisms and screw theory literature. In robotics, however, it is common to use the term to refer to a spatial velocity. We mostly use the term "twist" instead of "spatial velocity" to minimize verbiage, e.g., "body twist" versus "spatial velocity in the body frame."

[7] $se(3)$ is called the Lie algebra of the Lie group $SE(3)$. It consists of all possible $\dot{T}$ when $T = I$.

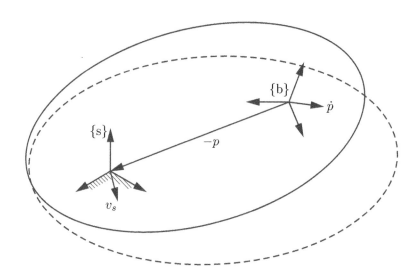

**Figure 3.17** Physical interpretation of $v_s$. The initial (solid line) and displaced (dashed line) configurations of a rigid body.

Observe that the skew-symmetric matrix $[\omega_s] = \dot{R}R^{\mathrm{T}}$ is the angular velocity expressed in fixed-frame coordinates but that $v_s = \dot{p} - \dot{R}R^{\mathrm{T}}p$ is **not** the linear velocity of the body-frame origin expressed in the fixed frame (that quantity would simply be $\dot{p}$). If we write $v_s$ as

$$v_s = \dot{p} - \omega_s \times p = \dot{p} + \omega_s \times (-p), \tag{3.73}$$

the physical meaning of $v_s$ can now be inferred: imagining the moving body to be infinitely large, $v_s$ is the instantaneous velocity of the point on this body currently at the fixed-frame origin, expressed in the fixed frame (see Figure 3.17).

As we did for $\omega_b$ and $v_b$, we assemble $\omega_s$ and $v_s$ into a six-dimensional twist,

$$\mathcal{V}_s = \begin{bmatrix} \omega_s \\ v_s \end{bmatrix} \in \mathbb{R}^6, \qquad [\mathcal{V}_s] = \begin{bmatrix} [\omega_s] & v_s \\ 0 & 0 \end{bmatrix} = \dot{T}T^{-1} \in se(3), \tag{3.74}$$

where $[\mathcal{V}_s]$ is the $4 \times 4$ matrix representation of $\mathcal{V}_s$. We call $\mathcal{V}_s$ the **spatial velocity in the space frame**, or simply the **spatial twist**.

If we regard the moving body as being infinitely large, there is an appealing and natural symmetry between $\mathcal{V}_s = (\omega_s, v_s)$ and $\mathcal{V}_b = (\omega_b, v_b)$:

(a) $\omega_b$ is the angular velocity expressed in {b}, and $\omega_s$ is the angular velocity expressed in {s}.
(b) $v_b$ is the linear velocity of a point at the origin of {b} expressed in {b}, and $v_s$ is the linear velocity of a point at the origin of {s} expressed in {s}.

We can obtain $\mathcal{V}_b$ from $\mathcal{V}_s$ as follows:

$$[\mathcal{V}_b] = T^{-1}\dot{T}$$
$$= T^{-1}[\mathcal{V}_s]T. \tag{3.75}$$

Going the other way,

$$[\mathcal{V}_s] = T\,[\mathcal{V}_b]\,T^{-1}. \tag{3.76}$$

Writing out the products in Equation (3.76), we get

$$[\mathcal{V}_s] = \begin{bmatrix} R[\omega_b]R^{\mathrm{T}} & -R[\omega_b]R^{\mathrm{T}}p + Rv_b \\ 0 & 0 \end{bmatrix}$$

which, using $R[\omega]R^{\mathrm{T}} = [R\omega]$ (Proposition 3.8) and $[\omega]p = -[p]\omega$ for $p, \omega \in \mathbb{R}^3$, can be manipulated into the following relation between $\mathcal{V}_b$ and $\mathcal{V}_s$:

$$\begin{bmatrix} \omega_s \\ v_s \end{bmatrix} = \begin{bmatrix} R & 0 \\ [p]R & R \end{bmatrix}\begin{bmatrix} \omega_b \\ v_b \end{bmatrix}.$$

Because the $6 \times 6$ matrix pre-multiplying $\mathcal{V}_b$ is useful for changing the frame of reference for twists and wrenches, as we will see shortly, we give it its own name.

**Definition 3.20**  Given $T = (R, p) \in SE(3)$, its **adjoint representation** $[\mathrm{Ad}_T]$ is

$$[\mathrm{Ad}_T] = \begin{bmatrix} R & 0 \\ [p]R & R \end{bmatrix} \in \mathbb{R}^{6\times 6}.$$

For any $\mathcal{V} \in \mathbb{R}^6$, the **adjoint map** associated with $T$ is

$$\mathcal{V}' = [\mathrm{Ad}_T]\mathcal{V},$$

which is sometimes also written as

$$\mathcal{V}' = \mathrm{Ad}_T(\mathcal{V}).$$

In terms of the matrix form $[\mathcal{V}] \in se(3)$ of $\mathcal{V} \in \mathbb{R}^6$,

$$[\mathcal{V}'] = T[\mathcal{V}]T^{-1}.$$

The adjoint map satisfies the following properties, verifiable by direct calculation:

**Proposition 3.21**  Let $T_1, T_2 \in SE(3)$ and $\mathcal{V} = (\omega, v)$. Then

$$\mathrm{Ad}_{T_1}\left(\mathrm{Ad}_{T_2}(\mathcal{V})\right) = \mathrm{Ad}_{T_1 T_2}(\mathcal{V}) \quad or \quad [\mathrm{Ad}_{T_1}][\mathrm{Ad}_{T_2}]\mathcal{V} = [\mathrm{Ad}_{T_1 T_2}]\mathcal{V}. \tag{3.77}$$

Also, for any $T \in SE(3)$ the following holds:

$$[\mathrm{Ad}_T]^{-1} = [\mathrm{Ad}_{T^{-1}}], \tag{3.78}$$

The second property follows from the first on choosing $T_1 = T^{-1}$ and $T_2 = T$, so that

$$\mathrm{Ad}_{T^{-1}}\left(\mathrm{Ad}_T(\mathcal{V})\right) = \mathrm{Ad}_{T^{-1}T}(\mathcal{V}) = \mathrm{Ad}_I(\mathcal{V}) = \mathcal{V}. \tag{3.79}$$

### 3.3.2.1    Summary of Results on Twists

The main results on twists derived thus far are summarized in the following proposition:

**Proposition 3.22**    *Given a fixed (space) frame {s}, a body frame {b}, and a differentiable $T_{sb}(t) \in SE(3)$, where*

$$T_{sb}(t) = \begin{bmatrix} R(t) & p(t) \\ 0 & 1 \end{bmatrix}, \tag{3.80}$$

*then*

$$T_{sb}^{-1}\dot{T}_{sb} = [\mathcal{V}_b] = \begin{bmatrix} [\omega_b] & v_b \\ 0 & 0 \end{bmatrix} \in se(3) \tag{3.81}$$

*is the matrix representation of the* **body twist**, *and*

$$\dot{T}_{sb}T_{sb}^{-1} = [\mathcal{V}_s] = \begin{bmatrix} [\omega_s] & v_s \\ 0 & 0 \end{bmatrix} \in se(3) \tag{3.82}$$

*is the matrix representation of the* **spatial twist**. *The twists $\mathcal{V}_s$ and $\mathcal{V}_b$ are related by*

$$\mathcal{V}_s = \begin{bmatrix} \omega_s \\ v_s \end{bmatrix} = \begin{bmatrix} R & 0 \\ [p]R & R \end{bmatrix} \begin{bmatrix} \omega_b \\ v_b \end{bmatrix} = [\mathrm{Ad}_{T_{sb}}]\mathcal{V}_b, \tag{3.83}$$

$$\mathcal{V}_b = \begin{bmatrix} \omega_b \\ v_b \end{bmatrix} = \begin{bmatrix} R^{\mathrm{T}} & 0 \\ -R^{\mathrm{T}}[p] & R^{\mathrm{T}} \end{bmatrix} \begin{bmatrix} \omega_s \\ v_s \end{bmatrix} = [\mathrm{Ad}_{T_{bs}}]\mathcal{V}_s. \tag{3.84}$$

*More generally, for any two frames {c} and {d}, a twist represented as $\mathcal{V}_c$ in {c} is related to its representation $\mathcal{V}_d$ in {d} by*

$$\mathcal{V}_c = [\mathrm{Ad}_{T_{cd}}]\mathcal{V}_d, \qquad \mathcal{V}_d = [\mathrm{Ad}_{T_{dc}}]\mathcal{V}_c.$$

Again analogously to the case of angular velocities, it is important to realize that, for a given twist, its fixed-frame representation $\mathcal{V}_s$ *does not depend on the choice of the body frame* {b}, and its body-frame representation $\mathcal{V}_b$ *does not depend on the choice of the fixed frame* {s}.

**Example 3.23**    Figure 3.18 shows a top view of a car, with a single steerable front wheel, driving on a plane. The $\hat{z}_b$-axis of the body frame {b} is into the page and the $\hat{z}_s$-axis of the fixed frame {s} is out of the page. The angle of the front wheel of the car causes the car's motion to be a pure angular velocity w = 2 rad/s about an axis out of the page at the point r in the plane. Inspecting the figure, we can write r as $r_s = (2, -1, 0)$ or $r_b = (2, -1.4, 0)$, w as $\omega_s = (0, 0, 2)$ or $\omega_b = (0, 0, -2)$, and $T_{sb}$ as

$$T_{sb} = \begin{bmatrix} R_{sb} & p_{sb} \\ 0 & 1 \end{bmatrix} = \begin{bmatrix} -1 & 0 & 0 & 4 \\ 0 & 1 & 0 & 0.4 \\ 0 & 0 & -1 & 0 \\ 0 & 0 & 0 & 1 \end{bmatrix}.$$

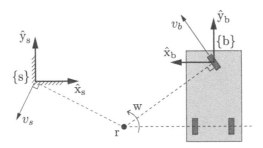

**Figure 3.18** The twist corresponding to the instantaneous motion of the chassis of a three-wheeled vehicle can be visualized as an angular velocity w about the point r.

From the figure and simple geometry, we get

$$v_s = \omega_s \times (-r_s) = r_s \times \omega_s = (-2, -4, 0),$$
$$v_b = \omega_b \times (-r_b) = r_b \times \omega_b = (2.8, 4, 0),$$

and thus obtain the twists $\mathcal{V}_s$ and $\mathcal{V}_b$:

$$\mathcal{V}_s = \begin{bmatrix} \omega_s \\ v_s \end{bmatrix} = \begin{bmatrix} 0 \\ 0 \\ 2 \\ -2 \\ -4 \\ 0 \end{bmatrix}, \qquad \mathcal{V}_b = \begin{bmatrix} \omega_b \\ v_b \end{bmatrix} = \begin{bmatrix} 0 \\ 0 \\ -2 \\ 2.8 \\ 4 \\ 0 \end{bmatrix}.$$

To confirm these results, try calculating $\mathcal{V}_s = [\mathrm{Ad}_{T_{sb}}]\mathcal{V}_b$.

### .3.2.2  The Screw Interpretation of a Twist

Just as an angular velocity $\omega$ can be viewed as $\hat{\omega}\dot{\theta}$, where $\hat{\omega}$ is the unit rotation axis and $\dot{\theta}$ is the rate of rotation about that axis, a twist $\mathcal{V}$ can be interpreted in terms of a **screw axis** $\mathcal{S}$ and a velocity $\dot{\theta}$ about the screw axis.

A screw axis represents the familiar motion of a screw: rotating about the axis while also translating along the axis. One representation of a screw axis $\mathcal{S}$ is the collection $\{q, \hat{s}, h\}$, where $q \in \mathbb{R}^3$ is any point on the axis, $\hat{s}$ is a unit vector in the direction of the axis, and $h$ is the **screw pitch**, which defines the ratio of the linear velocity along the screw axis to the angular velocity $\dot{\theta}$ about the screw axis (Figure 3.19).

Using Figure 3.19 and geometry, we can write the twist $\mathcal{V} = (\omega, v)$ corresponding to an angular velocity $\dot{\theta}$ about $\mathcal{S}$ (represented by $\{q, \hat{s}, h\}$) as

$$\mathcal{V} = \begin{bmatrix} \omega \\ v \end{bmatrix} = \begin{bmatrix} \hat{s}\dot{\theta} \\ -\hat{s}\dot{\theta} \times q + h\hat{s}\dot{\theta} \end{bmatrix}.$$

Note that the linear velocity $v$ is the sum of two terms: one due to translation along the screw axis, $h\hat{s}\dot{\theta}$, and the other due to the linear motion at the origin induced by rotation about the axis, $-\hat{s}\dot{\theta} \times q$. The first term is in the direction

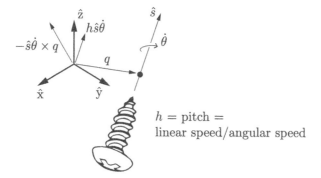

$h = \text{pitch} =$
linear speed/angular speed

**Figure 3.19** A screw axis $\mathcal{S}$ represented by a point $q$, a unit direction $\hat{s}$, and a pitch $h$.

of $\hat{s}$, while the second term is in the plane orthogonal to $\hat{s}$. It is not hard to show that, for any $\mathcal{V} = (\omega, v)$ where $\omega \neq 0$, there exists an equivalent screw axis $\{q, \hat{s}, h\}$ and velocity $\dot{\theta}$, where $\hat{s} = \omega/\|\omega\|$, $\dot{\theta} = \|\omega\|$, $h = \hat{\omega}^{\mathrm{T}} v/\dot{\theta}$, and $q$ is chosen so that the term $-\hat{s}\dot{\theta} \times q$ provides the portion of $v$ orthogonal to the screw axis.

If $\omega = 0$, then the pitch $h$ of the screw is infinite. In this case $\hat{s}$ is chosen as $v/\|v\|$, and $\dot{\theta}$ is interpreted as the linear velocity $\|v\|$ along $\hat{s}$.

Instead of representing the screw axis $\mathcal{S}$ using the cumbersome collection $\{q, \hat{s}, h\}$, with the possibility that $h$ may be infinite and with the nonuniqueness of $q$ (any $q$ along the screw axis may be used), we instead define the screw axis $\mathcal{S}$ using a normalized version of any twist $\mathcal{V} = (\omega, v)$ corresponding to motion along the screw:

(a) If $\omega \neq 0$ then $\mathcal{S} = \mathcal{V}/\|\omega\| = (\omega/\|\omega\|, v/\|\omega\|)$. The screw axis $\mathcal{S}$ is simply $\mathcal{V}$ normalized by the length of the angular velocity vector. The angular velocity about the screw axis is $\dot{\theta} = \|\omega\|$, such that $\mathcal{S}\dot{\theta} = \mathcal{V}$.

(b) If $\omega = 0$ then $\mathcal{S} = \mathcal{V}/\|v\| = (0, v/\|v\|)$. The screw axis $\mathcal{S}$ is simply $\mathcal{V}$ normalized by the length of the linear velocity vector. The linear velocity along the screw axis is $\dot{\theta} = \|v\|$, such that $\mathcal{S}\dot{\theta} = \mathcal{V}$.

This leads to the following definition of a "unit" (normalized) screw axis:

**Definition 3.24**    For a given reference frame, a **screw axis** $\mathcal{S}$ is written as

$$\mathcal{S} = \begin{bmatrix} \omega \\ v \end{bmatrix} \in \mathbb{R}^6,$$

where either (i) $\|\omega\| = 1$ or (ii) $\omega = 0$ and $\|v\| = 1$. If (i) holds then $v = -\omega \times q + h\omega$, where $q$ is a point on the axis of the screw and $h$ is the pitch of the screw ($h = 0$ for a pure rotation about the screw axis). If (ii) holds then the pitch of the screw is infinite and the twist is a translation along the axis defined by $v$.

**Important:** Although we use the pair $(\omega, v)$ for both a normalized screw axis $\mathcal{S}$ (where one of $\|\omega\|$ or $\|v\|$ must be unity) and a general twist $\mathcal{V}$ (where there are no constraints on $\omega$ and $v$), the meaning should be clear from the context.

Since a screw axis $\mathcal{S}$ is just a normalized twist, the $4 \times 4$ matrix representation $[\mathcal{S}]$ of $\mathcal{S} = (\omega, v)$ is

$$[\mathcal{S}] = \begin{bmatrix} [\omega] & v \\ 0 & 0 \end{bmatrix} \in se(3), \qquad [\omega] = \begin{bmatrix} 0 & -\omega_3 & \omega_2 \\ \omega_3 & 0 & -\omega_1 \\ -\omega_2 & \omega_1 & 0 \end{bmatrix} \in so(3), \qquad (3.85)$$

where the bottom row of $[\mathcal{S}]$ consists of all zeros. Also, a screw axis represented as $\mathcal{S}_a$ in a frame {a} is related to the representation $\mathcal{S}_b$ in a frame {b} by

$$\mathcal{S}_a = [\text{Ad}_{T_{ab}}]\mathcal{S}_b, \qquad \mathcal{S}_b = [\text{Ad}_{T_{ba}}]\mathcal{S}_a.$$

## 3.3 Exponential Coordinate Representation of Rigid-Body Motions

### 3.3.1 Exponential Coordinates of Rigid-Body Motions

In the planar example in Section 3.1, we saw that any planar rigid-body displacement can be achieved by rotating the rigid body about some fixed point in the plane (for a pure translation, this point lies at infinity). A similar result also exists for spatial rigid-body displacements: the **Chasles–Mozzi theorem** states that every rigid-body displacement can be expressed as a displacement along a fixed screw axis $\mathcal{S}$ in space.

By analogy to the exponential coordinates $\hat{\omega}\theta$ for rotations, we define the six-dimensional **exponential coordinates of a homogeneous transformation** $T$ as $\mathcal{S}\theta \in \mathbb{R}^6$, where $\mathcal{S}$ is the screw axis and $\theta$ is the distance that must be traveled along the screw axis to take a frame from the origin $I$ to $T$. If the pitch of the screw axis $\mathcal{S} = (\omega, v)$ is finite then $\|\omega\| = 1$ and $\theta$ corresponds to the angle of rotation about the screw axis. If the pitch of the screw is infinite then $\omega = 0$ and $\|v\| = 1$ and $\theta$ corresponds to the linear distance traveled along the screw axis.

Also by analogy to the rotation case, we define a matrix exponential (exp) and matrix logarithm (log):

$$\exp : \quad [\mathcal{S}]\theta \in se(3) \quad \rightarrow \quad T \in SE(3),$$
$$\log : \quad T \in SE(3) \quad \rightarrow \quad [\mathcal{S}]\theta \in se(3).$$

We begin by deriving a closed-form expression for the matrix exponential $e^{[\mathcal{S}]\theta}$. Expanding the matrix exponential in series form leads to

$$e^{[\mathcal{S}]\theta} = I + [\mathcal{S}]\theta + [\mathcal{S}]^2 \frac{\theta^2}{2!} + [\mathcal{S}]^3 \frac{\theta^3}{3!} + \cdots$$

$$= \begin{bmatrix} e^{[\omega]\theta} & G(\theta)v \\ 0 & 1 \end{bmatrix}, \qquad G(\theta) = I\theta + [\omega]\frac{\theta^2}{2!} + [\omega]^2 \frac{\theta^3}{3!} + \cdots . \quad (3.86)$$

Using the identity $[\omega]^3 = -[\omega]$, $G(\theta)$ can be simplified to

$$
\begin{aligned}
G(\theta) &= I\theta + [\omega]\frac{\theta^2}{2!} + [\omega]^2\frac{\theta^3}{3!} + \cdots \\
&= I\theta + \left(\frac{\theta^2}{2!} - \frac{\theta^4}{4!} + \frac{\theta^6}{6!} - \cdots\right)[\omega] + \left(\frac{\theta^3}{3!} - \frac{\theta^5}{5!} + \frac{\theta^7}{7!} - \cdots\right)[\omega]^2 \\
&= I\theta + (1 - \cos\theta)[\omega] + (\theta - \sin\theta)[\omega]^2.
\end{aligned}
\tag{3.87}
$$

Putting everything together leads to the following proposition:

**Proposition 3.25** *Let $\mathcal{S} = (\omega, v)$ be a screw axis. If $\|\omega\| = 1$ then, for any distance $\theta \in \mathbb{R}$ traveled along the axis,*

$$
e^{[\mathcal{S}]\theta} = \begin{bmatrix} e^{[\omega]\theta} & \left(I\theta + (1-\cos\theta)[\omega] + (\theta - \sin\theta)[\omega]^2\right)v \\ 0 & 1 \end{bmatrix}.
\tag{3.88}
$$

*If $\omega = 0$ and $\|v\| = 1$, then*

$$
e^{[\mathcal{S}]\theta} = \begin{bmatrix} I & v\theta \\ 0 & 1 \end{bmatrix}.
\tag{3.89}
$$

### 3.3.3.2 Matrix Logarithm of Rigid-Body Motions

The above derivation essentially provides a constructive proof of the Chasles–Mozzi theorem. That is, given an arbitrary $(R, p) \in SE(3)$, one can always find a screw axis $\mathcal{S} = (\omega, v)$ and a scalar $\theta$ such that

$$
e^{[\mathcal{S}]\theta} = \begin{bmatrix} R & p \\ 0 & 1 \end{bmatrix},
\tag{3.90}
$$

i.e., the matrix

$$
[\mathcal{S}]\theta = \begin{bmatrix} [\omega]\theta & v\theta \\ 0 & 0 \end{bmatrix} \in se(3)
$$

is the matrix logarithm of $T = (R, p)$.

**Algorithm:**
Given $(R, p)$ written as $T \in SE(3)$, find a $\theta \in [0, \pi]$ and a screw axis $\mathcal{S} = (\omega, v) \in \mathbb{R}^6$ (where at least one of $\|\omega\|$ and $\|v\|$ is unity) such that $e^{[\mathcal{S}]\theta} = T$. The vector $\mathcal{S}\theta \in \mathbb{R}^6$ comprises the exponential coordinates for $T$ and the matrix $[\mathcal{S}]\theta \in se(3)$ is the matrix logarithm of $T$.

(a) If $R = I$ then set $\omega = 0$, $v = p/\|p\|$, and $\theta = \|p\|$.
(b) Otherwise, use the matrix logarithm on $SO(3)$ to determine $\omega$ (written as $\hat{\omega}$ in the $SO(3)$ algorithm) and $\theta$ for $R$. Then $v$ is calculated as

$$
v = G^{-1}(\theta)p
\tag{3.91}
$$

where

$$
G^{-1}(\theta) = \frac{1}{\theta}I - \frac{1}{2}[\omega] + \left(\frac{1}{\theta} - \frac{1}{2}\cot\frac{\theta}{2}\right)[\omega]^2.
\tag{3.92}
$$

The verification of Equation (3.92) is left as an exercise.

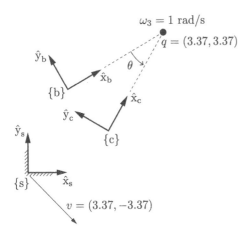

$\omega_3 = 1 \text{ rad/s}$

$q = (3.37, 3.37)$

$v = (3.37, -3.37)$

**Figure 3.20** Two frames in a plane.

**Example 3.26** In this example, the rigid-body motion is confined to the $\hat{x}_s$–$\hat{y}_s$-plane. The initial frame {b} and final frame {c} in Figure 3.20 can be represented by the $SE(3)$ matrices

$$T_{sb} = \begin{bmatrix} \cos 30° & -\sin 30° & 0 & 1 \\ \sin 30° & \cos 30° & 0 & 2 \\ 0 & 0 & 1 & 0 \\ 0 & 0 & 0 & 1 \end{bmatrix},$$

$$T_{sc} = \begin{bmatrix} \cos 60° & -\sin 60° & 0 & 2 \\ \sin 60° & \cos 60° & 0 & 1 \\ 0 & 0 & 1 & 0 \\ 0 & 0 & 0 & 1 \end{bmatrix}.$$

Because the motion occurs in the $\hat{x}_s$–$\hat{y}_s$-plane, the corresponding screw has an axis in the direction of the $\hat{z}_s$-axis and has zero pitch. The screw axis $\mathcal{S} = (\omega, v)$, expressed in {s}, therefore has the form

$$\omega = (0, 0, \omega_3),$$
$$v = (v_1, v_2, 0).$$

We seek the screw motion that displaces the frame at $T_{sb}$ to $T_{sc}$; i.e., $T_{sc} = e^{[\mathcal{S}]\theta} T_{sb}$ or

$$T_{sc} T_{sb}^{-1} = e^{[\mathcal{S}]\theta},$$

where

$$[\mathcal{S}] = \begin{bmatrix} 0 & -\omega_3 & 0 & v_1 \\ \omega_3 & 0 & 0 & v_2 \\ 0 & 0 & 0 & 0 \\ 0 & 0 & 0 & 0 \end{bmatrix}.$$

We can apply the matrix logarithm algorithm directly to $T_{sc}T_{sb}^{-1}$ to obtain $[\mathcal{S}]$ (and therefore $\mathcal{S}$) and $\theta$ as follows:

$$[\mathcal{S}] = \begin{bmatrix} 0 & -1 & 0 & 3.37 \\ 1 & 0 & 0 & -3.37 \\ 0 & 0 & 0 & 0 \\ 0 & 0 & 0 & 0 \end{bmatrix}, \quad \mathcal{S} = \begin{bmatrix} \omega_1 \\ \omega_2 \\ \omega_3 \\ v_1 \\ v_2 \\ v_3 \end{bmatrix} = \begin{bmatrix} 0 \\ 0 \\ 1 \\ 3.37 \\ -3.37 \\ 0 \end{bmatrix}, \quad \theta = \frac{\pi}{6} \text{ rad (or } 30°\text{)}.$$

The value of $\mathcal{S}$ means that the constant screw axis, expressed in the fixed frame $\{s\}$, is represented by an angular velocity of 1 rad/s about $\hat{z}_s$ and a linear velocity (of a point currently at the origin of $\{s\}$) of $(3.37, -3.37, 0)$ expressed in the $\{s\}$ frame.

Alternatively, we can observe that the displacement is not a pure translation – $T_{sb}$ and $T_{sc}$ have rotation components that differ by an angle of 30° – and we quickly determine that $\theta = 30°$ and $\omega_3 = 1$. We can also graphically determine the point $q = (q_x, q_y)$ in the $\hat{x}_s$–$\hat{y}_s$-plane through which the screw axis passes; for our example this point is given by $q = (3.37, 3.37)$.

For planar rigid-body motions such as this one, we could derive a planar matrix logarithm algorithm that maps elements of $SE(2)$ to elements of $se(2)$, which have the form

$$\begin{bmatrix} 0 & -\omega & v_1 \\ \omega & 0 & v_2 \\ 0 & 0 & 0 \end{bmatrix}.$$

## 3.4        Wrenches

Consider a linear force f acting on a rigid body at a point r. Defining a reference frame $\{a\}$, the point r can be represented as $r_a \in \mathbb{R}^3$ and the force f can be represented as $f_a \in \mathbb{R}^3$. This force creates a **torque** or **moment** $m_a \in \mathbb{R}^3$ in the $\{a\}$ frame:

$$m_a = r_a \times f_a.$$

Note that the point of application of the force along its line of action is immaterial.

Just as with twists, we can merge the moment and force into a single six-dimensional **spatial force**, or **wrench**, expressed in the $\{a\}$ frame, $\mathcal{F}_a$:

$$\mathcal{F}_a = \begin{bmatrix} m_a \\ f_a \end{bmatrix} \in \mathbb{R}^6. \tag{3.93}$$

If more than one wrench acts on a rigid body, the total wrench on the body is simply the vector sum of the individual wrenches, provided that the wrenches

are expressed in the same frame. A wrench with a zero linear component is called
a **pure moment**.

A wrench in the {a} frame can be represented in another frame {b} (Figure 3.21) if $T_{ba}$ is known. One way to derive the relationship between $\mathcal{F}_a$ and
$\mathcal{F}_b$ is to derive the appropriate transformations between the individual force and
moment vectors on the basis of techniques we have already used.

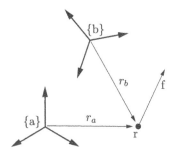

**Figure 3.21** Relation between wrench representations $\mathcal{F}_a$ and $\mathcal{F}_b$.

A simpler and more insightful way to derive the relationship between $\mathcal{F}_a$ and
$\mathcal{F}_b$ is, however, to (1) use the results we have already derived relating representations $\mathcal{V}_a$ and $\mathcal{V}_b$ of the same twist, and (2) use the fact that the power generated
(or dissipated) by an $(\mathcal{F}, \mathcal{V})$ pair must be the same regardless of the frame in
which it is represented. (Imagine if we could create power simply by changing our
choice of reference frame!) Recall that the dot product of a force and a velocity
is a power, and power is a coordinate-independent quantity. Because of this, we
know that

$$\mathcal{V}_b^{\mathrm{T}} \mathcal{F}_b = \mathcal{V}_a^{\mathrm{T}} \mathcal{F}_a. \tag{3.94}$$

From Proposition 3.22 we know that $\mathcal{V}_a = [\mathrm{Ad}_{T_{ab}}]\mathcal{V}_b$, and therefore Equation (3.94) can be rewritten as

$$\mathcal{V}_b^{\mathrm{T}} \mathcal{F}_b = ([\mathrm{Ad}_{T_{ab}}]\mathcal{V}_b)^{\mathrm{T}} \mathcal{F}_a$$
$$= \mathcal{V}_b^{\mathrm{T}} [\mathrm{Ad}_{T_{ab}}]^{\mathrm{T}} \mathcal{F}_a.$$

Since this must hold for all $\mathcal{V}_b$, this simplifies to

$$\mathcal{F}_b = [\mathrm{Ad}_{T_{ab}}]^{\mathrm{T}} \mathcal{F}_a. \tag{3.95}$$

Similarly,

$$\mathcal{F}_a = [\mathrm{Ad}_{T_{ba}}]^{\mathrm{T}} \mathcal{F}_b. \tag{3.96}$$

**Proposition 3.27**   *Given a wrench* F, *represented in* {a} *as* $\mathcal{F}_a$ *and in* {b} *as*
$\mathcal{F}_b$, *the two representations are related by*

$$\mathcal{F}_b = \mathrm{Ad}_{T_{ab}}^{\mathrm{T}}(\mathcal{F}_a) = [\mathrm{Ad}_{T_{ab}}]^{\mathrm{T}} \mathcal{F}_a, \tag{3.97}$$
$$\mathcal{F}_a = \mathrm{Ad}_{T_{ba}}^{\mathrm{T}}(\mathcal{F}_b) = [\mathrm{Ad}_{T_{ba}}]^{\mathrm{T}} \mathcal{F}_b. \tag{3.98}$$

Since we usually have a fixed space frame {s} and a body frame {b}, we can
define a **spatial wrench** $\mathcal{F}_s$ and a **body wrench** $\mathcal{F}_b$.

**Example 3.28** The robot hand in Figure 3.22 is holding an apple with a mass of 0.1 kg in a gravitational field $g = 10$ m/s$^2$ (rounded to keep the numbers simple) acting downward on the page. The mass of the hand is 0.5 kg. What is the force and torque measured by the six-axis force–torque sensor between the hand and the robot arm?

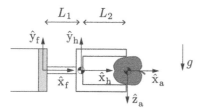

**Figure 3.22** A robot hand holding an apple subject to gravity.

We define frames {f} at the force–torque sensor, {h} at the center of mass of the hand, and {a} at the center of mass of the apple. According to the coordinate axes in Figure 3.22, the gravitational wrench on the hand in {h} is given by the column vector

$$\mathcal{F}_h = (0, 0, 0, 0, -5 \text{ N}, 0)$$

and the gravitational wrench on the apple in {a} is

$$\mathcal{F}_a = (0, 0, 0, 0, 0, 1 \text{ N}).$$

Given $L_1 = 10$ cm and $L_2 = 15$ cm, the transformation matrices $T_{hf}$ and $T_{af}$ are

$$T_{hf} = \begin{bmatrix} 1 & 0 & 0 & -0.1 \text{ m} \\ 0 & 1 & 0 & 0 \\ 0 & 0 & 1 & 0 \\ 0 & 0 & 0 & 1 \end{bmatrix}, \quad T_{af} = \begin{bmatrix} 1 & 0 & 0 & -0.25 \text{ m} \\ 0 & 0 & 1 & 0 \\ 0 & -1 & 0 & 0 \\ 0 & 0 & 0 & 1 \end{bmatrix}.$$

The wrench measured by the six-axis force–torque sensor is

$$\begin{aligned} \mathcal{F}_f &= [\mathrm{Ad}_{T_{hf}}]^{\mathrm{T}} \mathcal{F}_h + [\mathrm{Ad}_{T_{af}}]^{\mathrm{T}} \mathcal{F}_a \\ &= [0 \; 0 \; -0.5 \text{ Nm } 0 \; -5 \text{ N } 0]^{\mathrm{T}} + [0 \; 0 \; -0.25 \text{ Nm } 0 \; -1 \text{ N } 0]^{\mathrm{T}} \\ &= [0 \; 0 \; -0.75 \text{ Nm } 0 \; -6 \text{ N } 0]^{\mathrm{T}}. \end{aligned}$$

## 3.5 Summary

The following table succinctly summarizes some of the key concepts from the chapter, as well as the parallelism between rotations and rigid-body motions. For more details, consult the appropriate section of the chapter.

| Rotations | Rigid-Body Motions |
|---|---|
| $R \in SO(3) : 3 \times 3$ matrices<br><br>$R^{\mathrm{T}} R = I, \ \det R = 1.$ | $T \in SE(3) : 4 \times 4$ matrices<br><br>$T = \begin{bmatrix} R & p \\ 0 & 1 \end{bmatrix},$<br><br>where $R \in SO(3), p \in \mathbb{R}^3.$ |
| $R^{-1} = R^{\mathrm{T}}.$ | $T^{-1} = \begin{bmatrix} R^{\mathrm{T}} & -R^{\mathrm{T}} p \\ 0 & 1 \end{bmatrix}.$ |
| Change of coordinate frame:<br>$R_{ab} R_{bc} = R_{ac}, \ R_{ab} p_b = p_a.$ | Change of coordinate frame:<br>$T_{ab} T_{bc} = T_{ac}, \ T_{ab} p_b = p_a.$ |
| Rotating a frame $\{b\}$:<br><br>$R = \mathrm{Rot}(\hat{\omega}, \theta).$<br><br>$R_{sb'} = R R_{sb}$:<br>rotate by $\theta$ about $\hat{\omega}_s = \hat{\omega}$<br>$R_{sb''} = R_{sb} R$:<br>rotate by $\theta$ about $\hat{\omega}_b = \hat{\omega}.$ | Displacing a frame $\{b\}$:<br><br>$T = \begin{bmatrix} \mathrm{Rot}(\hat{\omega}, \theta) & p \\ 0 & 1 \end{bmatrix}.$<br><br>$T_{sb'} = T T_{sb}$: rotate by $\theta$ about $\hat{\omega}_s = \hat{\omega}$<br>(moves $\{b\}$ origin), translate $p$ in $\{s\}$.<br>$T_{sb''} = T_{sb} T$: translate $p$ in $\{b\}$,<br>rotate by $\theta$ about $\hat{\omega}$ in new body frame. |
| Unit rotation axis is $\hat{\omega} \in \mathbb{R}^3$,<br><br>where $\|\hat{\omega}\| = 1.$ | "Unit" screw axis is $\mathcal{S} = \begin{bmatrix} \omega \\ v \end{bmatrix} \in \mathbb{R}^6$,<br><br>where either (i) $\|\omega\| = 1$ or<br>(ii) $\omega = 0$ and $\|v\| = 1.$<br><br>For a screw axis $\{q, \hat{s}, h\}$ with finite $h$,<br><br>$\mathcal{S} = \begin{bmatrix} \omega \\ v \end{bmatrix} = \begin{bmatrix} \hat{s} \\ -\hat{s} \times q + h\hat{s} \end{bmatrix}.$ |
| Angular velocity is $\omega = \hat{\omega}\dot{\theta}.$ | Twist is $\mathcal{V} = \mathcal{S}\dot{\theta}.$ |
| For any 3-vector, e.g., $\omega \in \mathbb{R}^3$,<br><br>$[\omega] = \begin{bmatrix} 0 & -\omega_3 & \omega_2 \\ \omega_3 & 0 & -\omega_1 \\ -\omega_2 & \omega_1 & 0 \end{bmatrix} \in so(3)$<br>Identities for $\omega, x \in \mathbb{R}^3, R \in SO(3)$:<br>$[\omega] = -[\omega]^{\mathrm{T}}, \quad [\omega]x = -[x]\omega,$<br>$[\omega][x] = ([x][\omega])^{\mathrm{T}}, \quad R[\omega]R^{\mathrm{T}} = [R\omega].$ | For $\mathcal{V} = \begin{bmatrix} \omega \\ v \end{bmatrix} \in \mathbb{R}^6$,<br><br>$[\mathcal{V}] = \begin{bmatrix} [\omega] & v \\ 0 & 0 \end{bmatrix} \in se(3)$<br><br>(the pair $(\omega, v)$ can be a twist $\mathcal{V}$<br>or a "unit" screw axis $\mathcal{S}$,<br>depending on the context). |
| $\dot{R} R^{-1} = [\omega_s], \ R^{-1}\dot{R} = [\omega_b].$ | $\dot{T} T^{-1} = [\mathcal{V}_s], \ T^{-1}\dot{T} = [\mathcal{V}_b].$<br><br>$[\mathrm{Ad}_T] = \begin{bmatrix} R & 0 \\ [p]R & R \end{bmatrix} \in \mathbb{R}^{6 \times 6}$<br>Identities: $[\mathrm{Ad}_T]^{-1} = [\mathrm{Ad}_{T^{-1}}],$<br>$[\mathrm{Ad}_{T_1}][\mathrm{Ad}_{T_2}] = [\mathrm{Ad}_{T_1 T_2}].$ |
| Change of coordinate frame:<br>$\hat{\omega}_a = R_{ab}\hat{\omega}_b, \ \omega_a = R_{ab}\omega_b.$ | Change of coordinate frame:<br>$\mathcal{S}_a = [\mathrm{Ad}_{T_{ab}}]\mathcal{S}_b, \ \mathcal{V}_a = [\mathrm{Ad}_{T_{ab}}]\mathcal{V}_b.$ |

continued...

| Rotations | Rigid-Body Motions |
|---|---|
| Exp. coords. for $R \in SO(3)$: $\hat{\omega}\theta \in \mathbb{R}^3$: | Exp. coords. for $T \in SE(3)$: $\mathcal{S}\theta \in \mathbb{R}^6$: |
| $\exp : [\hat{\omega}]\theta \in so(3) \to R \in SO(3)$, | $\exp : [\mathcal{S}]\theta \in se(3) \to T \in SE(3)$, |
| $R = \text{Rot}(\hat{\omega}, \theta) = e^{[\hat{\omega}]\theta}$ | $T = e^{[\mathcal{S}]\theta} = \begin{bmatrix} e^{[\omega]\theta} & * \\ 0 & 1 \end{bmatrix}$, |
| $= I + \sin\theta[\hat{\omega}] + (1 - \cos\theta)[\hat{\omega}]^2$ | where $* =$ is given by $(I\theta + (1 - \cos\theta)[\omega] + (\theta - \sin\theta)[\omega]^2)v$. |
| $\log : R \in SO(3) \to [\hat{\omega}]\theta \in so(3)$. Algorithm in §3.2.3.3 | $\log : T \in SE(3) \to [\mathcal{S}]\theta \in se(3)$. Algorithm in §3.3.3.2 |
| Moment change of coord. frame: $m_a = R_{ab}m_b$. | Wrench change of coord. frame: $\mathcal{F}_a = (m_a, f_a) = [\text{Ad}_{T_{ba}}]^{\text{T}}\mathcal{F}_b$. |

## 3.6     Software

The following functions are included in the software distribution accompanying the book. The code below is in MATLAB format, but it is available in other languages. For more details on the software, consult the code and its documentation.

`invR = RotInv(R)`
Computes the inverse of the rotation matrix R.

`so3mat = VecToso3(omg)`
Returns the $3 \times 3$ skew-symmetric matrix corresponding to omg.

`omg = so3ToVec(so3mat)`
Returns the 3-vector corresponding to the $3 \times 3$ skew-symmetric matrix so3mat.

`[omghat,theta] = AxisAng3(expc3)`
Extracts the rotation axis $\hat{\omega}$ and the rotation amount $\theta$ from the 3-vector $\hat{\omega}\theta$ of exponential coordinates for rotation, expc3.

`R = MatrixExp3(so3mat)`
Computes the rotation matrix $R \in SO(3)$ corresponding to the matrix exponential of so3mat $\in so(3)$.

`so3mat = MatrixLog3(R)`
Computes the matrix logarithm so3mat $\in so(3)$ of the rotation matrix R $\in SO(3)$.

`T = RpToTrans(R,p)`
Builds the homogeneous transformation matrix T corresponding to a rotation matrix $R \in SO(3)$ and a position vector $p \in \mathbb{R}^3$.

`[R,p] = TransToRp(T)`

Extracts the rotation matrix and position vector from a homogeneous transformation matrix T.

`invT = TransInv(T)`
Computes the inverse of a homogeneous transformation matrix T.

`se3mat = VecTose3(V)`
Returns the $se(3)$ matrix corresponding to a 6-vector twist V.

`V = se3ToVec(se3mat)`
Returns the 6-vector twist corresponding to an $se(3)$ matrix se3mat.

`AdT = Adjoint(T)`
Computes the $6 \times 6$ adjoint representation $[\mathrm{Ad}_T]$ of the homogeneous transformation matrix T.

`S = ScrewToAxis(q,s,h)`
Returns a normalized screw axis representation $\mathcal{S}$ of a screw described by a unit vector s in the direction of the screw axis, located at the point q, with pitch h.

`[S,theta] = AxisAng6(expc6)`
Extracts the normalized screw axis $\mathcal{S}$ and the distance traveled along the screw $\theta$ from the 6-vector of exponential coordinates $\mathcal{S}\theta$.

`T = MatrixExp6(se3mat)`
Computes the homogeneous transformation matrix $T \in SE(3)$ corresponding to the matrix exponential of se3mat $\in se(3)$.

`se3mat = MatrixLog6(T)`
Computes the matrix logarithm se3mat $\in se(3)$ of the homogeneous transformation matrix $T \in SE(3)$.

## 3.7    Notes and References

The exponential coordinates for rotations introduced in this chapter are also referred to in the kinematics literature as the Euler–Rodrigues parameters. Other representations for rotations such as Euler angles, Cayley–Rodrigues parameters, and unit quaternions are described in Appendix B; further details on these and related parametrizations of the rotation group $SO(3)$ can be found in, e.g., Shuster (1993), McCarthy (1990), Tsiotras et al. (1997), Murray et al. (1994), and Park and Kang (1999).

Classical screw theory has its origins in the works of Mozzi and Chasles, who independently discovered that the motion of a rigid body can be obtained as a rotation about some axis followed by a translation about the same axis (Ceccarelli, 2000). Ball's 1900 treatise is often regarded as the classical reference on screw theory, while more modern treatments can be found in Bottema and Roth (1990), Angeles (2006), and McCarthy (1990).

The identification of the elements of classical screw theory with the Lie group structure of rigid-body motions $SE(3)$ was first made by Brockett (1983b), who went considerably further and showed that the forward kinematics of open chains can be expressed as the product of matrix exponentials (this is the subject of the next chapter). Derivations of the formulas for the matrix exponentials, logarithms, their derivatives, and other related formulas can be found in Lončarić (1985), Paden (1986), Park (1991), and Murray et al. (1994).

## 3.8    Exercises

**Exercise 3.1**   In terms of the $\hat{x}_s$, $\hat{y}_s$, $\hat{z}_s$ coordinates of a fixed space frame $\{s\}$, the frame $\{a\}$ has its $\hat{x}_a$-axis pointing in the direction $(0,0,1)$ and its $\hat{y}_a$-axis pointing in the direction $(-1,0,0)$, and the frame $\{b\}$ has its $\hat{x}_b$-axis pointing in the direction $(1,0,0)$ and its $\hat{y}_b$-axis pointing in the direction $(0,0,-1)$.

(a) Draw by hand the three frames, at different locations so that they are easy to see.

(b) Write down the rotation matrices $R_{sa}$ and $R_{sb}$.

(c) Given $R_{sb}$, how do you calculate $R_{sb}^{-1}$ without using a matrix inverse? Write down $R_{sb}^{-1}$ and verify its correctness using your drawing.

(d) Given $R_{sa}$ and $R_{sb}$, how do you calculate $R_{ab}$ (again without using matrix inverses)? Compute the answer and verify its correctness using your drawing.

(e) Let $R = R_{sb}$ be considered as a transformation operator consisting of a rotation about $\hat{x}$ by $-90°$. Calculate $R_1 = R_{sa}R$, and think of $R_{sa}$ as a representation of an orientation, $R$ as a rotation of $R_{sa}$, and $R_1$ as the new orientation after the rotation has been performed. Does the new orientation $R_1$ correspond to a rotation of $R_{sa}$ by $-90°$ about the world-fixed $\hat{x}_s$-axis or about the body-fixed $\hat{x}_a$-axis? Now calculate $R_2 = RR_{sa}$. Does the new orientation $R_2$ correspond to a rotation of $R_{sa}$ by $-90°$ about the world-fixed $\hat{x}_s$-axis or about the body-fixed $\hat{x}_a$-axis?

(f) Use $R_{sb}$ to change the representation of the point $p_b = (1,2,3)$ (which is in $\{b\}$ coordinates) to $\{s\}$ coordinates.

(g) Choose a point p represented by $p_s = (1,2,3)$ in $\{s\}$ coordinates. Calculate $p' = R_{sb}p_s$ and $p'' = R_{sb}^{T}p_s$. For each operation, should the result be interpreted as changing coordinates (from the $\{s\}$ frame to $\{b\}$) without moving the point p or as moving the location of the point without changing the reference frame of the representation?

(h) An angular velocity w is represented in $\{s\}$ as $\omega_s = (3,2,1)$. What is its representation $\omega_a$ in $\{a\}$?

(i) By hand, calculate the matrix logarithm $[\hat{\omega}]\theta$ of $R_{sa}$. (You may verify your answer with software.) Extract the unit angular velocity $\hat{\omega}$ and rotation amount $\theta$. Redraw the fixed frame {s} and in it draw $\hat{\omega}$.

(j) Calculate the matrix exponential corresponding to the exponential coordinates of rotation $\hat{\omega}\theta = (1, 2, 0)$. Draw the corresponding frame relative to {s}, as well as the rotation axis $\hat{\omega}$.

**Exercise 3.2** Let p be a point whose coordinates are $p = \left(\frac{1}{\sqrt{3}}, -\frac{1}{\sqrt{6}}, \frac{1}{\sqrt{2}}\right)$ with respect to the fixed frame $\hat{x}$–$\hat{y}$–$\hat{z}$. Suppose that p is rotated about the fixed-frame $\hat{x}$-axis by 30 degrees, then about the fixed-frame $\hat{y}$-axis by 135 degrees, and finally about the fixed-frame $\hat{z}$-axis by $-120$ degrees. Denote the coordinates of this newly rotated point by $p'$.

(a) What are the coordinates $p'$?
(b) Find the rotation matrix $R$ such that $p' = Rp$ for the $p'$ you obtained in (a).

**Exercise 3.3** Suppose that $p_i \in \mathbb{R}^3$ and $p_i' \in \mathbb{R}^3$ are related by $p_i' = Rp_i$, $i = 1, 2, 3$, for some unknown rotation matrix $R$. Find, if it exists, the rotation $R$ for the three input–output pairs $p_i \mapsto p_i'$, where

$$p_1 = (\sqrt{2}, 0, 2) \mapsto p_1' = (0, 2, \sqrt{2}),$$
$$p_2 = (1, 1, -1) \mapsto p_2' = \left(\frac{1}{\sqrt{2}}, \frac{1}{\sqrt{2}}, -\sqrt{2}\right),$$
$$p_3 = (0, 2\sqrt{2}, 0) \mapsto p_3' = (-\sqrt{2}, \sqrt{2}, -2).$$

**Exercise 3.4** In this exercise you are asked to prove the property $R_{ab}R_{bc} = R_{ac}$ of Equation (3.22). Define the unit axes of frames {a}, {b}, and {c} by the triplets of orthogonal unit vectors $\{\hat{x}_a, \hat{y}_a, \hat{z}_a\}$, $\{\hat{x}_b, \hat{y}_b, \hat{z}_b\}$, and $\{\hat{x}_c, \hat{y}_c, \hat{z}_c\}$, respectively. Suppose that the unit axes of frame {b} can be expressed in terms of the unit axes of frame {a} by

$$\begin{aligned}
\hat{x}_b &= r_{11}\hat{x}_a + r_{21}\hat{y}_a + r_{31}\hat{z}_a, \\
\hat{y}_b &= r_{12}\hat{x}_a + r_{22}\hat{y}_a + r_{32}\hat{z}_a, \\
\hat{z}_b &= r_{13}\hat{x}_a + r_{23}\hat{y}_a + r_{33}\hat{z}_a.
\end{aligned}$$

Similarly, suppose that the unit axes of frame {c} can be expressed in terms of the unit axes of frame {b} by

$$\begin{aligned}
\hat{x}_c &= s_{11}\hat{x}_b + s_{21}\hat{y}_b + s_{31}\hat{z}_b, \\
\hat{y}_c &= s_{12}\hat{x}_b + s_{22}\hat{y}_b + s_{32}\hat{z}_b, \\
\hat{z}_c &= s_{13}\hat{x}_b + s_{23}\hat{y}_b + s_{33}\hat{z}_b.
\end{aligned}$$

From the above prove that $R_{ab}R_{bc} = R_{ac}$.

**Exercise 3.5** Find the exponential coordinates $\hat{\omega}\theta \in \mathbb{R}^3$ for the $SO(3)$ matrix

$$\begin{bmatrix} 0 & -1 & 0 \\ 0 & 0 & -1 \\ 1 & 0 & 0 \end{bmatrix}.$$

**Exercise 3.6**   Given $R = \text{Rot}(\hat{x}, \pi/2)\text{Rot}(\hat{z}, \pi)$, find the unit vector $\hat{\omega}$ and angle $\theta$ such that $R = e^{[\hat{\omega}]\theta}$.

**Exercise 3.7**   (a) Given the rotation matrix

$$R = \begin{bmatrix} 0 & 0 & 1 \\ 0 & -1 & 0 \\ 1 & 0 & 0 \end{bmatrix},$$

find all possible values for $\hat{\omega} \in \mathbb{R}^3, \|\hat{\omega}\| = 1$, and $\theta \in [0, 2\pi)$ such that $e^{[\hat{\omega}]\theta} = R$.

(b) The two vectors $v_1, v_2 \in \mathbb{R}^3$ are related by

$$v_2 = Rv_1 = e^{[\hat{\omega}]\theta}v_1$$

where $\hat{\omega} \in \mathbb{R}^3$ has length 1, and $\theta \in [-\pi, \pi]$. Given $\hat{\omega} = (\frac{2}{3}, \frac{2}{3}, \frac{1}{3}), v_1 = (1, 0, 1), v_2 = (0, 1, 1)$, find all the angles $\theta$ that satisfy the above equation.

**Exercise 3.8**   (a) Suppose that we are seeking the logarithm of a rotation matrix $R$ whose trace is $-1$. From the exponential formula

$$e^{[\hat{\omega}]\theta} = I + \sin\theta\,[\hat{\omega}] + (1 - \cos\theta)[\hat{\omega}]^2, \qquad \|\omega\| = 1,$$

and recalling that $\text{tr}\,R = -1$ implies $\theta = \pi$, the above equation simplifies to

$$R = I + 2[\hat{\omega}]^2 = \begin{bmatrix} 1 - 2(\hat{\omega}_2^2 + \hat{\omega}_3^2) & 2\hat{\omega}_1\hat{\omega}_2 & 2\hat{\omega}_1\hat{\omega}_3 \\ 2\hat{\omega}_1\hat{\omega}_2 & 1 - 2(\hat{\omega}_1^2 + \hat{\omega}_3^2) & 2\hat{\omega}_2\hat{\omega}_3 \\ 2\hat{\omega}_1\hat{\omega}_2 & 2\hat{\omega}_2\hat{\omega}_3 & 1 - 2(\hat{\omega}_1^2 + \hat{\omega}_2^2) \end{bmatrix}.$$

Using the fact that $\hat{\omega}_1^2 + \hat{\omega}_2^2 + \hat{\omega}_3^2 = 1$, is it correct to conclude that

$$\hat{\omega}_1 = \sqrt{\frac{r_{11} + 1}{2}}, \qquad \hat{\omega}_2 = \sqrt{\frac{r_{22} + 1}{2}}, \qquad \hat{\omega}_3 = \sqrt{\frac{r_{33} + 1}{2}},$$

where $r_{ij}$ denotes the $(i, j)$th entry of $R$, is also a solution?

(b) Using the fact that $[\hat{\omega}]^3 = -[\hat{\omega}]$, the identity $R = I + 2[\hat{\omega}]^2$ can be written in the alternative forms

$$R - I = 2[\hat{\omega}]^2,$$
$$[\hat{\omega}](R - I) = 2[\hat{\omega}]^3 = -2[\hat{\omega}],$$
$$[\hat{\omega}](R + I) = 0.$$

The resulting equation consists of three linear equations in $(\hat{\omega}_1, \hat{\omega}_2, \hat{\omega}_3)$. What is the relation between the solution to this linear system and the logarithm of $R$?

**Exercise 3.9**   Exploiting the known properties of rotation matrices, determine the minimum number of arithmetic operations (multiplication and division, addition and subtraction) required to multiply two rotation matrices.

**Exercise 3.10** Because arithmetic precision is only finite, the numerically obtained product of two rotation matrices is not necessarily a rotation matrix; that is, the resulting rotation $A$ may not exactly satisfy $A^{\mathrm{T}}A = I$ as desired. Devise an iterative numerical procedure that takes an arbitrary matrix $A \in \mathbb{R}^{3\times3}$ and produces a matrix $R \in SO(3)$ that minimizes

$$\|A - R\|^2 = \operatorname{tr}(A - R)(A - R)^{\mathrm{T}}.$$

(Hint: See Appendix D for the relevant background on optimization.)

**Exercise 3.11** Properties of the matrix exponential.

(a) Under what conditions on general $A, B \in \mathbb{R}^{n\times n}$ does $e^A e^B = e^{A+B}$ hold?
(b) If $A = [\mathcal{V}_a]$ and $B = [\mathcal{V}_b]$, where $\mathcal{V}_a = (\omega_a, v_a)$ and $\mathcal{V}_b = (\omega_b, v_b)$ are arbitrary twists, then under what conditions on $\mathcal{V}_a$ and $\mathcal{V}_b$ does $e^A e^B = e^{A+B}$ hold? Try to give a physical description of this condition.

**Exercise 3.12** (a) Given a rotation matrix $A = \operatorname{Rot}(\hat{z}, \alpha)$, where $\operatorname{Rot}(\hat{z}, \alpha)$ indicates a rotation about the $\hat{z}$-axis by an angle $\alpha$, find all rotation matrices $R \in SO(3)$ that satisfy $AR = RA$.
(b) Given rotation matrices $A = \operatorname{Rot}(\hat{z}, \alpha)$ and $B = \operatorname{Rot}(\hat{z}, \beta)$, with $\alpha \neq \beta$, find all rotation matrices $R \in SO(3)$ that satisfy $AR = RB$.
(c) Given arbitrary rotation matrices $A, B \in SO(3)$, find all solutions $R \in SO(3)$ to the equation $AR = RB$.

**Exercise 3.13** (a) Show that the three eigenvalues of a rotation matrix $R \in SO(3)$ each have unit magnitude, and conclude that they can always be written $\{\mu + i\nu, \mu - i\nu, 1\}$, where $\mu^2 + \nu^2 = 1$.
(b) Show that a rotation matrix $R \in SO(3)$ can always be factored in the form

$$R = A \begin{bmatrix} \mu & \nu & 0 \\ -\nu & \mu & 0 \\ 0 & 0 & 1 \end{bmatrix} A^{-1},$$

where $A \in SO(3)$ and $\mu^2 + \nu^2 = 1$. (Hint: Denote the eigenvector associated with the eigenvalue $\mu + i\nu$ by $x + iy$, $x, y \in \mathbb{R}^3$, and the eigenvector associated with the eigenvalue 1 by $z \in \mathbb{R}^3$. For the purposes of this problem you may assume that the set of vectors $\{x, y, z\}$ can always be chosen to be linearly independent.)

**Exercise 3.14** Given $\omega \in \mathbb{R}^3$, $\|\omega\| = 1$, and $\theta$ a nonzero scalar, show that

$$\left(I\theta + (1 - \cos\theta)[\omega] + (\theta - \sin\theta)[\omega]^2\right)^{-1} = \frac{1}{\theta}I - \frac{1}{2}[\omega] + \left(\frac{1}{\theta} - \frac{1}{2}\cot\frac{\theta}{2}\right)[\omega]^2.$$

(Hint: From the identity $[\omega]^3 = -[\omega]$, express the inverse as a quadratic matrix polynomial in $[\omega]$.)

**Exercise 3.15** (a) Given a fixed frame $\{0\}$ and a moving frame $\{1\}$ initially aligned with $\{0\}$, perform the following sequence of rotations on $\{1\}$:

1. Rotate {1} about the {0} frame $\hat{x}$-axis by $\alpha$; call this new frame {2}.
2. Rotate {2} about the {0} frame $\hat{y}$-axis by $\beta$; call this new frame {3}.
3. Rotate {3} about the {0} frame $\hat{z}$-axis by $\gamma$; call this new frame {4}.
   What is the final orientation $R_{04}$?

(b) Suppose that the third step above is replaced by the following: "Rotate {3} about the $\hat{z}$-axis of frame {3} by $\gamma$; call this new frame {4}." What is the final orientation $R_{04}$?

(c) Find $T_{ca}$ for the following transformations:

$$
T_{ab} = \begin{bmatrix} \frac{1}{\sqrt{2}} & -\frac{1}{\sqrt{2}} & 0 & -1 \\ \frac{1}{\sqrt{2}} & \frac{1}{\sqrt{2}} & 0 & 0 \\ 0 & 0 & 1 & 1 \\ 0 & 0 & 0 & 1 \end{bmatrix}, \quad T_{cb} = \begin{bmatrix} \frac{1}{\sqrt{2}} & 0 & \frac{1}{\sqrt{2}} & 0 \\ 0 & 1 & 0 & 1 \\ -\frac{1}{\sqrt{2}} & 0 & \frac{1}{\sqrt{2}} & 0 \\ 0 & 0 & 0 & 1 \end{bmatrix}.
$$

**Exercise 3.16** In terms of the $\hat{x}_s$, $\hat{y}_s$, $\hat{z}_s$ coordinates of a fixed space frame {s}, frame {a} has its $\hat{x}_a$-axis pointing in the direction $(0, 0, 1)$ and its $\hat{y}_a$-axis pointing in the direction $(-1, 0, 0)$, and frame {b} has its $\hat{x}_b$-axis pointing in the direction $(1, 0, 0)$ and its $\hat{y}_b$-axis pointing in the direction $(0, 0, -1)$. The origin of {a} is at $(3, 0, 0)$ in {s} and the origin of {b} is at $(0, 2, 0)$ is {s}.

(a) Draw by hand a diagram showing {a} and {b} relative to {s}.

(b) Write down the rotation matrices $R_{sa}$ and $R_{sb}$ and the transformation matrices $T_{sa}$ and $T_{sb}$.

(c) Given $T_{sb}$, how do you calculate $T_{sb}^{-1}$ without using a matrix inverse? Write $T_{sb}^{-1}$ and verify its correctness using your drawing.

(d) Given $T_{sa}$ and $T_{sb}$, how do you calculate $T_{ab}$ (again without using matrix inverses)? Compute the answer and verify its correctness using your drawing.

(e) Let $T = T_{sb}$ be considered as a transformation operator consisting of a rotation about $\hat{x}$ by $-90°$ and a translation along $\hat{y}$ by 2 units. Calculate $T_1 = T_{sa}T$. Does $T_1$ correspond to a rotation and translation about $\hat{x}_s$ and $\hat{y}_s$, respectively (a world-fixed transformation of $T_{sa}$), or a rotation and translation about $\hat{x}_a$ and $\hat{y}_a$, respectively (a body-fixed transformation of $T_{sa}$)? Now calculate $T_2 = TT_{sa}$. Does $T_2$ correspond to a body-fixed or world-fixed transformation of $T_{sa}$?

(f) Use $T_{sb}$ to change the representation of the point $p_b = (1, 2, 3)$ in {b} coordinates to {s} coordinates.

(g) Choose a point p represented by $p_s = (1, 2, 3)$ in {s} coordinates. Calculate $p' = T_{sb}p_s$ and $p'' = T_{sb}^{-1}p_s$. For each operation, should the result be interpreted as changing coordinates (from the {s} frame to {b}) without moving the point p, or as moving the location of the point without changing the reference frame of the representation?

(h) A twist $\mathcal{V}$ is represented in {s} as $\mathcal{V}_s = (3, 2, 1, -1, -2, -3)$. What is its representation $\mathcal{V}_a$ in frame {a}?

(i) By hand, calculate the matrix logarithm $[\mathcal{S}]\theta$ of $T_{sa}$. (You may verify your answer with software.) Extract the normalized screw axis $\mathcal{S}$ and rotation amount $\theta$. Find a $\{q, \hat{s}, h\}$ representation of the screw axis. Redraw the fixed frame $\{s\}$ and in it draw $\mathcal{S}$.

(j) Calculate the matrix exponential corresponding to the exponential coordinates of rigid-body motion $\mathcal{S}\theta = (0, 1, 2, 3, 0, 0)$. Draw the corresponding frame relative to $\{s\}$, as well as the screw axis $\mathcal{S}$.

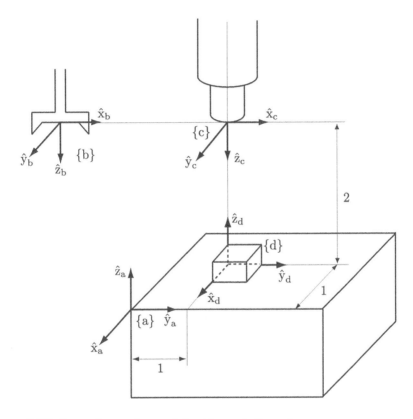

**Figure 3.23** Four reference frames defined in a robot's workspace.

**Exercise 3.17** Four reference frames are shown in the robot workspace of Figure 3.23: the fixed frame $\{a\}$, the end-effector frame effector $\{b\}$, the camera frame $\{c\}$, and the workpiece frame $\{d\}$.

(a) Find $T_{ad}$ and $T_{cd}$ in terms of the dimensions given in the figure.
(b) Find $T_{ab}$ given that

$$T_{bc} = \begin{bmatrix} 1 & 0 & 0 & 4 \\ 0 & 1 & 0 & 0 \\ 0 & 0 & 1 & 0 \\ 0 & 0 & 0 & 1 \end{bmatrix}.$$

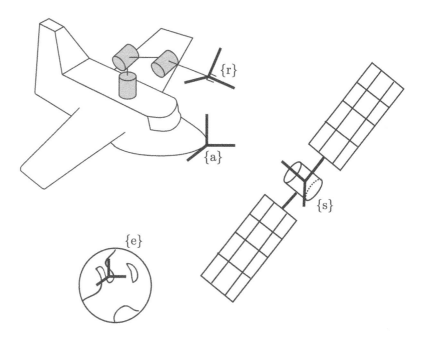

**Figure 3.24** A robot arm mounted on a spacecraft.

**Exercise 3.18**   Consider a robot arm mounted on a spacecraft as shown in Figure 3.24, in which frames are attached to the Earth {e}, a satellite {s}, the spacecraft {a}, and the robot arm {r}, respectively.

(a) Given $T_{ea}$, $T_{ar}$, and $T_{es}$, find $T_{rs}$.
(b) Suppose that the frame {s} origin as seen from {e} is $(1, 1, 1)$ and that

$$
T_{er} = \begin{bmatrix} -1 & 0 & 0 & 1 \\ 0 & 1 & 0 & 1 \\ 0 & 0 & -1 & 1 \\ 0 & 0 & 0 & 1 \end{bmatrix}.
$$

Write down the coordinates of the frame {s} origin as seen from frame {r}.

**Exercise 3.19**   Two satellites are circling the Earth as shown in Figure 3.25. Frames {1} and {2} are rigidly attached to the satellites in such a way that their $\hat{\text{x}}$-axes always point toward the Earth. Satellite 1 moves at a constant speed $v_1$, while satellite 2 moves at a constant speed $v_2$. To simplify matters, ignore the rotation of the Earth about its own axis. The fixed frame {0} is located at the center of the Earth. Figure 3.25 shows the position of the two satellites at $t = 0$.

(a) Derive the frames $T_{01}$, $T_{02}$ as a function of $t$.
(b) Using your results from part (a), find $T_{21}$ as a function of $t$.

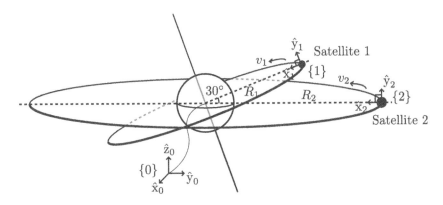

**Figure 3.25** Two satellites circling the Earth.

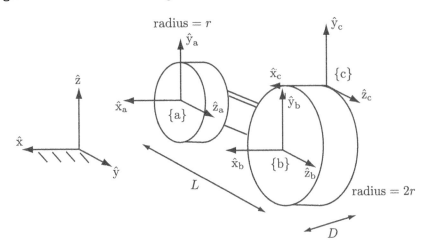

**Figure 3.26** A high-wheel bicycle.

**Exercise 3.20** Consider the high-wheel bicycle of Figure 3.26, in which the diameter of the front wheel is twice that of the rear wheel. Frames {a} and {b} are attached respectively to the centers of the wheels, and frame {c} is attached to the top of the front wheel. Assuming that the bike moves forward in the $\hat{y}$-direction, find $T_{ac}$ as a function of the front wheel's rotation angle $\theta$ (assume $\theta = 0$ at the instant shown in the figure).

**Exercise 3.21** The space station of Figure 3.27 moves in circular orbit around the Earth, and at the same time rotates about an axis always pointing toward the North Star. Owing to an instrument malfunction, a spacecraft heading toward the space station is unable to locate the docking port. An Earth-based ground

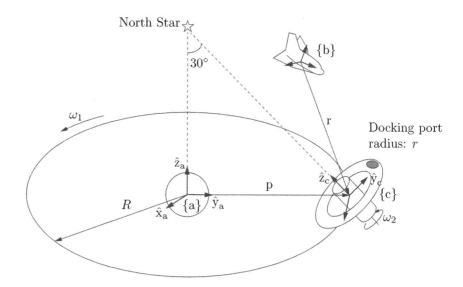

**Figure 3.27** A spacecraft and space station.

station sends the following information to the spacecraft:

$$T_{ab} = \begin{bmatrix} 0 & -1 & 0 & -100 \\ 1 & 0 & 0 & 300 \\ 0 & 0 & 1 & 500 \\ 0 & 0 & 0 & 1 \end{bmatrix}, \qquad p_a = \begin{bmatrix} 0 \\ 800 \\ 0 \end{bmatrix},$$

where $p_a$ is the vector p expressed in {a}-frame coordinates.

(a) From the given information, find $r_b$, the vector r expressed in {b}-frame coordinates.

(b) Determine $T_{bc}$ at the instant shown in the figure. Assume here that the $\hat{y}$- and $\hat{z}$-axes of the {a} and {c} frames are coplanar with the docking port.

**Exercise 3.22**   A target moves along a circular path at constant angular velocity $\omega$ rad/s in the $\hat{x}$–$\hat{y}$-plane, as shown in Figure 3.28. The target is tracked by a laser mounted on a moving platform, rising vertically at constant speed $v$. Assume that at $t = 0$ the laser and the platform start at $L_1$, while the target starts at frame $T_1$.

(a) Derive the frames $T_{01}, T_{12}, T_{03}$ as functions of $t$.

(b) Using your results from part (a), derive $T_{23}$ as a function of $t$.

**Exercise 3.23**   Two toy cars are moving on a round table as shown in Figure 3.29. Car 1 moves at a constant speed $v_1$ along the circumference of the table, while car 2 moves at a constant speed $v_2$ along a radius; the positions of the two vehicles at $t = 0$ are shown in the figures.

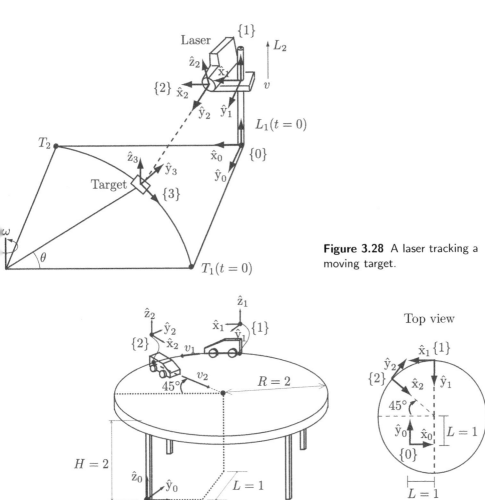

**Figure 3.28** A laser tracking a moving target.

**Figure 3.29** Two toy cars on a round table.

(a) Find $T_{01}$ and $T_{02}$ as a function of $t$.
(b) Find $T_{12}$ as a function of $t$.

**Exercise 3.24**  Figure 3.30 shows the configuration, at $t = 0$, of a robot arm whose first joint is a screw joint of pitch $h = 2$. The arm's link lengths are $L_1 = 10$, $L_2 = L_3 = 5$, and $L_4 = 3$. Suppose that all joint angular velocities are constant, with values $\omega_1 = \pi/4$, $\omega_2 = \pi/8$, $\omega_3 = -\pi/4$ rad/s. Find $T_{sb}(4) \in SE(3)$, i.e., the configuration of the end-effector frame $\{b\}$ relative to the fixed frame $\{s\}$ at time $t = 4$.

**Exercise 3.25**  A camera is rigidly attached to a robot arm, as shown in Fig-

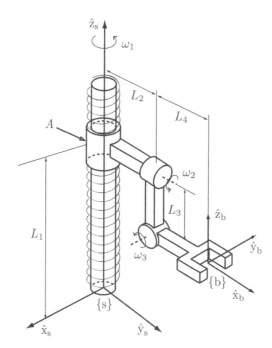

**Figure 3.30** A robot arm with a screw joint.

ure 3.31. The transformation $X \in SE(3)$ is constant. The robot arm moves from posture 1 to posture 2. The transformations $A \in SE(3)$ and $B \in SE(3)$ are measured and can be assumed to be known.

(a) Suppose that $X$ and $A$ are given as follows:

$$X = \begin{bmatrix} 1 & 0 & 0 & 1 \\ 0 & 1 & 0 & 0 \\ 0 & 0 & 1 & 0 \\ 0 & 0 & 0 & 1 \end{bmatrix}, \qquad A = \begin{bmatrix} 0 & 0 & 1 & 0 \\ 0 & 1 & 0 & 1 \\ -1 & 0 & 1 & 0 \\ 0 & 0 & 0 & 1 \end{bmatrix}.$$

What is $B$?

(b) Now suppose that

$$A = \begin{bmatrix} R_A & p_A \\ 0 & 1 \end{bmatrix}, \qquad B = \begin{bmatrix} R_B & p_B \\ 0 & 1 \end{bmatrix}$$

are known and we wish to find

$$X = \begin{bmatrix} R_X & p_X \\ 0 & 1 \end{bmatrix}.$$

Set $R_A = e^{[\alpha]}$ and $R_B = e^{[\beta]}$. What are the conditions on $\alpha \in \mathbb{R}^3$ and $\beta \in \mathbb{R}^3$ for a solution $R_X$ to exist?

(c) Now suppose that we have a set of $k$ equations

$$A_i X = X B_i \quad \text{for } i = 1, \dots, k.$$

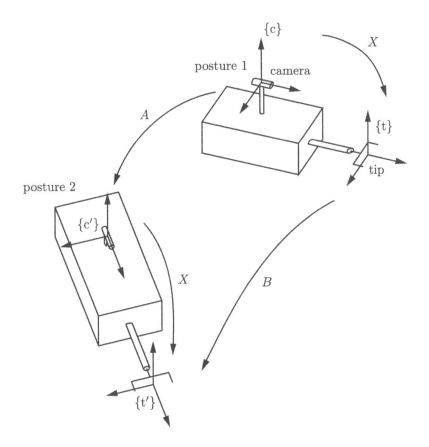

**Figure 3.31** A camera rigidly attached to a robot arm.

Assume that $A_i$ and $B_i$ are all known. What is the minimum number $k$ for which a unique solution exists?

**Exercise 3.26**   Draw the screw axis for which $q = (3, 0, 0)$, $\hat{s} = (0, 0, 1)$, and $h = 2$.

**Exercise 3.27**   Draw the screw axis for the twist $\mathcal{V} = (0, 2, 2, 4, 0, 0)$.

**Exercise 3.28**   Assume that the space-frame angular velocity is $\omega_s = (1, 2, 3)$ for a moving body with frame $\{b\}$ at

$$R = \begin{bmatrix} 0 & -1 & 0 \\ 0 & 0 & -1 \\ 1 & 0 & 0 \end{bmatrix}$$

relative to the space frame $\{s\}$. Calculate the body's angular velocity $\omega_b$ in $\{b\}$.

**Exercise 3.29**   Two frames $\{a\}$ and $\{b\}$ are attached to a moving rigid body. Show that the twist of $\{a\}$ in space-frame coordinates is the same as the twist of $\{b\}$ in space-frame coordinates.

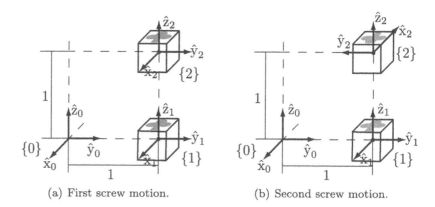

(a) First screw motion.          (b) Second screw motion.

**Figure 3.32** A cube undergoing two different screw motions.

**Exercise 3.30**   A cube undergoes two different screw motions from frame $\{1\}$ to frame $\{2\}$ as shown in Figure 3.32. In both cases, (a) and (b), the initial configuration of the cube is

$$T_{01} = \begin{bmatrix} 1 & 0 & 0 & 0 \\ 0 & 1 & 0 & 1 \\ 0 & 0 & 1 & 0 \\ 0 & 0 & 0 & 1 \end{bmatrix}.$$

(a) For each case, (a) and (b), find the exponential coordinates $\mathcal{S}\theta = (\omega, v)\theta$ such that $T_{02} = e^{[\mathcal{S}]\theta}T_{01}$, where no constraints are placed on $\omega$ or $v$.
(b) Repeat (a), this time with the constraint that $\|\omega\theta\| \in [-\pi, \pi]$.

**Exercise 3.31**   In Example 3.19 and Figure 3.16, the block that the robot must pick up weighs 1 kg, which means that the robot must provide approximately 10 N of force in the $\hat{z}_e$-direction of the block's frame $\{e\}$ (which you can assume is at the block's center of mass). Express this force as a wrench $\mathcal{F}_e$ in the $\{e\}$ frame. Given the transformation matrices in Example 3.19, express this same wrench in the end-effector frame $\{c\}$ as $\mathcal{F}_c$.

**Exercise 3.32**   Given two reference frames $\{a\}$ and $\{b\}$ in physical space, and a fixed frame $\{o\}$, define the distance between frames $\{a\}$ and $\{b\}$ as

$$\mathrm{dist}(T_{oa}, T_{ob}) \equiv \sqrt{\theta^2 + \|p_{ab}\|^2}$$

where $R_{ab} = e^{[\hat{\omega}]\theta}$. Suppose that the fixed frame is displaced to another frame $\{o'\}$ and that $T_{o'a} = ST_{oa}$, $T_{o'b} = ST_{o'b}$ for some constant $S = (R_s, p_s) \in SE(3)$.
(a) Evaluate $\mathrm{dist}(T_{o'a}, T_{o'b})$ using the above distance formula.
(b) Under what conditions on $S$ does $\mathrm{dist}(T_{oa}, T_{ob}) = \mathrm{dist}(T_{o'a}, T_{o'b})$?

**Exercise 3.33**   (a) Find the general solution to the differential equation $\dot{x} = Ax$,

where

$$A = \begin{bmatrix} -2 & 1 \\ 0 & -1 \end{bmatrix}.$$

What happens to the solution $x(t)$ as $t \to \infty$?

(b) Do the same for

$$A = \begin{bmatrix} 2 & -1 \\ 1 & 2 \end{bmatrix}.$$

What happens to the solution $x(t)$ as $t \to \infty$?

**Exercise 3.34**  Let $x \in \mathbb{R}^2$, $A \in \mathbb{R}^{2 \times 2}$, and consider the linear differential equation $\dot{x}(t) = Ax(t)$. Suppose that

$$x(t) = \begin{bmatrix} e^{-3t} \\ -3e^{-3t} \end{bmatrix}$$

is a solution for the initial condition $x(0) = (1, -3)$, and

$$x(t) = \begin{bmatrix} e^t \\ e^t \end{bmatrix}$$

is a solution for the initial condition $x(0) = (1, 1)$. Find $A$ and $e^{At}$.

**Exercise 3.35**  Given a differential equation of the form $\dot{x} = Ax + f(t)$, where $x \in \mathbb{R}^n$ and $f(t)$ is a given differentiable function of $t$, show that the general solution can be written

$$x(t) = e^{At} x(0) + \int_0^t e^{A(t-s)} f(s) \, ds.$$

(Hint: Define $z(t) = e^{-At} x(t)$ and evaluate $\dot{z}(t)$.)

**Exercise 3.36**  Referring to Appendix B, answer the following questions related to ZXZ Euler angles.

(a) Derive a procedure for finding the ZXZ Euler angles of a rotation matrix.
(b) Using the results of (a), find the ZXZ Euler angles for the following rotation matrix:

$$\begin{bmatrix} -\frac{1}{\sqrt{2}} & \frac{1}{\sqrt{2}} & 0 \\ -\frac{1}{2} & -\frac{1}{2} & \frac{1}{\sqrt{2}} \\ \frac{1}{2} & \frac{1}{2} & \frac{1}{\sqrt{2}} \end{bmatrix}.$$

**Exercise 3.37**  Consider a wrist mechanism with two revolute joints $\theta_1$ and $\theta_2$, in which the end-effector frame orientation $R \in SO(3)$ is given by

$$R = e^{[\hat{\omega}_1]\theta_1} e^{[\hat{\omega}_2]\theta_2},$$

with $\hat{\omega}_1 = (0, 0, 1)$ and $\hat{\omega}_2 = (0, \frac{1}{\sqrt{2}}, -\frac{1}{\sqrt{2}})$. Determine whether the following orientation is reachable (that is, find, if it exists, a solution $(\theta_1, \theta_2)$ for the following

$R$):

$$R = \begin{bmatrix} \frac{1}{\sqrt{2}} & 0 & -\frac{1}{\sqrt{2}} \\ 0 & 1 & 0 \\ \frac{1}{\sqrt{2}} & 0 & \frac{1}{\sqrt{2}} \end{bmatrix}$$

**Exercise 3.38**  Show that rotation matrices of the form

$$\begin{bmatrix} r_{11} & r_{12} & 0 \\ r_{21} & r_{22} & r_{23} \\ r_{31} & r_{32} & r_{33} \end{bmatrix}$$

can be parametrized using just two parameters $\theta$ and $\phi$ as follows:

$$\begin{bmatrix} \cos\theta & -\sin\theta & 0 \\ \sin\theta\cos\phi & \cos\theta\cos\phi & -\sin\phi \\ \sin\theta\sin\phi & \cos\theta\sin\phi & \cos\phi \end{bmatrix}.$$

What should the range of values be for $\theta$ and $\phi$?

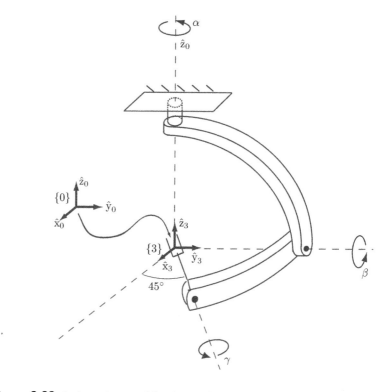

**Figure 3.33** A three-degree-of-freedom wrist mechanism.

**Exercise 3.39**  Figure 3.33 shows a three-dof wrist mechanism in its zero position (i.e., all joint angles are set to zero).

(a) Express the tool-frame orientation $R_{03} = R(\alpha, \beta, \gamma)$ as a product of three rotation matrices.

(b) Find all possible angles $(\alpha, \beta, \gamma)$ for the two values of $R_{03}$ given below. If no solution exists, explain why this is so in terms of the analogy between $SO(3)$ and a solid ball of radius $\pi$.

1. $R_{03} = \begin{bmatrix} 0 & 1 & 0 \\ 1 & 0 & 0 \\ 0 & 0 & -1 \end{bmatrix}$.

2. $R_{03} = e^{[\hat{\omega}]\pi/2}$, where $\hat{\omega} = (0, \frac{1}{\sqrt{5}}, \frac{2}{\sqrt{5}})$.

**Exercise 3.40** Refer to Appendix B.

(a) Verify formulas (B.10) and (B.11) for the unit quaternion representation of a rotation $R \in SO(3)$.

(b) Verify formula (B.12) for the rotation matrix $R$ representation of a unit quaternion $q \in S^3$.

(c) Verify the product rule for two unit quaternions. That is, given two unit quaternions $q, p \in S^3$ corresponding respectively to the rotations $R, Q \in SO(3)$, find a formula for the unit quaternion representation of the product $RQ \in SO(3)$.

**Exercise 3.41** The Cayley transform of Equation (B.18) in Appendix B can be generalized to higher orders as follows:

$$R = (I - [r])^k (I + [r])^{-k}. \tag{3.99}$$

(a) For the case $k = 2$, show that the rotation $R$ corresponding to $r$ can be computed from the formula

$$R = I - 4\frac{1 - r^{\mathrm{T}}r}{(1 + r^{\mathrm{T}}r)^2}[r] + \frac{8}{(1 + r^{\mathrm{T}}r)^2}[r]^2. \tag{3.100}$$

(b) Conversely, given a rotation matrix $R$, show that a vector $r$ that satisfies Equation (3.100) can be obtained as

$$r = -\hat{\omega}\tan\frac{\theta}{4}, \tag{3.101}$$

where, as before, $\hat{\omega}$ is the unit vector along the axis of rotation for $R$, and $\theta$ is the corresponding rotation angle. Is this solution unique?

(c) Show that the angular velocity in the body frame obeys the following relation:

$$\dot{r} = \frac{1}{4}\left((1 - r^{\mathrm{T}}r)I + 2[r] + 2rr^{\mathrm{T}}\right)\omega. \tag{3.102}$$

(d) Explain what happens to the singularity at $\pi$ that exists for the standard Cayley–Rodrigues parameters. Discuss the relative advantages and disadvantages of the modified Cayley–Rodrigues parameters, particularly for order $k = 4$ and higher.

(e) Compare the number of arithmetic operations needed for multiplying two rotation matrices, two unit quaternions, or two Cayley–Rodrigues representations. Which requires the fewest arithmetic operations?

**Exercise 3.42**   Rewrite the software for Chapter 3 in your favorite programming language.

**Exercise 3.43**   Write a function that returns "true" if a given $3 \times 3$ matrix is within $\epsilon$ of being a rotation matrix and "false" otherwise. It is up to you how to define the "distance" between a random $3 \times 3$ real matrix and the closest member of $SO(3)$. If the function returns "true," it should also return the "nearest" matrix in $SO(3)$. See, for example, Exercise 3.10.

**Exercise 3.44**   Write a function that returns "true" if a given $4 \times 4$ matrix is within $\epsilon$ of an element of $SE(3)$ and "false" otherwise.

**Exercise 3.45**   Write a function that returns "true" if a given $3 \times 3$ matrix is within $\epsilon$ of an element of $so(3)$ and "false" otherwise.

**Exercise 3.46**   Write a function that returns "true" if a given $4 \times 4$ matrix is within $\epsilon$ of an element of $se(3)$ and "false" otherwise.

**Exercise 3.47**   The primary purpose of the provided software is to be easy to read and educational, reinforcing the concepts in the book. The code is optimized neither for efficiency nor robustness, nor does it do full error-checking on its inputs.

Familiarize yourself with the whole code in your favorite language by reading the functions and their comments. This should help cement your understanding of the material in this chapter. Then:

(a) Rewrite one function to do full error-checking on its input, and have the function return a recognizable error value if the function is called with an improper input (e.g., an argument to the function is not an element of $SO(3)$, $SE(3)$, $so(3)$, or $se(3)$, as expected).
(b) Rewrite one function to improve its computational efficiency, perhaps by using what you know about properties of rotation or transformation matrices.
(c) Can you reduce the numerical sensitivity of either of the matrix logarithm functions?

**Exercise 3.48**   Use the provided software to write a program that allows the user to specify an initial configuration of a rigid body by $T$, a screw axis specified by $\{q, \hat{s}, h\}$ in the fixed frame $\{s\}$, and the total distance traveled along the screw axis, $\theta$. The program should calculate the final configuration $T_1 = e^{[\mathcal{S}]\theta}T$ attained when the rigid body follows the screw $\mathcal{S}$ a distance $\theta$, as well as the intermediate configurations at $\theta/4$, $\theta/2$, and $3\theta/4$. At the initial, intermediate, and final configurations, the program should plot the $\{b\}$-axes of the rigid body. The program should also calculate the screw axis $\mathcal{S}_1$ and the distance $\theta_1$ following

$S_1$ that takes the rigid body from $T_1$ to the origin and it should plot the screw axis $S_1$. Test the program with $q = (0, 2, 0)$, $\hat{s} = (0, 0, 1)$, $h = 2$, $\theta = \pi$, and

$$
T = \begin{bmatrix}
1 & 0 & 0 & 2 \\
0 & 1 & 0 & 0 \\
0 & 0 & 1 & 0 \\
0 & 0 & 0 & 1
\end{bmatrix}.
$$

**Exercise 3.49**   In this chapter, we developed expressions for the matrix exponential for spatial motions mapping elements of $so(3)$ to $SO(3)$ and elements of $se(3)$ to $SE(3)$. Similarly, we developed algorithms for the matrix logarithm going the other direction.

We could also develop matrix exponentials for planar motions, from $so(2)$ to $SO(2)$ and from $se(2)$ to $SE(2)$, as well as the matrix logarithms going from $SO(2)$ to $so(2)$ and $SE(2)$ to $se(2)$. For the $so(2)$ to $SO(2)$ case there is a single exponential coordinate. For the $se(2)$ to $SE(2)$ case there are three exponential coordinates, corresponding to a twist with three elements set to zero, $\mathcal{V} = (0, 0, \omega_z, v_x, v_y, 0)$.

For planar rotations and planar twists we could apply the matrix exponentials and logarithms that we derived for the spatial case by simply expressing the $so(2)$, $SO(2)$, $se(2)$, and $SE(2)$ elements as elements of $so(3)$, $SO(3)$, $se(3)$, and $SE(3)$. Instead, in this problem, write down explicitly the matrix exponential and logarithm for the $so(2)$ to $SO(2)$ case using a single exponential coordinate, and the matrix exponential and logarithm for the $se(2)$ to $SE(2)$ case using three exponential coordinates. Then provide software implementations of each of the four functions in your favorite programming language, and provide execution logs that show that they function as expected.

# 4 Forward Kinematics

The **forward kinematics** of a robot refers to the calculation of the position and orientation of its end-effector frame from its joint coordinates $\theta$. Figure 4.1 illustrates the forward kinematics problem for a 3R planar open chain. The link lengths are $L_1$, $L_2$, and $L_3$. Choose a fixed frame $\{0\}$ with origin located at the base joint as shown, and assume an end-effector frame $\{4\}$ has been attached to the tip of the third link. The Cartesian position $(x, y)$ and orientation $\phi$ of the end-effector frame as functions of the joint angles $(\theta_1, \theta_2, \theta_3)$ are then given by

$$x = L_1 \cos\theta_1 + L_2 \cos(\theta_1 + \theta_2) + L_3 \cos(\theta_1 + \theta_2 + \theta_3), \tag{4.1}$$

$$y = L_1 \sin\theta_1 + L_2 \sin(\theta_1 + \theta_2) + L_3 \sin(\theta_1 + \theta_2 + \theta_3), \tag{4.2}$$

$$\phi = \theta_1 + \theta_2 + \theta_3. \tag{4.3}$$

If one is only interested in the $(x, y)$ position of the end-effector, the robot's task space is then taken to be the $x$–$y$-plane, and the forward kinematics would consist of Equations (4.1) and (4.2) only. If the end-effector's position and orientation both matter, the forward kinematics would consist of the three equations (4.1)–(4.3).

While the above analysis can be done using only basic trigonometry, it is not

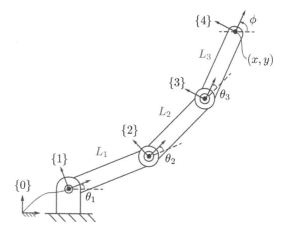

**Figure 4.1** Forward kinematics of a 3R planar open chain. For each frame, the $\hat{x}$- and $\hat{y}$-axis is shown; the $\hat{z}$-axes are parallel and out of the page.

difficult to imagine that for more general spatial chains the analysis can become considerably more complicated. A more systematic method of deriving the forward kinematics might involve attaching reference frames to each link; in Figure 4.1 the three link reference frames are respectively labeled $\{1\}$, $\{2\}$, and $\{3\}$. The forward kinematics can then be written as a product of four homogeneous transformation matrices:

$$T_{04} = T_{01}T_{12}T_{23}T_{34}, \tag{4.4}$$

where

$$T_{01} = \begin{bmatrix} \cos\theta_1 & -\sin\theta_1 & 0 & 0 \\ \sin\theta_1 & \cos\theta_1 & 0 & 0 \\ 0 & 0 & 1 & 0 \\ 0 & 0 & 0 & 1 \end{bmatrix}, \quad T_{12} = \begin{bmatrix} \cos\theta_2 & -\sin\theta_2 & 0 & L_1 \\ \sin\theta_2 & \cos\theta_2 & 0 & 0 \\ 0 & 0 & 1 & 0 \\ 0 & 0 & 0 & 1 \end{bmatrix},$$

$$T_{23} = \begin{bmatrix} \cos\theta_3 & -\sin\theta_3 & 0 & L_2 \\ \sin\theta_3 & \cos\theta_3 & 0 & 0 \\ 0 & 0 & 1 & 0 \\ 0 & 0 & 0 & 1 \end{bmatrix}, \quad T_{34} = \begin{bmatrix} 1 & 0 & 0 & L_3 \\ 0 & 1 & 0 & 0 \\ 0 & 0 & 1 & 0 \\ 0 & 0 & 0 & 1 \end{bmatrix}. \tag{4.5}$$

Observe that $T_{34}$ is constant and that each remaining $T_{i-1,i}$ depends only on the joint variable $\theta_i$.

As an alternative to this approach, let us define $M$ to be the position and orientation of frame $\{4\}$ when all joint angles are set to zero (the "home" or "zero" position of the robot). Then

$$M = \begin{bmatrix} 1 & 0 & 0 & L_1 + L_2 + L_3 \\ 0 & 1 & 0 & 0 \\ 0 & 0 & 1 & 0 \\ 0 & 0 & 0 & 1 \end{bmatrix}, \tag{4.6}$$

Now consider each revolute joint axis to be a zero-pitch screw axis. If $\theta_1$ and $\theta_2$ are held at their zero position then the screw axis corresponding to rotating about joint 3 can be expressed in the $\{0\}$ frame as

$$\mathcal{S}_3 = \begin{bmatrix} \omega_3 \\ v_3 \end{bmatrix} = \begin{bmatrix} 0 \\ 0 \\ 1 \\ 0 \\ -(L_1 + L_2) \\ 0 \end{bmatrix}.$$

You should be able to confirm this by simple visual inspection of Figure 4.1. When the arm is stretched out straight to the right at its zero configuration, imagine a turntable rotating with an angular velocity of $\omega_3 = 1$ rad/s about the axis of joint 3. The linear velocity $v_3$ of the point on the turntable at the origin of $\{0\}$ is in the $-\hat{y}_0$-direction at a rate of $L_1 + L_2$ units/s. Algebraically,

$v_3 = -\omega_3 \times q_3$, where $q_3$ is any point on the axis of joint 3 expressed in $\{0\}$, e.g., $q_3 = (L_1 + L_2, 0, 0)$.

The screw axis $\mathcal{S}_3$ can be expressed in $se(3)$ matrix form as

$$[\mathcal{S}_3] = \begin{bmatrix} [\omega] & v \\ 0 & 0 \end{bmatrix} = \begin{bmatrix} 0 & -1 & 0 & 0 \\ 1 & 0 & 0 & -(L_1 + L_2) \\ 0 & 0 & 0 & 0 \\ 0 & 0 & 0 & 0 \end{bmatrix}.$$

Therefore, for any $\theta_3$, the matrix exponential representation for screw motions from the previous chapter allows us to write

$$T_{04} = e^{[\mathcal{S}_3]\theta_3} M \qquad \text{(for } \theta_1 = \theta_2 = 0\text{)}. \tag{4.7}$$

Now, for $\theta_1 = 0$ and any fixed (but arbitrary) $\theta_3$, rotation about joint 2 can be viewed as applying a screw motion to the rigid (link 2)/(link 3) pair, i.e.,

$$T_{04} = e^{[\mathcal{S}_2]\theta_2} e^{[\mathcal{S}_3]\theta_3} M \qquad \text{(for } \theta_1 = 0\text{)}, \tag{4.8}$$

where $[\mathcal{S}_3]$ and $M$ are as defined previously, and

$$[\mathcal{S}_2] = \begin{bmatrix} 0 & -1 & 0 & 0 \\ 1 & 0 & 0 & -L_1 \\ 0 & 0 & 0 & 0 \\ 0 & 0 & 0 & 0 \end{bmatrix}. \tag{4.9}$$

Finally, keeping $\theta_2$ and $\theta_3$ fixed, rotation about joint 1 can be viewed as applying a screw motion to the entire rigid three-link assembly. We can therefore write, for arbitrary values of $(\theta_1, \theta_2, \theta_3)$,

$$T_{04} = e^{[\mathcal{S}_1]\theta_1} e^{[\mathcal{S}_2]\theta_2} e^{[\mathcal{S}_3]\theta_3} M, \tag{4.10}$$

where

$$[\mathcal{S}_1] = \begin{bmatrix} 0 & -1 & 0 & 0 \\ 1 & 0 & 0 & 0 \\ 0 & 0 & 0 & 0 \\ 0 & 0 & 0 & 0 \end{bmatrix}. \tag{4.11}$$

Thus the forward kinematics can be expressed as a product of matrix exponentials, each corresponding to a screw motion. Note that this latter derivation of the forward kinematics does not use any link reference frames; only $\{0\}$ and $M$ must be defined.

In this chapter we consider the forward kinematics of general open chains. One widely used representation for the forward kinematics of open chains relies on the **Denavit–Hartenberg parameters** (D–H parameters), and this representation uses Equation (4.4). Another representation relies on the **product of exponentials** (PoE) formula, which corresponds to Equation (4.10). The advantage of the D–H representation is that it requires the minimum number of parameters to describe the robot's kinematics: for an $n$-joint robot, it uses $3n$ numbers to describe the robot's structure and $n$ numbers to describe the joint values. The PoE representation is not minimal (it requires $6n$ numbers to describe the $n$ screw axes, in addition to the $n$ joint values), but it has advantages over the D–H representation (e.g., no link frames are necessary) and it is our preferred choice of forward kinematics representation. The D–H representation, and its relationship to the PoE representation, is described in Appendix C.

## Product of Exponentials Formula

To use the PoE formula, it is only necessary to assign a stationary frame {s} (e.g., at the fixed base of the robot or anywhere else that is convenient for defining a reference frame) and a frame {b} at the end-effector, described by $M$ when the robot is at its zero position. It is common to define a frame at each link, though, typically at the joint axis; these are needed for the D–H representation and they are useful for displaying a graphic rendering of a geometric model of the robot and for defining the mass properties of the link, which we will need starting in Chapter 8. Thus when we are defining the kinematics of an $n$-joint robot, we may either (1) minimally use the frames {s} and {b} if we are only interested in the kinematics, or (2) refer to {s} as frame {0}, use frames {$i$} for $i = 1, \ldots, n$ (the frames for links $i$ at joints $i$), and use one more frame {$n + 1$} (corresponding to {b}) at the end-effector. The frame {$n + 1$} (i.e., {b}) is fixed relative to {$n$}, but it is at a more convenient location to represent the configuration of the end-effector. In some cases we dispense with frame {$n + 1$} and simply refer to {$n$} as the end-effector frame {b}.

### First Formulation: Screw Axes in the Base Frame

The key concept behind the PoE formula is to regard each joint as applying a screw motion to all the outward links. To illustrate this consider a general spatial open chain like the one shown in Figure 4.2, consisting of $n$ one-dof joints that are connected serially. To apply the PoE formula, you must choose a fixed base frame {s} and an end-effector frame {b} attached to the last link. Place the robot in its zero position by setting all joint values to zero, with the direction of positive displacement (rotation for revolute joints, translation for prismatic joints) for each joint specified. Let $M \in SE(3)$ denote the configuration of the end-effector frame relative to the fixed base frame when the robot is in its zero position.

Now suppose that joint $n$ is displaced to some joint value $\theta_n$. The end-effector frame $M$ then undergoes a displacement of the form

$$T = e^{[\mathcal{S}_n]\theta_n} M, \tag{4.12}$$

where $T \in SE(3)$ is the new configuration of the end-effector frame and $\mathcal{S}_n = (\omega_n, v_n)$ is the screw axis of joint $n$ as expressed in the fixed base frame. If joint $n$ is revolute (corresponding to a screw motion of zero pitch) then $\omega_n \in \mathbb{R}^3$ is a

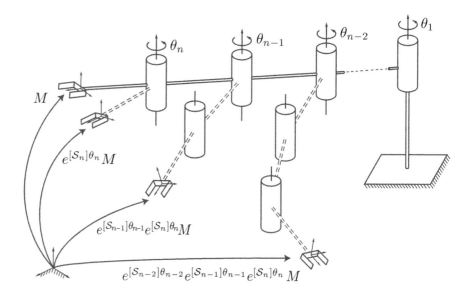

**Figure 4.2** Illustration of the PoE formula for an $n$-link spatial open chain.

unit vector in the positive direction of joint axis $n$; $v_n = -\omega_n \times q_n$, with $q_n$ any arbitrary point on joint axis $n$ as written in coordinates in the fixed base frame; and $\theta_n$ is the joint angle. If joint $n$ is prismatic then $\omega_n = 0$, $v_n \in \mathbb{R}^3$ is a unit vector in the direction of positive translation, and $\theta_n$ represents the prismatic extension/retraction.

If we assume that joint $n - 1$ is also allowed to vary then this has the effect of applying a screw motion to link $n - 1$ (and by extension to link $n$, since link $n$ is connected to link $n - 1$ via joint $n$). The end-effector frame thus undergoes a displacement of the form

$$T = e^{[\mathcal{S}_{n-1}]\theta_{n-1}} \left( e^{[\mathcal{S}_n]\theta_n} M \right). \tag{4.13}$$

Continuing with this reasoning and now allowing all the joints $(\theta_1, \ldots, \theta_n)$ to vary, it follows that

$$T(\theta) = e^{[\mathcal{S}_1]\theta_1} \cdots e^{[\mathcal{S}_{n-1}]\theta_{n-1}} e^{[\mathcal{S}_n]\theta_n} M. \tag{4.14}$$

This is the product of exponentials formula describing the forward kinematics of an $n$-dof open chain. Specifically, we call Equation (4.14) the **space form** of the product of exponentials formula, referring to the fact that the screw axes are expressed in the fixed space frame.

To summarize, to calculate the forward kinematics of an open chain using the space form of the PoE formula (4.14), we need the following elements:

(a) the end-effector configuration $M \in SE(3)$ when the robot is at its home position;

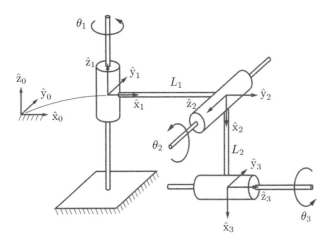

**Figure 4.3** A 3R spatial open chain.

(b) the screw axes $\mathcal{S}_1, \ldots, \mathcal{S}_n$ expressed in the fixed base frame, corresponding to the joint motions when the robot is at its home position;

(c) the joint variables $\theta_1, \ldots, \theta_n$.

Unlike the D–H representation, no link reference frames need to be defined. Further advantages will come to light when we examine the velocity kinematics in the next chapter.

### .1.2 Examples

We now derive the forward kinematics for some common spatial open chains using the PoE formula.

**Example 4.1** (3R spatial open chain) Consider the 3R open chain of Figure 4.3, shown in its home position (all joint variables set equal to zero). Choose the fixed frame $\{0\}$ and end-effector frame $\{3\}$ as indicated in the figure, and express all vectors and homogeneous transformations in terms of the fixed frame. The forward kinematics has the form

$$T(\theta) = e^{[\mathcal{S}_1]\theta_1} e^{[\mathcal{S}_2]\theta_2} e^{[\mathcal{S}_3]\theta_3} M,$$

where $M \in SE(3)$ is the end-effector frame configuration when the robot is in its zero position. By inspection $M$ can be obtained as

$$M = \begin{bmatrix} 0 & 0 & 1 & L_1 \\ 0 & 1 & 0 & 0 \\ -1 & 0 & 0 & -L_2 \\ 0 & 0 & 0 & 1 \end{bmatrix}.$$

The screw axis $\mathcal{S}_1 = (\omega_1, v_1)$ for joint axis 1 is then given by $\omega_1 = (0, 0, 1)$

and $v_1 = (0, 0, 0)$ (the fixed frame origin $(0,0,0)$ is a convenient choice for the point $q_1$ lying on joint axis 1). To determine the screw axis $\mathcal{S}_2$ for joint axis 2, observe that joint axis 2 points in the $-\hat{y}_0$-direction, so that $\omega_2 = (0, -1, 0)$. Choose $q_2 = (L_1, 0, 0)$, in which case $v_2 = -\omega_2 \times q_2 = (0, 0, -L_1)$. Finally, to determine the screw axis $\mathcal{S}_3$ for joint axis 3, note that $\omega_3 = (1, 0, 0)$. Choosing $q_3 = (0, 0, -L_2)$, it follows that $v_3 = -\omega_3 \times q_3 = (0, -L_2, 0)$.

In summary, we have the following $4 \times 4$ matrix representations for the three joint screw axes $\mathcal{S}_1$, $\mathcal{S}_2$, and $\mathcal{S}_3$:

$$[\mathcal{S}_1] = \begin{bmatrix} 0 & -1 & 0 & 0 \\ 1 & 0 & 0 & 0 \\ 0 & 0 & 0 & 0 \\ 0 & 0 & 0 & 0 \end{bmatrix},$$

$$[\mathcal{S}_2] = \begin{bmatrix} 0 & 0 & -1 & 0 \\ 0 & 0 & 0 & 0 \\ 1 & 0 & 0 & -L_1 \\ 0 & 0 & 0 & 0 \end{bmatrix},$$

$$[\mathcal{S}_3] = \begin{bmatrix} 0 & 0 & 0 & 0 \\ 0 & 0 & -1 & -L_2 \\ 0 & 1 & 0 & 0 \\ 0 & 0 & 0 & 0 \end{bmatrix}.$$

It will be more convenient to list the screw axes in the following tabular form:

| $i$ | $\omega_i$ | $v_i$ |
|---|---|---|
| 1 | $(0, 0, 1)$ | $(0, 0, 0)$ |
| 2 | $(0, -1, 0)$ | $(0, 0, -L_1)$ |
| 3 | $(1, 0, 0)$ | $(0, -L_2, 0)$ |

**Example 4.2** (3R planar open chain)   For the robot in Figure 4.1, we expressed the end-effector home configuration $M$ (Equation (4.6)) and the screw axes $\mathcal{S}_i$ as follows:

| $i$ | $\omega_i$ | $v_i$ |
|---|---|---|
| 1 | $(0, 0, 1)$ | $(0, 0, 0)$ |
| 2 | $(0, 0, 1)$ | $(0, -L_1, 0)$ |
| 3 | $(0, 0, 1)$ | $(0, -(L_1 + L_2), 0)$ |

Since the motion is in the $\hat{x}$–$\hat{y}$-plane, we could equivalently write each screw axis $\mathcal{S}_i$ as a 3-vector $(\omega_z, v_x, v_y)$:

| $i$ | $\omega_i$ | $v_i$ |
|---|---|---|
| 1 | 1 | $(0, 0)$ |
| 2 | 1 | $(0, -L_1)$ |
| 3 | 1 | $(0, -(L_1 + L_2))$ |

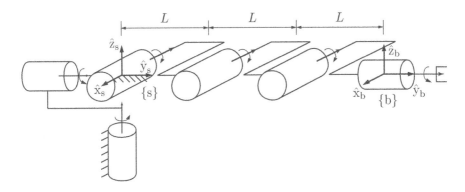

**Figure 4.4** PoE forward kinematics for the 6R open chain.

and $M$ as an element of $SE(2)$:

$$M = \begin{bmatrix} 1 & 0 & L_1 + L_2 + L_3 \\ 0 & 1 & 0 \\ 0 & 0 & 1 \end{bmatrix}.$$

In this case, the forward kinematics would use the simplified matrix exponential for planar motions (Exercise 3.49).

**Example 4.3** (6R spatial open chain)   We now derive the forward kinematics of the 6R open chain of Figure 4.4. Six-dof arms play an important role in robotics because they have the minimum number of joints that allows the end-effector to move a rigid body in all its degrees of freedom, subject only to limits on the robot's workspace. For this reason, six-dof robot arms are sometimes called general purpose manipulators.

The zero position and the direction of positive rotation for each joint axis are as shown in the figure. A fixed frame {s} and end-effector frame {b} are also assigned as shown. The end-effector frame $M$ in the zero position is then

$$M = \begin{bmatrix} 1 & 0 & 0 & 0 \\ 0 & 1 & 0 & 3L \\ 0 & 0 & 1 & 0 \\ 0 & 0 & 0 & 1 \end{bmatrix} \tag{4.15}$$

The screw axis for joint 1 is in the direction $\omega_1 = (0,0,1)$. The most convenient choice for point $q_1$ lying on joint axis 1 is the origin, so that $v_1 = (0,0,0)$. The screw axis for joint 2 is in the $\hat{y}$-direction of the fixed frame, so $\omega_2 = (0,1,0)$. Choosing $q_2 = (0,0,0)$, we have $v_2 = (0,0,0)$. The screw axis for joint 3 is in the direction $\omega_3 = (-1,0,0)$. Choosing $q_3 = (0,0,0)$ leads to $v_3 = (0,0,0)$. The screw axis for joint 4 is in the direction $\omega_4 = (-1,0,0)$. Choosing $q_4 = (0,L,0)$ leads to $v_4 = (0,0,L)$. The screw axis for joint 5 is in the direction $\omega_5 = (-1,0,0)$; choosing $q_5 = (0,2L,0)$ leads to $v_5 = (0,0,2L)$. The screw axis for joint 6 is

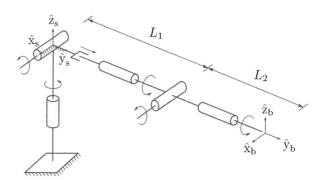

**Figure 4.5** The RRPRRR spatial open chain.

in the direction $\omega_6 = (0, 1, 0)$; choosing $q_6 = (0, 0, 0)$ leads to $v_6 = (0, 0, 0)$. In summary, the screw axes $\mathcal{S}_i = (\omega_i, v_i)$, $i = 1, \ldots, 6$, are as follows:

| $i$ | $\omega_i$ | $v_i$ |
|---|---|---|
| 1 | $(0, 0, 1)$ | $(0, 0, 0)$ |
| 2 | $(0, 1, 0)$ | $(0, 0, 0)$ |
| 3 | $(-1, 0, 0)$ | $(0, 0, 0)$ |
| 4 | $(-1, 0, 0)$ | $(0, 0, L)$ |
| 5 | $(-1, 0, 0)$ | $(0, 0, 2L)$ |
| 6 | $(0, 1, 0)$ | $(0, 0, 0)$ |

**Example 4.4** (An RRPRRR spatial open chain)   In this example we consider the six-degree-of-freedom RRPRRR spatial open chain of Figure 4.5. The end-effector frame in the zero position is given by

$$
M = \begin{bmatrix} 1 & 0 & 0 & 0 \\ 0 & 1 & 0 & L_1 + L_2 \\ 0 & 0 & 1 & 0 \\ 0 & 0 & 0 & 1 \end{bmatrix}.
$$

The screw axes $\mathcal{S}_i = (\omega_i, v_i)$ are listed in the following table:

| $i$ | $\omega_i$ | $v_i$ |
|---|---|---|
| 1 | $(0, 0, 1)$ | $(0, 0, 0)$ |
| 2 | $(1, 0, 0)$ | $(0, 0, 0)$ |
| 3 | $(0, 0, 0)$ | $(0, 1, 0)$ |
| 4 | $(0, 1, 0)$ | $(0, 0, 0)$ |
| 5 | $(1, 0, 0)$ | $(0, 0, -L_1)$ |
| 6 | $(0, 1, 0)$ | $(0, 0, 0)$ |

Note that the third joint is prismatic, so that $\omega_3 = 0$ and $v_3$ is a unit vector in the direction of positive translation.

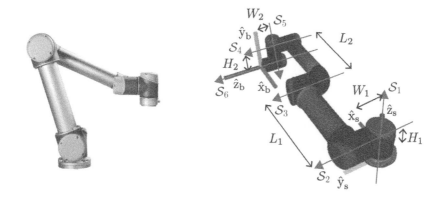

**Figure 4.6** (Left) Universal Robots' UR5 6R robot arm. (Right) Shown at its zero position. Positive rotations about the axes indicated are given by the usual right-hand rule. $W_1$ is the distance along the $\hat{y}_s$-direction between the anti-parallel axes of joints 1 and 5. $W_1 = 109$ mm, $W_2 = 82$ mm, $L_1 = 425$ mm, $L_2 = 392$ mm, $H_1 = 89$ mm, $H_2 = 95$ mm.

**Example 4.5** (Universal Robots' UR5 6R robot arm)   Universal Robots' UR5 6R robot arm is shown in Figure 4.6. Each joint is directly driven by a brushless motor combined with 100 : 1 zero-backlash harmonic drive gearing, which greatly increases the torque available at the joint while reducing its maximum speed.

Figure 4.6 shows the screw axes $S_1, \ldots, S_6$ when the robot is at its zero position. The end-effector frame {b} in the zero position is given by

$$
M = \begin{bmatrix} -1 & 0 & 0 & L_1 + L_2 \\ 0 & 0 & 1 & W_1 + W_2 \\ 0 & 1 & 0 & H_1 - H_2 \\ 0 & 0 & 0 & 1 \end{bmatrix}.
$$

The screw axes $S_i = (\omega_i, v_i)$ are listed in the following table:

| $i$ | $\omega_i$ | $v_i$ |
|---|---|---|
| 1 | $(0,0,1)$ | $(0,0,0)$ |
| 2 | $(0,1,0)$ | $(-H_1,0,0)$ |
| 3 | $(0,1,0)$ | $(-H_1,0,L_1)$ |
| 4 | $(0,1,0)$ | $(-H_1,0,L_1+L_2)$ |
| 5 | $(0,0,-1)$ | $(-W_1,L_1+L_2,0)$ |
| 6 | $(0,1,0)$ | $(H_2-H_1,0,L_1+L_2)$ |

As an example of the forward kinematics, set $\theta_2 = -\pi/2$ and $\theta_5 = \pi/2$, with all other joint angles equal to zero. Then the configuration of the end-effector is

$$
\begin{aligned}
T(\theta) &= e^{[S_1]\theta_1} e^{[S_2]\theta_2} e^{[S_3]\theta_3} e^{[S_4]\theta_4} e^{[S_5]\theta_5} e^{[S_6]\theta_6} M \\
&= I e^{-[S_2]\pi/2} I^2 e^{[S_5]\pi/2} I M \\
&= e^{-[S_2]\pi/2} e^{[S_5]\pi/2} M
\end{aligned}
$$

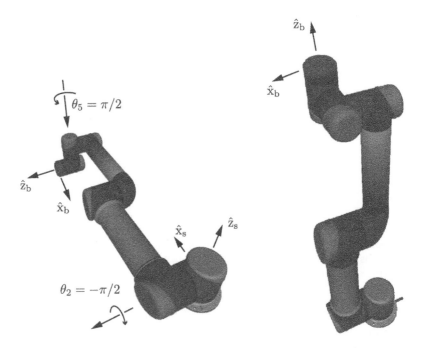

**Figure 4.7** (Left) The UR5 at its home position, with the axes of joints 2 and 5 indicated. (Right) The UR5 at joint angles $\theta = (\theta_1, \ldots, \theta_6) = (0, -\pi/2, 0, 0, \pi/2, 0)$.

since $e^0 = I$. Evaluating, we get

$$e^{-[\mathcal{S}_2]\pi/2} = \begin{bmatrix} 0 & 0 & -1 & 0.089 \\ 0 & 1 & 0 & 0 \\ 1 & 0 & 0 & 0.089 \\ 0 & 0 & 0 & 1 \end{bmatrix}, \qquad e^{[\mathcal{S}_5]\pi/2} = \begin{bmatrix} 0 & 1 & 0 & 0.708 \\ -1 & 0 & 0 & 0.926 \\ 0 & 0 & 1 & 0 \\ 0 & 0 & 0 & 1 \end{bmatrix},$$

where the linear units are meters, and

$$T(\theta) = e^{-[\mathcal{S}_2]\pi/2} e^{[\mathcal{S}_5]\pi/2} M = \begin{bmatrix} 0 & -1 & 0 & 0.095 \\ 1 & 0 & 0 & 0.109 \\ 0 & 0 & 1 & 0.988 \\ 0 & 0 & 0 & 1 \end{bmatrix}$$

as shown in Figure 4.7.

### 4.1.3     Second Formulation: Screw Axes in the End-Effector Frame

The matrix identity $e^{M^{-1}PM} = M^{-1}e^P M$ (Proposition 3.10) can also be expressed as $Me^{M^{-1}PM} = e^P M$. Beginning with the rightmost term of the previously derived product of exponentials formula, if we repeatedly apply this identity then after $n$ iterations we obtain

$$
\begin{aligned}
T(\theta) &= e^{[\mathcal{S}_1]\theta_1} \cdots e^{[\mathcal{S}_n]\theta_n} M \\
&= e^{[\mathcal{S}_1]\theta_1} \cdots M e^{M^{-1}[\mathcal{S}_n]M\theta_n} \\
&= e^{[\mathcal{S}_1]\theta_1} \cdots M e^{M^{-1}[\mathcal{S}_{n-1}]M\theta_{n-1}} e^{M^{-1}[\mathcal{S}_n]M\theta_n} \\
&= M e^{M^{-1}[\mathcal{S}_1]M\theta_1} \cdots e^{M^{-1}[\mathcal{S}_{n-1}]M\theta_{n-1}} e^{M^{-1}[\mathcal{S}_n]M\theta_n} \\
&= M e^{[\mathcal{B}_1]\theta_1} \cdots e^{[\mathcal{B}_{n-1}]\theta_{n-1}} e^{[\mathcal{B}_n]\theta_n},
\end{aligned}
\tag{4.16}
$$

where each $[\mathcal{B}_i]$ is given by $M^{-1}[\mathcal{S}_i]M$, i.e., $\mathcal{B}_i = [\mathrm{Ad}_{M^{-1}}]\mathcal{S}_i$, $i = 1, \ldots, n$. Equation (4.16) is an alternative form of the product of exponentials formula, representing the joint axes as screw axes $\mathcal{B}_i$ in the end-effector (body) frame when the robot is at its zero position. We call Equation (4.16) the **body form** of the product of exponentials formula.

It is worth thinking about the order of the transformations expressed in the space-form PoE formula (Equation (4.14)) and in the body-form formula (Equation (4.16)). In the space form, $M$ is first transformed by the most distal joint, progressively moving inward to more proximal joints. Note that the fixed space-frame representation of the screw axis for a more proximal joint is not affected by the joint displacement at a distal joint (e.g., joint 3's displacement does not affect joint 2's screw axis representation in the space frame). In the body form, $M$ is first transformed by the first joint, progressively moving outward to more distal joints. The body-frame representation of the screw axis for a more distal joint is not affected by the joint displacement at a proximal joint (e.g., joint 2's displacement does not affect joint 3's screw axis representation in the body frame.) Therefore, it makes sense that we need to determine the screw axes only at the robot's zero position: any $\mathcal{S}_i$ is unaffected by more distal transformations, and any $\mathcal{B}_i$ is unaffected by more proximal transformations.

**Example 4.6** (6R spatial open chain)   We now express the forward kinematics for the 6R open chain of Figure 4.4 in the second form,

$$
T(\theta) = M e^{[\mathcal{B}_1]\theta_1} e^{[\mathcal{B}_2]\theta_2} \cdots e^{[\mathcal{B}_6]\theta_6}.
$$

Assume the same fixed and end-effector frames and zero position as found previously; $M$ is still the same as in Equation (4.15), obtained as the end-effector frame as seen from the fixed frame with the chain in its zero position. The screw axis for each joint axis, expressed with respect to the end-effector frame in its zero position, is given in the following table:

| $i$ | $\omega_i$ | $v_i$ |
|---|---|---|
| 1 | $(0, 0, 1)$ | $(-3L, 0, 0)$ |
| 2 | $(0, 1, 0)$ | $(0, 0, 0)$ |
| 3 | $(-1, 0, 0)$ | $(0, 0, -3L)$ |
| 4 | $(-1, 0, 0)$ | $(0, 0, -2L)$ |
| 5 | $(-1, 0, 0)$ | $(0, 0, -L)$ |
| 6 | $(0, 1, 0)$ | $(0, 0, 0)$ |

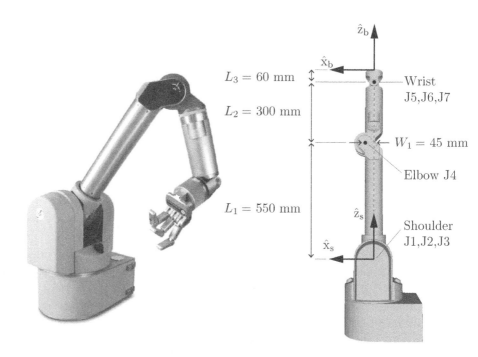

**Figure 4.8** Barrett Technology's WAM 7R robot arm at its zero configuration (right). At the zero configuration, axes 1, 3, 5, and 7 are along $\hat{z}_s$ and axes 2, 4, and 6 are aligned with $\hat{y}_s$, out of the page. Positive rotations are given by the right-hand rule. Axes 1, 2, and 3 intersect at the origin of $\{s\}$ and axes 5, 6, and 7 intersect at a point 60mm from $\{b\}$. The zero configuration is singular, as discussed in Section 5.3.

**Example 4.7** (Barrett Technology's WAM 7R robot arm)    Barrett Technology's WAM 7R robot arm is shown in Figure 4.8. The extra (seventh) joint means that the robot is **redundant** for the task of positioning its end-effector frame in $SE(3)$; in general, for a given end-effector configuration in the robot's workspace, there is a one-dimensional set of joint variables in the robot's seven-dimensional joint space that achieves that configuration. This extra degree of freedom can be used for obstacle avoidance or to optimize some objective function such as minimizing the motor power needed to hold the end-effector at that configuration.

Also, some joints of the WAM are driven by motors placed at the base of the robot, reducing the robot's moving mass. Torques are transferred from the motors to the joints by cables winding around drums at the joints and motors. Because the moving mass is reduced, the motor torque requirements are decreased, allowing low (cable) gear ratios and high speeds. This design is in contrast with that of the UR5, where the motor and harmonic drive gearing for each joint are directly at the joint.

Figure 4.8 illustrates the WAM's end-effector frame screw axes $\mathcal{B}_1, \ldots, \mathcal{B}_7$ when the robot is at its zero position. The end-effector frame $\{b\}$ in the zero position is given by

$$M = \begin{bmatrix} 1 & 0 & 0 & 0 \\ 0 & 1 & 0 & 0 \\ 0 & 0 & 1 & L_1 + L_2 + L_3 \\ 0 & 0 & 0 & 1 \end{bmatrix}.$$

The screw axes $\mathcal{B}_i = (\omega_i, v_i)$ are listed in the following table:

| $i$ | $\omega_i$ | $v_i$ |
|-----|------------|-------|
| 1 | $(0,0,1)$ | $(0,0,0)$ |
| 2 | $(0,1,0)$ | $(L_1 + L_2 + L_3, 0, 0)$ |
| 3 | $(0,0,1)$ | $(0,0,0)$ |
| 4 | $(0,1,0)$ | $(L_2 + L_3, 0, W_1)$ |
| 5 | $(0,0,1)$ | $(0,0,0)$ |
| 6 | $(0,1,0)$ | $(L_3, 0, 0)$ |
| 7 | $(0,0,1)$ | $(0,0,0)$ |

Figure 4.9 shows the WAM arm with $\theta_2 = 45°$, $\theta_4 = -45°$, $\theta_6 = -90°$ and all other joint angles equal to zero, giving

$$T(\theta) = M e^{[\mathcal{B}_2]\pi/4} e^{-[\mathcal{B}_4]\pi/4} e^{-[\mathcal{B}_6]\pi/2} = \begin{bmatrix} 0 & 0 & -1 & 0.3157 \\ 0 & 1 & 0 & 0 \\ 1 & 0 & 0 & 0.6571 \\ 0 & 0 & 0 & 1 \end{bmatrix}.$$

## The Universal Robot Description Format

The **Universal Robot Description Format** (URDF) is an XML (eXtensible Markup Language) file format used by the Robot Operating System (ROS) to describe the kinematics, inertial properties, and link geometry of robots. A URDF file describes the joints and links of a robot:

- **Joints.** Joints connect two links: a `parent` link and a `child` link. A few of the possible joint types include prismatic, revolute (including joint limits), continuous (revolute without joint limits), and fixed (a virtual joint that does not permit any motion). Each joint has an `origin` frame that defines the position and orientation of the `child` link frame relative to the `parent` link frame when the joint variable is zero. The `origin` is on the joint's axis. Each joint has an `axis` 3-vector, a unit vector expressed in the `child` link's frame, in the direction of positive rotation for a revolute joint or positive translation for a prismatic joint.
- **Links.** While the joints fully describe the kinematics of a robot, the links define its mass properties. These start to be needed in Chapter 8, when we begin to study the dynamics of robots. The elements of a link include

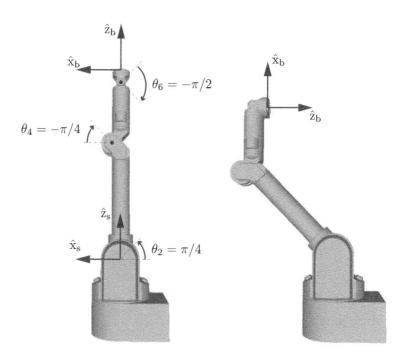

**Figure 4.9** (Left) The WAM at its home configuration, with the axes of joints 2, 4, and 6 indicated. (Right) The WAM at $\theta = (\theta_1, \ldots, \theta_7) = (0, \pi/4, 0, -\pi/4, 0, -\pi/2, 0)$.

its `mass`; an `origin` frame that defines the position and orientation of a frame at the link's center of mass relative to the link's joint frame described above; and an `inertia` matrix, relative to the link's center of mass frame, specified by the six elements on or above the diagonal. (As we will see in Chapter 8, the inertia matrix for a rigid body is a $3 \times 3$ symmetric positive-definite matrix. Since the inertia matrix is symmetric, it is only necessary to define the terms on and above the diagonal.)

Note that most links have two frames rigidly attached: a first frame at the joint (defined by the joint element that connects the link to its parent) and a second frame at the link's center of mass (defined by the link element).

A URDF file can represent any robot with a tree structure. This includes serial-chain robot arms and robot hands, but not a Stewart platform or other mechanisms with closed loops. An example of a robot with a tree structure is shown in Figure 4.10.

The orientation of a frame {b} relative to a frame {a} is represented using roll–pitch–yaw coordinates: first, a roll about the fixed $\hat{x}_a$-axis; then a pitch about the fixed $\hat{y}_a$-axis; then a yaw about the fixed $\hat{z}_a$-axis.

The kinematics and mass properties of the UR5 robot arm (Figure 4.11) are defined in the URDF file below, which demonstrates the syntax of the joint's elements (`parent`, `child`, `origin`, and `axis`) and the link's elements (`mass`,

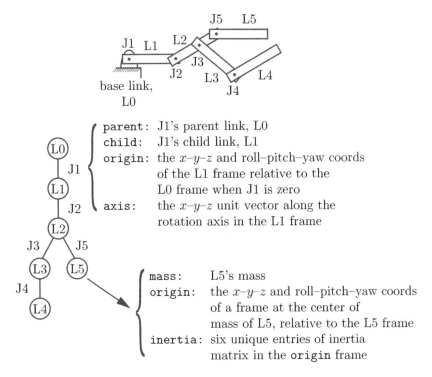

**Figure 4.10** A five-link robot represented as a tree, where nodes of the tree are the links and the edges of the tree are the joints.

origin, and inertia). A URDF requires a frame defined at every joint, so we define frames $\{1\}$ to $\{6\}$ in addition to the fixed base frame $\{0\}$ (i.e., $\{s\}$) and the end-effector frame $\{7\}$ (i.e., $\{b\}$). Figure 4.11 gives the extra information needed to fully write the URDF.

Although the joint types in the URDF are defined as "continuous," the UR5 joints do in fact have joint limits; they are omitted here for simplicity. The mass and inertial properties listed here are not exact.

---

**The UR5 URDF file (kinematics and inertial properties only).**

```
<?xml version="1.0" ?>
<robot name="ur5">

<!-- ********** KINEMATIC PROPERTIES (JOINTS) ********** -->
  <joint name="world_joint" type="fixed">
    <parent link="world"/>
    <child link="base_link"/>
    <origin rpy="0.0 0.0 0.0" xyz="0.0 0.0 0.0"/>
  </joint>
  <joint name="joint1" type="continuous">
    <parent link="base_link"/>
    <child link="link1"/>
    <origin rpy="0.0 0.0 0.0" xyz="0.0 0.0 0.089159"/>
```

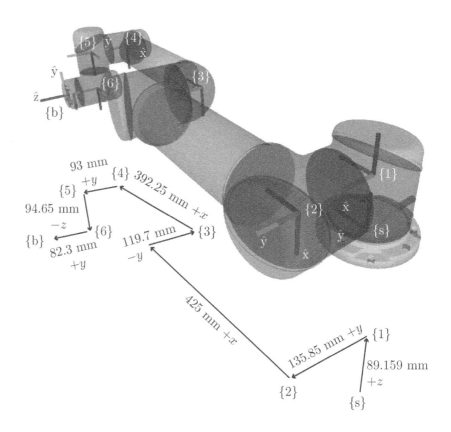

**Figure 4.11** The orientations of the frames {s} (also called {0}), {b} (also called {7}), and {1} through {6} are illustrated on the translucent UR5. The frames {s} and {1} are aligned with each other; frames {2} and {3} are aligned with each other; and frames {4}, {5}, and {6} are aligned with each other. Therefore, only the axes of frames {s}, {2}, {4}, and {b} are labeled. Just below the image of the robot is a skeleton indicating how the frames are offset from each other, including distances and directions (expressed in the {s} frame).

```
    <axis xyz="0 0 1"/>
  </joint>
  <joint name="joint2" type="continuous">
    <parent link="link1"/>
    <child link="link2"/>
    <origin rpy="0.0 1.570796325 0.0" xyz="0.0 0.13585 0.0"/>
    <axis xyz="0 1 0"/>
  </joint>
  <joint name="joint3" type="continuous">
    <parent link="link2"/>
    <child link="link3"/>
    <origin rpy="0.0 0.0 0.0" xyz="0.0 -0.1197 0.425"/>
    <axis xyz="0 1 0"/>
  </joint>
  <joint name="joint4" type="continuous">
    <parent link="link3"/>
```

```
    <child link="link4"/>
    <origin rpy="0.0 1.570796325 0.0" xyz="0.0 0.0 0.39225"/>
    <axis xyz="0 1 0"/>
  </joint>
  <joint name="joint5" type="continuous">
    <parent link="link4"/>
    <child link="link5"/>
    <origin rpy="0.0 0.0 0.0" xyz="0.0 0.093 0.0"/>
    <axis xyz="0 0 1"/>
  </joint>
  <joint name="joint6" type="continuous">
    <parent link="link5"/>
    <child link="link6"/>
    <origin rpy="0.0 0.0 0.0" xyz="0.0 0.0 0.09465"/>
    <axis xyz="0 1 0"/>
  </joint>
  <joint name="ee_joint" type="fixed">
    <origin rpy="-1.570796325 0 0" xyz="0 0.0823 0"/>
    <parent link="link6"/>
    <child link="ee_link"/>
  </joint>

<!-- ********** INERTIAL PROPERTIES (LINKS) ********** -->
  <link name="world"/>
  <link name="base_link">
    <inertial>
      <mass value="4.0"/>
      <origin rpy="0 0 0" xyz="0.0 0.0 0.0"/>
      <inertia ixx="0.00443333156" ixy="0.0" ixz="0.0"
               iyy="0.00443333156" iyz="0.0" izz="0.0072"/>
    </inertial>
  </link>
  <link name="link1">
    <inertial>
      <mass value="3.7"/>
      <origin rpy="0 0 0" xyz="0.0 0.0 0.0"/>
      <inertia ixx="0.010267495893" ixy="0.0" ixz="0.0"
               iyy="0.010267495893" iyz="0.0" izz="0.00666"/>
    </inertial>
  </link>
  <link name="link2">
    <inertial>
      <mass value="8.393"/>
      <origin rpy="0 0 0" xyz="0.0 0.0 0.28"/>
      <inertia ixx="0.22689067591" ixy="0.0" ixz="0.0"
               iyy="0.22689067591" iyz="0.0" izz="0.0151074"/>
    </inertial>
  </link>
  <link name="link3">
    <inertial>
      <mass value="2.275"/>
      <origin rpy="0 0 0" xyz="0.0 0.0 0.25"/>
      <inertia ixx="0.049443313556" ixy="0.0" ixz="0.0"
               iyy="0.049443313556" iyz="0.0" izz="0.004095"/>
    </inertial>
  </link>
```

```
<link name="link4">
  <inertial>
    <mass value="1.219"/>
    <origin rpy="0 0 0" xyz="0.0 0.0 0.0"/>
    <inertia ixx="0.111172755531" ixy="0.0" ixz="0.0"
             iyy="0.111172755531" iyz="0.0" izz="0.21942"/>
  </inertial>
</link>
<link name="link5">
  <inertial>
    <mass value="1.219"/>
    <origin rpy="0 0 0" xyz="0.0 0.0 0.0"/>
    <inertia ixx="0.111172755531" ixy="0.0" ixz="0.0"
             iyy="0.111172755531" iyz="0.0" izz="0.21942"/>
  </inertial>
</link>
<link name="link6">
  <inertial>
    <mass value="0.1879"/>
    <origin rpy="0 0 0" xyz="0.0 0.0 0.0"/>
    <inertia ixx="0.0171364731454" ixy="0.0" ixz="0.0"
             iyy="0.0171364731454" iyz="0.0" izz="0.033822"/>
  </inertial>
</link>
<link name="ee_link"/>
</robot>
```

Beyond the properties described above, a URDF can describe other properties of a robot, such as its visual appearance (including geometric models of the links) as well as simplified representations of link geometries that can be used for collision detection in motion planning algorithms.

## 4.3    Summary

- Given an open chain with a fixed reference frame {s} and a reference frame {b} attached to some point on its last link – this frame is denoted the end-effector frame – the forward kinematics is the mapping $T(\theta)$ from the joint values $\theta$ to the position and orientation of {b} in {s}.
- In the Denavit–Hartenberg representation the forward kinematics of an open chain is described in terms of the relative displacements between reference frames attached to each link. If the link frames are sequentially labeled $\{0\}, \ldots, \{n+1\}$, where $\{0\}$ is the fixed frame {s}, $\{i\}$ is a frame attached to link $i$ at joint $i$ (with $i = 1, \ldots, n$), and $\{n+1\}$ is the end-effector frame {b} then the forward kinematics is expressed as

$$T_{0,n+1}(\theta) = T_{01}(\theta_1) \cdots T_{n-1,n}(\theta_n) T_{n,n+1}$$

where $\theta_i$ denotes the joint $i$ variable and $T_{n,n+1}$ indicates the (fixed) configuration of the end-effector frame in $\{n\}$. If the end-effector frame {b} is chosen to be coincident with $\{n\}$ then we can dispense with the frame $\{n+1\}$.

- The Denavit–Hartenberg convention requires that reference frames assigned to each link obey a strict convention (see Appendix C). Following this convention, the link frame transformation $T_{i-1,i}$ between link frames $\{i-1\}$ and $\{i\}$ can be parametrized using only four parameters, the Denavit–Hartenberg parameters. Three of these parameters describe the kinematic structure, while the fourth is the joint value. Four numbers is the minimum needed to represent the displacement between two link frames.

- The forward kinematics can also be expressed as the following product of exponentials (the space form),

$$T(\theta) = e^{[\mathcal{S}_1]\theta_1} \cdots e^{[\mathcal{S}_n]\theta_n} M,$$

where $\mathcal{S}_i = (\omega_i, v_i)$ denotes the screw axis associated with positive motion along joint $i$ expressed in fixed-frame $\{s\}$ coordinates, $\theta_i$ is the joint-$i$ variable, and $M \in SE(3)$ denotes the position and orientation of the end-effector frame $\{b\}$ when the robot is in its zero position. It is not necessary to define individual link frames; it is only necessary to define $M$ and the screw axes $\mathcal{S}_1, \ldots, \mathcal{S}_n$.

- The product of exponentials formula can also be written in the equivalent body form,

$$T(\theta) = M e^{[\mathcal{B}_1]\theta_1} \cdots e^{[\mathcal{B}_n]\theta_n},$$

where $\mathcal{B}_i = [\mathrm{Ad}_{M^{-1}}]\mathcal{S}_i$, $i = 1, \ldots, n$; $\mathcal{B}_i = (\omega_i, v_i)$ is the screw axis corresponding to joint axis $i$, expressed in $\{b\}$, with the robot in its zero position.

- The Universal Robot Description Format (URDF) is a file format used by the Robot Operating System and other software for representing the kinematics, inertial properties, visual properties, and other information for general tree-like robot mechanisms, including serial chains. A URDF file includes descriptions of joints, which connect a parent link and a child link and fully specify the kinematics of the robot, as well as descriptions of links, which specify its inertial properties.

## .4     Software

Software functions associated with this chapter are listed in MATLAB format below.

```
T = FKinBody(M,Blist,thetalist)
```
Computes the end-effector frame given the zero position of the end-effector M,

the list of joint screws `Blist` expressed in the end-effector frame, and the list of joint values `thetalist`.

`T = FKinSpace(M,Slist,thetalist)`
Computes the end-effector frame given the zero position of the end-effector `M`, the list of joint screws `Slist` expressed in the space frame, and the list of joint values `thetalist`.

## 4.5    Notes and References

The literature on robot kinematics is quite extensive, and with very few exceptions most approaches are based on the Denavit–Hartenberg (D–H) parameters originally presented in Denavit and Hartenberg (1955) and summarized in Appendix C. Our approach is based on the product of exponentials (PoE) formula first presented by Brockett (1983b). Computational aspects of the PoE formula are discussed in Park (1994).

Appendix C also elucidates in some detail the many advantages of the PoE formula over the D–H parameters, e.g., the elimination of link reference frames, the uniform treatment of revolute and prismatic joints, and the intuitive geometric interpretation of the joint axes as screws. These advantages more than offset the lone advantage of the D–H parameters, namely that they constitute a minimal set. Moreover, it should be noted that, when using D–H parameters, there are differing conventions for assigning link frames, e.g., some methods align the joint axis with the $\hat{x}$-axis rather than with the $\hat{z}$-axis of the link frame as we have done. Both the link frames and the accompanying D–H parameters need to be specified together in order to have a complete description of the robot's forward kinematics.

In summary, unless using a minimal set of parameters to represent a joint's spatial motion is critical, there is no compelling reason to prefer the D–H parameters over the PoE formula. In the next chapter, an even stronger case will be made for preferring the PoE formula to model the forward kinematics.

## 4.6    Exercises

**Exercise 4.1**  Familiarize yourself with the functions `FKinBody` and `FKinSpace` in your favorite programming language. Can you make these functions more computationally efficient? If so, indicate how. If not, explain why not.

**Exercise 4.2**  The RRRP SCARA robot of Figure 4.12 is shown in its zero position. Determine the end-effector zero position configuration $M$, the screw axes $\mathcal{S}_i$ in $\{0\}$, and the screw axes $\mathcal{B}_i$ in $\{b\}$. For $\ell_0 = \ell_1 = \ell_2 = 1$ and the joint variable values $\theta = (0, \pi/2, -\pi/2, 1)$, use both the `FKinSpace` and the `FKinBody`

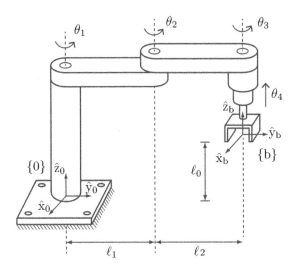

**Figure 4.12** An RRRP SCARA robot for performing pick-and-place operations.

functions to find the end-effector configuration $T \in SE(3)$. Confirm that they agree with each other.

**Exercise 4.3**   Determine the end-effector frame screw axes $\mathcal{B}_i$ for the 3R robot in Figure 4.3.

**Exercise 4.4**   Determine the end-effector frame screw axes $\mathcal{B}_i$ for the RRPRRR robot in Figure 4.5.

**Exercise 4.5**   Determine the end-effector frame screw axes $\mathcal{B}_i$ for the UR5 robot in Figure 4.6.

**Exercise 4.6**   Determine the space frame screw axes $\mathcal{S}_i$ for the WAM robot in Figure 4.8.

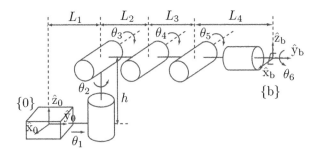

**Figure 4.13** A PRRRRR spatial open chain in its zero configuration.

**Exercise 4.7**   The PRRRRR spatial open chain of Figure 4.13 is shown in its zero position. Determine the end-effector zero position configuration $M$, the screw axes $\mathcal{S}_i$ in $\{0\}$, and the screw axes $\mathcal{B}_i$ in $\{b\}$.

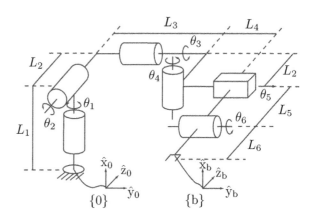

**Figure 4.14** A spatial RRRRPR open chain.

**Exercise 4.8** The spatial RRRRPR open chain of Figure 4.14 is shown in its zero position, with fixed and end-effector frames chosen as indicated. Determine the end-effector zero position configuration $M$, the screw axes $\mathcal{S}_i$ in $\{0\}$, and the screw axes $\mathcal{B}_i$ in $\{b\}$.

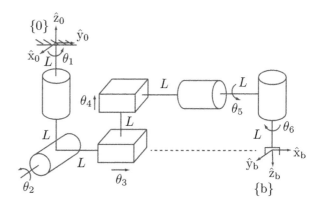

**Figure 4.15** A spatial RRPPRR open chain with prescribed fixed and end-effector frames.

**Exercise 4.9** The spatial RRPPRR open chain of Figure 4.15 is shown in its zero position. Determine the end-effector zero position configuration $M$, the screw axes $\mathcal{S}_i$ in $\{0\}$, and the screw axes $\mathcal{B}_i$ in $\{b\}$.

**Exercise 4.10** The URRPR spatial open chain of Figure 4.16 is shown in its zero position. Determine the end-effector zero position configuration $M$, the screw axes $\mathcal{S}_i$ in $\{0\}$, and the screw axes $\mathcal{B}_i$ in $\{b\}$.

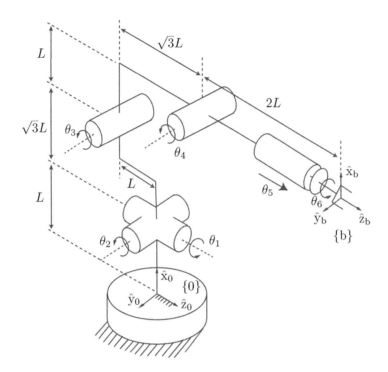

**Figure 4.16** A URRPR spatial open-chain robot.

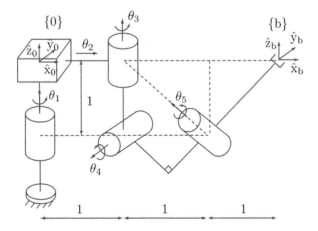

**Figure 4.17** An RPRRR spatial open chain.

**Exercise 4.11**    The spatial RPRRR open chain of Figure 4.17 is shown in its zero position. Determine the end-effector zero position configuration $M$, the screw axes $\mathcal{S}_i$ in $\{0\}$, and the screw axes $\mathcal{B}_i$ in $\{b\}$.

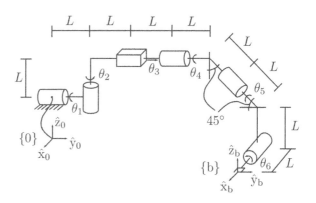

**Figure 4.18** An RRPRRR spatial open chain.

**Exercise 4.12**    The RRPRRR spatial open chain of Figure 4.18 is shown in its zero position (all joints lie on the same plane). Determine the end-effector zero position configuration $M$, the screw axes $\mathcal{S}_i$ in $\{0\}$, and the screw axes $\mathcal{B}_i$ in $\{b\}$. Setting $\theta_5 = \pi$ and all other joint variables to zero, find $T_{06}$ and $T_{60}$.

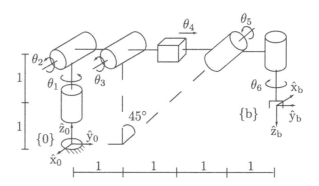

**Figure 4.19** A spatial RRRPRR open chain with prescribed fixed and end-effector frames.

**Exercise 4.13**    The spatial RRRPRR open chain of Figure 4.19 is shown in its zero position. Determine the end-effector zero position configuration $M$, the screw axes $\mathcal{S}_i$ in $\{0\}$, and the screw axes $\mathcal{B}_i$ in $\{b\}$.

**Exercise 4.14**    The RPH robot of Figure 4.20 is shown in its zero position. Determine the end-effector zero position configuration $M$, the screw axes $\mathcal{S}_i$ in $\{s\}$, and the screw axes $\mathcal{B}_i$ in $\{b\}$. Use both the FKinSpace and the FKinBody

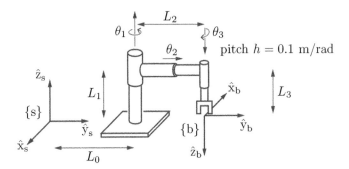

**Figure 4.20** An RPH open chain shown at its zero position. All arrows along/about the joint axes are drawn in the positive direction (i.e., in the direction of increasing joint value). The pitch of the screw joint is 0.1 m/rad, i.e., it advances linearly by 0.1 m for every radian rotated. The link lengths are $L_0 = 4$, $L_1 = 3$, $L_2 = 2$, and $L_3 = 1$ (figure not drawn to scale).

functions to find the end-effector configuration $T \in SE(3)$ when $\theta = (\pi/2, 3, \pi)$. Confirm that the results agree.

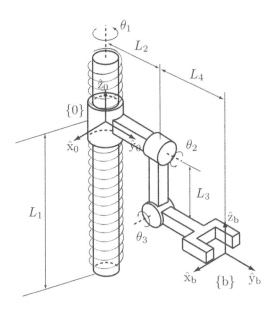

**Figure 4.21** HRR robot. The pitch of the screw joint is denoted by $h$.

**Exercise 4.15** The HRR robot in Figure 4.21 is shown in its zero position. Determine the end-effector zero position configuration $M$, the screw axes $\mathcal{S}_i$ in $\{0\}$, and the screw axes $\mathcal{B}_i$ in $\{b\}$.

**Exercise 4.16** The forward kinematics of a four-dof open chain in its zero position is written in the following exponential form:

$$T(\theta) = e^{[A_1]\theta_1} e^{[A_2]\theta_2} M e^{[A_3]\theta_3} e^{[A_4]\theta_4}.$$

Suppose that the manipulator's zero position is redefined as follows:

$$(\theta_1, \theta_2, \theta_3, \theta_4) = (\alpha_1, \alpha_2, \alpha_3, \alpha_4).$$

Defining $\theta'_i = \theta_i - \alpha_i$, $i = 1, \ldots, 4$, the forward kinematics can then be written

$$T_{04}(\theta'_1, \theta'_2, \theta'_3, \theta'_4) = e^{[A'_1]\theta'_1} e^{[A'_2]\theta'_2} M' e^{[A'_3]\theta'_3} e^{[A'_4]\theta'_4}.$$

Find $M'$ and each of the $A'_i$.

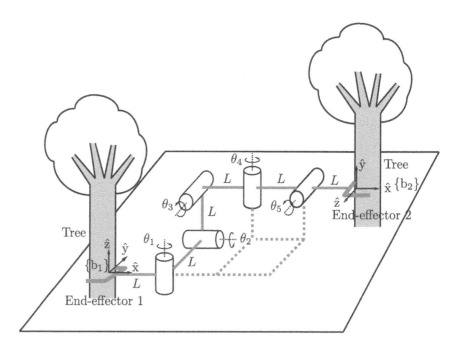

**Figure 4.22** Snake robot.

**Exercise 4.17**   Figure 4.22 shows a snake robot with end-effectors at each end. Reference frames $\{b_1\}$ and $\{b_2\}$ are attached to the two end-effectors, as shown.

(a) Suppose that end-effector 1 is grasping a tree (which can be thought of as "ground") and end-effector 2 is free to move. Assume that the robot is in its zero configuration. Then $T_{b_1 b_2} \in SE(3)$ can be expressed in the following product of exponentials form:

$$T_{b_1 b_2} = e^{[\mathcal{S}_1]\theta_1} e^{[\mathcal{S}_2]\theta_2} \cdots e^{[\mathcal{S}_5]\theta_5} M.$$

Find $\mathcal{S}_3, \mathcal{S}_5$, and $M$.

(b) Now suppose that end-effector 2 is rigidly grasping a tree and end-effector 1 is free to move. Then $T_{b_2b_1} \in SE(3)$ can be expressed in the following product of exponentials form:

$$T_{b_2b_1} = e^{[\mathcal{A}_5]\theta_5}e^{[\mathcal{A}_4]\theta_4}e^{[\mathcal{A}_3]\theta_3}Ne^{[\mathcal{A}_2]\theta_2}e^{[\mathcal{A}_1]\theta_1}.$$

Find $\mathcal{A}_2, \mathcal{A}_4$, and $N$.

**Exercise 4.18** The two identical PUPR open chains of Figure 4.23 are shown in their zero position.

(a) In terms of the given fixed frame {A} and end-effector frame {a}, the forward kinematics for the robot on the left (robot A) can be expressed in the following product of exponentials form:

$$T_{Aa} = e^{[\mathcal{S}_1]\theta_1}e^{[\mathcal{S}_2]\theta_2}\cdots e^{[\mathcal{S}_5]\theta_5}M_a.$$

Find $\mathcal{S}_2$ and $\mathcal{S}_4$.

(b) Suppose that the end-effector of robot A is inserted into the end-effector of robot B in such a way that the origins of the end-effectors coincide; the two robots then form a single-loop closed chain. Then the configuration space of the single-loop closed chain can be expressed in the form

$$M = e^{-[\mathcal{B}_5]\phi_5}e^{-[\mathcal{B}_4]\phi_4}e^{-[\mathcal{B}_3]\phi_3}e^{-[\mathcal{B}_2]\phi_2}e^{-[\mathcal{B}_1]\phi_1}e^{[\mathcal{S}_1]\theta_1}e^{[\mathcal{S}_2]\theta_2}e^{[\mathcal{S}_3]\theta_3}e^{[\mathcal{S}_4]\theta_4}e^{[\mathcal{S}_5]\theta_5}$$

for some constant $M \in SE(3)$ and $\mathcal{B}_i = (\omega_i, v_i)$, for $i = 1, \ldots, 5$. Find $\mathcal{B}_3$, $\mathcal{B}_5$, and $M$. (Hint: Given any $A \in \mathbb{R}^{n \times n}, (e^A)^{-1} = e^{-A}$.)

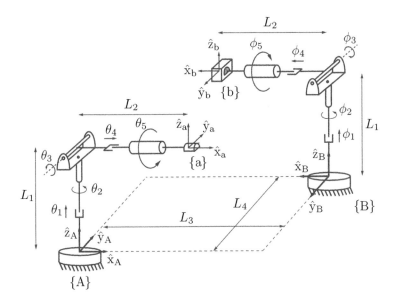

**Figure 4.23** Two PUPR open chains.

**Exercise 4.19**  The RRPRR spatial open chain of Figure 4.24 is shown in its zero position.

(a)  The forward kinematics can be expressed in the form

$$T_{sb} = M_1 e^{[\mathcal{A}_1]\theta_1} M_2 e^{[\mathcal{A}_2]\theta_2} \cdots M_5 e^{[\mathcal{A}_5]\theta_5}.$$

Find $M_2, M_3, \mathcal{A}_2$, and $\mathcal{A}_3$. (Hint: Appendix C may be helpful.)

(b)  Expressing the forward kinematics in the form

$$T_{sb} = e^{[\mathcal{S}_1]\theta_1} e^{[\mathcal{S}_2]\theta_2} \cdots e^{[\mathcal{S}_5]\theta_5} M,$$

find $M$ and $\mathcal{S}_1, \ldots, \mathcal{S}_5$ in terms of the quantities $M_1, \ldots, M_5$, $\mathcal{A}_1, \ldots, \mathcal{A}_5$ appearing in (a).

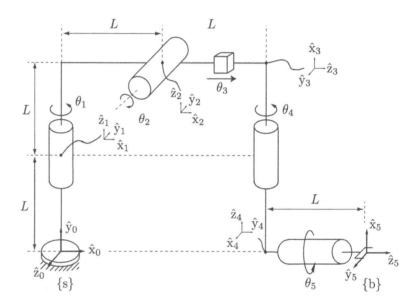

**Figure 4.24**  A spatial RRPRR open chain.

**Exercise 4.20**  The spatial PRRPRR open chain of Figure 4.25 is shown in its zero position, with space and end-effector frames chosen as indicated. Derive its forward kinematics in the form

$$T_{0n} = e^{[\mathcal{S}_1]\theta_1} e^{[\mathcal{S}_2]\theta_2} e^{[\mathcal{S}_3]\theta_3} e^{[\mathcal{S}_4]\theta_4} e^{[\mathcal{S}_5]\theta_5} M e^{[\mathcal{S}_6]\theta_6},$$

where $M \in SE(3)$.

**Exercise 4.21**  (Refer to Appendix C.) For each $T \in SE(3)$ below, find, if they exist, values for the four parameters $(\alpha, a, d, \phi)$ that satisfy

$$T = \text{Rot}(\hat{x}, \alpha) \, \text{Trans}(\hat{x}, a) \, \text{Trans}(\hat{z}, d) \, \text{Rot}(\hat{z}, \phi).$$

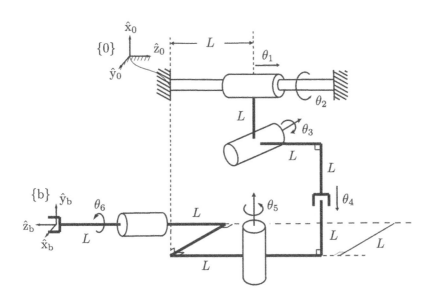

**Figure 4.25** A spatial PRRPRR open chain.

(a) $T = \begin{bmatrix} 0 & 1 & 1 & 3 \\ 1 & 0 & 0 & 0 \\ 0 & 1 & 0 & 1 \\ 0 & 0 & 0 & 1 \end{bmatrix}$.

(b) $T = \begin{bmatrix} \cos\beta & \sin\beta & 0 & 1 \\ \sin\beta & -\cos\beta & 0 & 0 \\ 0 & 0 & -1 & -2 \\ 0 & 0 & 0 & 1 \end{bmatrix}$.

(c) $T = \begin{bmatrix} 0 & -1 & 0 & -1 \\ 0 & 0 & -1 & 0 \\ 1 & 0 & 0 & 2 \\ 0 & 0 & 0 & 1 \end{bmatrix}$.

# 5 Velocity Kinematics and Statics

In the previous chapter we saw how to calculate the robot end-effector frame's position and orientation for a given set of joint positions. In this chapter we examine the related problem of calculating the twist of the end-effector of an open chain from a given set of joint positions and velocities.

Before we reach the representation of the end-effector twist as $\mathcal{V} \in \mathbb{R}^6$, starting in Section 5.1, let us consider the case where the end-effector configuration is represented by a minimal set of coordinates $x \in \mathbb{R}^m$ and the velocity is given by $\dot{x} = dx/dt \in \mathbb{R}^m$. In this case, the forward kinematics can be written as

$$x(t) = f(\theta(t)),$$

where $\theta \in \mathbb{R}^n$ is a set of joint variables. By the chain rule, the time derivative at time $t$ is

$$\dot{x} = \frac{\partial f(\theta)}{\partial \theta} \frac{d\theta(t)}{dt} = \frac{\partial f(\theta)}{\partial \theta} \dot{\theta}$$
$$= J(\theta)\dot{\theta},$$

where $J(\theta) \in \mathbb{R}^{m \times n}$ is called the **Jacobian**. The Jacobian matrix represents the linear sensitivity of the end-effector velocity $\dot{x}$ to the joint velocity $\dot{\theta}$, and it is a function of the joint variables $\theta$.

To provide a concrete example, consider a 2R planar open chain (left-hand side of Figure 5.1) with forward kinematics given by

$$x_1 = L_1 \cos\theta_1 + L_2 \cos(\theta_1 + \theta_2)$$
$$x_2 = L_1 \sin\theta_1 + L_2 \sin(\theta_1 + \theta_2).$$

Differentiating both sides with respect to time yields

$$\dot{x}_1 = -L_1\dot{\theta}_1 \sin\theta_1 - L_2(\dot{\theta}_1 + \dot{\theta}_2)\sin(\theta_1 + \theta_2)$$
$$\dot{x}_2 = L_1\dot{\theta}_1 \cos\theta_1 + L_2(\dot{\theta}_1 + \dot{\theta}_2)\cos(\theta_1 + \theta_2),$$

which can be rearranged into an equation of the form $\dot{x} = J(\theta)\dot{\theta}$:

$$\begin{bmatrix} \dot{x}_1 \\ \dot{x}_2 \end{bmatrix} = \begin{bmatrix} -L_1 \sin\theta_1 - L_2 \sin(\theta_1 + \theta_2) & -L_2 \sin(\theta_1 + \theta_2) \\ L_1 \cos\theta_1 + L_2 \cos(\theta_1 + \theta_2) & L_2 \cos(\theta_1 + \theta_2) \end{bmatrix} \begin{bmatrix} \dot{\theta}_1 \\ \dot{\theta}_2 \end{bmatrix}. \quad (5.1)$$

Writing the two columns of $J(\theta)$ as $J_1(\theta)$ and $J_2(\theta)$, and the tip velocity $\dot{x}$ as

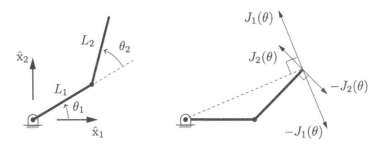

**Figure 5.1** (Left) A 2R robot arm. (Right) Columns 1 and 2 of the Jacobian correspond to the endpoint velocity when $\dot{\theta}_1 = 1$ (and $\dot{\theta}_2 = 0$) and when $\dot{\theta}_2 = 1$ (and $\dot{\theta}_1 = 0$), respectively.

$v_{\text{tip}}$, Equation (5.1) becomes

$$v_{\text{tip}} = J_1(\theta)\dot{\theta}_1 + J_2(\theta)\dot{\theta}_2. \tag{5.2}$$

As long as $J_1(\theta)$ and $J_2(\theta)$ are not collinear, it is possible to generate a tip velocity $v_{\text{tip}}$ in any arbitrary direction in the $x_1$–$x_2$-plane by choosing appropriate joint velocities $\dot{\theta}_1$ and $\dot{\theta}_2$. Since $J_1(\theta)$ and $J_2(\theta)$ depend on the joint values $\theta_1$ and $\theta_2$, one may ask whether there are any configurations at which $J_1(\theta)$ and $J_2(\theta)$ become collinear. For our example the answer is yes: if $\theta_2$ is $0°$ or $180°$ then, regardless of the value of $\theta_1$, $J_1(\theta)$ and $J_2(\theta)$ will be collinear and the Jacobian $J(\theta)$ becomes a singular matrix. Such configurations are therefore called **singularities**; they are characterized by a situation where the robot tip is unable to generate velocities in certain directions.

Now let's substitute $L_1 = L_2 = 1$ and consider the robot at two different nonsingular postures: $\theta = (0, \pi/4)$ and $\theta = (0, 3\pi/4)$. The Jacobians $J(\theta)$ at these two configurations are

$$J\left(\begin{bmatrix} 0 \\ \pi/4 \end{bmatrix}\right) = \begin{bmatrix} -0.71 & -0.71 \\ 1.71 & 0.71 \end{bmatrix} \quad \text{and} \quad J\left(\begin{bmatrix} 0 \\ 3\pi/4 \end{bmatrix}\right) = \begin{bmatrix} -0.71 & -0.71 \\ 0.29 & -0.71 \end{bmatrix}.$$

The right-hand side of Figure 5.1 illustrates the robot at the $\theta_2 = \pi/4$ configuration. Column $i$ of the Jacobian matrix, $J_i(\theta)$, corresponds to the tip velocity when $\dot{\theta}_i = 1$ and the other joint velocity is zero. These tip velocities (and therefore columns of the Jacobian) are indicated in Figure 5.1.

The Jacobian can be used to map bounds on the rotational speed of the joints to bounds on $v_{\text{tip}}$, as illustrated in Figure 5.2. Rather than mapping a polygon of joint velocities through the Jacobian as in Figure 5.2, we could instead map a unit circle of joint velocities in the $\theta_1$–$\theta_2$-plane. This circle represents an "iso-effort" contour in the joint velocity space, where total actuator effort is considered to be the sum of squares of the joint velocities. This circle maps through the Jacobian to an ellipse in the space of tip velocities, and this ellipse is referred to as the **manipulability ellipsoid**.[1] Figure 5.3 shows examples of this mapping

---

[1] A two-dimensional ellipsoid, as in our example, is commonly referred to as an ellipse.

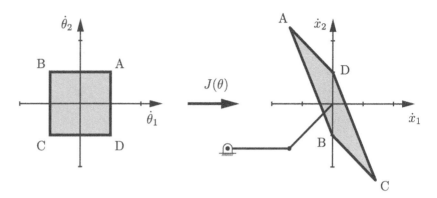

**Figure 5.2** Mapping the set of possible joint velocities, represented as a square in the $\dot{\theta}_1$–$\dot{\theta}_2$ space, through the Jacobian to find the parallelogram of possible end-effector velocities. The extreme points A, B, C, and D in the joint velocity space map to the extreme points A, B, C, and D in the end-effector velocity space.

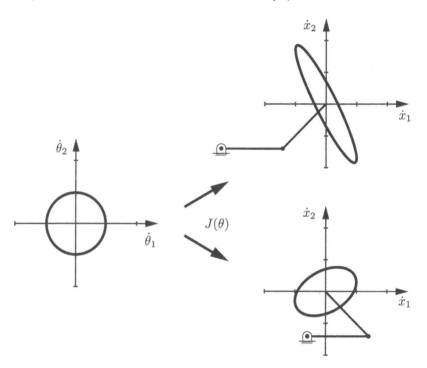

**Figure 5.3** Manipulability ellipsoids for two different postures of the 2R planar open chain.

for the two different postures of the 2R arm. As the manipulator configuration approaches a singularity, the ellipse collapses to a line segment, since the ability of the tip to move in one direction is lost.

Using the manipulability ellipsoid one can quantify how close a given posture

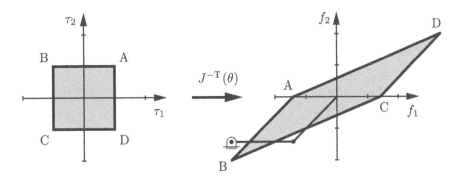

**Figure 5.4** Mapping joint torque bounds to tip force bounds.

is to a singularity. For example, we can compare the lengths of the major and minor principal semi-axes of the manipulability ellipsoid, respectively denoted $\ell_{max}$ and $\ell_{min}$. The closer the ellipsoid is to a circle, i.e., the closer the ratio $\ell_{max}/\ell_{min}$ is to 1, the more easily can the tip move in arbitrary directions and thus the more removed it is from a singularity.

The Jacobian also plays a central role in static analysis. Suppose that an external force is applied to the robot tip. What are the joint torques required to resist this external force?

This question can be answered via a conservation of power argument. Assuming that negligible power is used to move the robot,[2] the power measured at the robot's tip must equal the power generated at the joints. Denoting the tip force vector generated by the robot as $f_{tip}$ and the joint torque vector by $\tau$, the conservation of power then requires that

$$f_{tip}^{T} v_{tip} = \tau^{T} \dot{\theta},$$

for all arbitrary joint velocities $\dot{\theta}$. Since $v_{tip} = J(\theta)\dot{\theta}$, the equality

$$f_{tip}^{T} J(\theta)\dot{\theta} = \tau^{T}\dot{\theta}$$

must hold for all possible $\dot{\theta}$. This can only be true if

$$\tau = J^{T}(\theta)f_{tip}. \tag{5.3}$$

The joint torque $\tau$ needed to create the tip force $f_{tip}$ is calculated from the equation above.

For our two-link planar chain example, $J(\theta)$ is a square matrix dependent on $\theta$. If the configuration $\theta$ is not a singularity then both $J(\theta)$ and its transpose are invertible, and Equation (5.3) can be written

$$f_{tip} = ((J(\theta))^{T})^{-1}\tau = J^{-T}(\theta)\tau. \tag{5.4}$$

---

[2] Since the robot is at equilibrium, the joint velocity $\dot{\theta}$ is technically zero. This can be considered the limiting case as $\dot{\theta}$ approaches zero. To be more formal, we could invoke the "principle of virtual work" which deals with infinitesimal joint displacements instead of joint velocities.

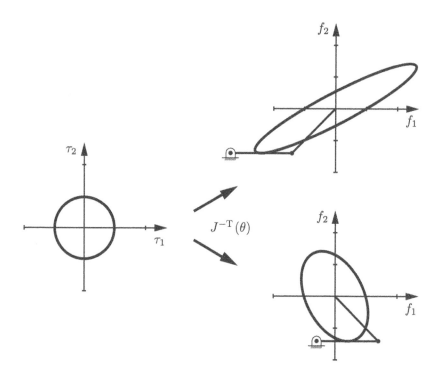

**Figure 5.5** Force ellipsoids for two different postures of the 2R planar open chain.

Using the equation above one can now determine, under the same static equilibrium assumption, what input torques are needed to generate a desired tip force, e.g., the joint torques needed for the robot tip to push against a wall with a specified normal force. For a given posture $\theta$ of the robot at equilibrium and a set of joint torque limits such as

$$-1 \, \mathrm{Nm} \leq \tau_1 \leq 1 \, \mathrm{Nm},$$
$$-1 \, \mathrm{Nm} \leq \tau_2 \leq 1 \, \mathrm{Nm},$$

then Equation (5.4) can be used to generate the set of all possible tip forces as indicated in Figure 5.4.

As for the manipulability ellipsoid, a **force ellipsoid** can be drawn by mapping a unit circle "iso-effort" contour in the $\tau_1$–$\tau_2$-plane to an ellipsoid in the $f_1$–$f_2$ tip-force plane via the Jacobian transpose inverse $J^{-\mathrm{T}}(\theta)$ (see Figure 5.5). The force ellipsoid illustrates how easily the robot can generate forces in different directions. As is evident from the manipulability and force ellipsoids, if it is easy to generate a tip velocity in a given direction then it is difficult to generate a force in that same direction, and vice versa (Figure 5.6). In fact, for a given robot configuration, the principal axes of the manipulability ellipsoid and force ellipsoid are aligned, and the lengths of the principal semi-axes of the force ellipsoid are

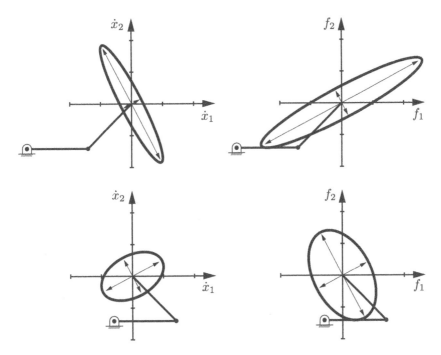

**Figure 5.6** Left-hand column: Manipulability ellipsoids at two different arm configurations. Right-hand column: The force ellipsoids for the same two arm configurations.

the reciprocals of the lengths of the principal semi-axes of the manipulability ellipsoid.

At a singularity, the manipulability ellipsoid collapses to a line segment. The force ellipsoid, on the other hand, becomes infinitely long in a direction orthogonal to the manipulability ellipsoid line segment (i.e., the direction of the aligned links) and skinny in the orthogonal direction. Consider, for example, carrying a heavy suitcase with your arm. It is much easier if your arm hangs straight down under gravity (with your elbow fully straightened at a singularity), because the force you must support passes directly through your joints, therefore requiring no torques about the joints. Only the joint structure bears the load, not the muscles generating torques. The manipulability ellipsoid loses dimension at a singularity and therefore its area drops to zero, but the force ellipsoid's area goes to infinity (assuming that the joints can support the load!).

In this chapter we present methods for deriving the Jacobian for general open chains, where the configuration of the end-effector is expressed as $T \in SE(3)$ and its velocity is expressed as a twist $\mathcal{V}$ in the fixed base frame or the end-effector body frame. We also examine how the Jacobian can be used for velocity and static analysis, including identifying kinematic singularities and determining the manipulability and force ellipsoids. Later chapters on inverse kinematics, motion planning, dynamics, and control make extensive use of the Jacobian and related notions introduced in this chapter.

## 5.1     Manipulator Jacobian

In the 2R planar open chain example, we saw that, for any joint configuration $\theta$, the tip velocity vector $v_{\text{tip}}$ and joint velocity vector $\dot{\theta}$ are linearly related via the Jacobian matrix $J(\theta)$, i.e., $v_{\text{tip}} = J(\theta)\dot{\theta}$. The tip velocity $v_{\text{tip}}$ depends on the coordinates of interest for the tip, which in turn determine the specific form of the Jacobian. For example, in the most general case $v_{\text{tip}}$ can be taken to be a six-dimensional twist, while, for pure orienting devices such as a wrist, $v_{\text{tip}}$ is usually taken to be the angular velocity of the end-effector frame. Other choices for $v_{\text{tip}}$ lead to different formulations for the Jacobian. We begin with the general case where $v_{\text{tip}}$ is taken to be a six-dimensional twist $\mathcal{V}$.

All the derivations below are mathematical expressions of the same simple idea, embodied in Equation (5.2): given the configuration $\theta$ of the robot, the 6-vector $J_i(\theta)$, which is column $i$ of $J(\theta)$, is the twist $\mathcal{V}$ when $\dot{\theta}_i = 1$ and all other joint velocities are zero. This twist is determined in the same way as the joint screw axes were determined in the previous chapter, using a point $q_i$ on joint axis $i$ for revolute joints. The only difference is that the screw axes of the Jacobian depend on the joint variables $\theta$ whereas the screw axes for the forward kinematics of Chapter 4 were always for the case $\theta = 0$.

The two standard types of Jacobian that we will consider are: the space Jacobian $J_s(\theta)$ satisfying $\mathcal{V}_s = J_s(\theta)\dot{\theta}$, where each column $J_{si}(\theta)$ corresponds to a screw axis expressed in the fixed space frame {s}; and the body Jacobian $J_b(\theta)$ satisfying $\mathcal{V}_b = J_b(\theta)\dot{\theta}$, where each column $J_{bi}(\theta)$ corresponds to a screw axis expressed in the end-effector frame {b}. We start with the space Jacobian.

### 5.1.1     Space Jacobian

In this section we derive the relationship between an open chain's joint velocity vector $\dot{\theta}$ and the end-effector's spatial twist $\mathcal{V}_s$. We first recall a few basic properties from linear algebra and linear differential equations: (i) if $A, B \in \mathbb{R}^{n \times n}$ are both invertible then $(AB)^{-1} = B^{-1}A^{-1}$; (ii) if $A \in \mathbb{R}^{n \times n}$ is constant and $\theta(t)$ is a scalar function of $t$ then $d(e^{A\theta})/dt = Ae^{A\theta}\dot{\theta} = e^{A\theta}A\dot{\theta}$; (iii) $(e^{A\theta})^{-1} = e^{-A\theta}$.

Consider an $n$-link open chain whose forward kinematics is expressed in the following product of exponentials form:

$$T(\theta_1, \ldots, \theta_n) = e^{[\mathcal{S}_1]\theta_1} e^{[\mathcal{S}_2]\theta_2} \cdots e^{[\mathcal{S}_n]\theta_n} M. \tag{5.5}$$

The spatial twist $\mathcal{V}_s$ is given by $[\mathcal{V}_s] = \dot{T}T^{-1}$, where

$$\dot{T} = \left(\frac{d}{dt}e^{[\mathcal{S}_1]\theta_1}\right)\cdots e^{[\mathcal{S}_n]\theta_n}M + e^{[\mathcal{S}_1]\theta_1}\left(\frac{d}{dt}e^{[\mathcal{S}_2]\theta_2}\right)\cdots e^{[\mathcal{S}_n]\theta_n}M + \cdots$$

$$= [\mathcal{S}_1]\dot{\theta}_1 e^{[\mathcal{S}_1]\theta_1}\cdots e^{[\mathcal{S}_n]\theta_n}M + e^{[\mathcal{S}_1]\theta_1}[\mathcal{S}_2]\dot{\theta}_2 e^{[\mathcal{S}_2]\theta_2}\cdots e^{[\mathcal{S}_n]\theta_n}M + \cdots$$

Also,
$$T^{-1} = M^{-1} e^{-[\mathcal{S}_n]\theta_n} \cdots e^{-[\mathcal{S}_1]\theta_1}.$$

Calculating $\dot{T}T^{-1}$ we obtain

$$[\mathcal{V}_s] = [\mathcal{S}_1]\dot{\theta}_1 + e^{[\mathcal{S}_1]\theta_1}[\mathcal{S}_2]e^{-[\mathcal{S}_1]\theta_1}\dot{\theta}_2 + e^{[\mathcal{S}_1]\theta_1}e^{[\mathcal{S}_2]\theta_2}[\mathcal{S}_3]e^{-[\mathcal{S}_2]\theta_2}e^{-[\mathcal{S}_1]\theta_1}\dot{\theta}_3 + \cdots.$$

The above can also be expressed in vector form by means of the adjoint mapping:

$$\mathcal{V}_s = \underbrace{\mathcal{S}_1}_{J_{s1}}\dot{\theta}_1 + \underbrace{\mathrm{Ad}_{e^{[\mathcal{S}_1]\theta_1}}(\mathcal{S}_2)}_{J_{s2}}\dot{\theta}_2 + \underbrace{\mathrm{Ad}_{e^{[\mathcal{S}_1]\theta_1}e^{[\mathcal{S}_2]\theta_2}}(\mathcal{S}_3)}_{J_{s3}}\dot{\theta}_3 + \cdots \qquad (5.6)$$

Observe that $\mathcal{V}_s$ is a sum of $n$ spatial twists of the form

$$\mathcal{V}_s = J_{s1}\dot{\theta}_1 + J_{s2}(\theta)\dot{\theta}_2 + \cdots + J_{sn}(\theta)\dot{\theta}_n, \qquad (5.7)$$

where each $J_{si}(\theta) = (\omega_{si}(\theta), v_{si}(\theta))$ depends explictly on the joint values $\theta \in \mathbb{R}^n$ for $i = 2, \ldots, n$. In matrix form,

$$\begin{aligned}
\mathcal{V}_s &= \begin{bmatrix} J_{s1} & J_{s2}(\theta) & \cdots & J_{sn}(\theta) \end{bmatrix} \begin{bmatrix} \dot{\theta}_1 \\ \vdots \\ \dot{\theta}_n \end{bmatrix} \\
&= J_s(\theta)\dot{\theta}.
\end{aligned} \qquad (5.8)$$

The matrix $J_s(\theta)$ is said to be the **Jacobian** in fixed (space) frame coordinates, or more simply the **space Jacobian**.

**Definition 5.1** Let the forward kinematics of an $n$-link open chain be expressed in the following product of exponentials form:

$$T = e^{[\mathcal{S}_1]\theta_1} \cdots e^{[\mathcal{S}_n]\theta_n} M. \qquad (5.9)$$

The **space Jacobian** $J_s(\theta) \in \mathbb{R}^{6 \times n}$ relates the joint rate vector $\dot{\theta} \in \mathbb{R}^n$ to the spatial twist $\mathcal{V}_s$ via

$$\mathcal{V}_s = J_s(\theta)\dot{\theta}. \qquad (5.10)$$

The $i$th column of $J_s(\theta)$ is

$$J_{si}(\theta) = \mathrm{Ad}_{e^{[\mathcal{S}_1]\theta_1}\cdots e^{[\mathcal{S}_{i-1}]\theta_{i-1}}}(\mathcal{S}_i), \qquad (5.11)$$

for $i = 2, \ldots, n$, with the first column $J_{s1} = \mathcal{S}_1$.

To understand the physical meaning behind the columns of $J_s(\theta)$, observe that the $i$th column is of the form $\mathrm{Ad}_{T_{i-1}}(\mathcal{S}_i)$, where $T_{i-1} = e^{[\mathcal{S}_1]\theta_1} \cdots e^{[\mathcal{S}_{i-1}]\theta_{i-1}}$; recall that $\mathcal{S}_i$ is the screw axis describing the $i$th joint axis in terms of the fixed frame with the robot in its zero position. $\mathrm{Ad}_{T_{i-1}}(\mathcal{S}_i)$ is therefore the screw axis describing the $i$th joint axis after it undergoes the rigid body displacement $T_{i-1}$. Physically this is the same as moving the first $i-1$ joints from their zero position to the current values $\theta_1, \ldots, \theta_{i-1}$. Therefore, the $i$th column $J_{si}(\theta)$ of $J_s(\theta)$ is simply the screw vector describing joint axis $i$, expressed in fixed-frame coordinates, as a function of the joint variables $\theta_1, \ldots, \theta_{i-1}$.

In summary, the procedure for determining the columns $J_{si}$ of $J_s(\theta)$ is similar to the procedure for deriving the joint screws $\mathcal{S}_i$ in the product of exponentials formula $e^{[\mathcal{S}_1]\theta_1}\cdots e^{[\mathcal{S}_n]\theta_n}M$: each column $J_{si}(\theta)$ is the screw vector describing joint axis $i$, expressed in fixed-frame coordinates, but for arbitrary $\theta$ rather than $\theta = 0$.

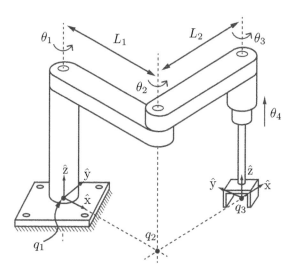

**Figure 5.7** Space Jacobian for a spatial RRRP chain.

**Example 5.2** (Space Jacobian for a spatial RRRP chain)  We now illustrate the procedure for finding the space Jacobian for the spatial RRRP chain of Figure 5.7. Denote the $i$th column of $J_s(\theta)$ by $J_{si} = (\omega_{si}, v_{si})$. The $[\text{Ad}_{T_{i-1}}]$ matrices are implicit in our calculations of the screw axes of the displaced joint axes.

- Observe that $\omega_{s1}$ is constant and in the $\hat{z}_s$-direction: $\omega_{s1} = (0,0,1)$. Choosing $q_1$ as the origin, $v_{s1} = (0,0,0)$.
- $\omega_{s2}$ is also constant in the $\hat{z}_s$-direction, so $\omega_{s2} = (0,0,1)$. Choose $q_2$ as the point $(L_1 c_1, L_1 s_1, 0)$, where $c_1 = \cos\theta_1$, $s_1 = \sin\theta_1$. Then $v_{s2} = -\omega_2 \times q_2 = (L_1 s_1, -L_1 c_1, 0)$.
- The direction of $\omega_{s3}$ is always fixed in the $\hat{z}_s$-direction regardless of the values of $\theta_1$ and $\theta_2$, so $\omega_{s3} = (0,0,1)$. Choosing $q_3 = (L_1 c_1 + L_2 c_{12}, L_1 s_1 + L_2 s_{12}, 0)$, where $c_{12} = \cos(\theta_1 + \theta_2)$, $s_{12} = \sin(\theta_1 + \theta_2)$, it follows that $v_{s3} = (L_1 s_1 + L_2 s_{12}, -L_1 c_1 - L_2 c_{12}, 0)$.
- Since the final joint is prismatic, $\omega_{s4} = (0,0,0)$, and the joint-axis direction is given by $v_{s4} = (0,0,1)$.

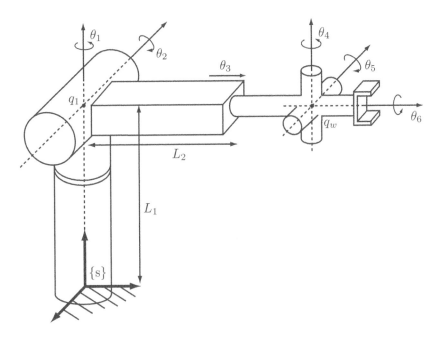

**Figure 5.8** Space Jacobian for the spatial RRPRRR chain.

The space Jacobian is therefore

$$
J_s(\theta) =
\begin{bmatrix}
0 & 0 & 0 & 0 \\
0 & 0 & 0 & 0 \\
1 & 1 & 1 & 0 \\
0 & L_1 s_1 & L_1 s_1 + L_2 s_{12} & 0 \\
0 & -L_1 c_1 & -L_1 c_1 - L_2 c_{12} & 0 \\
0 & 0 & 0 & 1
\end{bmatrix}.
$$

**Example 5.3** (Space Jacobian for spatial RRPRRR chain)    We now derive the space Jacobian for the spatial RRPRRR chain of Figure 5.8. The base frame is chosen as shown in the figure.

- The first joint axis is in the direction $\omega_{s1} = (0, 0, 1)$. Choosing $q_1 = (0, 0, L_1)$, we get $v_{s1} = -\omega_1 \times q_1 = (0, 0, 0)$.
- The second joint axis is in the direction $\omega_{s2} = (-c_1, -s_1, 0)$. Choosing $q_2 = (0, 0, L_1)$, we get $v_{s2} = -\omega_2 \times q_2 = (L_1 s_1, -L_1 c_1, 0)$.
- The third joint is prismatic, so $\omega_{s3} = (0, 0, 0)$. The direction of the prismatic joint axis is given by

$$
v_{s3} = \mathrm{Rot}(\hat{z}, \theta_1)\mathrm{Rot}(\hat{x}, -\theta_2)
\begin{bmatrix}
0 \\
1 \\
0
\end{bmatrix}
=
\begin{bmatrix}
-s_1 c_2 \\
c_1 c_2 \\
-s_2
\end{bmatrix}.
$$

- Now consider the wrist portion of the chain. The wrist center is located at the point

$$
q_w = \begin{bmatrix} 0 \\ 0 \\ L_1 \end{bmatrix} + \mathrm{Rot}(\hat{z}, \theta_1)\mathrm{Rot}(\hat{x}, -\theta_2)\begin{bmatrix} 0 \\ L_2 + \theta_3 \\ 0 \end{bmatrix} = \begin{bmatrix} -(L_2 + \theta_3)s_1c_2 \\ (L_2 + \theta_3)c_1c_2 \\ L_1 - (L_2 + \theta_3)s_2 \end{bmatrix}.
$$

Observe that the directions of the wrist axes depend on $\theta_1$, $\theta_2$, and the preceding wrist axes. These are

$$
\omega_{s4} = \mathrm{Rot}(\hat{z}, \theta_1)\mathrm{Rot}(\hat{x}, -\theta_2)\begin{bmatrix} 0 \\ 0 \\ 1 \end{bmatrix} = \begin{bmatrix} -s_1s_2 \\ c_1s_2 \\ c_2 \end{bmatrix},
$$

$$
\omega_{s5} = \mathrm{Rot}(\hat{z}, \theta_1)\mathrm{Rot}(\hat{x}, -\theta_2)\mathrm{Rot}(\hat{z}, \theta_4)\begin{bmatrix} -1 \\ 0 \\ 0 \end{bmatrix} = \begin{bmatrix} -c_1c_4 + s_1c_2s_4 \\ -s_1c_4 - c_1c_2s_4 \\ s_2s_4 \end{bmatrix},
$$

$$
\omega_{s6} = \mathrm{Rot}(\hat{z}, \theta_1)\mathrm{Rot}(\hat{x}, -\theta_2)\mathrm{Rot}(\hat{z}, \theta_4)\mathrm{Rot}(\hat{x}, -\theta_5)\begin{bmatrix} 0 \\ 1 \\ 0 \end{bmatrix}
$$

$$
= \begin{bmatrix} -c_5(s_1c_2c_4 + c_1s_4) + s_1s_2s_5 \\ c_5(c_1c_2c_4 - s_1s_4) - c_1s_2s_5 \\ -s_2c_4c_5 - c_2s_5 \end{bmatrix}.
$$

The space Jacobian can now be computed and written in matrix form as follows:

$$
J_s(\theta) = \begin{bmatrix} \omega_{s1} & \omega_{s2} & 0 & \omega_{s4} & \omega_{s5} & \omega_{s6} \\ 0 & -\omega_{s2} \times q_2 & v_{s3} & -\omega_{s4} \times q_w & -\omega_{s5} \times q_w & -\omega_{s6} \times q_w \end{bmatrix}.
$$

Note that we were able to obtain the entire Jacobian directly, without having to explicitly differentiate the forward kinematic map.

### 5.1.2    Body Jacobian

In the previous section we derived the relationship between the joint rates and $[\mathcal{V}_s] = \dot{T}T^{-1}$, the end-effector's twist expressed in fixed-frame coordinates. Here we derive the relationship between the joint rates and $[\mathcal{V}_b] = T^{-1}\dot{T}$, the end-effector twist in end-effector-frame coordinates. For this purpose it will be more convenient to express the forward kinematics in the alternative product of exponentials form:

$$
T(\theta) = Me^{[\mathcal{B}_1]\theta_1}e^{[\mathcal{B}_2]\theta_2}\cdots e^{[\mathcal{B}_n]\theta_n}. \tag{5.12}
$$

Computing $\dot{T}$,

$$\dot{T} = M e^{[\mathcal{B}_1]\theta_1} \cdots e^{[\mathcal{B}_{n-1}]\theta_{n-1}} \left( \frac{d}{dt} e^{[\mathcal{B}_n]\theta_n} \right)$$

$$+ M e^{[\mathcal{B}_1]\theta_1} \cdots \left( \frac{d}{dt} e^{[\mathcal{B}_{n-1}]\theta_{n-1}} \right) e^{[\mathcal{B}_n]\theta_n} + \cdots$$

$$= M e^{[\mathcal{B}_1]\theta_1} \cdots e^{[\mathcal{B}_n]\theta_n} [\mathcal{B}_n] \dot{\theta}_n$$

$$+ M e^{[\mathcal{B}_1]\theta_1} \cdots e^{[\mathcal{B}_{n-1}]\theta_{n-1}} [\mathcal{B}_{n-1}] e^{[\mathcal{B}_n]\theta_n} \dot{\theta}_{n-1} + \cdots$$

$$+ M e^{[\mathcal{B}_1]\theta_1} [\mathcal{B}_1] e^{[\mathcal{B}_2]\theta_2} \cdots e^{[\mathcal{B}_n]\theta_n} \dot{\theta}_1.$$

Also,

$$T^{-1} = e^{-[\mathcal{B}_n]\theta_n} \cdots e^{-[\mathcal{B}_1]\theta_1} M^{-1}.$$

Evaluating $T^{-1}\dot{T}$,

$$[\mathcal{V}_b] = [\mathcal{B}_n] \dot{\theta}_n + e^{-[\mathcal{B}_n]\theta_n} [\mathcal{B}_{n-1}] e^{[\mathcal{B}_n]\theta_n} \dot{\theta}_{n-1} + \cdots$$

$$+ e^{-[\mathcal{B}_n]\theta_n} \cdots e^{-[\mathcal{B}_2]\theta_2} [\mathcal{B}_1] e^{[\mathcal{B}_2]\theta_2} \cdots e^{[\mathcal{B}_n]\theta_n} \dot{\theta}_1$$

or, in vector form,

$$\mathcal{V}_b = \underbrace{\mathcal{B}_n}_{J_{bn}} \dot{\theta}_n + \underbrace{\text{Ad}_{e^{-[\mathcal{B}_n]\theta_n}}(\mathcal{B}_{n-1})}_{J_{b,n-1}} \dot{\theta}_{n-1} + \cdots + \underbrace{\text{Ad}_{e^{-[\mathcal{B}_n]\theta_n} \cdots e^{-[\mathcal{B}_2]\theta_2}}(\mathcal{B}_1)}_{J_{b1}} \dot{\theta}_1. \quad (5.13)$$

The twist $\mathcal{V}_b$ can therefore be expressed as a sum of $n$ body twists:

$$\mathcal{V}_b = J_{b1}(\theta)\dot{\theta}_1 + \cdots + J_{bn-1}(\theta)\dot{\theta}_{n-1} + J_{bn}\dot{\theta}_n, \quad (5.14)$$

where each $J_{bi}(\theta) = (\omega_{bi}(\theta), v_{bi}(\theta))$ depends explictly on the joint values $\theta$ for $i = 1, \ldots, n-1$. In matrix form,

$$\mathcal{V}_b = \begin{bmatrix} J_{b1}(\theta) & \cdots & J_{bn-1}(\theta) & J_{bn} \end{bmatrix} \begin{bmatrix} \dot{\theta}_1 \\ \vdots \\ \dot{\theta}_n \end{bmatrix} = J_b(\theta)\dot{\theta}. \quad (5.15)$$

The matrix $J_b(\theta)$ is the Jacobian in the end-effector- (or body-) frame coordinates and is called, more simply, the **body Jacobian**.

**Definition 5.4** Let the forward kinematics of an $n$-link open chain be expressed in the following product of exponentials form:

$$T = M e^{[\mathcal{B}_1]\theta_1} \cdots e^{[\mathcal{B}_n]\theta_n}. \quad (5.16)$$

The **body Jacobian** $J_b(\theta) \in \mathbb{R}^{6 \times n}$ relates the joint rate vector $\dot{\theta} \in \mathbb{R}^n$ to the end-effector twist $\mathcal{V}_b = (\omega_b, v_b)$ via

$$\mathcal{V}_b = J_b(\theta)\dot{\theta}. \quad (5.17)$$

The $i$th column of $J_b(\theta)$ is

$$J_{bi}(\theta) = \text{Ad}_{e^{-[\mathcal{B}_n]\theta_n} \cdots e^{-[\mathcal{B}_{i+1}]\theta_{i+1}}}(\mathcal{B}_i), \quad (5.18)$$

for $i = n-1, \ldots, 1$, with $J_{bn} = \mathcal{B}_n$.

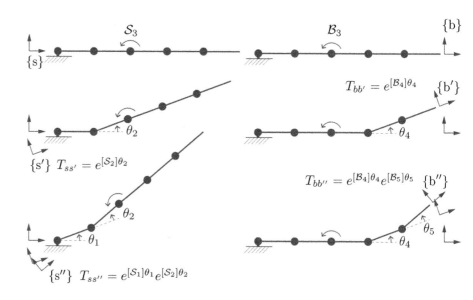

**Figure 5.9** A 5R robot. (Left-hand column) Derivation of $J_{s3}$, the third column of the space Jacobian. (Right-hand column) Derivation of $J_{b3}$, the third column of the body Jacobian.

A physical interpretation can be given to the columns of $J_b(\theta)$: each column $J_{bi}(\theta) = (\omega_{bi}(\theta), v_{bi}(\theta))$ of $J_b(\theta)$ is the screw vector for joint axis $i$, expressed in the coordinates of the end-effector frame rather than those of the fixed frame. The procedure for determining the columns of $J_b(\theta)$ is similar to the procedure for deriving the forward kinematics in the product of exponentials form $Me^{[\mathcal{B}_1]\theta_1} \cdots e^{[\mathcal{B}_n]\theta_n}$, the only difference being that each of the end-effector-frame joint screws $J_{bi}(\theta)$ are expressed for arbitrary $\theta$ rather than $\theta = 0$.

### 5.1.3    Visualizing the Space and Body Jacobian

Another, perhaps simpler, way to derive the formulas for the $i$th column of the space Jacobian (5.11) and the $i$th column of the body Jacobian (5.18) comes from inspecting the 5R robot in Figure 5.9. Let's start with the third column, $J_{s3}$, of the space Jacobian using the left-hand column of Figure 5.9.

The screw corresponding to joint axis 3 is written as $\mathcal{S}_3$ in $\{s\}$ when the robot is at its zero configuration. Clearly the joint variables $\theta_3$, $\theta_4$, and $\theta_5$ have no impact on the spatial twist resulting from the joint velocity $\dot{\theta}_3$, because they do not displace axis 3 relative to $\{s\}$. So we fix those joint variables at zero, making the robot from joint 2 outward a rigid body $B$. If $\theta_1 = 0$ and $\theta_2$ is arbitrary then the frame $\{s'\}$ at $T_{ss'} = e^{[\mathcal{S}_2]\theta_2}$ is at the same position and orientation relative to $B$ as frame $\{s\}$ when $\theta_1 = \theta_2 = 0$. Now, if $\theta_1$ is also arbitrary then the frame $\{s''\}$ at $T_{ss''} = e^{[\mathcal{S}_1]\theta_1}e^{[\mathcal{S}_2]\theta_2}$ is at the same position and orientation relative to $B$ as frame $\{s\}$ when $\theta_1 = \theta_2 = 0$. Thus $\mathcal{S}_3$ represents the screw relative to $\{s''\}$ for arbitrary joint angles $\theta_1$ and $\theta_2$. The column $J_{s3}$, though, is the screw relative to {s}. The mapping that changes the frame of representation of $\mathcal{S}_3$ from {s''} to {s} is $[\mathrm{Ad}_{T_{ss''}}] = [\mathrm{Ad}_{e^{[\mathcal{S}_1]\theta_1}e^{[\mathcal{S}_2]\theta_2}}]$, i.e., $J_{s3} = [\mathrm{Ad}_{T_{ss''}}]\mathcal{S}_3$, precisely Equation (5.11) for joint $i = 3$. Equation (5.11) is the generalization of the reasoning above for any joint $i = 2, \ldots, n$.

Now let's derive the third column, $J_{b3}$, of the body Jacobian by inspecting the right-hand column of Figure 5.9. The screw corresponding to joint 3 is written $\mathcal{B}_3$ in {b} when the robot is at its zero configuration. Clearly the joint variables $\theta_1$, $\theta_2$, and $\theta_3$ have no impact on the body twist resulting from the joint velocity $\dot\theta_3$, because they do not displace axis 3 relative to {b}. So we fix those joint variables at zero, making the robot a rigid body $B$ from the base to joint 4. If $\theta_5 = 0$ and $\theta_4$ is arbitrary, then the frame {b'} at $T_{bb'} = e^{[\mathcal{B}_4]\theta_4}$ is the new end-effector frame. Now if $\theta_5$ is also arbitrary, then the frame {b''} at $T_{bb''} = e^{[\mathcal{B}_4]\theta_4}e^{[\mathcal{B}_5]\theta_5}$ is the new end-effector frame. The column $J_{b3}$ is simply the screw axis of joint 3 expressed in {b''}. Since $\mathcal{B}_3$ is expressed in {b}, we have

$$
\begin{aligned}
J_{b3} &= [\mathrm{Ad}_{T_{b''b}}]\mathcal{B}_3 \\
&= [\mathrm{Ad}_{T_{bb''}^{-1}}]\mathcal{B}_3 \\
&= [\mathrm{Ad}_{e^{-[\mathcal{B}_5]\theta_5}e^{-[\mathcal{B}_4]\theta_4}}]\mathcal{B}_3.
\end{aligned}
$$

where we have made use of the fact that $(T_1 T_2)^{-1} = T_2^{-1}T_1^{-1}$. This formula for $J_{b3}$ is precisely Equation (5.18) for joint $i = 3$. Equation (5.18) is the generalization of the reasoning above for any joint $i = 1, \ldots, n - 1$.

## .1.4   Relationship between the Space and Body Jacobian

Denoting the fixed frame by {s} and the end-effector frame by {b}, the forward kinematics can be written as $T_{sb}(\theta)$. The twist of the end-effector frame can be written in terms of the fixed- and end-effector-frame coordinates as

$$
\begin{aligned}
[\mathcal{V}_s] &= \dot T_{sb}T_{sb}^{-1}, \\
[\mathcal{V}_b] &= T_{sb}^{-1}\dot T_{sb},
\end{aligned}
$$

with $\mathcal{V}_s$ and $\mathcal{V}_b$ related by $\mathcal{V}_s = \mathrm{Ad}_{T_{sb}}(\mathcal{V}_b)$ and $\mathcal{V}_b = \mathrm{Ad}_{T_{bs}}(\mathcal{V}_s)$. The twists $\mathcal{V}_s$ and $\mathcal{V}_b$ are also related to their respective Jacobians via

$$
\mathcal{V}_s = J_s(\theta)\dot\theta, \tag{5.19}
$$
$$
\mathcal{V}_b = J_b(\theta)\dot\theta. \tag{5.20}
$$

Equation (5.19) can therefore be written

$$
\mathrm{Ad}_{T_{sb}}(\mathcal{V}_b) = J_s(\theta)\dot\theta. \tag{5.21}
$$

Applying $[\mathrm{Ad}_{T_{bs}}]$ to both sides of Equation (5.21) and using the general property $[\mathrm{Ad}_X][\mathrm{Ad}_Y] = [\mathrm{Ad}_{XY}]$ of the adjoint map, we obtain

$$
\mathrm{Ad}_{T_{bs}}(\mathrm{Ad}_{T_{sb}}(\mathcal{V}_b)) = \mathrm{Ad}_{T_{bs}T_{sb}}(\mathcal{V}_b) = \mathcal{V}_b = \mathrm{Ad}_{T_{bs}}(J_s(q)\dot\theta).
$$

Since we also have $\mathcal{V}_b = J_b(\theta)\dot{\theta}$ for all $\dot{\theta}$, it follows that $J_s(\theta)$ and $J_b(\theta)$ are related by

$$J_b(\theta) = \mathrm{Ad}_{T_{bs}}\left(J_s(\theta)\right) = [\mathrm{Ad}_{T_{bs}}]J_s(\theta). \tag{5.22}$$

The space Jacobian can in turn be obtained from the body Jacobian via

$$J_s(\theta) = \mathrm{Ad}_{T_{sb}}\left(J_b(\theta)\right) = [\mathrm{Ad}_{T_{sb}}]J_b(\theta). \tag{5.23}$$

The fact that the space and body Jacobians, and the space and body twists, are similarly related by the adjoint map should not be surprising since each column of the space or body Jacobian corresponds to a twist.

An important implication of Equations (5.22) and (5.23) is that $J_b(\theta)$ and $J_s(\theta)$ always have the same rank; this is shown explicitly in Section 5.3 on singularity analysis.

### 5.1.5    Alternative Notions of the Jacobian

The space and body Jacobians derived above are matrices that relate joint rates to the twist of the end-effector. There exist alternative notions of the Jacobian that are based on a representation of the end-effector configuration using a minimum set of coordinates $q$. Such representations are particularly relevant when the task space is considered to be a subspace of $SE(3)$. For example, the configuration of the end-effector of a planar robot could be treated as $q = (x, y, \theta) \in \mathbb{R}^3$ instead of as an element of $SE(2)$.

When using a minimum set of coordinates, the end-effector velocity is not given by a twist $\mathcal{V}$ but by the time derivative of the coordinates $\dot{q}$, and the Jacobian $J_a$ in the velocity kinematics $\dot{q} = J_a(\theta)\dot{\theta}$ is sometimes called the **analytic Jacobian** as opposed to the **geometric Jacobian** in space and body form, described above.[3]

For an $SE(3)$ task space, a typical choice of the minimal coordinates $q \in \mathbb{R}^6$ includes three coordinates for the origin of the end-effector frame in the fixed frame and three coordinates for the orientation of the end-effector frame in the fixed frame. Example coordinates for the orientation include the Euler angles (see Appendix B) and exponential coordinates for rotation.

**Example 5.5** (Analytic Jacobian with exponential coordinates for rotation)    In this example, we find the relationship between the geometric Jacobian $J_b$ in the body frame and an analytic Jacobian $J_a$ that uses exponential coordinates $r = \hat{\omega}\theta$ to represent the orientation. (Recall that $\|\hat{\omega}\| = 1$ and $\theta \in [0, \pi]$.)

---

[3] The term "geometric Jacobian" has also been used to describe the relationship between joint rates and a representation of the end-effector velocity that combines the rate of change of the position coordinates of the end-effector (which is neither the linear portion of a body twist nor the linear portion of a spatial twist) and a representation of the angular velocity. Unlike a body or spatial twist, which depends only on the body or space frame, respectively, this "hybrid" notion of a spatial velocity depends on the definition of both frames.

First, consider an open chain with $n$ joints and the body Jacobian

$$\mathcal{V}_b = J_b(\theta)\dot{\theta},$$

where $J_b(\theta) \in \mathbb{R}^{6 \times n}$. The angular and linear velocity components of $\mathcal{V}_b = (\omega_b, v_b)$ can be written explicitly as

$$\mathcal{V}_b = \begin{bmatrix} \omega_b \\ v_b \end{bmatrix} = J_b(\theta)\dot{\theta} = \begin{bmatrix} J_\omega(\theta) \\ J_v(\theta) \end{bmatrix} \dot{\theta},$$

where $J_\omega$ is the $3 \times n$ matrix corresponding to the top three rows of $J_b$ and $J_v$ is the $3 \times n$ matrix corresponding to the bottom three rows of $J_b$.

Now suppose that our minimal set of coordinates $q \in \mathbb{R}^6$ is given by $q = (r, x)$, where $x \in \mathbb{R}^3$ is the position of the origin of the end-effector frame and $r = \hat{\omega}\theta \in \mathbb{R}^3$ is the exponential coordinate representation for the rotation. The coordinate time derivative $\dot{x}$ is related to $v_b$ by a rotation that gives $v_b$ in the fixed coordinates:

$$\dot{x} = R_{sb}(\theta)v_b = R_{sb}(\theta)J_v(\theta)\dot{\theta},$$

where $R_{sb}(\theta) = e^{[r]} = e^{[\hat{\omega}]\theta}$.

The time-derivative $\dot{r}$ is related to the body angular velocity $\omega_b$ by

$$\omega_b = A(r)\dot{r},$$

where

$$A(r) = I - \frac{1 - \cos\|r\|}{\|r\|^2}[r] + \frac{\|r\| - \sin\|r\|}{\|r\|^3}[r]^2.$$

(The derivation of this formula is explored in Exercise 5.10.) Provided that the matrix $A(r)$ is invertible, $\dot{r}$ can be obtained from $\omega_b$:

$$\dot{r} = A^{-1}(r)\omega_b = A^{-1}(r)J_\omega(\theta)\dot{\theta}.$$

Putting these together, we obtain

$$\dot{q} = \begin{bmatrix} \dot{r} \\ \dot{x} \end{bmatrix} = \begin{bmatrix} A^{-1}(r) & 0 \\ 0 & R_{sb} \end{bmatrix} \begin{bmatrix} \omega_b \\ v_b \end{bmatrix}, \tag{5.24}$$

i.e., the analytic Jacobian $J_a$ is related to the body Jacobian $J_b$ by

$$J_a(\theta) = \begin{bmatrix} A^{-1}(r) & 0 \\ 0 & R_{sb}(\theta) \end{bmatrix} \begin{bmatrix} J_\omega(\theta) \\ J_v(\theta) \end{bmatrix} = \begin{bmatrix} A^{-1}(r) & 0 \\ 0 & R_{sb}(\theta) \end{bmatrix} J_b(\theta). \tag{5.25}$$

## .1.6 Looking Ahead to Inverse Velocity Kinematics

In the above sections we asked the question "What twist results from a given set of joint velocities?" The answer, written independently of the frame in which the twists are represented, was given by

$$\mathcal{V} = J(\theta)\dot{\theta}.$$

Often we are interested in the inverse question: given a desired twist $\mathcal{V}$, what joint velocities $\dot{\theta}$ are needed? This is a question of inverse velocity kinematics, which is discussed in more detail in Section 6.3. Briefly, if $J(\theta)$ is square (so that the number of joints $n$ is equal to six, the number of elements of a twist) and of full rank then $\dot{\theta} = J^{-1}(\theta)\mathcal{V}$. If $n \neq 6$ or the robot is at a singularity, however, then $J(\theta)$ is not invertible. In the case $n < 6$, arbitrary twists $\mathcal{V}$ cannot be achieved – the robot does not have enough joints. If $n > 6$ then we call the robot **redundant**. In this case, a desired twist $\mathcal{V}$ places six constraints on the joint rates, and the remaining $n - 6$ freedoms correspond to internal motions of the robot that are not evident in the motion of the end-effector. For example, if you consider your arm from your shoulder to your palm as a seven-joint open chain, when you place your palm at a fixed configuration in space (e.g., on the surface of a table), you still have one internal degree of freedom corresponding to the position of your elbow.

## 5.2     Statics of Open Chains

Using our familiar principle of conservation of power, we have

power at the joints = (power to move the robot) + (power at the end-effector)

and, considering the robot to be at static equilibrium (no power is being used to move the robot), we can equate the power at the joints to the power at the end-effector,[4]

$$\tau^{\mathrm{T}}\dot{\theta} = \mathcal{F}_b^{\mathrm{T}}\mathcal{V}_b,$$

where $\tau$ is the column vector of the joint torques. Using the identity $\mathcal{V}_b = J_b(\theta)\dot{\theta}$, we get

$$\tau = J_b^{\mathrm{T}}(\theta)\mathcal{F}_b,$$

relating the joint torques to the wrench written in the end-effector frame. Similarly,

$$\tau = J_s^{\mathrm{T}}(\theta)\mathcal{F}_s$$

in the fixed space frame. Independently of the choice of the frame, we can simply write

$$\tau = J^{\mathrm{T}}(\theta)\mathcal{F}. \tag{5.26}$$

If an external wrench $-\mathcal{F}$ is applied to the end-effector when the robot is at equilibrium with joint values $\theta$, Equation (5.26) calculates the joint torques needed

---

[4] We are considering the limiting case as $\dot{\theta}$ goes to zero, consistent with our assumption that the robot is at equilibrium.

to generate the opposing wrench $\mathcal{F}$, keeping the robot at equilibrium.[5] This is important in force control of a robot, for example.

One could also ask the opposite question, namely, what is the end-effector wrench generated by a given set of joint torques? If $J^{\mathrm{T}}$ is a $6 \times 6$ invertible matrix, then clearly $\mathcal{F} = J^{-\mathrm{T}}(\theta)\tau$. If the number of joints $n$ is not equal to six then $J^{\mathrm{T}}$ is not invertible, and the question is not well posed.

If the robot is redundant ($n > 6$) then, even if the end-effector is embedded in concrete, the robot is not immobilized and the joint torques may cause internal motions of the links. The static equilibrium assumption is no longer satisfied, and we need to include dynamics to know what will happen to the robot.

If $n \leq 6$ and $J^{\mathrm{T}} \in \mathbb{R}^{n \times 6}$ has rank $n$ then embedding the end-effector in concrete will immobilize the robot. If $n < 6$, no matter what $\tau$ we choose, the robot cannot *actively* generate forces in the $6 - n$ wrench directions defined by the null space of $J^{\mathrm{T}}$,

$$\mathrm{Null}(J^{\mathrm{T}}(\theta)) = \{\mathcal{F} \mid J^{\mathrm{T}}(\theta)\mathcal{F} = 0\},$$

since no actuators act in these directions. The robot can, however, resist arbitrary externally applied wrenches in the space $\mathrm{Null}(J^{\mathrm{T}}(\theta))$ without moving, owing to the lack of joints that would allow motions due to these forces. For example, consider a motorized rotating door with a single revolute joint ($n = 1$) and an end-effector frame at the door knob. The door can only actively generate a force at the knob that is tangential to the allowed circle of motion of the knob (defining a single direction in the wrench space), but it can resist arbitrary wrenches in the orthogonal five-dimensional wrench space without moving.

## .3    Singularity Analysis

The Jacobian allows us to identify postures at which the robot's end-effector loses the ability to move instantaneously in one or more directions. Such a posture is called a **kinematic singularity**, or simply a **singularity**. Mathematically, a singular posture is one in which the Jacobian $J(\theta)$ fails to be of maximal rank. To understand why, consider the body Jacobian $J_b(\theta)$, whose columns are denoted $J_{bi}$, $i = 1, \ldots, n$. Then

$$\mathcal{V}_b = \begin{bmatrix} J_{b1}(\theta) & J_{b2}(\theta) & \cdots & J_{bn}(\theta) \end{bmatrix} \begin{bmatrix} \dot{\theta}_1 \\ \vdots \\ \dot{\theta}_n \end{bmatrix}$$

$$= J_{b1}(\theta)\dot{\theta}_1 + \cdots + J_{bn}(\theta)\dot{\theta}_n.$$

Thus, the tip frame can achieve twists that are linear combinations of the $J_{bi}$. As long as $n \geq 6$, the maximum rank that $J_b(\theta)$ can attain is six. Singular

---

[5] If the robot has to support itself against gravity to maintain static equilibrium, the torques $\tau$ must be added to the torques that offset gravity.

postures correspond to those values of $\theta$ at which the rank of $J_b(\theta)$ drops below the maximum possible value; at such postures the tip frame loses the ability to generate instantaneous spatial velocities in in one or more dimensions. This loss of mobility at a singularity is accompanied by the ability to resist arbitrary wrenches in the direction corresponding to the lost mobility.

The mathematical definition of a kinematic singularity is independent of the choice of body or space Jacobian. To see why, recall the relationship between $J_s(\theta)$ and $J_b(\theta)$: $J_s(\theta) = \text{Ad}_{T_{sb}}(J_b(\theta)) = [\text{Ad}_{T_{sb}}]J_b(\theta)$ or, more explicitly,

$$J_s(\theta) = \begin{bmatrix} R_{sb} & 0 \\ [p_{sb}]R_{sb} & R_{sb} \end{bmatrix} J_b(\theta).$$

We now claim that the matrix $[\text{Ad}_{T_{sb}}]$ is always invertible. This can be established by examining the linear equation

$$\begin{bmatrix} R_{sb} & 0 \\ [p_{sb}]R_{sb} & R_{sb} \end{bmatrix}\begin{bmatrix} x \\ y \end{bmatrix} = 0.$$

Its unique solution is $x = y = 0$, implying that the matrix $[\text{Ad}_{T_{sb}}]$ is invertible. Since multiplying any matrix by an invertible matrix does not change its rank, it follows that

$$\text{rank } J_s(\theta) = \text{rank } J_b(\theta),$$

as claimed; singularities of the space and body Jacobian are one and the same.

Kinematic singularities are also independent of the choice of fixed frame and end-effector frame. Choosing a different fixed frame is equivalent to simply relocating the robot arm, which should have absolutely no effect on whether a particular posture is singular. This obvious fact can be verified by referring to Figure 5.10(a). The forward kinematics with respect to the original fixed frame is denoted $T(\theta)$, while the forward kinematics with respect to the relocated fixed frame is denoted $T'(\theta) = PT(\theta)$, where $P \in SE(3)$ is constant. Then the body Jacobian of $T'(\theta)$, denoted $J_b'(\theta)$, is obtained from $(T')^{-1}\dot{T}'$. A simple calculation reveals that

$$(T')^{-1}\dot{T}' = (T^{-1}P^{-1})(P\dot{T}) = T^{-1}\dot{T},$$

i.e., $J_b'(\theta) = J_b(\theta)$, so that the singularities of the original and relocated robot arms are the same.

To see that singularities are independent of the end-effector frame, refer to Figure 5.10(b) and suppose that the forward kinematics for the original end-effector frame is given by $T(\theta)$ while the forward kinematics for the relocated end-effector frame is $T'(\theta) = T(\theta)Q$, where $Q \in SE(3)$ is constant. This time, looking at the space Jacobian – recall that singularities of $J_b(\theta)$ coincide with those of $J_s(\theta)$ – let $J_s'(\theta)$ denote the space Jacobian of $T'(\theta)$. A simple calculation reveals that

$$\dot{T}'(T')^{-1} = (\dot{T}Q)(Q^{-1}T^{-1}) = \dot{T}T^{-1}.$$

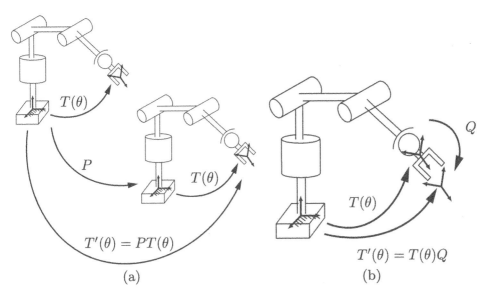

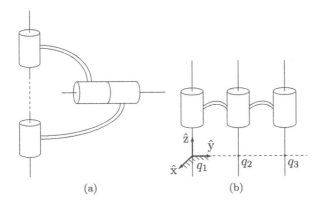

**Figure 5.10** Kinematic singularities are invariant with respect to the choice of fixed and end-effector frames. (a) Choosing a different fixed frame, which is equivalent to relocating the base of the robot arm; (b) choosing a different end-effector frame.

**Figure 5.11** (a) A kinematic singularity in which two joint axes are collinear. (b) A kinematic singularity in which three revolute joint axes are parallel and coplanar.

That is, $J'_s(\theta) = J_s(\theta)$, so that the kinematic singularities are invariant with respect to the choice of end-effector frame.

In the remainder of this section we consider some common kinematic singularities that occur in six-dof open chains with revolute and prismatic joints. We now know that either the space or body Jacobian can be used for our analysis; we use the space Jacobian in the examples below.

### Case I: Two Collinear Revolute Joint Axes

The first case we consider is one in which two revolute joint axes are collinear (see Figure 5.11(a)). Without loss of generality these joint axes can be labeled 1 and 2. The corresponding columns of the Jacobian are

$$J_{s1}(\theta) = \begin{bmatrix} \omega_{s1} \\ -\omega_{s1} \times q_1 \end{bmatrix} \quad \text{and} \quad J_{s2}(\theta) = \begin{bmatrix} \omega_{s2} \\ -\omega_{s2} \times q_2 \end{bmatrix}.$$

Since the two joint axes are collinear, we must have $\omega_{s1} = \pm\omega_{s2}$; let us assume the positive sign. Also, $\omega_{si} \times (q_1 - q_2) = 0$ for $i = 1, 2$. Then $J_{s1} = J_{s2}$, the set $\{J_{s1}, J_{s2}, \ldots, J_{s6}\}$ cannot be linearly independent, and the rank of $J_s(\theta)$ must be less than six.

### Case II: Three Coplanar and Parallel Revolute Joint Axes

The second case we consider is one in which three revolute joint axes are parallel and also lie on the same plane (three coplanar axes: see Figure 5.11(b)). Without loss of generality we label these as joint axes 1, 2, and 3. In this case we choose the fixed frame as shown in the figure; then

$$J_s(\theta) = \begin{bmatrix} \omega_{s1} & \omega_{s1} & \omega_{s1} & \cdots \\ 0 & -\omega_{s1} \times q_2 & -\omega_{s1} \times q_3 & \cdots \end{bmatrix}.$$

Since $q_2$ and $q_3$ are points on the same unit axis, it is not difficult to verify that the first three columns cannot be linearly independent.

### Case III: Four Revolute Joint Axes Intersecting at a Common Point

Here we consider the case where four revolute joint axes intersect at a common point (Figure 5.12). Again, without loss of generality, label these axes from 1 to 4. In this case we choose the fixed-frame origin to be the common point of intersection, so that $q_1 = \cdots = q_4 = 0$, and therefore

$$J_s(\theta) = \begin{bmatrix} \omega_{s1} & \omega_{s2} & \omega_{s3} & \omega_{s4} & \cdots \\ 0 & 0 & 0 & 0 & \cdots \end{bmatrix}.$$

The first four columns clearly cannot be linearly independent; one can be written as a linear combination of the other three. Such a singularity occurs, for example, when the wrist center of an elbow-type robot arm is directly above the shoulder.

### Case IV: Four Coplanar Revolute Joints

Here we consider the case in which four revolute joint axes are coplanar. Again, without loss of generality, label these axes from 1 to 4. Choose a fixed frame such that the joint axes all lie on the $x$–$y$-plane; in this case the unit vector $\omega_{si} \in \mathbb{R}^3$ in the direction of joint axis $i$ is of the form

$$\omega_{si} = \begin{bmatrix} \omega_{six} \\ \omega_{siy} \\ 0 \end{bmatrix}.$$

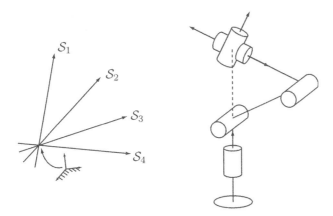

**Figure 5.12** A kinematic singularity in which four revolute joint axes intersect at a common point.

Similarly, any reference point $q_i \in \mathbb{R}^3$ lying on joint axis $i$ is of the form

$$q_i = \begin{bmatrix} q_{ix} \\ q_{iy} \\ 0 \end{bmatrix}$$

and consequently

$$v_{si} = -\omega_{si} \times q_i = \begin{bmatrix} 0 \\ 0 \\ \omega_{siy}q_{ix} - \omega_{six}q_{iy} \end{bmatrix}.$$

The first four columns of the space Jacobian $J_s(\theta)$ are

$$\begin{bmatrix} \omega_{s1x} & \omega_{s2x} & \omega_{s3x} & \omega_{s4x} \\ \omega_{s1y} & \omega_{s2y} & \omega_{s3y} & \omega_{s4y} \\ 0 & 0 & 0 & 0 \\ 0 & 0 & 0 & 0 \\ 0 & 0 & 0 & 0 \\ \omega_{s1y}q_{1x} - \omega_{s1x}q_{1y} & \omega_{s2y}q_{2x} - \omega_{s2x}q_{2y} & \omega_{s3y}q_{3x} - \omega_{s3x}q_{3y} & \omega_{s4y}q_{4x} - \omega_{s4x}q_{4y} \end{bmatrix}$$

and cannot be linearly independent since they only have three nonzero components.

### Case V: Six Revolute Joints Intersecting a Common Line

The final case we consider is six revolute joint axes intersecting a common line. Choose a fixed frame such that the common line lies along the $\hat{z}$-axis, and select the intersection between this common line and joint axis $i$ as the reference point $q_i \in \mathbb{R}^3$ for joint axis $i$; each $q_i$ is thus of the form $q_i = (0, 0, q_{iz})$, and

$$v_{si} = -\omega_{si} \times q_i = (\omega_{siy}q_{iz}, -\omega_{six}q_{iz}, 0),$$

for $i = 1, \ldots, 6$. The space Jacobian $J_s(\theta)$ thus becomes

$$
\begin{bmatrix}
\omega_{s1x} & \omega_{s2x} & \omega_{s3x} & \omega_{s4x} & \omega_{s5x} & \omega_{s6x} \\
\omega_{s1y} & \omega_{s2y} & \omega_{s3y} & \omega_{s4y} & \omega_{s5y} & \omega_{s6y} \\
\omega_{s1z} & \omega_{s2z} & \omega_{s3z} & \omega_{s4z} & \omega_{s5z} & \omega_{s6z} \\
\omega_{s1y}q_{1z} & \omega_{s2y}q_{2z} & \omega_{s3y}q_{3z} & \omega_{s4y}q_{4z} & \omega_{s5y}q_{5z} & \omega_{s6y}q_{6z} \\
-\omega_{s1x}q_{1z} & -\omega_{s2x}q_{2z} & -\omega_{s3x}q_{3z} & -\omega_{s4x}q_{4z} & -\omega_{s5x}q_{5z} & -\omega_{s6x}q_{6z} \\
0 & 0 & 0 & 0 & 0 & 0
\end{bmatrix},
$$

which is clearly singular.

## 5.4 Manipulability

In the previous section we saw that, at a kinematic singularity, a robot's end-effector loses the ability to translate or rotate in one or more directions. A kinematic singularity presents a binary proposition – a particular configuration is either kinematically singular or it is not – and it is reasonable to ask if a non-singular configuration is "close" to being singular. The answer is yes; in fact, one can even determine the directions in which the end-effector's ability to move is diminished, and to what extent. The manipulability ellipsoid allows one to visualize geometrically the directions in which the end-effector moves with least effort or with greatest effort.

Manipulability ellipsoids are illustrated for a 2R planar arm in Figure 5.3. The Jacobian is given by Equation (5.1).

For a general $n$-joint open chain and a task space with coordinates $q \in \mathbb{R}^m$, where $m \leq n$, the manipulability ellipsoid corresponds to the end-effector velocities for joint rates $\dot{\theta}$ satisfying $\|\dot{\theta}\| = 1$, a unit sphere in the $n$-dimensional joint-velocity space.[6] Assuming $J$ is invertible, the unit joint-velocity condition can be written

$$
\begin{aligned}
1 &= \dot{\theta}^{\mathrm{T}}\dot{\theta} \\
&= (J^{-1}\dot{q})^{\mathrm{T}}(J^{-1}\dot{q}) \\
&= \dot{q}^{\mathrm{T}}J^{-\mathrm{T}}J^{-1}\dot{q} \\
&= \dot{q}^{\mathrm{T}}(JJ^{\mathrm{T}})^{-1}\dot{q} = \dot{q}^{\mathrm{T}}A^{-1}\dot{q}.
\end{aligned}
\tag{5.27}
$$

If $J$ is full rank (i.e., of rank $m$), the matrix $A = JJ^{\mathrm{T}} \in \mathbb{R}^{m \times m}$ is square, symmetric, and positive definite, as is $A^{-1}$.

Consulting a textbook on linear algebra, we see that for any symmetric positive-definite $A^{-1} \in \mathbb{R}^{m \times m}$, the set of vectors $\dot{q} \in \mathbb{R}^m$ satisfying

$$
\dot{q}^{\mathrm{T}}A^{-1}\dot{q} = 1
$$

---

[6] A two-dimensional ellipsoid is usually referred to as an "ellipse," and an ellipsoid in more than three dimensions is often referred to as a "hyperellipsoid," but here we use the term ellipsoid independently of the dimension. Similarly, we refer to a "sphere" independently of the dimension, instead of using "circle" for two dimensions and "hypersphere" for more than three dimensions.

defines an ellipsoid in the $m$-dimensional space. Letting $v_i$ and $\lambda_i$ be the eigenvectors and eigenvalues of $A$, the directions of the principal axes of the ellipsoid are $v_i$ and the lengths of the principal semi-axes are $\sqrt{\lambda_i}$, as illustrated in Figure 5.13. Furthermore, the volume $V$ of the ellipsoid is proportional to the product of the semi-axis lengths:

$$V \text{ is proportional to } \sqrt{\lambda_1 \lambda_2 \cdots \lambda_m} = \sqrt{\det(A)} = \sqrt{\det(JJ^{\mathrm{T}})}.$$

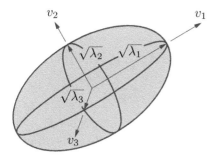

**Figure 5.13** An ellipsoid visualization of $\dot{q}^{\mathrm{T}} A^{-1} \dot{q} = 1$ in the $\dot{q}$ space $\mathbb{R}^3$, where the principal semi-axis lengths are the square roots of the eigenvalues $\lambda_i$ of $A$ and the directions of the principal semi-axes are the eigenvectors $v_i$.

For the geometric Jacobian $J$ (either $J_b$ in the end-effector frame or $J_s$ in the fixed frame), we can express the $6 \times n$ Jacobian as

$$J(\theta) = \begin{bmatrix} J_\omega(\theta) \\ J_v(\theta) \end{bmatrix},$$

where $J_\omega$ comprises the top three rows of $J$ and $J_v$ the bottom three rows of $J$. It makes sense to separate the two because the units of angular velocity and linear velocity are different. This leads to two three-dimensional manipulability ellipsoids, one for angular velocities and one for linear velocities. These manipulability ellipsoids have principal semi-axes aligned with the eigenvectors of $A$, with lengths given by the square roots of the eigenvalues, where $A = J_\omega J_\omega^{\mathrm{T}}$ for the angular velocity manipulability ellipsoid and $A = J_v J_v^{\mathrm{T}}$ for the linear velocity manipulability ellipsoid.

When calculating the linear-velocity manipulability ellipsoid, it generally makes more sense to use the body Jacobian $J_b$ instead of the space Jacobian $J_s$, since we are usually interested in the linear velocity of a point at the origin of the end-effector frame rather than that of a point at the origin of the fixed-space frame.

Apart from the geometry of the manipulability ellipsoid, it can be useful to assign a single scalar measure defining how easily the robot can move at a given posture. One measure is the ratio of the longest and shortest semi-axes of the manipulability ellipsoid,

$$\mu_1(A) = \frac{\sqrt{\lambda_{\max}(A)}}{\sqrt{\lambda_{\min}(A)}} = \sqrt{\frac{\lambda_{\max}(A)}{\lambda_{\min}(A)}} \geq 1,$$

where $A = JJ^{\mathrm{T}}$. When $\mu_1(A)$ is low (i.e., close to 1) then the manipulability ellipsoid is nearly spherical or **isotropic**, meaning that it is equally easy to move in any direction. This situation is generally desirable. As the robot approaches a singularity, however, $\mu_1(A)$ goes to infinity.

A similar measure $\mu_2(A)$ is just the square of $\mu_1(A)$, and is known as the **condition number** of the matrix $A = JJ^{\mathrm{T}}$,

$$\mu_2(A) = \frac{\lambda_{\max}(A)}{\lambda_{\min}(A)} \geq 1.$$

Again, smaller values (close to 1) are preferred. The condition number of a matrix is commonly used to characterize the sensitivity of the result of multiplying that matrix by a vector to small errors in the vector.

A final measure is simply proportional to the volume of the manipulability ellipsoid,

$$\mu_3(A) = \sqrt{\lambda_1 \lambda_2 \cdots} = \sqrt{\det(A)}.$$

In this case, unlike the first two measures, a larger value is better.

Just like the manipulability ellipsoid, a force ellipsoid can be drawn for joint torques $\tau$ satisfying $\|\tau\| = 1$. Beginning from $\tau = J^{\mathrm{T}}(\theta)\mathcal{F}$, we arrive at similar results to those above, except that now the ellipsoid satisfies

$$1 = f^{\mathrm{T}} J J^{\mathrm{T}} f = f^{\mathrm{T}} B^{-1} f,$$

where $B = (JJ^{\mathrm{T}})^{-1} = A^{-1}$. For the force ellipsoid, the matrix $B$ plays the same role as $A$ in the manipulability ellipsoid; it is the eigenvectors and the square roots of eigenvalues of $B$ that define the shape of the force ellipsoid.

Since eigenvectors of any invertible matrix $A$ are also eigenvectors of $B = A^{-1}$, the principal axes of the force ellipsoid are aligned with the principal axes of the manipulability ellipsoid. Furthermore, since the eigenvalues of $B = A^{-1}$ associated with each principal axis are the reciprocals of the corresponding eigenvalues of $A$, the lengths of the principal semi-axes of the force ellipsoid are given by $1/\sqrt{\lambda_i}$, where $\lambda_i$ are the eigenvalues of $A$. Thus the force ellipsoid is obtained from the manipulability ellipsoid simply by stretching the manipulability ellipsoid along each principal axis $i$ by a factor $1/\lambda_i$. Furthermore, since the volume $V_A$ of the manipulability ellipsoid is proportional to the product of the semi-axes, $\sqrt{\lambda_1 \lambda_2 \cdots}$, and the volume $V_B$ of the force ellipsoid is proportional to $1/\sqrt{\lambda_1 \lambda_2 \cdots}$, the product of the two volumes $V_A V_B$ is constant *independently* of the joint variables $\theta$. Therefore, positioning the robot to increase the manipulability-ellipsoid volume measure $\mu_3(A)$ simultaneously reduces the force-ellipsoid volume measure $\mu_3(B)$. This also explains the observation made at the start of the chapter that, as the robot approaches a singularity, $V_A$ goes to zero while $V_B$ goes to infinity.

## .5    Summary

- Let the forward kinematics of an $n$-link open chain be expressed in the following product of exponentials form:

$$T(\theta) = e^{[\mathcal{S}_1]\theta_1} \cdots e^{[\mathcal{S}_n]\theta_n} M.$$

  The space Jacobian $J_s(\theta) \in \mathbb{R}^{6 \times n}$ relates the joint rate vector $\dot{\theta} \in \mathbb{R}^n$ to the spatial twist $\mathcal{V}_s$, via $\mathcal{V}_s = J_s(\theta)\dot{\theta}$. The $i$th column of $J_s(\theta)$ is given by

$$J_{si}(\theta) = \mathrm{Ad}_{e^{[\mathcal{S}_1]\theta_1} \cdots [\mathcal{S}_{i-1}]\theta_{i-1}}(\mathcal{S}_i),$$

  for $i = 2, \ldots, n$, with the first column $J_{s1} = \mathcal{S}_1$. The screw vector $J_{si}$ for joint $i$ is expressed in space-frame coordinates, with the joint values $\theta$ assumed to be arbitrary rather than zero.

- Let the forward kinematics of an $n$-link open chain be expressed in the following product of exponentials form:

$$T(\theta) = M e^{[\mathcal{B}_1]\theta_1} \cdots e^{[\mathcal{B}_n]\theta_n}.$$

  The body Jacobian $J_b(\theta) \in \mathbb{R}^{6 \times n}$ relates the joint rate vector $\dot{\theta} \in \mathbb{R}^n$ to the end-effector body twist $\mathcal{V}_b = (\omega_b, v_b)$ via $\mathcal{V}_b = J_b(\theta)\dot{\theta}$. The $i$th column of $J_b(\theta)$ is given by

$$J_{bi}(\theta) = \mathrm{Ad}_{e^{-[\mathcal{B}_n]\theta_n} \cdots e^{-[\mathcal{B}_{i+1}]\theta_{i+1}}}(\mathcal{B}_i),$$

  for $i = n-1, \ldots, 1$, with $J_{bn} = \mathcal{B}_n$. The screw vector $J_{bi}$ for joint $i$ is expressed in body-frame coordinates, with the joint values $\theta$ assumed to be arbitrary rather than zero.

- The body and space Jacobians are related via

$$J_s(\theta) = [\mathrm{Ad}_{T_{sb}}]J_b(\theta),$$
$$J_b(\theta) = [\mathrm{Ad}_{T_{bs}}]J_s(\theta),$$

  where $T_{sb} = T(\theta)$.

- Consider a spatial open chain with $n$ one-dof joints that is assumed to be in static equilibrium. Let $\tau \in \mathbb{R}^n$ denote the vector of the joint torques and forces and $\mathcal{F} \in \mathbb{R}^6$ be the wrench applied at the end-effector, in either space- or body-frame coordinates. Then $\tau$ and $\mathcal{F}$ are related by

$$\tau = J_b^{\mathrm{T}}(\theta)\mathcal{F}_b = J_s^{\mathrm{T}}(\theta)\mathcal{F}_s.$$

- A kinematically singular configuration for an open chain, or more simply a kinematic singularity, is any configuration $\theta \in \mathbb{R}^n$ at which the rank of the Jacobian is not maximal. For six-dof spatial open chains consisting of revolute and prismatic joints, some common singularities include (i) two collinear revolute joint axes; (ii) three coplanar and parallel revolute joint axes; (iii) four revolute joint axes intersecting at a common point; (iv) four coplanar revolute joints; and (v) six revolute joints intersecting a common line.

- The manipulability ellipsoid describes how easily the robot can move in different directions. For a Jacobian $J$, the principal axes of the manipulability ellipsoid are defined by the eigenvectors of $JJ^{\mathrm{T}}$ and the corresponding lengths of the principal semi-axes are the square roots of the eigenvalues.
- The force ellipsoid describes how easily the robot can generate forces in different directions. For a Jacobian $J$, the principal axes of the force ellipsoid are defined by the eigenvectors of $(JJ^{\mathrm{T}})^{-1}$ and the corresponding lengths of the principal semi-axes are the square roots of the eigenvalues.
- Measures of the manipulability and force ellipsoids include the ratio of the longest principal semi-axis to the shortest; the square of this measure; and the volume of the ellipsoid. The first two measures indicate that the robot is far from being singular if they are small (close to 1).

## 5.6     Software

Software functions associated with this chapter are listed below.

`Jb = JacobianBody(Blist,thetalist)`
Computes the body Jacobian $J_b(\theta) \in \mathbb{R}^{6 \times n}$ given a list of joint screws $\mathcal{B}_i$ expressed in the body frame and a list of joint angles.

`Js = JacobianSpace(Slist,thetalist)`
Computes the space Jacobian $J_s(\theta) \in \mathbb{R}^{6 \times n}$ given a list of joint screws $\mathcal{S}_i$ expressed in the fixed space frame and a list of joint angles.

## 5.7     Notes and References

One of the key advantages of the PoE formulation is in the derivation of the Jacobian; the columns of the Jacobian are simply the (configuration-dependent) screws for the joint axes. Compact closed-form expressions for the columns of the Jacobian are also obtained because taking the derivatives of matrix exponentials is particularly straightforward.

There is extensive literature on the singularity analysis of 6R open chains. In addition to the three cases presented in this chapter, other cases are examined in Murray et al. (1994) and in some of the exercises at the end of this chapter, including the case when some of the revolute joints are replaced by prismatic joints. Many of the mathematical techniques and analyses used in open-chain singularity analysis can also be used to determine the singularities of parallel mechanisms; this is discussed in Chapter 7.

The concept of a robot's manipulability was first formulated in a quantitative way by Yoshikawa (1985). There is now a vast literature on the manipulability analysis of open chains; see, e.g., Klein and Blaho (1987) and Park and Brockett (1994).

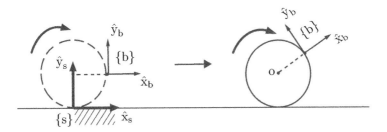

**Figure 5.14** A rolling wheel.

## .8     Exercises

**Exercise 5.1**   A wheel of unit radius is rolling to the right at a rate of 1 rad/s (see Figure 5.14; the dashed circle shows the wheel at $t = 0$).

(a) Find the spatial twist $\mathcal{V}_s(t)$ as a function of $t$.
(b) Find the linear velocity of the {b}-frame origin expressed in {s}-frame coordinates.

**Exercise 5.2**   The 3R planar open chain of Figure 5.15(a) is shown in its zero position.

(a) Suppose that the last link must apply a wrench corresponding to a force of 5 N in the $\hat{x}_s$-direction of the {s} frame, with zero component in the $\hat{y}_s$-direction and zero moment about the $\hat{z}_s$-axis. What torques should be applied at each joint?
(b) Suppose that now the last link must apply a force of 5 N in the $\hat{y}_s$-direction, with zero components in other wrench directions. What torques should be applied at each joint?

**Exercise 5.3**   Answer the following questions for the 4R planar open chain of Figure 5.15(b).

(a) For the forward kinematics of the form

$$T(\theta) = e^{[\mathcal{S}_1]\theta_1} e^{[\mathcal{S}_2]\theta_2} e^{[\mathcal{S}_3]\theta_3} e^{[\mathcal{S}_4]\theta_4} M,$$

write down $M \in SE(2)$ and each $\mathcal{S}_i = (\omega_{zi}, v_{xi}, v_{yi}) \in \mathbb{R}^3$.
(b) Write down the body Jacobian.
(c) Suppose that the chain is in static equilibrium at the configuration $\theta_1 = \theta_2 = 0, \theta_3 = \pi/2, \theta_4 = -\pi/2$ and that a force $f = (10, 10, 0)$ and a moment $m = (0, 0, 10)$ are applied to the tip (both $f$ and $m$ are expressed with respect to the fixed frame). What are the torques experienced at each joint?

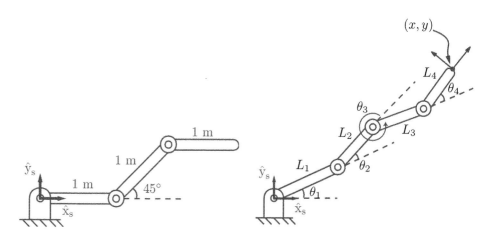

**Figure 5.15** (a) A 3R planar open chain. The length of each link is 1 m. (b) A 4R planar open chain.

(d) Under the same conditions as (c), suppose that a force $f = (-10, 10, 0)$ and a moment $m = (0, 0, -10)$, also expressed in the fixed frame, are applied to the tip. What are the torques experienced at each joint?

(e) Find all kinematic singularities for this chain.

**Exercise 5.4** Figure 5.16 shows two fingers grasping a can. Frame {b} is attached to the center of the can. Frames {b$_1$} and {b$_2$} are attached to the can at the two contact points as shown. The force $f_1 = (f_{1,x}, f_{1,y}, f_{1,z})$ is the force applied by fingertip 1 to the can, expressed in {b$_1$} coordinates. Similarly, $f_2 = (f_{2,x}, f_{2,y}, f_{2,z})$ is the force applied by fingertip 2 to the can, expressed in {b$_2$} coordinates.

(a) Assume that the system is in static equilibrium, and find the total wrench $\mathcal{F}_b$ applied by the two fingers to the can. Express your result in {b} coordinates.

(b) Suppose that $\mathcal{F}_{\text{ext}}$ is an arbitrary external wrench applied to the can ($\mathcal{F}_{\text{ext}}$ is also expressed in frame-{b} coordinates). Find all $\mathcal{F}_{\text{ext}}$ that cannot be resisted by the fingertip forces.

**Exercise 5.5** Referring to Figure 5.17, a rigid body, shown at the top right, rotates about the point $(L, L)$ with angular velocity $\dot{\theta} = 1$.

(a) Find the position of point $P$ on the moving body relative to the fixed reference frame {s} in terms of $\theta$.

(b) Find the velocity of point $P$ in terms of the fixed frame.

(c) What is $T_{sb}$, the configuration of frame {b}, as seen from the fixed frame {s}?

(d) Find the twist of $T_{sb}$ in body coordinates.

(e) Find the twist of $T_{sb}$ in space coordinates.

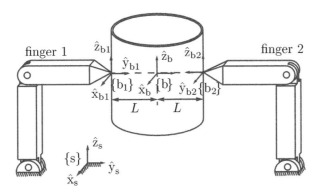

**Figure 5.16** Two fingers grasping a can.

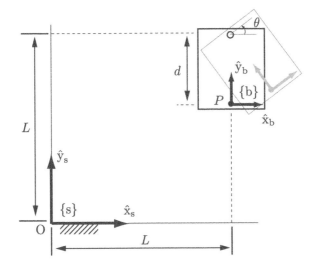

**Figure 5.17** A rigid body rotating in the plane.

(f) What is the relationship between the twists from (d) and (e)?

(g) What is the relationship between the twist from (d) and $\dot{P}$ from (b)?

(h) What is the relationship between the twist from (e) and $\dot{P}$ from (b)?

**Exercise 5.6** Figure 5.18 shows a design for a new amusement park ride. A rider sits at the location indicated by the moving frame {b}. The fixed frame {s} is attached to the top shaft as shown. The dimensions indicated in the figure are $R = 10$ m and $L = 20$ m, and the two joints each rotate at a constant angular velocity of 1 rad/s.

(a) Suppose $t = 0$ at the instant shown in the figure. Find the linear velocity $v_b$ and angular velocity $\omega_b$ of the rider as functions of time $t$. Express your answer in frame-{b} coordinates.

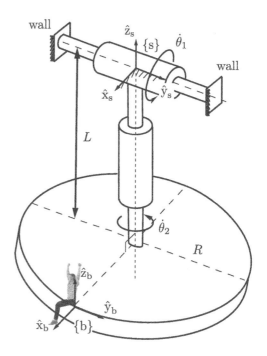

**Figure 5.18** A new amusement park ride.

(b) Let $p$ be the linear coordinates expressing the position of the rider in $\{s\}$. Find the linear velocity $\dot{p}(t)$.

**Exercise 5.7**  The RRP robot in Figure 5.19 is shown in its zero position.

(a) Write down the screw axes in the space frame. Evaluate the forward kinematics when $\theta = (90°, 90°, 1)$. Hand-draw or use a computer to show the arm and the end-effector frame in this configuration. Obtain the space Jacobian $J_s$ for this configuration.
(b) Write down the screw axes in the end-effector body frame. Evaluate the forward kinematics when $\theta = (90°, 90°, 1)$ and confirm that you get the same result as in part (a). Obtain the body Jacobian $J_b$ for this configuration.

**Exercise 5.8**  The RPR robot of Figure 5.20 is shown in its zero position. The fixed and end-effector frames are respectively denoted $\{s\}$ and $\{b\}$.

(a) Find the space Jacobian $J_s(\theta)$ for arbitrary configurations $\theta \in \mathbb{R}^3$.
(b) Assume the manipulator is in its zero position. Suppose that an external force $f \in \mathbb{R}^3$ is applied to the $\{b\}$ frame origin. Find all the directions in which $f$ can be resisted by the manipulator with $\tau = 0$.

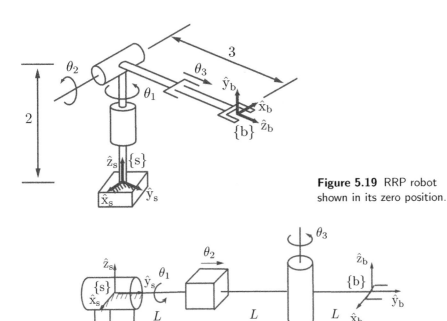

**Figure 5.19** RRP robot shown in its zero position.

**Figure 5.20** RPR robot.

**Exercise 5.9** Find the kinematic singularities of the 3R wrist given the forward kinematics

$$R = e^{[\hat{\omega}_1]\theta_1}e^{[\hat{\omega}_2]\theta_2}e^{[\hat{\omega}_3]\theta_3},$$

where $\hat{\omega}_1 = (0,0,1)$, $\hat{\omega}_2 = (1/\sqrt{2}, 0, 1/\sqrt{2})$, and $\hat{\omega}_3 = (1,0,0)$.

**Exercise 5.10** In this exercise, for an $n$-link open chain we derive the analytic Jacobian corresponding to the exponential coordinates on $SO(3)$.

(a) Given an $n \times n$ matrix $A(t)$ parametrized by $t$ that is also differentiable with respect to $t$, its exponential $X(t) = e^{A(t)}$ is then an $n \times n$ matrix that is always nonsingular. Prove the following:

$$X^{-1}\dot{X} = \int_0^1 e^{-A(t)s}\dot{A}(t)e^{A(t)s}ds,$$

$$\dot{X}X^{-1} = \int_0^1 e^{A(t)s}\dot{A}(t)e^{-A(t)s}ds.$$

(Hint: The formula

$$\frac{d}{d\epsilon}e^{(A+\epsilon B)t}|_{\epsilon=0} = \int_0^t e^{As}Be^{A(t-s)}ds$$

may be useful.)

(b) Use the result above to show that, for $r(t) \in \mathbb{R}^3$ and $R(t) = e^{[r(t)]}$, the angular velocity in the body frame, $[\omega_b] = R^T \dot{R}$, is related to $\dot{r}$ by

$$\omega_b = A(r)\dot{r},$$

$$A(r) = I - \frac{1 - \cos \|r\|}{\|r\|^2}[r] + \frac{\|r\| - \sin \|r\|}{\|r\|^3}[r]^2.$$

(c) Derive the corresponding formula relating the angular velocity in the space frame, $[\omega_s] = \dot{R}R^T$, to $\dot{r}$.

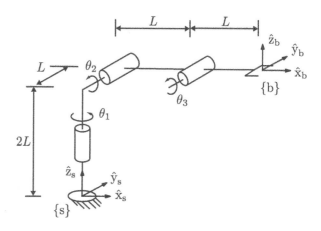

**Figure 5.21** A spatial 3R open chain.

**Exercise 5.11**   The spatial 3R open chain of Figure 5.21 is shown in its zero position. Let $p$ be the coordinates of the origin of {b} expressed in {s}.

(a) In its zero position, suppose we wish to make the end-effector move with linear velocity $\dot{p} = (10, 0, 0)$. Is this motion possible? If so, what are the required input joint velocities $\dot{\theta}_1, \dot{\theta}_2$, and $\dot{\theta}_3$?

(b) Suppose that the robot is in the configuration $\theta_1 = 0, \theta_2 = 45°, \theta_3 = -45°$. Assuming static equilibrium, suppose that we wish to generate an end-effector force $f_b = (10, 0, 0)$, where $f_b$ is expressed with respect to the end-effector frame {b}. What are the required input joint torques $\tau_1, \tau_2$, and $\tau_3$?

(c) Under the same conditions as in (b), suppose that we now seek to generate an end-effector moment $m_b = (10, 0, 0)$, where $m_b$ is expressed with respect to the end-effector frame {b}. What are the required input joint torques $\tau_1, \tau_2, \tau_3$?

(d) Suppose that the maximum allowable torques for each joint motor are

$$\|\tau_1\| \le 10, \quad \|\tau_2\| \le 20, \quad \text{and} \quad \|\tau_3\| \le 5.$$

In the zero position, what is the maximum force that can be applied by the tip in the end-effector-frame $\hat{x}$-direction?

**Exercise 5.12**    The RRRP chain of Figure 5.22 is shown in its zero position. Let $p$ be the coordinates of the origin of $\{b\}$ expressed in $\{s\}$.

(a) Determine the body Jacobian $J_b(\theta)$ when $\theta_1 = \theta_2 = 0, \theta_3 = \pi/2, \theta_4 = L$.
(b) Find $\dot{p}$ when $\theta_1 = \theta_2 = 0, \theta_3 = \pi/2, \theta_4 = L$ and $\dot{\theta}_1 = \dot{\theta}_2 = \dot{\theta}_3 = \dot{\theta}_4 = 1$.

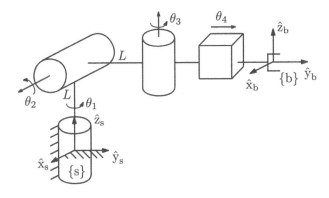

**Figure 5.22** An RRRP spatial open chain.

**Exercise 5.13**    For the 6R spatial open chain of Figure 5.23,

(a) Determine its space Jacobian $J_s(\theta)$.
(b) Find its kinematic singularities. Explain each singularity in terms of the alignment of the joint screws and of the directions in which the end-effector loses one or more degrees of freedom of motion.

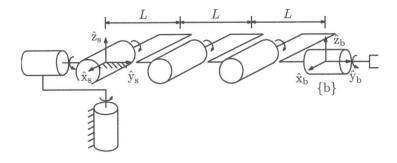

**Figure 5.23** Singularities of a 6R open chain.

**Exercise 5.14**    Show that a six-dof spatial open chain is at a kinematic singularity when any two of its revolute joint axes are parallel, and any prismatic joint axis is normal to the plane spanned by the two parallel revolute joint axes (see Figure 5.24).

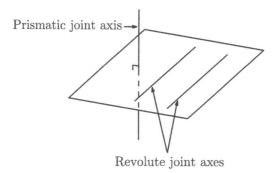

Prismatic joint axis

Revolute joint axes

**Figure 5.24** A kinematic singularity involving prismatic and revolute joints.

**Exercise 5.15** The spatial PRRRRP open chain of Figure 5.25 is shown in its zero position.

(a) At the zero position, find the first three columns of the space Jacobian.
(b) Find all configurations for which the first three columns of the space Jacobian become linearly dependent.
(c) Suppose that the chain is in the configuration $\theta_1 = \theta_2 = \theta_3 = \theta_5 = \theta_6 = 0$, $\theta_4 = 90°$. Assuming static equilibrium, suppose that a pure force $f_b = (10, 0, 10)$, where $f_b$ is expressed in terms of the end-effector frame, is applied to the origin of the end-effector frame. Find the torques $\tau_1, \tau_2$, and $\tau_3$ experienced at the first three joints.

**Exercise 5.16** Consider the PRPRRR spatial open chain of Figure 5.26, shown in its zero position. The distance from the origin of the fixed frame to the origin of the end-effector frame at the home position is $L$.

(a) Determine the first three columns of the space Jacobian $J_s$.

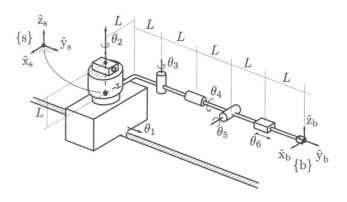

**Figure 5.25** A spatial PRRRRP open chain.

(b) Determine the last two columns of the body Jacobian $J_b$.

(c) For what value of $L$ is the home position a singularity?

(d) In the zero position, what joint forces and torques $\tau$ must be applied in order to generate a pure end-effector force of 100 N in the $-\hat{z}_b$-direction?

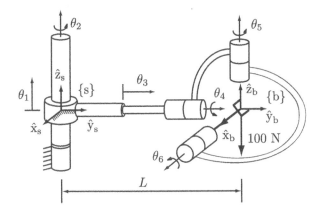

Figure 5.26 A PRPRRR spatial open chain.

**Exercise 5.17** The PRRRRP robot of Figure 5.27 is shown in its zero position.

(a) Find the first three columns of the space Jacobian $J_s(\theta)$.

(b) Assuming the robot is in its zero position and $\dot{\theta} = (1, 0, 1, -1, 2, 0)$, find the spatial twist $\mathcal{V}_s$.

(c) Is the zero position a kinematic singularity? Explain your answer.

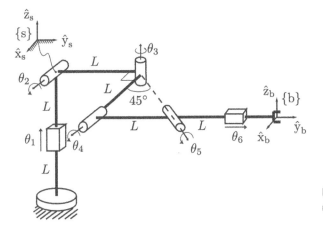

Figure 5.27 A PRRRRP robot.

**Exercise 5.18** The six-dof RRPRPR open chain of Figure 5.28 has a fixed frame {s} and an end-effector frame {b} attached as shown. At its zero position, joint axes 1, 2, and 6 lie in the $\hat{y}$–$\hat{z}$-plane of the fixed frame, and joint axis 4 is aligned along the fixed-frame $\hat{x}$-axis.

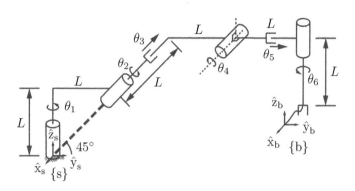

**Figure 5.28** An RRPRPR open chain shown at its zero position.

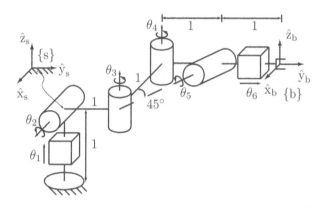

**Figure 5.29** A spatial PRRRRP open chain with a skewed joint axis.

(a) Find the first three columns of the space Jacobian $J_s(\theta)$.
(b) At the zero position, let $\dot{\theta} = (1, 0, 1, -1, 2, 0)$. Find the spatial twist $\mathcal{V}_s$.
(c) Is the zero position a kinematic singularity? Explain your answer.

**Exercise 5.19**   The spatial PRRRRP open chain of Figure 5.29 is shown in its zero position.

(a) Determine the first four columns of the space Jacobian $J_s(\theta)$.
(b) Determine whether the zero position is a kinematic singularity.
(c) Calculate the joint forces and torques required for the tip to apply the following end-effector wrenches:
  1. $\mathcal{F}_s = (0, 1, -1, 1, 0, 0)$.
  2. $\mathcal{F}_s = (1, -1, 0, 1, 0, -1)$.

**Exercise 5.20**   The spatial RRPRRR open chain of Figure 5.30 is shown in its zero position.

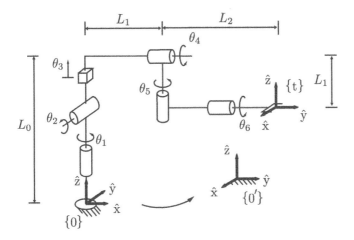

**Figure 5.30** A spatial RRPRRR open chain.

(a) For the fixed frame {0} and tool (end-effector) frame {t} as shown, express the forward kinematics in the product of exponentials form

$$T(\theta) = e^{[\mathcal{S}_1]\theta_1} e^{[\mathcal{S}_2]\theta_2} e^{[\mathcal{S}_3]\theta_3} e^{[\mathcal{S}_4]\theta_4} e^{[\mathcal{S}_5]\theta_5} e^{[\mathcal{S}_6]\theta_6} M.$$

(b) Find the first three columns of the space Jacobian $J_s(\theta)$.
(c) Suppose that the fixed frame {0} is moved to another location {0′} as shown in the figure. Find the first three columns of the space Jacobian $J_s(\theta)$ with respect to this new fixed frame.
(d) Determine whether the zero position is a kinematic singularity and, if so, provide a geometric description in terms of the joint screw axes.

**Exercise 5.21**    Figure 5.31 shows an RRPRRR exercise robot used for stroke patient rehabilitation.

(a) Assume the manipulator is in its zero position. Suppose that $M_{0c} \in SE(3)$ is the displacement from frame {0} to frame {c} and $M_{ct} \in SE(3)$ is the displacement from frame {c} to frame {t}. Express the forward kinematics $T_{0t}$ in the form

$$T_{0t} = e^{[\mathcal{A}_1]\theta_1} e^{[\mathcal{A}_2]\theta_2} M_{0c} e^{[\mathcal{A}_3]\theta_3} e^{[\mathcal{A}_4]\theta_4} M_{ct} e^{[\mathcal{A}_5]\theta_5} e^{[\mathcal{A}_6]\theta_6}.$$

Find $\mathcal{A}_2, \mathcal{A}_4$, and $\mathcal{A}_5$.
(b) Suppose that $\theta_2 = 90°$ and all the other joint variables are fixed at zero. Set the joint velocities to $(\dot{\theta}_1, \dot{\theta}_2, \dot{\theta}_3, \dot{\theta}_4, \dot{\theta}_5, \dot{\theta}_6) = (1, 0, 1, 0, 0, 1)$, and find the spatial twist $\mathcal{V}_s$ in frame-{0} coordinates.
(c) Is the configuration described in part (b) a kinematic singularity? Explain your answer.

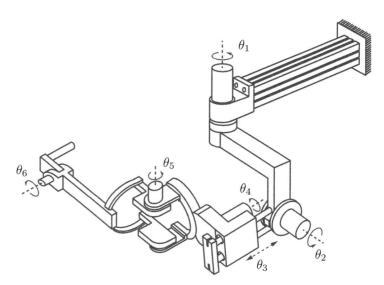

(a) Rehabilitation robot ARMin III (Nef et al., 2009). Figure courtesy of ETH Zürich.

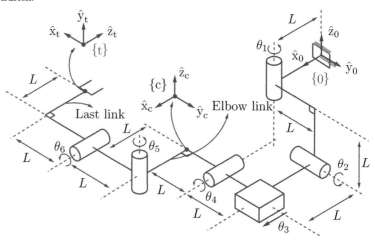

(b) Kinematic model of the ARMin III.

**Figure 5.31** The ARMin III rehabilitation robot.

(d) Suppose that a person now operates the rehabilitation robot. At the configuration described in part (b), a wrench $\mathcal{F}_{\text{elbow}}$ is applied to the elbow link, and a wrench $\mathcal{F}_{\text{tip}}$ is applied to the last link. Both $\mathcal{F}_{\text{elbow}}$ and $\mathcal{F}_{\text{tip}}$ are expressed in frame $\{0\}$ coordinates and are given by $\mathcal{F}_{\text{elbow}} = (1, 0, 0, 0, 0, 1)$ and $\mathcal{F}_{\text{tip}} = (0, 1, 0, 1, 1, 0)$. Find the joint forces and torques $\tau$ that must be applied for the robot to maintain static equilibrium.

**Exercise 5.22**   Consider an $n$-link open chain, with reference frames attached

**Figure 5.32** Left: The 2R robot arm. Right: The arm at four different configurations.

to each link. Let

$$T_{0k} = e^{[\mathcal{S}_1]\theta_1} \cdots e^{[\mathcal{S}_k]\theta_k} M_k, \qquad k = 1, \ldots, n$$

be the forward kinematics up to link frame $\{k\}$. Let $J_s(\theta)$ be the space Jacobian for $T_{0n}$; $J_s(\theta)$ has columns $J_{si}$ as shown below:

$$J_s(\theta) = \begin{bmatrix} J_{s1}(\theta) & \cdots & J_{sn}(\theta) \end{bmatrix}.$$

Let $[\mathcal{V}_k] = \dot{T}_{0k} T_{0k}^{-1}$ be the twist of link frame $\{k\}$ in fixed frame $\{0\}$ coordinates.

(a) Derive explicit expressions for $\mathcal{V}_2$ and $\mathcal{V}_3$.
(b) On the basis of your results from (a), derive a recursive formula for $\mathcal{V}_{k+1}$ in terms of $\mathcal{V}_k$, $J_{s1}, \ldots, J_{s,k+1}$, and $\dot{\theta}$.

**Exercise 5.23** Write a program that allows the user to enter the link lengths $L_1$ and $L_2$ of a 2R planar robot (Figure 5.32) and a list of robot configurations (each defined by the joint angles $(\theta_1, \theta_2)$) and plots the manipulability ellipse at each of those configurations. The program should plot the arm (as two line segments) at each configuration and the manipulability ellipse centered at the endpoint of the arm. Choose the same scaling for all the ellipses so that they can be easily visualized (e.g., the ellipse should usually be shorter than the arm but not so small that you cannot easily see it). The program should also print the three manipulability measures $\mu_1, \mu_2$, and $\mu_3$ for each configuration.

(a) Choose $L_1 = L_2 = 1$ and plot the arm and its manipulability ellipse at the four configurations $(-10°, 20°), (60°, 60°), (135°, 90°)$, and $(190°, 160°)$. At which of these configurations does the manipulability ellipse appear most isotropic? Does this agree with the manipulability measures calculated by the program?
(b) Does the ratio of the length of the major axis of the manipulability ellipse and the length of the minor axis depend on $\theta_1$? On $\theta_2$? Explain your answers.
(c) Choose $L_1 = L_2 = 1$. Hand-draw the following: the arm at $(-45°, 90°)$; the endpoint linear velocity vector arising from $\dot{\theta}_1 = 1$ rad/s and $\dot{\theta}_2 = 0$; the endpoint linear velocity vector arising from $\dot{\theta}_1 = 0$ and $\dot{\theta}_2 = 1$ rad/s; and the vector sum of these two vectors to get the endpoint linear velocity when $\dot{\theta}_1 = 1$ rad/s and $\dot{\theta}_2 = 1$ rad/s.

**Exercise 5.24** Modify the program in the previous exercise to plot the force ellipse. Demonstrate it at the same four configurations as in the first part of the previous exercise.

**Exercise 5.25** The kinematics of the 6R UR5 robot are given in Section 4.1.2.

(a) Give the numerical space Jacobian $J_s$ when all joint angles are $\pi/2$. Separate the Jacobian matrix into an angular velocity portion $J_\omega$ (the joint rates act on the angular velocity) and a linear velocity portion $J_v$ (the joint rates act on the linear velocity).

(b) For this configuration, calculate the directions and lengths of the principal semi-axes of the three-dimensional angular-velocity manipulability ellipsoid (based on $J_\omega$) and the directions and lengths of the principal semi-axes of the three-dimensional linear-velocity manipulability ellipsoid (based on $J_v$). Comment on why it is usually preferred to use the body Jacobian instead of the space Jacobian for the manipulability ellipsoid.

(c) For this configuration, calculate the directions and lengths of the principal semi-axes of the three-dimensional moment (torque) force ellipsoid (based on $J_\omega$) and the directions and lengths of the principal semi-axes of the three-dimensional linear force ellipsoid (based on $J_v$).

**Exercise 5.26** The kinematics of the 7R WAM robot are given in Section 4.1.3.

(a) Give the numerical body Jacobian $J_b$ when all joint angles are $\pi/2$. Separate the Jacobian matrix into an angular-velocity portion $J_\omega$ (the joint rates act on the angular velocity) and a linear-velocity portion $J_v$ (the joint rates act on the linear velocity).

(b) For this configuration, calculate the directions and lengths of the principal semi-axes of the three-dimensional angular-velocity manipulability ellipsoid (based on $J_\omega$) and the directions and lengths of the principal semi-axes of the three-dimensional linear-velocity manipulability ellipsoid (based on $J_v$).

(c) For this configuration, calculate the directions and lengths of the principal semi-axes of the three-dimensional moment (torque) force ellipsoid (based on $J_\omega$) and the directions and lengths of the principal semi-axes of the three-dimensional linear force ellipsoid (based on $J_v$).

**Exercise 5.27** Examine the software functions for this chapter in your favorite programming language. Verify that they work in the way that you expect. Can you make them more computationally efficient?

# Inverse Kinematics

For a general $n$ degree-of-freedom open chain with forward kinematics $T(\theta)$, $\theta \in \mathbb{R}^n$, the inverse kinematics problem can be stated as follows: given a homogeneous transform $X \in SE(3)$, find solutions $\theta$ that satisfy $T(\theta) = X$. To highlight the main features of the inverse kinematics problem, let us examine the two-link planar open chain of Figure 6.1(a) as a motivational example. Considering only the position of the end-effector and ignoring its orientation, the forward kinematics can be expressed as

$$\left[ \begin{array}{c} x \\ y \end{array} \right] = \left[ \begin{array}{c} L_1 \cos \theta_1 + L_2 \cos(\theta_1 + \theta_2) \\ L_1 \sin \theta_1 + L_2 \sin(\theta_1 + \theta_2) \end{array} \right]. \tag{6.1}$$

Assuming $L_1 > L_2$, the set of reachable points, or the workspace, is an annulus of inner radius $L_1 - L_2$ and outer radius $L_1 + L_2$. Given some end-effector position $(x, y)$, it is not hard to see that there will be either zero, one, or two solutions depending on whether $(x, y)$ lies in the exterior, boundary, or interior of this annulus, respectively. When there are two solutions, the angle at the second joint (the "elbow" joint) may be positive or negative. These two solutions are sometimes called "lefty" and "righty" solutions, or "elbow-up" and "elbow-down" solutions.

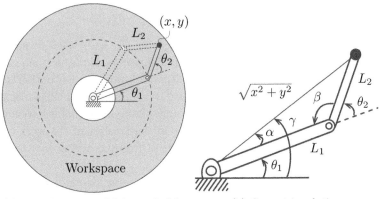

(a) A workspace, and lefty and righty configurations.

(b) Geometric solution.

**Figure 6.1** Inverse kinematics of a 2R planar open chain.

Finding an explicit solution $(\theta_1, \theta_2)$ for a given $(x, y)$ is also not difficult. For this purpose, we will find it useful to introduce the two-argument arctangent function $\text{atan2}(y, x)$, which returns the angle from the origin to a point $(x, y)$ in the plane. It is similar to the inverse tangent $\tan^{-1}(y/x)$, but whereas $\tan^{-1}(y/x)$ is equal to $\tan^{-1}(-y/-x)$, and therefore $\tan^{-1}$ only returns angles in the range $[-\pi/2, \pi/2]$, the atan2 function returns angles in the range $(-\pi, \pi]$. For this reason, atan2 is sometimes called the four-quadrant arctangent.

We also recall the law of cosines,

$$c^2 = a^2 + b^2 - 2ab \cos C,$$

where $a$, $b$, and $c$ are the lengths of the three sides of a triangle and $C$ is the interior angle of the triangle opposite the side of length $c$.

Referring to Figure 6.1(b), angle $\beta$, restricted to lie in the interval $[0, \pi]$, can be determined from the law of cosines,

$$L_1^2 + L_2^2 - 2L_1L_2 \cos \beta = x^2 + y^2,$$

from which it follows that

$$\beta = \cos^{-1}\left(\frac{L_1^2 + L_2^2 - x^2 - y^2}{2L_1L_2}\right).$$

Also from the law of cosines,

$$\alpha = \cos^{-1}\left(\frac{x^2 + y^2 + L_1^2 - L_2^2}{2L_1\sqrt{x^2 + y^2}}\right).$$

The angle $\gamma$ is determined using the two-argument arctangent function, $\gamma = \text{atan2}(y, x)$. With these angles, the righty solution to the inverse kinematics is

$$\theta_1 = \gamma - \alpha, \qquad \theta_2 = \pi - \beta$$

and the lefty solution is

$$\theta_1 = \gamma + \alpha, \qquad \theta_2 = \beta - \pi.$$

If $x^2 + y^2$ lies outside the range $[L_1 - L_2, L_1 + L_2]$ then no solution exists.

This simple motivational example illustrates that, for open chains, the inverse kinematics problem may have multiple solutions; this situation is in contrast with the forward kinematics, where a unique end-effector displacement $T$ exists for given joint values $\theta$. In fact, three-link planar open chains have an infinite number of solutions for points $(x, y)$ lying in the interior of the workspace; in this case the chain possesses an extra degree of freedom and is said to be kinematically redundant.

In this chapter we first consider the inverse kinematics of spatial open chains with six degrees of freedom. At most a finite number of solutions exists in this case, and we consider two popular structures – the PUMA and Stanford robot arms – for which analytic inverse kinematic solutions can be easily obtained. For more general open chains, we adapt the Newton–Raphson method to the

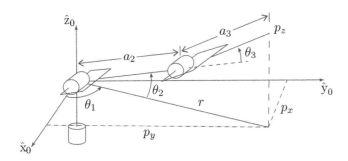

**Figure 6.2** Inverse position kinematics of a 6R PUMA-type arm.

inverse kinematics problem. The result is an iterative numerical algorithm which, provided that an initial guess of the joint variables is sufficiently close to a true solution, converges quickly to that solution.

## .1    Analytic Inverse Kinematics

We begin by writing the forward kinematics of a spatial six-dof open chain in the following product of exponentials form:

$$T(\theta) = e^{[\mathcal{S}_1]\theta_1} e^{[\mathcal{S}_2]\theta_2} e^{[\mathcal{S}_3]\theta_3} e^{[\mathcal{S}_4]\theta_4} e^{[\mathcal{S}_5]\theta_5} e^{[\mathcal{S}_6]\theta_6} M.$$

Given some end-effector frame $X \in SE(3)$, the inverse kinematics problem is to find solutions $\theta \in \mathbb{R}^6$ satisfying $T(\theta) = X$. In the following subsections we derive analytic inverse kinematic solutions for the PUMA and Stanford arms.

### .1.1    6R PUMA-Type Arm

We first consider a 6R arm of the PUMA type. Referring to Figure 6.2, when the arm is placed in its zero position: (i) the two shoulder joint axes intersect orthogonally at a common point, with joint axis 1 aligned in the $\hat{z}_0$-direction and joint axis 2 aligned in the $-\hat{y}_0$-direction; (ii) joint axis 3 (the elbow joint) lies in the $\hat{x}_0$–$\hat{y}_0$-plane and is aligned parallel with joint axis 2; (iii) joint axes 4, 5, and 6 (the wrist joints) intersect orthogonally at a common point (the wrist center) to form an orthogonal wrist and, for the purposes of this example, we assume that these joint axes are aligned in the $\hat{z}_0$-, $\hat{y}_0$-, and $\hat{x}_0$-directions, respectively. The lengths of links 2 and 3 are $a_2$ and $a_3$, respectively. The arm may also have an offset at the shoulder (see Figure 6.3). The inverse kinematics problem for PUMA-type arms can be decoupled into inverse-position and inverse-orientation subproblems, as we now show.

We first consider the simple case of a zero-offset PUMA-type arm. Referring to Figure 6.2 and expressing all vectors in terms of fixed-frame coordinates, denote

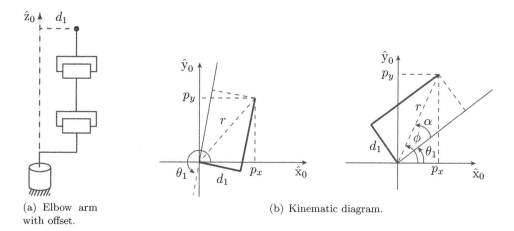

(a) Elbow arm with offset.

(b) Kinematic diagram.

**Figure 6.3** A 6R PUMA-type arm with a shoulder offset.

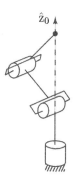

**Figure 6.4** Singular configuration of the zero-offset 6R PUMA-type arm.

the components of the wrist center $p \in \mathbb{R}^3$ by $p = (p_x, p_y, p_z)$. Projecting $p$ onto the $\hat{x}_0$–$\hat{y}_0$-plane, it can be seen that

$$\theta_1 = \operatorname{atan2}(p_y, p_x).$$

Note that a second valid solution for $\theta_1$ is given by

$$\theta_1 = \operatorname{atan2}(p_y, p_x) + \pi,$$

when the original solution for $\theta_2$ is replaced by $\pi - \theta_2$. As long as $p_x, p_y \neq 0$ both these solutions are valid. When $p_x = p_y = 0$ the arm is in a singular configuration (see Figure 6.4), and there are infinitely many possible solutions for $\theta_1$.

If there is an offset $d_1 \neq 0$ as shown in Figure 6.3, then in general there will be two solutions for $\theta_1$, the righty and lefty solutions (Figure 6.3). As seen from the figure, $\theta_1 = \phi - \alpha$ where $\phi = \operatorname{atan2}(p_y, p_x)$ and $\alpha = \operatorname{atan2}(d_1, \sqrt{r^2 - d_1^2})$. The second solution is given by

$$\theta_1 = \pi + \operatorname{atan2}(p_y, p_x) + \operatorname{atan2}\left(-\sqrt{p_x^2 + p_y^2 - d_1^2}, d_1\right).$$

Determining angles $\theta_2$ and $\theta_3$ for the PUMA-type arm now reduces to the inverse

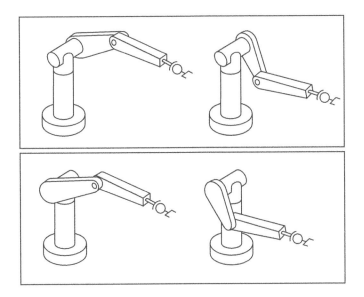

**Figure 6.5** Four possible inverse kinematics solutions for the 6R PUMA type arm with shoulder offset.

kinematics problem for a planar two-link chain:

$$\cos\theta_3 = \frac{r^2 - d_1^2 + p_z^2 - a_2^2 - a_3^2}{2a_2a_3}$$

$$= \frac{p_x^2 + p_y^2 + p_z^2 - d_1^2 - a_2^2 - a_3^2}{2a_2a_3} = D.$$

Using $D$ defined above, $\theta_3$ is given by

$$\theta_3 = \text{atan2}\left(\pm\sqrt{1 - D^2}, D\right)$$

and $\theta_2$ can be obtained in a similar fashion as

$$\theta_2 = \text{atan2}\left(p_z, \sqrt{r^2 - d_1^2}\right) - \text{atan2}\left(a_3 s_3, a_2 + a_3 c_3\right)$$

$$= \text{atan2}\left(p_z, \sqrt{p_x^2 + p_y^2 - d_1^2}\right) - \text{atan2}\left(a_3 s_3, a_2 + a_3 c_3\right),$$

where $s_3 = \sin\theta_3$ and $c_3 = \cos\theta_3$. The two solutions for $\theta_3$ correspond to the well-known elbow-up and elbow-down configurations for the two-link planar arm. In general, a PUMA-type arm with an offset will have four solutions to the inverse position problem, as shown in Figure 6.5; the postures in the upper panel are lefty solutions (elbow-up and elbow-down), while those in the lower panel are righty solutions (elbow-up and elbow-down).

We now solve the inverse orientation problem; that is, that of finding $(\theta_4, \theta_5, \theta_6)$ given the end-effector orientation. This problem is completely straightforward:

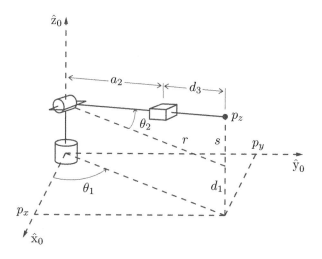

**Figure 6.6** The first three joints of a Stanford-type arm.

having found $(\theta_1, \theta_2, \theta_3)$, the forward kinematics can be manipulated into the form

$$e^{[\mathcal{S}_4]\theta_4}e^{[\mathcal{S}_5]\theta_5}e^{[\mathcal{S}_6]\theta_6} = e^{-[\mathcal{S}_3]\theta_3}e^{-[\mathcal{S}_2]\theta_2}e^{-[\mathcal{S}_1]\theta_1}XM^{-1}, \qquad (6.2)$$

where the right-hand side is now known, and the $\omega_i$-components of $\mathcal{S}_4$, $\mathcal{S}_5$, and $\mathcal{S}_6$ are

$$\omega_4 = (0, 0, 1),$$
$$\omega_5 = (0, 1, 0),$$
$$\omega_6 = (1, 0, 0).$$

Denoting the $SO(3)$ component of the right-hand side of Equation (6.2) by $R$, the wrist joint angles $(\theta_4, \theta_5, \theta_6)$ can be determined as the solution to

$$\mathrm{Rot}(\hat{z}, \theta_4)\mathrm{Rot}(\hat{y}, \theta_5)\mathrm{Rot}(\hat{x}, \theta_6) = R,$$

which correspond exactly to the ZYX Euler angles, derived in Appendix B.

### 6.1.2    Stanford-Type Arms

If the elbow joint in a 6R PUMA-type arm is replaced by a prismatic joint, as shown in Figure 6.6, we then have an RRPRRR Stanford-type arm. Here we consider the inverse position kinematics for the arm of Figure 6.6; the inverse orientation kinematics is identical to that for the PUMA-type arm and so is not repeated here.

The first joint variable $\theta_1$ can be found in similar fashion to the PUMA-type arm: $\theta_1 = \mathrm{atan2}(p_y, p_x)$ (provided that $p_x$ and $p_y$ are not both zero). The variable

$\theta_2$ is then found from Figure 6.6 to be

$$\theta_2 = \text{atan2}(s, r),$$

where $r^2 = p_x^2 + p_y^2$ and $s = p_z - d_1$. Similarly to the case of the PUMA-type arm, a second solution for $\theta_1$ and $\theta_2$ is given by

$$\theta_1 = \pi + \text{atan2}(p_y, p_x),$$
$$\theta_2 = \pi - \text{atan2}(s, r).$$

The translation distance $\theta_3$ is found from the relation

$$(\theta_3 + a_2)^2 = r^2 + s^2$$

as

$$\theta_3 = \sqrt{r^2 + s^2} = \sqrt{p_x^2 + p_y^2 + (p_z - d_1)^2} - a_2.$$

Ignoring the negative square root solution for $\theta_3$, we obtain two solutions to the inverse position kinematics as long as the wrist center $p$ does not intersect the $\hat{z}_0$-axis of the fixed frame. If there is an offset then, as in the case of the PUMA-type arm, there will be lefty and righty solutions.

## .2     Numerical Inverse Kinematics

Iterative numerical methods can be applied if the inverse kinematics equations do not admit analytic solutions. Even in cases where an analytic solution does exist, numerical methods are often used to improve the accuracy of these solutions. For example, in a PUMA-type arm, the last three axes may not exactly intersect at a common point, and the shoulder joint axes may not be exactly orthogonal. In such cases, rather than throw away any analytic inverse kinematic solutions that are available, such solutions can be used as the initial guess in an iterative numerical procedure for solving the inverse kinematics.

There exist a variety of iterative methods for finding the roots of a nonlinear equation, and our aim is not to discuss these in detail – any text on numerical analysis will cover these methods in depth – but rather to develop ways in which to transform the inverse kinematics equations so that they become amenable to existing numerical methods. We will make use of an approach fundamental to nonlinear root-finding, the Newton–Raphson method. Also, methods of optimization are needed in situations where an exact solution may not exist and we seek the closest approximate solution; or, conversely, an infinity of inverse kinematics solutions exists (i.e., if the robot is kinematically redundant) and we seek a solution that is optimal with respect to some criterion. We now therefore discuss the Newton–Raphson method for nonlinear root-finding and also the first-order necessary conditions for optimization.

### 6.2.1      Newton–Raphson Method

To solve the equation $g(\theta) = 0$ numerically for a given differentiable function $g : \mathbb{R} \to \mathbb{R}$, assume $\theta^0$ is an initial guess for the solution. Write the Taylor expansion of $g(\theta)$ at $\theta^0$ and truncate it at first order:

$$g(\theta) = g(\theta^0) + \frac{\partial g}{\partial \theta}(\theta^0)(\theta - \theta^0) + \text{higher-order terms (h.o.t.)}.$$

Keeping only the terms up to first order, set $g(\theta) = 0$ and solve for $\theta$ to obtain

$$\theta = \theta^0 - \left(\frac{\partial g}{\partial \theta}(\theta^0)\right)^{-1} g(\theta^0).$$

Using this value of $\theta$ as the new guess for the solution and repeating the above, we get the following iteration:

$$\theta^{k+1} = \theta^k - \left(\frac{\partial g}{\partial \theta}(\theta^k)\right)^{-1} g(\theta^k).$$

The above iteration is repeated until some stopping criterion is satisfied, e.g., $|g(\theta^k) - g(\theta^{k+1})|/|g(\theta^k)| \leq \epsilon$ for some user-prescribed threshold value $\epsilon$.

The same formula applies for the case when $g$ is multi-dimensional, i.e., $g : \mathbb{R}^n \to \mathbb{R}^n$, in which case

$$\frac{\partial g}{\partial \theta}(\theta) = \begin{bmatrix} \frac{\partial g_1}{\partial \theta_1}(\theta) & \cdots & \frac{\partial g_1}{\partial \theta_n}(\theta) \\ \vdots & \ddots & \vdots \\ \frac{\partial g_n}{\partial \theta_1}(\theta) & \cdots & \frac{\partial g_n}{\partial \theta_n}(\theta) \end{bmatrix} \in \mathbb{R}^{n \times n}.$$

The case where the above matrix fails to be invertible is discussed in Section 6.2.2.

### 6.2.2      Numerical Inverse Kinematics Algorithm

Suppose we express the end-effector frame using a coordinate vector $x$ governed by the forward kinematics $x = f(\theta)$, a nonlinear vector equation mapping the $n$ joint coordinates to the $m$ end-effector coordinates. Assume that $f : \mathbb{R}^n \to \mathbb{R}^m$ is differentiable, and let $x_d$ be the desired end-effector coordinates. Then $g(\theta)$ for the Newton–Raphson method is defined as $g(\theta) = x_d - f(\theta)$, and the goal is to find joint coordinates $\theta_d$ such that

$$g(\theta_d) = x_d - f(\theta_d) = 0.$$

Given an initial guess $\theta^0$ which is "close to" a solution $\theta_d$, the kinematics can be expressed as the Taylor expansion

$$x_d = f(\theta_d) = f(\theta^0) + \underbrace{\frac{\partial f}{\partial \theta}\bigg|_{\theta^0}}_{J(\theta^0)} \underbrace{(\theta_d - \theta^0)}_{\Delta\theta} + \text{h.o.t.}, \tag{6.3}$$

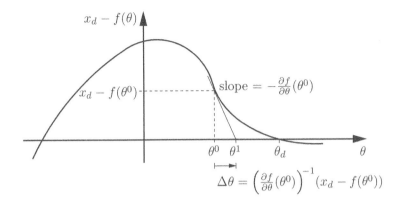

**Figure 6.7** The first step of the Newton–Raphson method for nonlinear root-finding for a scalar $x$ and $\theta$. In the first step, the slope $-\partial f/\partial\theta$ is evaluated at the point $(\theta^0, x_d - f(\theta^0))$. In the second step, the slope is evaluated at the point $(\theta^1, x_d - f(\theta^1))$ and eventually the process converges to $\theta_d$. Note that an initial guess to the left of the plateau of $x_d - f(\theta)$ would be likely to result in convergence to the other root of $x_d - f(\theta)$, and an initial guess at or near the plateau would result in a large initial $|\Delta\theta|$ and the iterative process might not converge at all.

where $J(\theta^0) \in \mathbb{R}^{m \times n}$ is the coordinate Jacobian evaluated at $\theta^0$. Truncating the Taylor expansion at first order, we can approximate Equation (6.3) as

$$J(\theta^0)\Delta\theta = x_d - f(\theta^0). \tag{6.4}$$

Assuming that $J(\theta^0)$ is square ($m = n$) and invertible, we can solve for $\Delta\theta$ as

$$\Delta\theta = J^{-1}(\theta^0)\left(x_d - f(\theta^0)\right). \tag{6.5}$$

If the forward kinematics is linear in $\theta$, i.e., the higher-order terms in Equation (6.3) are zero, then the new guess $\theta^1 = \theta^0 + \Delta\theta$ exactly satisfies $x_d = f(\theta^1)$. If the forward kinematics is not linear in $\theta$, as is usually the case, the new guess $\theta^1$ should still be closer to the root than $\theta^0$, and the process is then repeated, producing a sequence $\{\theta^0, \theta^1, \theta^2, \ldots\}$ converging to $\theta_d$ (Figure 6.7).

As indicated in Figure 6.7, if there are multiple inverse kinematics solutions, the iterative process tends to converge to the solution that is "closest" to the initial guess $\theta^0$. You can think of each solution as having its own basin of attraction. If the initial guess is not in one of these basins (e.g., the initial guess is not sufficiently close to a solution), the iterative process may not converge.

In practice, for computational efficiency reasons, Equation (6.4) is often solved without directly calculating the inverse $J^{-1}(\theta^0)$. More efficient techniques exist for solving a set of linear equations $Ax = b$ for $x$. For example, for invertible square matrices $A$, the LU decomposition of $A$ can be used to solve for $x$ with fewer operations. In MATLAB, for example, the syntax

```
x = A\b
```

solves $Ax = b$ for $x$ without computing $A^{-1}$.

If $J$ is not invertible, either because it is not square or because it is singular, then $J^{-1}$ in Equation (6.5) does not exist. Equation (6.4) can still be solved (or approximately solved) for $\Delta\theta$ by replacing $J^{-1}$ in Equation (6.5) with the Moore–Penrose **pseudoinverse** $J^\dagger$. For any equation of the form $Jy = z$, where $J \in \mathbb{R}^{m \times n}$, $y \in \mathbb{R}^n$, and $z \in \mathbb{R}^m$, the solution

$$y^* = J^\dagger z$$

falls into one of two categories:

- The solution $y^*$ exactly satisfies $Jy^* = z$ and, for any solution $y$ exactly satisfying $Jy = z$, we have $\|y^*\| \leq \|y\|$. In other words, among all solutions, $y^*$ minimizes the two-norm. There can be an infinite number of solutions $y$ to $Jy = z$ if the robot has more joints $n$ than end-effector coordinates $m$, i.e., the Jacobian $J$ is "fat."
- If there is no $y$ that exactly satisfies $Jy = z$ then $y^*$ minimizes the two-norm of the error, i.e., $\|Jy^* - z\| \leq \|Jy - z\|$ for any $y \in \mathbb{R}^n$. This case corresponds to rank $J < m$, i.e., the robot has fewer joints $n$ than end-effector coordinates $m$ (a "tall" Jacobian $J$) or it is at a singularity.

Many programming languages provide functions to calculate the pseudoinverse; for example, the usage in MATLAB is

```
y = pinv(J) * z
```

In the case where $J$ is full rank (rank $m$ for $n > m$ or rank $n$ for $n < m$), i.e., the robot is not at a singularity, the pseudoinverse can be calculated as

$$J^\dagger = J^{\mathrm{T}}(JJ^{\mathrm{T}})^{-1} \quad \text{if } J \text{ is fat, } n > m \text{ (called a right inverse, since } JJ^\dagger = I)$$
$$J^\dagger = (J^{\mathrm{T}}J)^{-1}J^{\mathrm{T}} \quad \text{if } J \text{ is tall, } n < m \text{ (called a left inverse, since } J^\dagger J = I).$$

Replacing the Jacobian inverse with the pseudoinverse, Equation (6.5) becomes

$$\Delta\theta = J^\dagger(\theta^0)\left(x_d - f(\theta^0)\right). \tag{6.6}$$

If rank$(J) < m$ then the solution $\Delta\theta$ calculated in Equation (6.6) may not exactly satisfy Equation (6.4), but it satisfies this condition as closely as possible in a least-squares sense. If $n > m$ then the solution is the smallest joint variable change (in the two-norm sense) that exactly satisfies Equation (6.4).

Equation (6.6) suggests using the Newton–Raphson iterative algorithm for finding $\theta_d$:

(a) **Initialization**: Given $x_d \in \mathbb{R}^m$ and an initial guess $\theta^0 \in \mathbb{R}^n$, set $i = 0$.
(b) Set $e = x_d - f(\theta^i)$. While $\|e\| > \epsilon$ for some small $\epsilon$:
  - Set $\theta^{i+1} = \theta^i + J^\dagger(\theta^i)e$.
  - Increment $i$.

To modify this algorithm to work with a desired end-effector configuration represented as $T_{sd} \in SE(3)$ instead of as a coordinate vector $x_d$, we can replace the coordinate Jacobian $J$ with the end-effector body Jacobian $J_b \in \mathbb{R}^{6 \times n}$. Note, however, that the vector $e = x_d - f(\theta^i)$, representing the direction from the current guess (evaluated through the forward kinematics) to the desired end-effector configuration, cannot simply be replaced by $T_{sd} - T_{sb}(\theta^i)$; the pseudoinverse of $J_b$ should act on a body twist $\mathcal{V}_b \in \mathbb{R}^6$. To find the right analogy, we should think of $e = x_d - f(\theta^i)$ as a velocity vector which, if followed for unit time, would cause a motion from $f(\theta^i)$ to $x_d$. Similarly, we should look for a body twist $\mathcal{V}_b$ which, if followed for unit time, would cause a motion from $T_{sb}(\theta^i)$ to the desired configuration $T_{sd}$.

To find this $\mathcal{V}_b$, we first calculate the desired configuration in the body frame,

$$T_{bd}(\theta^i) = T_{sb}^{-1}(\theta^i)T_{sd} = T_{bs}(\theta^i)T_{sd}.$$

Then $\mathcal{V}_b$ is determined using the matrix logarithm,

$$[\mathcal{V}_b] = \log T_{bd}(\theta^i).$$

This leads to the following inverse kinematics algorithm, which is analogous to the above coordinate-vector algorithm:

(a) **Initialization**: Given $T_{sd}$ and an initial guess $\theta^0 \in \mathbb{R}^n$, set $i = 0$.
(b) Set $[\mathcal{V}_b] = \log\left(T_{sb}^{-1}(\theta^i)T_{sd}\right)$. While $\|\omega_b\| > \epsilon_\omega$ or $\|v_b\| > \epsilon_v$ for small $\epsilon_\omega, \epsilon_v$:
   - Set $\theta^{i+1} = \theta^i + J_b^\dagger(\theta^i)\mathcal{V}_b$.
   - Increment $i$.

An equivalent form can be derived in the space frame, using the space Jacobian $J_s(\theta)$ and the spatial twist $\mathcal{V}_s = [\mathrm{Ad}_{T_{sb}}]\mathcal{V}_b$.

For this numerical inverse kinematics method to converge, the initial guess $\theta^0$ should be sufficiently close to a solution $\theta_d$. This condition can be satisfied by starting the robot from an initial home configuration where both the actual end-effector configuration and the joint angles are known and ensuring that the requested end-effector position $T_{sd}$ changes slowly relative to the frequency of the calculation of the inverse kinematics. Then, for the rest of the robot's run, the calculated $\theta_d$ at the previous timestep serves as the initial guess $\theta^0$ for the new $T_{sd}$ at the next timestep.

**Example 6.1** (Planar 2R robot)   Now we apply the body Jacobian Newton–Raphson inverse kinematics algorithm to the 2R robot in Figure 6.8. Each link is 1 m in length, and we would like to find the joint angles that place the tip of the robot at $(x, y) = (0.366 \text{ m}, 1.366 \text{ m})$, which corresponds to to $\theta_d = (30°, 90°)$ and

$$T_{sd} = \begin{bmatrix} -0.5 & -0.866 & 0 & 0.366 \\ 0.866 & -0.5 & 0 & 1.366 \\ 0 & 0 & 1 & 0 \\ 0 & 0 & 0 & 1 \end{bmatrix}$$

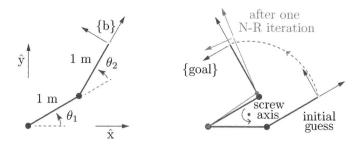

**Figure 6.8** (Left) A 2R robot. (Right) The goal is to find the joint angles yielding the end-effector frame {goal} corresponding to $\theta_1 = 30°$ and $\theta_2 = 90°$. The initial guess is $(0°, 30°)$. After one Newton–Raphson iteration, the calculated joint angles are $(34.23°, 79.18°)$. The screw axis that takes the initial frame to the goal frame (by means of the curved dashed line) is also indicated.

as shown by the frame {goal} in Figure 6.8. The forward kinematics, expressed in the end-effector frame, is given by

$$
M = \begin{bmatrix} 1 & 0 & 0 & 2 \\ 0 & 1 & 0 & 0 \\ 0 & 0 & 1 & 0 \\ 0 & 0 & 0 & 1 \end{bmatrix}, \qquad \mathcal{B}_1 = \begin{bmatrix} 0 \\ 0 \\ 1 \\ 0 \\ 2 \\ 0 \end{bmatrix}, \qquad \mathcal{B}_2 = \begin{bmatrix} 0 \\ 0 \\ 1 \\ 0 \\ 1 \\ 0 \end{bmatrix}.
$$

Our initial guess at the solution is $\theta^0 = (0, 30°)$, and we specify an error tolerance of $\epsilon_\omega = 0.001$ rad (or $0.057°$) and $\epsilon_v = 10^{-4}$ m (100 microns). The progress of the Newton–Raphson method is illustrated in the table below, where only the $(\omega_{zb}, v_{xb}, v_{yb})$-components of the body twist $\mathcal{V}_b$ are given since the robot's motion is restricted to the $x$–$y$-plane:

| $i$ | $(\theta_1, \theta_2)$ | $(x, y)$ | $\mathcal{V}_b = (\omega_{zb}, v_{xb}, v_{yb})$ | $\|\omega_b\|$ | $\|v_b\|$ |
|---|---|---|---|---|---|
| 0 | $(0.00, 30.00°)$ | $(1.866, 0.500)$ | $(1.571, 0.498, 1.858)$ | 1.571 | 1.924 |
| 1 | $(34.23°, 79.18°)$ | $(0.429, 1.480)$ | $(0.115, -0.074, 0.108)$ | 0.115 | 0.131 |
| 2 | $(29.98°, 90.22°)$ | $(0.363, 1.364)$ | $(-0.004, 0.000, -0.004)$ | 0.004 | 0.004 |
| 3 | $(30.00°, 90.00°)$ | $(0.366, 1.366)$ | $(0.000, 0.000, 0.000)$ | 0.000 | 0.000 |

The iterative procedure converges to within the tolerances after three iterations. Figure 6.8 shows the initial guess, the goal configuration, and the configuration after one iteration. Notice that the first $v_{xb}$ calculated is positive, even though the origin of the goal frame is in the $-\hat{x}_b$-direction of the initial guess. The reason is that the constant body velocity $\mathcal{V}_b$ that takes the initial guess to {goal} in one second is a rotation about the screw axis indicated in the figure.

**.3   Inverse Velocity Kinematics**

One solution for controlling a robot so that it follows a desired end-effector trajectory $T_{sd}(t)$ is to calculate the inverse kinematics $\theta_d(k\Delta t)$ at each discrete timestep $k$, then control the joint velocities $\dot{\theta}$ as follows

$$\dot{\theta} = \big(\theta_d(k\Delta t) - \theta((k-1)\Delta t)\big)/\Delta t$$

during the time interval $[(k-1)\Delta t, k\Delta t]$. This amounts to a feedback controller since the desired new joint angles $\theta_d(k\Delta t)$ are being compared with the most recently measured actual joint angles $\theta((k-1)\Delta t)$ in order to calculate the required joint velocities.

Another option that avoids the computation of inverse kinematics is to calculate the required joint velocities $\dot{\theta}$ directly from the relationship $J\dot{\theta} = \mathcal{V}_d$, where the desired end-effector twist $\mathcal{V}_d$ and $J$ are expressed with respect to the same frame:

$$\dot{\theta} = J^{\dagger}(\theta)\mathcal{V}_d. \tag{6.7}$$

The matrix form $[\mathcal{V}_d(t)]$ of the desired twist is either $T_{sd}^{-1}(t)\dot{T}_{sd}(t)$ (the matrix form of the body twist of the desired trajectory at time $t$) or $\dot{T}_{sd}(t)T_{sd}^{-1}(t)$ (the matrix form of the spatial twist), depending on whether the body Jacobian or space Jacobian is used; however small velocity errors are likely to accumulate over time, resulting in increasing position error. Thus, a position feedback controller should choose $\mathcal{V}_d(t)$ so as to keep the end-effector following $T_{sd}(t)$ with little position error. Feedback control is discussed in Chapter 11.

In the case of a redundant robot with $n > 6$ joints, of the $(n-6)$-dimensional set of joint velocities satisfying Equation (6.7), the use of the pseudoinverse $J^{\dagger}(\theta)$ returns joint velocities $\dot{\theta}$ minimizing the two-norm $\|\dot{\theta}\| = \sqrt{\dot{\theta}^{\mathrm{T}}\dot{\theta}}$.

The use of the pseudoinverse in Equation (6.7) implicitly weights the cost of each joint velocity identically. We could instead give the joint velocities different weights; for example, the velocity at the first joint, which moves a lot of the robot's mass, could be weighted more heavily than the velocity at the last joint, which moves little of the robot's mass. As we will see later, the kinetic energy of a robot can be written

$$\frac{1}{2}\dot{\theta}^{\mathrm{T}}M(\theta)\dot{\theta},$$

where $M(\theta)$ is the symmetric, positive-definite, configuration-dependent mass matrix of the robot. The mass matrix $M(\theta)$ can be used as a weighting function in the inverse velocity kinematics, and the goal is to find the $\dot{\theta}$ that minimizes the kinetic energy while also satisfying $J(\theta)\dot{\theta} = \mathcal{V}_d$.

Another possibility is to find the $\dot{\theta}$ that causes the robot to minimize a configuration-dependent potential energy function $h(\theta)$ while satisfying $J(\theta)\dot{\theta} = \mathcal{V}_d$. For example, $h(\theta)$ could be the gravitational potential energy, or an artificial potential function whose value increases as the robot approaches an obstacle.

Then the rate of change of $h(\theta)$ is

$$\frac{d}{dt}h(\theta) = \frac{dh(\theta)}{d\theta}\frac{d\theta}{dt} = \nabla h(\theta)^{\mathrm{T}}\dot{\theta},$$

where $\nabla h(\theta)$ points in the direction of maximum ascent of $h(\theta)$.

More generally, we may wish to minimize the sum of the kinetic energy and the rate of change of the potential energy:

$$\min_{\dot{\theta}} \frac{1}{2}\dot{\theta}^{\mathrm{T}}M(\theta)\dot{\theta} + \nabla h(\theta)^{\mathrm{T}}\dot{\theta},$$

subject to the constraint $J(\theta)\dot{\theta} = \mathcal{V}_d$. From the first-order necessary conditions for optimality (Appendix D)

$$J^{\mathrm{T}}\lambda = M\dot{\theta} + \nabla h,$$
$$\mathcal{V}_d = J\dot{\theta},$$

the optimal $\dot{\theta}$ and $\lambda$ can be derived as follows:

$$\dot{\theta} = G\mathcal{V}_d + (I - GJ)M^{-1}\nabla h,$$
$$\lambda = B\mathcal{V}_d + BJM^{-1}\nabla h,$$

where $B \in \mathbb{R}^{m \times m}$ and $G \in \mathbb{R}^{n \times m}$ are defined by

$$B = (JM^{-1}J^{\mathrm{T}})^{-1},$$
$$G = M^{-1}J^{\mathrm{T}}(JM^{-1}J^{\mathrm{T}})^{-1} = M^{-1}J^{\mathrm{T}}B.$$

Recalling the static relation $\tau = J^{\mathrm{T}}\mathcal{F}$ from the previous chapter, the Lagrange multiplier $\lambda$ (see Appendix D) can be interpreted as a wrench in task space. Moreover, in the expression $\lambda = B\mathcal{V}_d + BJM^{-1}\nabla h$, the first term, $B\mathcal{V}_d$, can be interpreted as a dynamic force generating the end-effector velocity $\mathcal{V}_d$ while the second term, $BJM^{-1}\nabla h$, can be interpreted as the static wrench counteracting gravity.

If the potential function $h(\theta)$ is zero or unspecified, the kinetic-energy-minimizing solution is

$$\dot{\theta} = M^{-1}J^{\mathrm{T}}(JM^{-1}J^{\mathrm{T}})^{-1}\mathcal{V}_d,$$

where $M^{-1}J^{\mathrm{T}}(JM^{-1}J^{\mathrm{T}})^{-1}$ is the weighted pseudoinverse according to the mass matrix $M(\theta)$.

## 6.4     A Note on Closed Loops

A desired end-effector trajectory over a time interval $[0, t_f]$ is a closed loop if $T_{sd}(0) = T_{sd}(t_f)$. It should be noted that numerical methods for calculating inverse kinematics for redundant robots, at either the configuration or velocity levels, are likely to yield motions that are not closed loops in the joint space, i.e., $\theta(0) \neq \theta(t_f)$. If closed-loop motions in joint space are required, an extra set of conditions on the inverse kinematics must be satisfied.

## .5  Summary

- Given a spatial open chain with forward kinematics $T(\theta)$, $\theta \in \mathbb{R}^n$, in the inverse kinematics problem one seeks to find, for a desired end-effector configuration $X \in SE(3)$, solutions $\theta$ that satisfy $X = T(\theta)$. Unlike the forward kinematics problem, the inverse kinematics problem can possess multiple solutions, or no solutions in the event that $X$ lies outside the workspace. For a spatial open chain with $n$ joints and an $X$ in the workspace, $n = 6$ typically leads to a finite number of inverse kinematic solutions while $n > 6$ leads to an infinite number of solutions.

- The inverse kinematics can be solved analytically for the six-dof PUMA-type robot arm, a popular 6R design consisting of a 3R orthogonal axis wrist connected to a 2R orthogonal axis shoulder by an elbow joint.

- Stanford-type arms also admit analytic inverse kinematics solutions. These arms are obtained by replacing the elbow joint in the generalized 6R PUMA-type arm by a prismatic joint. Geometric inverse kinematic algorithms similar to those for PUMA-type arms have been developed.

- Iterative numerical methods are used in cases where analytic inverse kinematic solutions are unavailable. These methods typically involve solving the inverse kinematics equations using an iterative procedure like the Newton–Raphson method, and they require an initial guess at the joint variables. The performance of the iterative procedure depends to a large extent on the quality of the initial guess and, in the case where there are several possible inverse kinematic solutions, the method finds the solution that is "closest" to the initial guess. Each iteration is of the form

$$\dot{\theta}^{i+1} = \theta^i + J^\dagger(\theta^i)\mathcal{V},$$

where $J^\dagger(\theta)$ is the pseudoinverse of the Jacobian $J(\theta)$ and $\mathcal{V}$ is the twist that takes $T(\theta^i)$ to $T_{sd}$ in one second.

## .6  Software

Software functions associated with this chapter are listed below.

`[thetalist,success] = IKinBody(Blist,M,T,thetalist0,eomg,ev)`
This function uses iterative Newton–Raphson to calculate the inverse kinematics given the list of joint screws $\mathcal{B}_i$ expressed in the end-effector frame, the end-effector home configuration $M$, the desired end-effector configuration $T$, an initial guess at the joint angles $\theta^0$, and the tolerances $\epsilon_\omega$ and $\epsilon_v$ on the final error. If a solution is not found within a maximum number of iterations, then `success` is false.

`[thetalist,success] = IKinSpace(Slist,M,T,thetalist0,eomg,ev)`

This is similar to `IKinBody`, except that the joint screws $\mathcal{S}_i$ are expressed in the space frame and the tolerances are interpreted in the space frame, also.

## 6.7     Notes and References

The inverse kinematics of the most general 6R open chain is known to have up to 16 solutions; this result was proved by Lee and Liang (1988) and by Raghavan and Roth (1990). Procedures for finding closed-form inverse kinematics solutions to somewhat more general six-dof open chains than those treated in this chapter are described in Paden (1986) and Murray et al. (1994); these procedures use solutions to a collection of some basic screw-theoretic subproblems, called Paden–Kahan subproblems, e.g., finding the angle of rotation for a zero-pitch screw motion between a pair of given points. Iterative numerical procedures for finding all 16 solutions of a general 6R open chain were reported in Manocha and Canny (1989).

A comprehensive summary of inverse kinematics methods for kinematically redundant robot arms is given in Chiaverini et al. (2016). Many of these methods rely on results and solution techniques from least-squares optimization, and for this reason we provide a brief review of the basics of optimization in Appendix D; a classic reference for optimization is Luenberger and Ye (2008). The repeatability (or cyclicity) conditions for a general class of inverse kinematic redundancy resolution schemes are examined in Shamir and Yomdin (1988).

## 6.8     Exercises

**Exercise 6.1**   Write a program that solves the analytical inverse kinematics for a planar 3R robot with link lengths $L_1 = 3$, $L_2 = 2$, and $L_3 = 1$, given the desired position $(x, y)$ and orientation $\theta$ of a frame fixed to the tip of the robot. Each joint has no joint limits. Your program should find all the solutions (how many are there in the general case?), give the joint angles for each, and draw the robot in these configurations. Test the code for the case of $(x, y, \theta) = (4, 2, 0)$.

**Exercise 6.2**   Solve the inverse position kinematics (you do not need to solve the orientation kinematics) of the 6R open chain shown in Figure 6.9.

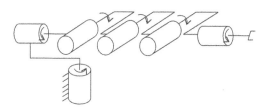

**Figure 6.9**  A 6R open chain.

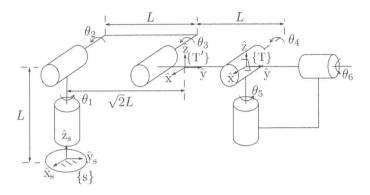

**Figure 6.10** A 6R open chain.

**Exercise 6.3**  Find the inverse kinematics solutions when the end-effector frame {T} of the 6R open chain shown in Figure 6.10 is set to {T′} as shown. The orientation of {T} at the zero position is the same as that of the fixed frame {s}, and {T′} is the result of a pure translation of {T} along the $\hat{y}_s$-axis.

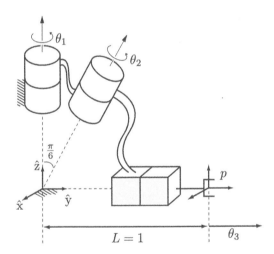

**Figure 6.11** An RRP open chain.

**Exercise 6.4**  The RRP open chain of Figure 6.11 is shown in its zero position. Joint axes 1 and 2 intersect at the fixed frame origin, and the end-effector frame origin $p$ is located at $(0, 1, 0)$ when the robot is in its zero position.

(a) Suppose that $\theta_1 = 0$. Solve for $\theta_2$ and $\theta_3$ when the end-effector frame origin $p$ is at $(-6, 5, \sqrt{3})$.
(b) If joint 1 is not fixed to zero but instead allowed to vary, find all the inverse kinematic solutions $(\theta_1, \theta_2, \theta_3)$ for the $p$ given in (a).

**Exercise 6.5**  The four-dof robot of Figure 6.12 is shown in its zero position.

Joint 1 is a screw joint of pitch $h$. Given the end-effector position $p = (p_x, p_y, p_z)$ and orientation $R = e^{[\hat{z}]\alpha}$, where $\hat{z} = (0, 0, 1)$ and $\alpha \in [0, 2\pi]$, find the inverse kinematics solution $(\theta_1, \theta_2, \theta_3, \theta_4)$ as a function of $p$ and $\alpha$.

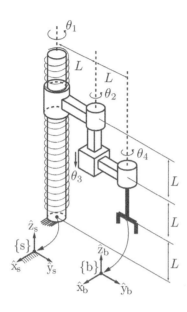

**Figure 6.12** An open chain with a screw joint.

**Exercise 6.6**  Figure 6.13(a) shows a surgical robot, which can be modeled as an RRPRRP open chain as shown in Figure 6.13(b).

(a) In the general case, how many inverse kinematic solutions will exist for a given end-effector frame?
(b) Consider points $A$ and $B$ on the surgical robot shown in Figure 6.13(b). Given coordinates $(x_A, y_A, z_A)$ and $(x_B, y_B, z_B)$ for the points $A$ and $B$ in the fixed frame, find the joint variables $\theta_1$, $\theta_2$, $\theta_3$, $\theta_4$, and $\theta_5$. You should find an explicit formula for $(\theta_1, \theta_2, \theta_3)$ while for $(\theta_4, \theta_5)$ you can just describe the procedure.

**Exercise 6.7**  In this exercise you are asked to draw a plot of a scalar $x_d - f(\theta)$ versus a scalar $\theta$ (similar to Figure 6.7) with two roots. Draw it so that, for some initial guess $\theta^0$, the iterative process actually jumps over the closest root and eventually converges to the further root. Hand-draw the plot and show the iteration process that results in convergence to the further root. Comment on the basins of attraction of the two roots in your plot.

**Exercise 6.8**  Use Newton–Raphson iterative numerical root finding to perform two steps in finding the root of

$$g(x, y) = \begin{bmatrix} x^2 - 4 \\ y^2 - 9 \end{bmatrix}$$

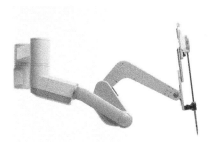

(a) da Vinci S Surgical System instrument arm,
© 2016 Intuitive Surgical, Inc.

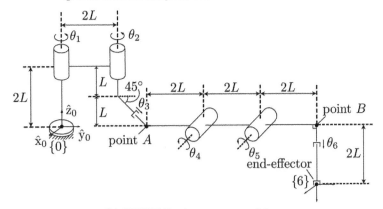

(b) RRPRRP robot at zero position.

**Figure 6.13** Surgical robot and kinematic model.

when your initial guess is $(x^0, y^0) = (1, 1)$. Write the general form of the gradient (for any guess $(x, y)$) and compute the results of the first two iterations. You can do this by hand or write a program. Also, give all the correct roots, not just the one that would be found from your initial guess. How many are there?

**Exercise 6.9** Modify the function IKinBody to print out the results of each Newton–Raphson iteration, in a table similar to that for the 2R robot example in Section 6.2. Show the table produced when the initial guess for the 2R robot of Figure 6.8 is $(0, 30°)$ and the goal configuration corresponds to $(90°, 120°)$.

**Exercise 6.10** The 3R orthogonal axis wrist mechanism of Figure 6.14 is shown in its zero position, with joint axes 1 and 3 collinear.

(a) Given a desired wrist orientation $R \in SO(3)$, derive an iterative numerical procedure for solving its inverse kinematics.
(b) Perform a single iteration of Newton–Raphson root-finding using body-frame numerical inverse kinematics. First write down the forward kinematics and Jacobian for general configurations of the wrist. Then apply your results for

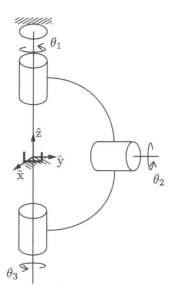

**Figure 6.14** A 3R wrist.

the specific case of an initial guess of $\theta_1 = \theta_3 = 0$, $\theta_2 = \pi/6$, with a desired end-effector frame at

$$R = \begin{bmatrix} \frac{1}{\sqrt{2}} & -\frac{1}{\sqrt{2}} & 0 \\ \frac{1}{\sqrt{2}} & \frac{1}{\sqrt{2}} & 0 \\ 0 & 0 & 1 \end{bmatrix} \in SO(3).$$

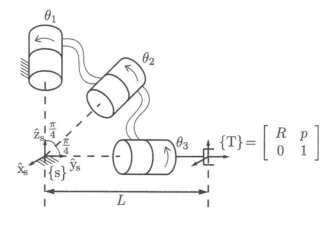

$$\{T\} = \begin{bmatrix} R & p \\ 0 & 1 \end{bmatrix}$$

**Figure 6.15** A 3R nonorthogonal chain.

**Exercise 6.11**   The 3R nonorthogonal chain of Figure 6.15 is shown in its zero position.

(a) Derive a numerical procedure for solving the inverse position kinematics; that is, given some end-effector position $p$ as indicated in the figure, find $(\theta_1, \theta_2, \theta_3)$.

(b) Given an end-effector orientation $R \in SO(3)$, find all inverse kinematic solutions $(\theta_1, \theta_2, \theta_3)$.

**Exercise 6.12** Use the function IKinSpace to find joint variables $\theta_d$ of the UR5 (Section 4.1.2) satisfying

$$T(\theta_d) = T_{sd} = \begin{bmatrix} 0 & 1 & 0 & -0.5 \\ 0 & 0 & -1 & 0.1 \\ -1 & 0 & 0 & 0.1 \\ 0 & 0 & 0 & 1 \end{bmatrix}.$$

The distances are in meters. Use $\epsilon_\omega = 0.001$ rad ($0.057°$) and $\epsilon_v = 0.0001$ (0.1 mm). For your initial guess $\theta^0$, choose all joint angles as 0.1 rad. If $T_{sd}$ is outside the workspace, or if you find that your initial guess is too far from a solution to converge, you may demonstrate IKinBody using another $T_{sd}$.

Note that numerical inverse kinematics is intended to find a solution close to the initial guess. Since your initial guess is not close to a solution (and remember that there are generally multiple solutions), the procedure may thrash about before finding a solution far from the initial guess. This solution may not respect joint limits. You can post-process the solution so that all joint angles are in the range $[0, 2\pi)$.

**Exercise 6.13** Use the function IKinBody to find joint variables $\theta_d$ of the WAM (Section 4.1.3) satisfying

$$T(\theta_d) = T_{sd} = \begin{bmatrix} 1 & 0 & 0 & 0.5 \\ 0 & 1 & 0 & 0 \\ 0 & 0 & 1 & 0.4 \\ 0 & 0 & 0 & 1 \end{bmatrix}.$$

Distances are in meters. Use $\epsilon_\omega = 0.001$ rad ($0.057°$) and $\epsilon_v = 0.0001$ (0.1 mm). For your initial guess $\theta^0$, choose all joint angles as 0.1 rad. If $T_{sd}$ is outside the workspace, or if you find that your initial guess is too far from a solution to converge, you may demonstrate IKinBody using another $T_{sd}$.

Note that numerical inverse kinematics is intended to find a solution close to the initial guess. Since your initial guess is not close to a solution (and remember that there are generally multiple solutions), the procedure may thrash about before finding a solution far away from the initial guess. This solution may not respect joint limits. You can post-process the solution so that all joint angles are in the range $[0, 2\pi)$.

**Exercise 6.14** The fundamental theorem of linear algebra (FTLA) states that, given a matrix $A \in \mathbb{R}^{m \times n}$,

$$\text{null}(A) = \text{range}(A^{\mathrm{T}})^{\perp},$$
$$\text{null}(A^{\mathrm{T}}) = \text{range}(A)^{\perp},$$

where $\text{null}(A)$ denotes the null space of $A$ (i.e., the subspace of $\mathbb{R}^n$ of vectors

$x$ that satisfy $Ax = 0$), range($A$) denotes the range or column space of $A$ (i.e., the subspace of $\mathbb{R}^m$ spanned by the columns of $A$), and range($A$)$^\perp$ denotes the orthogonal complement to range($A$) (i.e., the set of all vectors in $\mathbb{R}^m$ that are orthogonal to every vector in range($A$)).

In this problem you are asked to use the FTLA to prove the existence of Lagrange multipliers (see Appendix D) for the equality-constrained optimization problem. Let $f : \mathbb{R}^n \to \mathbb{R}$, assumed differentiable, be the objective function to be minimized. The vector $x$ must satisfy the equality constraint $g(x) = 0$ for given differentiable $g : \mathbb{R}^n \to \mathbb{R}^m$.

Suppose that $x^*$ is a local minimum. Let $x(t)$ be any arbitrary curve on the surface parametrized implicitly by $g(x) = 0$ (implying that $g(x(t)) = 0$ for all $t$) such that $x(0) = x^*$. Further, assume that $x^*$ is a regular point of the surface. Taking the time derivative of both sides of $g(x(t)) = 0$ at $t = 0$ then leads to

$$\frac{\partial g}{\partial x}(x^*)\dot{x}(0) = 0. \tag{6.8}$$

At the same time, because $x(0) = x^*$ is a local minimum it follows that $f(x(t))$ (viewed as an objective function in $t$) has a local minimum at $t = 0$, implying that

$$\frac{d}{dt}f(x(t))\bigg|_{t=0} = \frac{\partial f}{\partial x}(x^*)\dot{x}(0) = 0. \tag{6.9}$$

Since (6.8) and (6.9) must hold for all arbitrary curves $x(t)$ on the surface defined by $g(x) = 0$, use the FTLA to prove the existence of a Lagrange multiplier $\lambda^* \in \mathbb{R}^m$ such that the first-order necessary condition,

$$\nabla f(x^*) + \frac{\partial g}{\partial x}(x^*)^\mathrm{T}\lambda^* = 0,$$

holds.

**Exercise 6.15**   (a) For matrices $A$, $B$, $C$, and $D$, if $A^{-1}$ exists show that

$$\begin{bmatrix} A & D \\ C & B \end{bmatrix}^{-1} = \begin{bmatrix} A^{-1} + EG^{-1}F & -EG^{-1} \\ -G^{-1}F & G^{-1} \end{bmatrix},$$

where $G = B - CA^{-1}D$, $E = A^{-1}D$, and $F = CA^{-1}$.

(b) Use the above result to solve for the first-order necessary conditions for the equality-constrained optimization problem

$$\min_{x\in\mathbb{R}^n} \frac{1}{2}x^\mathrm{T}Qx + c^\mathrm{T}x$$

subject to $Hx = b$, where $Q \in \mathbb{R}^{n\times n}$ is symmetric and positive definite and $H \in \mathbb{R}^{m\times n}$ is some matrix of maximal rank $m$. See Appendix D.

# Kinematics of Closed Chains

Any kinematic chain that contains one or more loops is called a **closed chain**. Several examples of closed chains were encountered in Chapter 2, from the planar four-bar linkage to spatial mechanisms like the Stewart–Gough platform and the Delta robot (Figure 7.1). These mechanisms are examples of **parallel mechanisms**: closed chains consisting of fixed and moving platforms connected by a set of "legs." The legs themselves are typically open chains but sometimes can also be other closed chains (like the Delta robot in Figure 7.1(b)). In this chapter we analyze the kinematics of closed chains, paying special attention to parallel mechanisms.

The Stewart–Gough platform is used widely as both a motion simulator and a six-axis force–torque sensor. When used as a force–torque sensor, the six prismatic joints experience internal linear forces whenever any external force is applied to the moving platform; by measuring these internal linear forces one can estimate the applied external force. The Delta robot is a three-dof mechanism whose moving platform moves in such a way that it always remains parallel to the fixed platform. Because the three actuators are all attached to the three revolute joints of the fixed platform, the moving parts are relatively light; this allows the Delta to achieve very fast motions.

Closed chains admit a much greater variety of designs than open chains, and their kinematic and static analysis is consequently more complicated. This complexity can be traced to two defining features of closed chains: (i) not all joints are actuated, and (ii) the joint variables must satisfy a number of loop-closure constraint equations, which may or may not be independent depending on the configuration of the mechanism. The presence of unactuated (or passive) joints, together with the fact that the number of actuated joints may deliberately be designed to exceed the mechanism's kinematic degrees of freedom – such mechanisms are said to be **redundantly actuated** – not only makes the kinematics analysis more challenging but also introduces new types of singularities not present in open chains.

Recall also that, for open chains, the kinematic analysis proceeds in a more or less straightforward fashion, with the formulation of the forward kinematics (e.g., via the product of exponentials formalism) followed by that of the inverse kinematics. For general closed chains it is usually difficult to obtain an explicit set of equations for the forward kinematics in the form $X = T(\theta)$, where $X \in SE(3)$

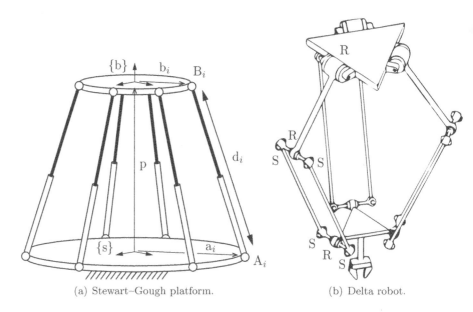

(a) Stewart–Gough platform.                    (b) Delta robot.

**Figure 7.1** Two popular parallel mechanisms.

is the end-effector frame and $\theta \in \mathbb{R}^n$ are the joint coordinates. The more effective approaches exploit, as much as possible, any kinematic symmetries and other special features of the mechanism.

In this chapter we begin with a series of case studies involving some well-known parallel mechanisms and eventually build up a repertoire of kinematic analysis tools and methodologies for handling more general closed chains. Our focus will be on parallel mechanisms that are exactly actuated, i.e., the number of actuated degrees of freedom is equal to the number of degrees of freedom of the mechanism. Methods for the forward and inverse position kinematics of parallel mechanisms are discussed; this is followed by the characterization and derivation of the constraint Jacobian, and the Jacobians of both the inverse and forward kinematics. The chapter concludes with a discussion of the different types of kinematic singularities that arise in closed chains.

## 7.1    Inverse and Forward Kinematics

One general observation that can be made for serial mechanisms versus parallel mechanisms is the following: for serial chains, the forward kinematics is generally straightforward while inverse kinematics may be complex (e.g., there may be multiple solutions or no solution). For parallel mechanisms, the inverse kinematics is often relatively straightforward (e.g., given the configuration of a platform, it may not be hard to determine the joint variables), while the forward kinematics

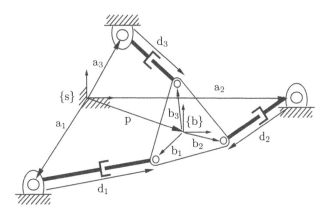

**Figure 7.2** 3×RPR planar parallel mechanism.

may be quite complex: an arbitrarily chosen set of joint values may be infeasible or it may correspond to multiple possible configurations of the platform.

We now continue with two case studies, the 3×RPR planar parallel mechanism and its spatial counterpart, the 3×SPS Stewart–Gough platform. The analysis of these two mechanisms draws upon some simplification techniques that result in a reduced form of the governing kinematic equations, which in turn can be applied to the analysis of more general parallel mechanisms.

## .1.1   3×RPR Planar Parallel Mechanism

The first example we consider is the 3-dof planar 3×RPR parallel mechanism shown in Figure 7.2. A fixed frame {s} and body frame {b} are assigned to the platform as shown. The three prismatic joints are typically actuated while the six revolute joints are passive. Denote the lengths of each of the three legs by $s_i$, $i = 1, 2, 3$. The forward kinematics problem is to determine, for given values of $s = (s_1, s_2, s_3)$, the body frame's position and orientation. Conversely, the inverse kinematics problem is to determine $s$ from $T_{sb} \in SE(2)$.

Let p be the vector from the origin of the {s} frame to the origin of the {b} frame. Let $\phi$ denote the angle measured from the $\hat{x}_s$-axis of the {s} frame to the $\hat{x}_b$-axis of the {b} frame. Further, define the vectors $a_i$, $b_i$, $d_i$, $i = 1, 2, 3$, as shown in the figure. From these definitions, clearly

$$d_i = p + b_i - a_i, \tag{7.1}$$

for $i = 1, 2, 3$. Let

$$\begin{bmatrix} p_x \\ p_y \end{bmatrix} = \mathrm{p} \quad \text{in } \{s\}\text{-frame coordinates,}$$

$$\begin{bmatrix} a_{ix} \\ a_{iy} \end{bmatrix} = \mathrm{a}_i \quad \text{in } \{s\}\text{-frame coordinates,}$$

$$\begin{bmatrix} d_{ix} \\ d_{iy} \end{bmatrix} = \mathrm{d}_i \quad \text{in } \{s\}\text{-frame coordinates,}$$

$$\begin{bmatrix} b_{ix} \\ b_{iy} \end{bmatrix} = \mathrm{b}_i \quad \text{in } \{b\}\text{-frame coordinates.}$$

Note that $(a_{ix}, a_{iy})$ and $(b_{ix}, b_{iy})$ for $i = 1, 2, 3$ are all constant, and that, with the exception of the $(b_{ix}, b_{iy})$, all other vectors are expressed in $\{s\}$-frame coordinates. To express Equation (7.1) in terms of $\{s\}$-frame coordinates, $\mathrm{b}_i$ must be expressed in $\{s\}$-frame coordinates. This is straightforward: defining

$$R_{sb} = \begin{bmatrix} \cos\phi & -\sin\phi \\ \sin\phi & \cos\phi \end{bmatrix},$$

it follows that

$$\begin{bmatrix} d_{ix} \\ d_{iy} \end{bmatrix} = \begin{bmatrix} p_x \\ p_y \end{bmatrix} + R_{sb} \begin{bmatrix} b_{ix} \\ b_{iy} \end{bmatrix} - \begin{bmatrix} a_{ix} \\ a_{iy} \end{bmatrix},$$

for $i = 1, 2, 3$. Also, since $s_i^2 = d_{ix}^2 + d_{iy}^2$, we have

$$\begin{aligned} s_i^2 = {} & (p_x + b_{ix}\cos\phi - b_{iy}\sin\phi - a_{ix})^2 \\ & + (p_y + b_{ix}\sin\phi + b_{iy}\cos\phi - a_{iy})^2, \end{aligned} \tag{7.2}$$

for $i = 1, 2, 3$.

Formulated as above, the inverse kinematics is trivial to compute: given values for $(p_x, p_y, \phi)$, the leg lengths $(s_1, s_2, s_3)$ can be directly calculated from the above equations (negative values of $s_i$ will not be physically realizable in most cases and can be ignored). In contrast, the forward kinematics problem of determining the body frame's position and orientation $(p_x, p_y, \phi)$ from the leg lengths $(s_1, s_2, s_3)$ is not trivial. The following tangent half-angle substitution transforms the three equations in (7.2) into a system of polynomials in $t$, where

$$t = \tan\frac{\phi}{2},$$

$$\sin\phi = \frac{2t}{1 + t^2},$$

$$\cos\phi = \frac{1 - t^2}{1 + t^2}.$$

After some algebraic manipulation, the system of polynomials (7.2) can eventually be reduced to a single sixth-order polynomial in $t$; this effectively shows that the $3 \times$RPR mechanism may have up to six forward kinematics solutions.

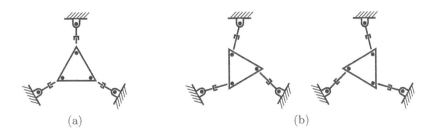

(a)  (b)

**Figure 7.3** (a) The 3×RPR at a singular configuration. From this configuration, extending the legs may cause the platform to snap to a counterclockwise rotation or a clockwise rotation. (b) Two solutions to the forward kinematics when all prismatic joint extensions are identical.

Showing that all six mathematical solutions are physically realizable requires further verification.

Figure 7.3(a) shows the mechanism at a singular configuration, where each leg length is identical and as short as possible. This configuration is a singularity, because extending the legs from this symmetric configuration causes the platform to rotate either clockwise or counterclockwise; we cannot predict which.[1] Singularities are covered in greater detail in Section 7.3. Figure 7.3(b) shows two solutions to the forward kinematics when all leg lengths are identical.

## .1.2 Stewart–Gough Platform

We now examine the inverse and forward kinematics of the 6×SPS Stewart–Gough platform of Figure 7.1(a). In this design the fixed and moving platforms are connected by six serial SPS structures, with the spherical joints passive and the prismatic joints actuated. The derivation of the kinematic equations is close to that for the 3×RPR planar mechanism discussed above. Let {s} and {b} denote the fixed and body frames, respectively, and let $d_i$ be the vector directed from joint $A_i$ to joint $B_i$, $i = 1, \ldots, 6$. Referring to Figure 7.1(a), we make the following definitions:

$$p \in \mathbb{R}^3 = \text{p in \{s\}-frame coordinates,}$$
$$a_i \in \mathbb{R}^3 = \text{a}_i \text{ in \{s\}-frame coordinates,}$$
$$b_i \in \mathbb{R}^3 = \text{b}_i \text{ in \{b\}-frame coordinates,}$$
$$d_i \in \mathbb{R}^3 = \text{d}_i \text{ in \{s\}-frame coordinates,}$$
$$R \in SO(3) \text{ is the orientation of \{b\} as seen from \{s\}.}$$

In order to derive the kinematic constraint equations, note that, vectorially,

$$d_i = p + b_i - a_i, \qquad i = 1, \ldots, 6.$$

Writing the above equations explicitly in {s}-frame coordinates yields

$$d_i = p + Rb_i - a_i, \qquad i = 1, \ldots, 6.$$

[1] A third possibility is that the extending legs crush the platform!

Denoting the length of leg $i$ by $s_i$, we have

$$s_i^2 = d_i^{\mathrm{T}} d_i = (p + Rb_i - a_i)^{\mathrm{T}}(p + Rb_i - a_i),$$

for $i = 1, \ldots, 6$. Note that $a_i$ and $b_i$ are all known constant vectors. Writing the constraint equations in this form, the inverse kinematics becomes straightforward: given $p$ and $R$, the six leg lengths $s_i$ can be determined directly from the above equations.

The forward kinematics is not as straightforward: given each leg length $s_i$, $i = 1, \ldots, 6$, we must solve for $p \in \mathbb{R}^3$ and $R \in SO(3)$. These six constraint equations, together with six further constraints imposed by the condition $R^{\mathrm{T}} R = I$, constitute a set of 12 equations in 12 unknowns (three for $p$, nine for $R$).

### 7.1.3 General Parallel Mechanisms

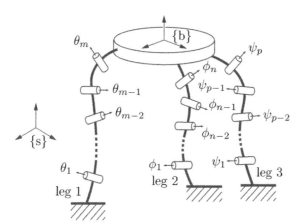

**Figure 7.4** A general parallel mechanism.

For both the $3\times$RPR mechanism and the Stewart–Gough platform, we were able to exploit certain features of the mechanism that resulted in a reduced set of equations; for example, the fact that the legs of the Stewart–Gough platform can be modeled as straight lines considerably simplified the analysis. In this section we briefly consider the case when the legs are general open chains.

Consider such a parallel mechanism, as shown in Figure 7.4; here the fixed and moving platforms are connected by three open chains. Let the configuration of the moving platform be given by $T_{sb}$. Denote the forward kinematics of the three chains by $T_1(\theta)$, $T_2(\phi)$, and $T_3(\psi)$, respectively, where $\theta \in \mathbb{R}^m$, $\phi \in \mathbb{R}^n$, and $\psi \in \mathbb{R}^p$. The loop-closure conditions can be written $T_{sb} = T_1(\theta) = T_2(\phi) = T_3(\psi)$. Eliminating $T_{sb}$, we get

$$T_1(\theta) = T_2(\phi), \tag{7.3}$$
$$T_2(\phi) = T_3(\psi). \tag{7.4}$$

Equations (7.3) and (7.4) each consist of 12 equations (nine for the rotation component and three for the position component), six of which are independent:

from the rotation matrix constraint $R^\mathrm{T}R = I$, the nine equations for the rotation component can be reduced to a set of three independent equations. Thus there are 24 constraint equations, 12 of which are independent, with $n+m+p$ unknown variables. The mechanism therefore has $d = n + m + p - 12$ degrees of freedom.

In the forward kinematics problem, given $d$ values for the joint variables $(\theta, \phi, \psi)$, Equations (7.3) and (7.4) can be solved for the remaining joint variables. Generally this is not trivial and multiple solutions are likely. Once the joint values for any one of the open chain legs are known, the forward kinematics of that leg can then be evaluated to determine the forward kinematics of the closed chain.

In the inverse kinematics problem, given the body-frame displacement $T_{sb} \in SE(3)$, we set $T = T_1 = T_2 = T_3$ and solve Equations (7.3) and (7.4) for the joint variables $(\theta, \phi, \psi)$. As suggested by these case studies, for most parallel mechanisms there are features of the mechanism that can be exploited to eliminate some of these equations and simplify them to a more computationally amenable form.

## .2     Differential Kinematics

We now consider the differential kinematics of parallel mechanisms. Unlike the case for open chains, in which the objective is to relate the input joint velocities to the twist of the end-effector frame, the analysis for closed chains is complicated by the fact that not all the joints are actuated. Only the actuated joints can be prescribed input velocities; the velocities of the remaining passive joints must then be determined from the kinematic constraint equations. These passive joint velocities are usually required in order to eventually determine the twist of the closed chain's end-effector frame.

For open chains, the Jacobian of the forward kinematics is central to the velocity and static analysis. For closed chains, in addition to the forward kinematics Jacobian, the Jacobian defined by the kinematic constraint equations – we will call this the **constraint Jacobian** – also plays a central role in the velocity and static analysis. Usually there are features of the mechanism that can be exploited to simplify and reduce the procedure for obtaining the two Jacobians. We illustrate this with a case study of the Stewart–Gough platform, and show that the Jacobian of the inverse kinematics can be obtained straightforwardly via static analysis. The velocity analysis for more general parallel mechanisms is then detailed.

## .2.1     Stewart–Gough Platform

Earlier we saw that the inverse kinematics for the Stewart–Gough platform can be solved analytically. That is, given the body-frame orientation $R \in SO(3)$ and position $p \in \mathbb{R}^3$, the leg lengths $s \in \mathbb{R}^6$ can be obtained analytically in the functional form $s = g(R, p)$. In principle one could differentiate this equation and manipulate it into the form

$$\dot{s} = G(R, p)\mathcal{V}_s, \tag{7.5}$$

where $\dot{s} \in \mathbb{R}^6$ denotes the leg velocities, $\mathcal{V}_s \in \mathbb{R}^6$ is the spatial twist, and $G(R, p) \in \mathbb{R}^{6 \times 6}$ is the Jacobian of the inverse kinematics. In most cases this procedure will require considerable algebraic manipulation.

Here we take a different approach, based on the static analysis via the conservation of power principles used to determine the static relationship $\tau = J^{\mathrm{T}}\mathcal{F}$ for open chains. The static relationship for closed chains can be expressed in exactly the same form. We illustrate this with an analysis of the Stewart–Gough platform.

In the absence of external forces, the only forces applied to the moving platform occur at the spherical joints. In what follows, all vectors are expressed in {s}-frame coordinates. Let

$$f_i = \hat{n}_i \tau_i$$

be the three-dimensional linear force applied by leg $i$, where $\hat{n}_i \in \mathbb{R}^3$ is a unit vector indicating the direction of the applied force and $\tau_i \in \mathbb{R}$ is the magnitude of the linear force. The moment $m_i$ generated by $f_i$ is

$$m_i = r_i \times f_i,$$

where $r_i \in \mathbb{R}^3$ denotes the vector from the {s}-frame origin to the point of application of the force (the location of spherical joint $i$ in this case). Since neither the spherical joint at the moving platform nor the spherical joint at the fixed platform can resist any torques about them, the force $f_i$ must be along the line of the leg. Therefore, instead of calculating the moment $m_i$ using the spherical joint at the moving platform, we can calculate the moment using the spherical joint at the fixed platform:

$$m_i = q_i \times f_i,$$

where $q_i \in \mathbb{R}^3$ denotes the vector from the fixed-frame origin to the base joint of leg $i$. Since $q_i$ is constant, expressing the moment as $q_i \times f_i$ is preferable.

Combining $f_i$ and $m_i$ into the six-dimensional wrench $\mathcal{F}_i = (m_i, f_i)$, the resultant wrench $\mathcal{F}_s$ on the moving platform is given by

$$\mathcal{F}_s = \sum_{i=1}^{6} \mathcal{F}_i = \sum_{i=1}^{6} \begin{bmatrix} r_i \times \hat{n}_i \\ \hat{n}_i \end{bmatrix} \tau_i$$

$$= \begin{bmatrix} -\hat{n}_1 \times q_1 & \cdots & -\hat{n}_6 \times q_6 \\ \hat{n}_1 & \cdots & \hat{n}_6 \end{bmatrix} \begin{bmatrix} \tau_1 \\ \vdots \\ \tau_6 \end{bmatrix}$$

$$= J_s^{-\mathrm{T}} \tau,$$

where $J_s$ is the spatial Jacobian of the forward kinematics, with inverse given by

$$J_s^{-1} = \left[ \begin{array}{ccc} -\hat{n}_1 \times q_1 & \cdots & -\hat{n}_6 \times q_6 \\ \hat{n}_1 & \cdots & \hat{n}_6 \end{array} \right]^{\mathrm{T}}.$$

## 7.2.2  General Parallel Mechanisms

Because of its kinematic structure, the Stewart–Gough platform lends itself particularly well to a static analysis, as each of the six joint forces are directed along their respective legs. The Jacobian (or more precisely, the inverse Jacobian) can therefore be derived in terms of the screws associated with each straight-line leg. In this subsection we consider more general parallel mechanisms where the static analysis is less straightforward. Using the previous three-legged spatial parallel mechanism of Figure 7.4 as an illustrative example, we derive a procedure for determining the forward kinematics Jacobian that can be generalized to other types of parallel mechanisms.

The mechanism of Figure 7.4 consists of two platforms connected by three legs with $m$, $n$, and $p$ joints, respectively. For simplicity, we will take $m = n = p = 5$, so that the mechanism has $d = n + m + p - 12 = 3$ degrees of freedom (generalizing what follows to different types and numbers of legs is completely straightforward). For the fixed and body frames indicated in the figure, we can write the forward kinematics for the three chains as follows:

$$T_1(\theta_1, \theta_2, \ldots, \theta_5) = e^{[S_1]\theta_1} e^{[S_2]\theta_2} \cdots e^{[S_5]\theta_5} M_1,$$
$$T_2(\phi_1, \phi_2, \ldots, \phi_5) = e^{[P_1]\phi_1} e^{[P_2]\phi_2} \cdots e^{[P_5]\phi_5} M_2,$$
$$T_3(\psi_1, \psi_2, \ldots, \psi_5) = e^{[Q_1]\psi_1} e^{[Q_2]\psi_2} \cdots e^{[Q_5]\psi_5} M_3.$$

The kinematic loop constraints can be expressed as

$$T_1(\theta) = T_2(\phi), \tag{7.6}$$
$$T_2(\phi) = T_3(\psi). \tag{7.7}$$

Since these constraints must be satisfied at all times, we can express their time derivatives in terms of their spatial twists, using

$$\dot{T}_1 T_1^{-1} = \dot{T}_2 T_2^{-1}, \tag{7.8}$$
$$\dot{T}_2 T_2^{-1} = \dot{T}_3 T_3^{-1}. \tag{7.9}$$

Since $\dot{T}_i T_i^{-1} = [\mathcal{V}_i]$, where $\mathcal{V}_i$ is the spatial twist of chain $i$'s end-effector frame, the above identities can also be expressed in terms of the forward kinematics Jacobian for each chain:

$$J_1(\theta)\dot{\theta} = J_2(\phi)\dot{\phi}, \tag{7.10}$$
$$J_2(\phi)\dot{\phi} = J_3(\psi)\dot{\psi}, \tag{7.11}$$

which can be rearranged as

$$
\begin{bmatrix} J_1(\theta) & -J_2(\phi) & 0 \\ 0 & -J_2(\phi) & J_3(\psi) \end{bmatrix} \begin{bmatrix} \dot{\theta} \\ \dot{\phi} \\ \dot{\psi} \end{bmatrix} = 0. \tag{7.12}
$$

Now we rearrange the 15 joints into those that are actuated and those that are passive. Assume without loss of generality that the three actuated joints are $(\theta_1, \phi_1, \psi_1)$. Define the vector of the actuated joints $q_a \in \mathbb{R}^3$ and the vector of the passive joints $q_p \in \mathbb{R}^{12}$ as

$$
q_a = \begin{bmatrix} \theta_1 \\ \phi_1 \\ \psi_1 \end{bmatrix}, \qquad q_p = \begin{bmatrix} \theta_2 \\ \vdots \\ \psi_5 \end{bmatrix},
$$

and we have $q = (q_a, q_p) \in \mathbb{R}^{15}$. Equation (7.12) can now be rearranged into the form

$$
\begin{bmatrix} H_a(q) & H_p(q) \end{bmatrix} \begin{bmatrix} \dot{q}_a \\ \dot{q}_p \end{bmatrix} = 0, \tag{7.13}
$$

or, equivalently,

$$
H_a \dot{q}_a + H_p \dot{q}_p = 0, \tag{7.14}
$$

where $H_a \in \mathbb{R}^{12 \times 3}$ and $H_p \in \mathbb{R}^{12 \times 12}$. If $H_p$ is invertible, we have

$$
\dot{q}_p = -H_p^{-1} H_a \dot{q}_a. \tag{7.15}
$$

So, assuming that $H_p$ is invertible, once the velocities of the actuated joints are given, then the velocities of the remaining passive joints can be obtained uniquely via Equation (7.15).

It still remains to derive the forward kinematics Jacobian with respect to the actuated joints, i.e., to find $J_a(q) \in \mathbb{R}^{6 \times 3}$ satisfying $\mathcal{V}_s = J_a(q)\dot{q}_a$, where $\mathcal{V}_s$ is the spatial twist of the end-effector frame. For this purpose we can use the forward kinematics for any of the three open chains: for example, in terms of chain 1, $J_1(\theta)\dot{\theta} = \mathcal{V}_s$, and from Equation (7.15) we can write

$$
\dot{\theta}_2 = g_2^{\mathrm{T}} \dot{q}_a, \tag{7.16}
$$
$$
\dot{\theta}_3 = g_3^{\mathrm{T}} \dot{q}_a, \tag{7.17}
$$
$$
\dot{\theta}_4 = g_4^{\mathrm{T}} \dot{q}_a, \tag{7.18}
$$
$$
\dot{\theta}_5 = g_5^{\mathrm{T}} \dot{q}_a, \tag{7.19}
$$

where each $g_i(q) \in \mathbb{R}^3$, for $i = 2, \ldots, 5$, can be obtained from Equation (7.15). Defining the row vector $e_1^{\mathrm{T}} = [1\ 0\ 0]$, the differential forward kinematics for chain

1 can now be written

$$
\mathcal{V}_s = J_1(\theta) \begin{bmatrix} e_1^{\mathrm{T}} \\ g_2^{\mathrm{T}} \\ g_3^{\mathrm{T}} \\ g_4^{\mathrm{T}} \\ g_5^{\mathrm{T}} \end{bmatrix} \begin{bmatrix} \dot{\theta}_1 \\ \dot{\phi}_1 \\ \dot{\psi}_1 \end{bmatrix}. \tag{7.20}
$$

Since we are seeking $J_a(q)$ in $\mathcal{V}_s = J_a(q)\dot{q}_a$, and since $\dot{q}_a^{\mathrm{T}} = [\dot{\theta}_1 \ \dot{\phi}_1 \ \dot{\psi}_1]$, from the above it now follows that

$$
J_a(q) = J_1(q_1, \ldots, q_5) \begin{bmatrix} e_1^{\mathrm{T}} \\ g_2(q)^{\mathrm{T}} \\ g_3(q)^{\mathrm{T}} \\ g_4(q)^{\mathrm{T}} \\ g_5(q)^{\mathrm{T}} \end{bmatrix}; \tag{7.21}
$$

this equation could also have been derived using either chain 2 or chain 3.

Given values for the actuated joints $q_a$, we still need to solve for the passive joints $q_p$ from the loop-constraint equations. Eliminating in advance as many elements of $q_p$ as possible will obviously simplify matters. The second point to note is that $H_p(q)$ may become singular, in which case $\dot{q}_p$ cannot be obtained from $\dot{q}_a$. Configurations in which $H_p(q)$ becomes singular correspond to **actuator singularities**, which are discussed in the next section.

## .3     Singularities

Characterizing the singularities of closed chains involves many more subtleties than for open chains. In this section we highlight the essential features of closed-chain singularities via two planar examples: a four-bar linkage (see Figure 7.5) and a five-bar linkage (see Figure 7.6). On the basis of these examples we classify closed-chain singularities into three basic types: **actuator singularities**, **configuration space singularities**, and **end-effector singularities**.

We begin with the four-bar linkage of Figure 7.5. Recall from Chapter 2 that its C-space is a one-dimensional curve embedded in a four-dimensional ambient space (each dimension is parametrized by one of the four joints). Projecting the C-space onto the joint angles $(\theta, \phi)$ leads to the bold curve shown in Figure 7.5. In terms of $\theta$ and $\phi$, the kinematic loop constraint equations for the four-bar linkage can be expressed as

$$
\phi = \tan^{-1}\left(\frac{\beta}{\alpha}\right) \pm \cos^{-1}\left(\frac{\gamma}{\sqrt{\alpha^2 + \beta^2}}\right), \tag{7.22}
$$

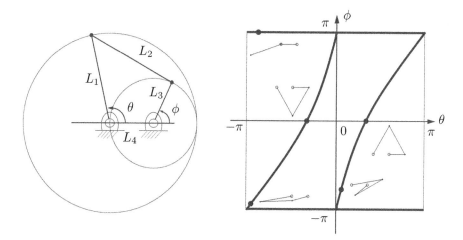

**Figure 7.5** (Left) A planar four-bar linkage and (right) its one-dimensional C-space, represented in bold in the $\theta$–$\phi$ space. Also shown on the right are five sample configurations (bold dots), three of which are near bifurcation points and two of which are far removed from a bifurcation point.

where

$$\alpha = 2L_3L_4 - 2L_1L_3 \cos\theta, \tag{7.23}$$

$$\beta = -2L_1L_3 \sin\theta, \tag{7.24}$$

$$\gamma = L_2^2 - L_4^2 - L_3^2 - L_1^2 + 2L_1L_4 \cos\theta. \tag{7.25}$$

The existence and uniqueness of solutions to the equations above depend on the link lengths $L_1, \ldots, L_4$. In particular, a solution will fail to exist if $\gamma^2 \leq \alpha^2 + \beta^2$. Figure 7.5 depicts the feasible configurations for the choice of link lengths $L_1 = L_2 = 4$ and $L_3 = L_4 = 2$. For this set of link lengths, $\theta$ and $\phi$ both range from 0 to $2\pi$.

A distinctive feature of Figure 7.5 is the presence of **bifurcation points** where branches of the curve meet. As the mechanism approaches these configurations, it has a choice of which branch to follow. Figure 7.5 shows sample configurations on the different branches near, and also far from, the bifurcation points.

We now turn to the five-bar linkage of Figure 7.6. The kinematic loop-constraint equations can be written

$$L_1 \cos\theta_1 + \cdots + L_4 \cos(\theta_1 + \theta_2 + \theta_3 + \theta_4) = L_5, \tag{7.26}$$

$$L_1 \sin\theta_1 + \cdots + L_4 \sin(\theta_1 + \theta_2 + \theta_3 + \theta_4) = 0, \tag{7.27}$$

where we have eliminated in advance the joint variable $\theta_5$ from the loop-closure conditions. Writing these two equations in the form $f(\theta_1, \ldots, \theta_4) = 0$, where $f : \mathbb{R}^4 \to \mathbb{R}^2$, the configuration space can be regarded as a two-dimensional surface in $\mathbb{R}^4$. Like the bifurcation points of the four-bar linkage, self-intersections of the surface can also occur. At such points the constraint Jacobian loses rank.

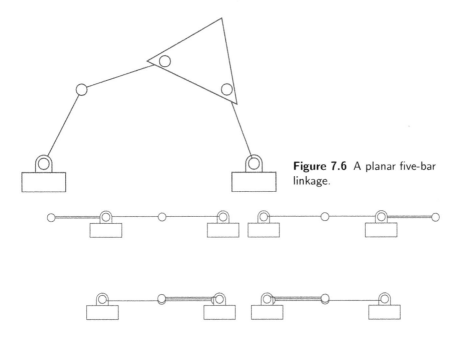

**Figure 7.6** A planar five-bar linkage.

**Figure 7.7** Configuration space singularities of the planar five-bar linkage.

For the five-bar linkage, any point $\theta$ at which

$$\operatorname{rank}\left(\frac{\partial f}{\partial \theta}(\theta)\right) < 2 \tag{7.28}$$

corresponds to what we call a **configuration space singularity**. Figure 7.7 illustrates the possible configuration space singularities of the five-bar linkage. Notice that so far we have made no mention of which joints of the five-bar linkage are actuated, or where the end-effector frame is placed. The notion of a configuration space singularity is completely independent of the choice of actuated joints or where the end-effector frame is placed.

We now consider the case when two joints of the five-bar linkage are actuated. Referring to Figure 7.8, the two revolute joints fixed to ground are the actuated joints. Under normal operating conditions, the motions of the actuated joints can be independently controlled. Alternatively, locking the actuated joints should immobilize the five-bar linkage and turn it into a rigid structure.

For the **nondegenerate actuator singularity** shown in the left-hand panel of Figure 7.8, rotating the two actuated joints oppositely and outward will pull the mechanism apart; rotating them oppositely and inward would either crush the inner two links or cause the center joint to unpredictably buckle upward or downward. For the **degenerate actuator singularity** shown on the right, even when the actuated joints are locked in place the inner two links are free to rotate.

The reason for classifying these singularities as **actuator singularities** is that, by relocating the actuators to a different set of joints, such singularities can

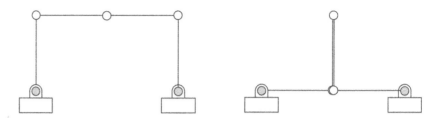

**Figure 7.8** Actuator singularities of the planar five-bar linkage, where in each case the two actuated joints are shaded gray. The singularity on the left is nondegenerate, while the singularity on the right is degenerate.

be eliminated. For both the degenerate and nondegenerate actuator singularities of the five-bar linkage, relocating one actuator to one of the other three passive joints eliminates the singularity.

Visualizing the actuator singularities of the planar five-bar linkage is straightforward enough but, for more complex spatial closed chains, visualization may be difficult. Actuator singularities can be characterized mathematically by the rank of the constraint Jacobian. As before, write the kinematic loop constraints in differential form:

$$H(q)\dot{q} = \begin{bmatrix} H_a(q) & H_p(q) \end{bmatrix} \begin{bmatrix} \dot{q}_a \\ \dot{q}_p \end{bmatrix} = 0, \tag{7.29}$$

where $q_a \in \mathbb{R}^a$ is the vector of the $a$ actuated joints and $q_p \in \mathbb{R}^p$ is the vector of the $p$ passive joints. It follows that $H(q) \in \mathbb{R}^{p \times (a+p)}$ and that $H_p(q)$ is a $p \times p$ matrix.

With the above definitions, we have the following:

- If rank $H_p(q) < p$ then $q$ is an **actuator singularity**. Distinguishing between **degenerate** and **nondegenerate** singularities involves additional mathematical subtleties and relies on second-order derivative information; we do not pursue this further here.
- If rank $H(q) < p$ then $q$ is a **configuration space singularity**. Note that under this condition $H_p(q)$ is also singular (the converse is not true, however). The configuration space singularities can therefore be regarded as the intersection of all possible actuator singularities obtained over all possible combinations of actuated joints.

The final class of singularities depends on the choice of end-effector frame. For the five-bar linkage, let us ignore the orientation of the end-effector frame and focus exclusively on the $x$–$y$ location of the end-effector frame. Figure 7.9 shows the five-bar linkage in an **end-effector singularity** for a given choice of end-effector location. Note that velocities along the dashed line are not possible in this configuration, similarly to the case of singularities for open chains. To see why these velocities are not possible, consider the effective 2R open chain created by the rightmost joint, the link connecting it to the platform, the joint on the

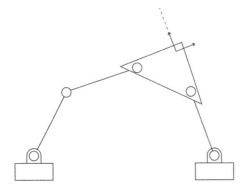

**Figure 7.9** End-effector singularity of the planar five-bar linkage.

platform, and the effective link connecting the platform joint to the end-effector frame. Since the two links of the 2R robot are aligned, the end-effector frame can have no component of motion along the direction of the links.

End-effector singularities are independent of the choice of actuated joints. They can be mathematically characterized as follows. Choose any valid set of actuated joints $q_a$ such that the mechanism is not at an actuator singularity. Write the forward kinematics in the form

$$f(q_a) = T_{sb}. \tag{7.30}$$

One can then check for rank deficiencies in the Jacobian of $f$, as was done for open chains, to determine the presence of an end-effector singularity.

## 7.4 Summary

- Any kinematic chain that contains one or more loops is called a **closed chain**. **Parallel mechanisms** are a class of closed chains that are characterized by two platforms – one moving and one stationary – connected by several legs; the legs are typically open chains, but can themselves be closed chains. The kinematic analysis of closed chains is complicated compared with that of open chains because only a subset of joints is actuated and because the joint variables must satisfy a number of loop-closure constraint equations which may or may not be independent, depending on the configuration of the mechanism.
- For a parallel mechanism with equal numbers of actuators and degrees of freedom, the inverse kinematics problem involves finding, from the given position and orientation of the moving platform, the joint coordinates of the actuated joints. For well-known parallel mechanisms like the planar $3 \times$ RPR and the spatial Stewart–Gough platform, the inverse kinematics admits unique solutions.
- For a parallel mechanism with equal numbers of actuators and degrees of freedom, the forward kinematics problem involves finding the position and

orientation of the moving platform given coordinates for all the actuated joints. For well-known parallel mechanisms like the $3 \times$ RPR and the spatial Stewart–Gough platform, the forward kinematics usually admits multiple solutions. In the case of the most general Stewart–Gough platform, a maximum of 40 solutions is possible.

- The differential kinematics of a closed chain relates the velocities of the actuated joints to the linear and angular velocities of the moving platform's end-effector frame. For an $m$-dof closed chain consisting of $n$ one-dof joints, let $q_a \in \mathbb{R}^m$ and $q_p \in \mathbb{R}^{n-m}$ respectively denote the vector of actuated and passive joints. The kinematic loop-closure constraints can then be expressed in differential form as $H_a \dot{q}_a + H_p \dot{q}_p = 0$, where $H_a \in \mathbb{R}^{(n-m) \times m}$ and $H_p \in \mathbb{R}^{(n-m) \times (n-m)}$ are configuration-dependent matrices. If $H_p$ is invertible then $\dot{q}_p = -H_p^{-1} H_a \dot{q}_a$; the differential forward kinematics can then be expressed in the form $\mathcal{V} = J(q_a, q_p) \dot{q}_a$, where $\mathcal{V}$ is the twist of the end-effector frame and $J(q_a, q_p) \in \mathbb{R}^{6 \times m}$ is a configuration-dependent Jacobian matrix. For closed chains like the Stewart–Gough platform, the differential forward kinematics can also be obtained from a static analysis by exploiting the fact that, just as for open chains, the wrench $\mathcal{F}$ applied by the end-effector is related to the joint forces or torques $\tau$ by $\tau = J^{\mathrm{T}} \mathcal{F}$.

- Singularities for closed chains can be classified into three types: (i) configuration space singularities at self-intersections of the configuration space surface (also called bifurcation points for one-dimensional configuration spaces); (ii) nondegenerate actuator singularities, when the actuated joints cannot be independently actuated, and degenerate actuator singularities when locking all joints fails to make the mechanism a rigid structure; and (iii) end-effector singularities when the end-effector loses one or more degrees of freedom of motion. Configuration space singularities are independent of the choice of actuated joints, while actuator singularities depend on which joints are actuated. End-effector singularities depend on the placement of the end-effector frame but do not depend on the choice of actuated joints.

## 7.5 Notes and References

A comprehensive reference for all aspects of parallel robots is Merlet (2006); Merlet et al. (2016) provides a more compact summary but with more recent references. A major outstanding problem in parallel mechanism kinematics in the 1990s was the question of how many forward kinematics solutions can exist for the general 6–6 platform consisting of six SPS legs (with the prismatic joints actuated) connecting a fixed platform to a moving platform. Raghavan and Roth (1990) showed that there can be at most 40 solutions, while Husty (1996) developed an algorithm for finding all 40 of them.

Singularities of closed chains have also received considerable attention in the literature. The terminology for closed-chain singularities used in this chapter was introduced in Park and Kim (1999); in particular, the distinction between degenerate and nondegenerate actuator singularities derives in part from similar terminology used in Morse theory to identify those critical points where the Hessian is singular (i.e., degenerate). The $3 \times$UPU mechanism, which is addressed in the exercises in both Chapter 2 and the current chapter, can exhibit rather unusual singularity behavior; a more detailed singularity analysis of this mechanism can be found in Han et al. (2002) or di Gregorio and Parenti-Castelli (2002).

## .6 Exercises

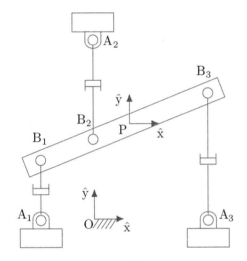

**Figure 7.10** $3 \times$RPR planar parallel mechanism.

**Exercise 7.1** In the $3 \times$RPR planar parallel mechanism of Figure 7.10 the prismatic joints are actuated. Define $a_i \in \mathbb{R}^2$ to be the vector from the fixed-frame origin O to joint $A_i$, $i = 1, 2, 3$, expressed in fixed-frame coordinates. Define $b_i \in \mathbb{R}^2$ to be the vector from the moving-platform-frame origin P to joint $B_i$, $i = 1, 2, 3$, defined in terms of the moving-platform-frame coordinates.

(a) Solve the inverse kinematics.
(b) Derive a procedure to solve the forward kinematics.
(c) Is the configuration shown in the figure an end-effector singularity? Explain your answer by examining the inverse kinematics Jacobian. Is this also an actuator singularity?

**Exercise 7.2** For the $3 \times$RPR planar parallel mechanism in Figure 7.11(a), let $\phi$ be the angle measured from the {s}-frame $\hat{x}$-axis to the {b}-frame $\hat{x}$-axis, and $p \in \mathbb{R}^2$ be the vector from the {s}-frame origin to the {b}-frame origin, expressed in {s}-frame coordinates. Let $a_i \in \mathbb{R}^2$ be the vector from the {s}-frame origin to the three joints fixed to ground, $i = 1, 2, 3$ (note that two of the joints are overlapping), expressed in {s}-frame coordinates. Let $b_i \in \mathbb{R}^2$ be the vector from the {b}-frame origin to the three joints attached to the moving platform, $i = 1, 2, 3$ (note that two of the joints are overlapping), expressed in {b}-frame coordinates. The three prismatic joints are actuated, and the leg lengths are $\theta_1$, $\theta_2$, and $\theta_3$, as shown.

(a) Derive a set of independent equations relating $(\phi, p)$ and $(\theta_1, \theta_2, \theta_3)$.
(b) What is the maximum possible number of forward kinematics solutions?
(c) Assuming static equilibrium, given joint forces $\tau = (1, 0, -1)$ applied at joints $(\theta_1, \theta_2, \theta_3)$, find the planar wrench $(m_{bz}, f_{bx}, f_{by})$ in the end-effector frame {b}.
(d) Now construct a mechanism with three connected 3×RPR parallel mechanisms as shown in Figure 7.11(b). How many degrees of freedom does this mechanism have?

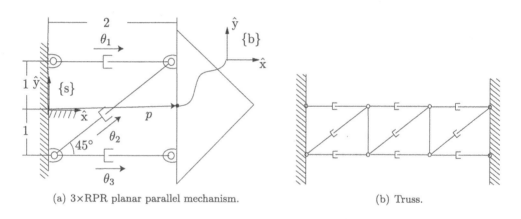

(a) 3×RPR planar parallel mechanism.        (b) Truss.

**Figure 7.11** 3×RPR planar parallel mechanism and truss structure.

**Exercise 7.3**   For the 3×RRR planar parallel mechanism shown in Figure 7.12, let $\phi$ be the orientation of the end-effector frame and $p \in \mathbb{R}^2$ be the vector p expressed in fixed-frame coordinates. Let $a_i \in \mathbb{R}^2$ be the vector $a_i$ expressed in fixed-frame coordinates and $b_i \in \mathbb{R}^2$ be the vector $b_i$ expressed in the moving body-frame coordinates.

(a) Derive a set of independent equations relating $(\phi, p)$ and $(\theta_1, \theta_2, \theta_3)$.
(b) What is the maximum possible number of inverse and forward kinematic solutions for this mechanism?

**Exercise 7.4**   Figure 7.13 shows a six-bar linkage in its zero position. Let $(p_x, p_y)$ be the position of the {b}-frame origin expressed in {s}-frame coordinates,

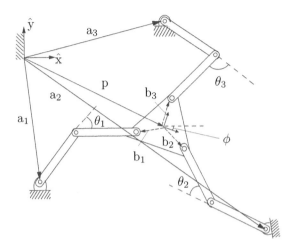

**Figure 7.12** 3×RRR planar parallel mechanism.

and let $\phi$ be the orientation of the {b} frame. The inverse kinematics problem is defined as that of finding the joint variables $(\theta, \psi)$ given $(p_x, p_y, \phi)$.

(a) In order to solve the inverse kinematics problem, how many equations are needed? Derive these equations.

(b) Assume that joints A, D, and E are actuated. Determine whether the configuration shown in Figure 7.13 is an actuator singularity by analyzing an equation of the form

$$\begin{bmatrix} H_a & H_p \end{bmatrix} \begin{bmatrix} \dot{q}_a \\ \dot{q}_p \end{bmatrix} = 0,$$

where $q_a$ is the vector of the actuated joints and $q_p$ is the vector of the passive joints.

(c) Suppose instead that joints A, B, and D are actuated. Find the forward kinematics Jacobian $J_a$ from $\mathcal{V}_s = J_a \dot{q}_a$, where $\mathcal{V}_s$ is the twist in {s}-frame coordinates and $\dot{q}_a$ is the vector of actuated joint rates.

**Exercise 7.5**   Consider the 3×PSP spatial parallel mechanism of Figure 7.14.

(a) How many degrees of freedom does this mechanism have?

(b) Let $R_{sb} = \mathrm{Rot}(\hat{z}, \theta)\mathrm{Rot}(\hat{y}, \phi)\mathrm{Rot}(\hat{x}, \psi)$ be the orientation of the body frame {b}, and let $p_{sb} = (x, y, z) \in \mathbb{R}^3$ be the vector from the {s}-frame origin to the {b}-frame origin (both $R_{sb}$ and $p_{sb}$ are expressed in {s}-frame coordinates). The vectors $a_i, b_i, d_i$, $i = 1, 2, 3$, are defined as shown in the figure. Derive a set of independent kinematic constraint equations relating $(\theta, \phi, \psi, x, y, z)$ and the defined vectors.

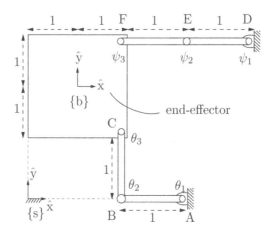

**Figure 7.13** A six-bar linkage.

(c) Given values for $(x, y, z)$, is it possible to solve for the vertical prismatic joint values $s_i$, where $s_i = \|d_i\|$ for $i = 1, 2, 3$? If so, derive an algorithm for doing so.

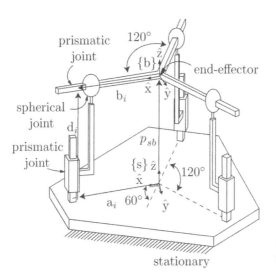

**Figure 7.14** 3×PSP spatial parallel manipulator.

**Exercise 7.6**  The Eclipse mechanism of Figure 7.15 is a six-dof parallel mechanism whose moving platform can tilt by ±90° with respect to ground and can also rotate by 360° about the vertical axis. Assume that the six sliding joints are actuated.

(a) Derive the forward and inverse kinematics. How many forward kinematic solutions are there for general nonsingular configurations?
(b) Find and classify all singularities of this mechanism.

**Exercise 7.7**  For the Delta robot of Figure 7.1(b), obtain the following:

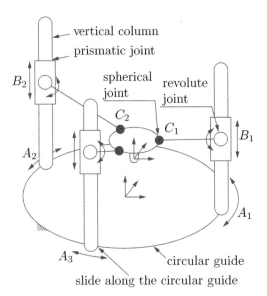

Figure 7.15 The Eclipse mechanism.

(a) the forward kinematics,
(b) the inverse kinematics,
(c) the Jacobian $J_a$ (assume that the revolute joints at the fixed base are actuated).
(d) Identify all actuator singularities of the Delta robot.

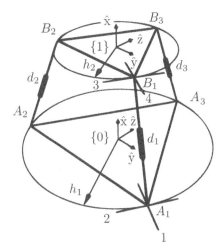

Figure 7.16 The 3×UPU mechanism.

**Exercise 7.8** In the 3×UPU platform of Figure 7.16, the axes of the universal joints are attached to the fixed and moving platforms in the sequence indicated, i.e., axis 1 is attached orthogonally to the fixed base, while axis 4 is attached orthogonally to the moving platform. Obtain the following:

(a) the forward kinematics,
(b) the inverse kinematics,
(c) the Jacobian $J_a$ (assume that the revolute joints at the fixed base are actuated).
(d) Identify all actuator singularities of this robot.
(e) If you can, build a mechanical prototype and see whether the mechanism behaves as predicted by your analysis.

# 8 Dynamics of Open Chains

In this chapter we study once again the motions of open-chain robots, but this time taking into account the forces and torques that cause them; this is the subject of **robot dynamics**. The associated dynamic equations – also referred to as the **equations of motion** – are a set of second-order differential equations of the form

$$\tau = M(\theta)\ddot{\theta} + h(\theta, \dot{\theta}), \tag{8.1}$$

where $\theta \in \mathbb{R}^n$ is the vector of joint variables, $\tau \in \mathbb{R}^n$ is the vector of joint forces and torques, $M(\theta) \in \mathbb{R}^{n \times n}$ is a symmetric positive-definite **mass matrix**, and $h(\theta, \dot{\theta}) \in \mathbb{R}^n$ are forces that lump together centripetal, Coriolis, gravity, and friction terms that depend on $\theta$ and $\dot{\theta}$. One should not be deceived by the apparent simplicity of these equations; even for "simple" open chains, e.g., those with joint axes that are either orthogonal or parallel to each other, $M(\theta)$ and $h(\theta, \dot{\theta})$ can be extraordinarily complex.

Just as a distinction was made between a robot's forward and inverse kinematics, it is also customary to distinguish between a robot's **forward** and **inverse dynamics**. The forward problem is the problem of determining the robot's acceleration $\ddot{\theta}$ given the state $(\theta, \dot{\theta})$ and the joint forces and torques,

$$\ddot{\theta} = M^{-1}(\theta)\left(\tau - h(\theta, \dot{\theta})\right), \tag{8.2}$$

and the inverse problem is finding the joint forces and torques $\tau$ corresponding to the robot's state and a desired acceleration, i.e., Equation (8.1).

A robot's dynamic equations are typically derived in one of two ways: by a direct application of Newton's and Euler's dynamic equations for a rigid body (often called the **Newton–Euler formulation**) or by the **Lagrangian dynamics** formulation derived from the kinetic and potential energy of the robot. The Lagrangian formalism is conceptually elegant and quite effective for robots with simple structures, e.g., with three or fewer degrees of freedom. The calculations can quickly become cumbersome for robots with more degrees of freedom, however. For general open chains, the Newton–Euler formulation leads to efficient recursive algorithms for both the inverse and forward dynamics that can also be assembled into closed-form analytic expressions for, e.g., the mass matrix $M(\theta)$ and the other terms in the dynamics equation (8.1). The Newton–Euler formulation also takes advantage of tools we have already developed in this book.

In this chapter we study both the Lagrangian and Newton–Euler dynamics formulations for an open-chain robot. While we usually express the dynamics in terms of the joint space variables $\theta$, it is sometimes convenient to express it in terms of the configuration, twist, and rate of change of the twist of the end-effector. This is the task-space dynamics, studied in Section 8.6. Sometimes robots are subject to a set of constraints on their motion, such as when the robot makes contact with a rigid environment. This leads to a formulation of the constrained dynamics (Section 8.7), whereby the space of joint torques and forces is divided into a subspace that causes motion of the robot and a subspace that causes forces against the constraints. The URDF file format for specifying robot inertial properties is described in Section 8.8. Finally, some practical issues that arise in the derivation of robot dynamics, such as the effect of motor gearing and friction, are described in Section 8.9.

## 8.1    Lagrangian Formulation

### 8.1.1    Basic Concepts and Motivating Examples

The first step in the Lagrangian formulation of dynamics is to choose a set of independent coordinates $q \in \mathbb{R}^n$ that describes the system's configuration. The coordinates $q$ are called **generalized coordinates**. Once generalized coordinates have been chosen, these then define the **generalized forces** $f \in \mathbb{R}^n$. The forces $f$ and the coordinate rates $\dot{q}$ are dual to each other in the sense that the inner product $f^{\mathrm{T}}\dot{q}$ corresponds to power. A Lagrangian function $\mathcal{L}(q, \dot{q})$ is then defined as the overall system's kinetic energy $\mathcal{K}(q, \dot{q})$ minus the potential energy $\mathcal{P}(q)$,

$$\mathcal{L}(q, \dot{q}) = \mathcal{K}(q, \dot{q}) - \mathcal{P}(q).$$

The equations of motion can now be expressed in terms of the Lagrangian as follows:

$$f = \frac{d}{dt}\frac{\partial \mathcal{L}}{\partial \dot{q}} - \frac{\partial \mathcal{L}}{\partial q}, \tag{8.3}$$

These equations are also referred to as the **Euler–Lagrange equations with external forces**.[1] The derivation can be found in dynamics texts.

We illustrate the Lagrangian dynamics formulation through two examples. In the first example, consider a particle of mass $\mathrm{m}$ constrained to move on a vertical line. The particle's configuration space is this vertical line, and a natural choice for a generalized coordinate is the height of the particle, which we denote by the scalar variable $x \in \mathbb{R}$. Suppose that the gravitational force $\mathrm{m}g$ acts downward, and an external force $f$ is applied upward. By Newton's second law, the equation of motion for the particle is

$$f - \mathrm{m}g = \mathrm{m}\ddot{x}. \tag{8.4}$$

---

[1] The external force $f$ is zero in the standard form of the Euler–Lagrange equations.

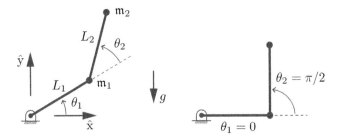

**Figure 8.1** (Left) A 2R open chain under gravity. (Right) At $\theta = (0, \pi/2)$.

We now apply the Lagrangian formalism to derive the same result. The kinetic energy is $m\dot{x}^2/2$, the potential energy is $mgx$, and the Lagrangian is

$$\mathcal{L}(x, \dot{x}) = \mathcal{K}(x, \dot{x}) - \mathcal{P}(x) = \frac{1}{2}m\dot{x}^2 - mgx. \tag{8.5}$$

The equation of motion is then given by

$$f = \frac{d}{dt}\frac{\partial \mathcal{L}}{\partial \dot{x}} - \frac{\partial \mathcal{L}}{\partial x} = m\ddot{x} + mg, \tag{8.6}$$

which matches Equation (8.4).

We now derive the dynamic equations for a planar 2R open chain moving in the presence of gravity (Figure 8.1). The chain moves in the $\hat{x}$–$\hat{y}$-plane, with gravity $g$ acting in the $-\hat{y}$-direction. Before the dynamics can be derived, the mass and inertial properties of all the links must be specified. To keep things simple the two links are modeled as point masses $m_1$ and $m_2$ concentrated at the ends of each link. The position and velocity of the link-1 mass are then given by

$$\begin{bmatrix} x_1 \\ y_1 \end{bmatrix} = \begin{bmatrix} L_1 \cos\theta_1 \\ L_1 \sin\theta_1 \end{bmatrix},$$

$$\begin{bmatrix} \dot{x}_1 \\ \dot{y}_1 \end{bmatrix} = \begin{bmatrix} -L_1 \sin\theta_1 \\ L_1 \cos\theta_1 \end{bmatrix} \dot{\theta}_1,$$

while those of the link-2 mass are given by

$$\begin{bmatrix} x_2 \\ y_2 \end{bmatrix} = \begin{bmatrix} L_1 \cos\theta_1 + L_2 \cos(\theta_1 + \theta_2) \\ L_1 \sin\theta_1 + L_2 \sin(\theta_1 + \theta_2) \end{bmatrix},$$

$$\begin{bmatrix} \dot{x}_2 \\ \dot{y}_2 \end{bmatrix} = \begin{bmatrix} -L_1 \sin\theta_1 - L_2 \sin(\theta_1 + \theta_2) & -L_2 \sin(\theta_1 + \theta_2) \\ L_1 \cos\theta_1 + L_2 \cos(\theta_1 + \theta_2) & L_2 \cos(\theta_1 + \theta_2) \end{bmatrix} \begin{bmatrix} \dot{\theta}_1 \\ \dot{\theta}_2 \end{bmatrix}.$$

We choose the joint coordinates $\theta = (\theta_1, \theta_2)$ as the generalized coordinates. The generalized forces $\tau = (\tau_1, \tau_2)$ then correspond to joint torques (since $\tau^{\mathrm{T}}\dot{\theta}$ corresponds to power). The Lagrangian $\mathcal{L}(\theta, \dot{\theta})$ is of the form

$$\mathcal{L}(\theta, \dot{\theta}) = \sum_{i=1}^{2}(\mathcal{K}_i - \mathcal{P}_i), \tag{8.7}$$

where the link kinetic energy terms $\mathcal{K}_1$ and $\mathcal{K}_2$ are

$$\mathcal{K}_1 = \frac{1}{2}\mathsf{m}_1(\dot{x}_1^2 + \dot{y}_1^2) = \frac{1}{2}\mathsf{m}_1 L_1^2 \dot{\theta}_1^2$$

$$\mathcal{K}_2 = \frac{1}{2}\mathsf{m}_2(\dot{x}_2^2 + \dot{y}_2^2)$$

$$= \frac{1}{2}\mathsf{m}_2\left((L_1^2 + 2L_1 L_2 \cos\theta_2 + L_2^2)\dot{\theta}_1^2 + 2(L_2^2 + L_1 L_2 \cos\theta_2)\dot{\theta}_1\dot{\theta}_2 + L_2^2\dot{\theta}_2^2\right),$$

and the link potential energy terms $\mathcal{P}_1$ and $\mathcal{P}_2$ are

$$\mathcal{P}_1 = \mathsf{m}_1 g y_1 = \mathsf{m}_1 g L_1 \sin\theta_1,$$

$$\mathcal{P}_2 = \mathsf{m}_2 g y_2 = \mathsf{m}_2 g(L_1 \sin\theta_1 + L_2 \sin(\theta_1 + \theta_2)).$$

The Euler–Lagrange equations (8.3) for this example are of the form

$$\tau_i = \frac{d}{dt}\frac{\partial\mathcal{L}}{\partial\dot{\theta}_i} - \frac{\partial\mathcal{L}}{\partial\theta_i}, \qquad i = 1, 2. \tag{8.8}$$

The dynamic equations for the 2R planar chain follow from explicit evaluation of the right-hand side of (8.8) (we omit the detailed calculations, which are straightforward but tedious):

$$
\left.
\begin{aligned}
\tau_1 &= \left(\mathsf{m}_1 L_1^2 + \mathsf{m}_2(L_1^2 + 2L_1 L_2 \cos\theta_2 + L_2^2)\right)\ddot{\theta}_1 \\
&\quad + \mathsf{m}_2(L_1 L_2 \cos\theta_2 + L_2^2)\ddot{\theta}_2 - \mathsf{m}_2 L_1 L_2 \sin\theta_2(2\dot{\theta}_1\dot{\theta}_2 + \dot{\theta}_2^2) \\
&\quad + (\mathsf{m}_1 + \mathsf{m}_2)L_1 g \cos\theta_1 + \mathsf{m}_2 g L_2 \cos(\theta_1 + \theta_2), \\
\tau_2 &= \mathsf{m}_2(L_1 L_2 \cos\theta_2 + L_2^2)\ddot{\theta}_1 + \mathsf{m}_2 L_2^2 \ddot{\theta}_2 + \mathsf{m}_2 L_1 L_2 \dot{\theta}_1^2 \sin\theta_2 \\
&\quad + \mathsf{m}_2 g L_2 \cos(\theta_1 + \theta_2).
\end{aligned}
\right\} \tag{8.9}
$$

We can gather terms together into an equation of the form

$$\tau = M(\theta)\ddot{\theta} + \underbrace{c(\theta,\dot{\theta}) + g(\theta)}_{h(\theta,\dot{\theta})}, \tag{8.10}$$

with

$$M(\theta) = \begin{bmatrix} \mathsf{m}_1 L_1^2 + \mathsf{m}_2(L_1^2 + 2L_1 L_2 \cos\theta_2 + L_2^2) & \mathsf{m}_2(L_1 L_2 \cos\theta_2 + L_2^2) \\ \mathsf{m}_2(L_1 L_2 \cos\theta_2 + L_2^2) & \mathsf{m}_2 L_2^2 \end{bmatrix},$$

$$c(\theta,\dot{\theta}) = \begin{bmatrix} -\mathsf{m}_2 L_1 L_2 \sin\theta_2(2\dot{\theta}_1\dot{\theta}_2 + \dot{\theta}_2^2) \\ \mathsf{m}_2 L_1 L_2 \dot{\theta}_1^2 \sin\theta_2 \end{bmatrix},$$

$$g(\theta) = \begin{bmatrix} (\mathsf{m}_1 + \mathsf{m}_2)L_1 g \cos\theta_1 + \mathsf{m}_2 g L_2 \cos(\theta_1 + \theta_2) \\ \mathsf{m}_2 g L_2 \cos(\theta_1 + \theta_2) \end{bmatrix},$$

where $M(\theta)$ is the symmetric positive-definite mass matrix, $c(\theta,\dot{\theta})$ is the vector containing the Coriolis and centripetal torques, and $g(\theta)$ is the vector containing the gravitational torques. These reveal that the equations of motion are linear in $\ddot{\theta}$, quadratic in $\dot{\theta}$, and trigonometric in $\theta$. This is true in general for serial chains containing revolute joints, not just for the 2R robot.

The $M(\theta)\ddot{\theta} + c(\theta,\dot{\theta})$ terms in Equation (8.10) could have been derived by writing $f_i = m_i a_i$ for each point mass, where the accelerations $a_i$ are written in terms of $\theta$, by differentiating the expressions for $(\dot{x}_1, \dot{y}_1)$ and $(\dot{x}_2, \dot{y}_2)$ given above:

$$f_1 = \begin{bmatrix} f_{x1} \\ f_{y1} \\ f_{z1} \end{bmatrix} = m_1 \begin{bmatrix} \ddot{x}_1 \\ \ddot{y}_1 \\ \ddot{z}_1 \end{bmatrix} = m_1 \begin{bmatrix} -L_1\dot{\theta}_1^2 c_1 - L_1\ddot{\theta}_1 s_1, \\ -L_1\dot{\theta}_1^2 s_1 + L_1\ddot{\theta}_1 c_1 \\ 0 \end{bmatrix}, \tag{8.11}$$

$$f_2 = m_2 \begin{bmatrix} -L_1\dot{\theta}_1^2 c_1 - L_2(\dot{\theta}_1 + \dot{\theta}_2)^2 c_{12} - L_1\ddot{\theta}_1 s_1 - L_2(\ddot{\theta}_1 + \ddot{\theta}_2)s_{12} \\ -L_1\dot{\theta}_1^2 s_1 - L_2(\dot{\theta}_1 + \dot{\theta}_2)^2 s_{12} + L_1\ddot{\theta}_1 c_1 + L_2(\ddot{\theta}_1 + \ddot{\theta}_2)c_{12} \\ 0 \end{bmatrix}, \tag{8.12}$$

where $s_{12}$ indicates $\sin(\theta_1 + \theta_2)$, etc. Defining $r_{11}$ as the vector from joint 1 to $m_1$, $r_{12}$ as the vector from joint 1 to $m_2$, and $r_{22}$ as the vector from joint 2 to $m_2$, the moments in world-aligned frames $\{i\}$ attached to joints 1 and 2 can be expressed as $m_1 = r_{11} \times f_1 + r_{12} \times f_2$ and $m_2 = r_{22} \times f_2$. (Note that joint 1 must provide torques to move both $m_1$ and $m_2$, but joint 2 only needs to provide torque to move $m_2$.) The joint torques $\tau_1$ and $\tau_2$ are just the third elements of $m_1$ and $m_2$, i.e., the moments about the $\hat{z}_i$-axes out of the page, respectively.

In $(x, y)$ coordinates, the accelerations of the masses are written simply as second time-derivatives of the coordinates, e.g., $(\ddot{x}_2, \ddot{y}_2)$. This is because the $\hat{x}$-$\hat{y}$ frame is an inertial frame. The joint coordinates $(\theta_1, \theta_2)$ are not in an inertial frame, however, so accelerations are expressed as a sum of terms that are linear in the second derivatives of joint variables, $\ddot{\theta}$, and quadratic of the first derivatives of joint variables, $\dot{\theta}^T \dot{\theta}$, as seen in Equations (8.11) and (8.12). Quadratic terms containing $\dot{\theta}_i^2$ are called **centripetal** terms, and quadratic terms containing $\dot{\theta}_i \dot{\theta}_j, i \neq j$, are called **Coriolis** terms. In other words, $\ddot{\theta} = 0$ does not mean zero acceleration of the masses, due to the centripetal and Coriolis terms.

To better understand the centripetal and Coriolis terms, consider the arm at the configuration $(\theta_1, \theta_2) = (0, \pi/2)$, i.e., $\cos\theta_1 = \sin(\theta_1 + \theta_2) = 1$, $\sin\theta_1 = \cos(\theta_1 + \theta_2) = 0$. Assuming $\ddot{\theta} = 0$, the acceleration $(\ddot{x}_2, \ddot{y}_2)$ of $m_2$ from Equation (8.12) can be written

$$\begin{bmatrix} \ddot{x}_2 \\ \ddot{y}_2 \end{bmatrix} = \underbrace{\begin{bmatrix} -L_1\dot{\theta}_1^2 \\ -L_2\dot{\theta}_1^2 - L_2\dot{\theta}_2^2 \end{bmatrix}}_{\text{centripetal terms}} + \underbrace{\begin{bmatrix} 0 \\ -2L_2\dot{\theta}_1\dot{\theta}_2 \end{bmatrix}}_{\text{Coriolis terms}}.$$

Figure 8.2 shows the centripetal acceleration $a_{\text{cent}1} = (-L_1\dot{\theta}_1^2, -L_2\dot{\theta}_1^2)$ when $\dot{\theta}_2 = 0$, the centripetal acceleration $a_{\text{cent}2} = (0, -L_2\dot{\theta}_2^2)$ when $\dot{\theta}_1 = 0$, and the Coriolis acceleration $a_{\text{cor}} = (0, -2L_2\dot{\theta}_1\dot{\theta}_2)$ when both $\dot{\theta}_1$ and $\dot{\theta}_2$ are positive. As illustrated in Figure 8.2, each centripetal acceleration $a_{\text{cent}i}$ pulls $m_2$ toward joint $i$ to keep $m_2$ rotating about the center of the circle defined by joint $i$.[2] Therefore $a_{\text{cent}i}$ creates zero torque about joint $i$. The Coriolis acceleration $a_{\text{cor}}$ in this example passes through joint 2, so it creates zero torque about joint 2

---

[2] Without this centripetal acceleration, and therefore centripetal force, the mass $m_2$ would fly off along a tangent to the circle.

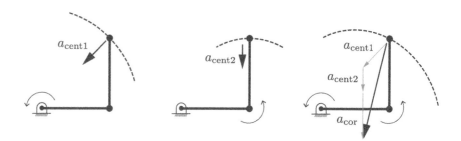

**Figure 8.2** Accelerations of $\mathsf{m}_2$ when $\theta = (0, \pi/2)$ and $\ddot{\theta} = 0$. (Left) The centripetal acceleration $a_{\text{cent1}} = (-L_1\dot{\theta}_1^2, -L_2\dot{\theta}_1^2)$ of $\mathsf{m}_2$ when $\dot{\theta}_2 = 0$. (Middle) The centripetal acceleration $a_{\text{cent2}} = (0, -L_2\dot{\theta}_2^2)$ of $\mathsf{m}_2$ when $\dot{\theta}_1 = 0$. (Right) When both joints are rotating with $\dot{\theta}_i > 0$, the acceleration is the vector sum of $a_{\text{cent1}}$, $a_{\text{cent2}}$, and the Coriolis acceleration $a_{\text{cor}} = (0, -2L_2\dot{\theta}_1\dot{\theta}_2)$.

but it creates negative torque about joint 1; the torque about joint 1 is negative because $\mathsf{m}_2$ gets closer to joint 1 (due to joint 2's motion). Therefore the inertia due to $\mathsf{m}_2$ about the $\hat{z}_1$-axis is dropping, meaning that the positive momentum about joint 1 drops while joint 1's speed $\dot{\theta}_1$ is constant. Therefore joint 1 must apply a negative torque, since torque is defined as the rate of change of angular momentum. Otherwise $\dot{\theta}_1$ would increase as $\mathsf{m}_2$ gets closer to joint 1, just as a skater's rotation speed increases as she pulls in her outstretched arms while doing a spin.

### 8.1.2 General Formulation

We now describe the Lagrangian dynamics formulation for general $n$-link open chains. The first step is to select a set of generalized coordinates $\theta \in \mathbb{R}^n$ for the configuration space of the system. For open chains all of whose joints are actuated, it is convenient and always possible to choose $\theta$ to be the vector of the joint values. The generalized forces will be denoted $\tau \in \mathbb{R}^n$. If $\theta_i$ is a revolute joint then $\tau_i$ will correspond to a torque, while if $\theta_i$ is a prismatic joint then $\tau_i$ will correspond to a force.

Once $\theta$ has been chosen and the generalized forces $\tau$ identified, the next step is to formulate the Lagrangian $\mathcal{L}(\theta, \dot{\theta})$ as follows:

$$\mathcal{L}(\theta, \dot{\theta}) = \mathcal{K}(\theta, \dot{\theta}) - \mathcal{P}(\theta), \qquad (8.13)$$

where $\mathcal{K}(\theta, \dot{\theta})$ is the kinetic energy and $\mathcal{P}(\theta)$ is the potential energy of the overall system. For rigid-link robots the kinetic energy can always be written in the form

$$\mathcal{K}(\theta, \dot{\theta}) = \frac{1}{2}\sum_{i=1}^{n}\sum_{j=1}^{n} m_{ij}(\theta)\dot{\theta}_i\dot{\theta}_j = \frac{1}{2}\dot{\theta}^{\text{T}}M(\theta)\dot{\theta}, \qquad (8.14)$$

where $m_{ij}(\theta)$ is the $(i, j)$th element of the $n \times n$ mass matrix $M(\theta)$; a constructive proof of this assertion is provided when we examine the Newton–Euler formulation.

The dynamic equations are analytically obtained by evaluating the right-hand side of

$$\tau_i = \frac{d}{dt}\frac{\partial \mathcal{L}}{\partial \dot{\theta}_i} - \frac{\partial \mathcal{L}}{\partial \theta_i}, \qquad i = 1, \ldots, n. \tag{8.15}$$

With the kinetic energy expressed as in Equation (8.14), the dynamics can be written explicitly as

$$\tau_i = \sum_{j=1}^{n} m_{ij}(\theta)\ddot{\theta}_j + \sum_{j=1}^{n}\sum_{k=1}^{n} \Gamma_{ijk}(\theta)\dot{\theta}_j\dot{\theta}_k + \frac{\partial \mathcal{P}}{\partial \theta_i}, \qquad i = 1, \ldots, n, \tag{8.16}$$

where the $\Gamma_{ijk}(\theta)$, known as the **Christoffel symbols of the first kind**, are defined as follows:

$$\Gamma_{ijk}(\theta) = \frac{1}{2}\left(\frac{\partial m_{ij}}{\partial \theta_k} + \frac{\partial m_{ik}}{\partial \theta_j} - \frac{\partial m_{jk}}{\partial \theta_i}\right). \tag{8.17}$$

This shows that the Christoffel symbols, which generate the Coriolis and centripetal terms $c(\theta, \dot{\theta})$, are derived from the mass matrix $M(\theta)$.

As we have already seen, the equations (8.16) are often gathered together in the form

$$\tau = M(\theta)\ddot{\theta} + c(\theta, \dot{\theta}) + g(\theta) \qquad \text{or} \qquad M(\theta)\ddot{\theta} + h(\theta, \dot{\theta}),$$

where $g(\theta)$ is simply $\partial \mathcal{P}/\partial \theta$.

We can see explicitly that the Coriolis and centripetal terms are quadratic in the velocity by using the form

$$\tau = M(\theta)\ddot{\theta} + \dot{\theta}^{\mathrm{T}}\Gamma(\theta)\dot{\theta} + g(\theta), \tag{8.18}$$

where $\Gamma(\theta)$ is an $n \times n \times n$ matrix and the product $\dot{\theta}^{\mathrm{T}}\Gamma(\theta)\dot{\theta}$ should be interpreted as follows:

$$\dot{\theta}^{\mathrm{T}}\Gamma(\theta)\dot{\theta} = \begin{bmatrix} \dot{\theta}^{\mathrm{T}}\Gamma_1(\theta)\dot{\theta} \\ \dot{\theta}^{\mathrm{T}}\Gamma_2(\theta)\dot{\theta} \\ \vdots \\ \dot{\theta}^{\mathrm{T}}\Gamma_n(\theta)\dot{\theta} \end{bmatrix},$$

where $\Gamma_i(\theta)$ is an $n \times n$ matrix with $(j,k)$th entry $\Gamma_{ijk}$.

It is also common to see the dynamics written as

$$\tau = M(\theta)\ddot{\theta} + C(\theta, \dot{\theta})\dot{\theta} + g(\theta),$$

where $C(\theta, \dot{\theta}) \in \mathbb{R}^{n \times n}$ is called the **Coriolis matrix**, with $(i,j)$th entry

$$c_{ij}(\theta, \dot{\theta}) = \sum_{k=1}^{n} \Gamma_{ijk}(\theta)\dot{\theta}_k. \tag{8.19}$$

The Coriolis matrix is used to prove the following **passivity property** (Proposition 8.1), which can be used to prove the stability of certain robot control laws, as we will see in Section 11.4.2.2.

**Proposition 8.1**   *The matrix $\dot{M}(\theta) - 2C(\theta, \dot{\theta}) \in \mathbb{R}^{n \times n}$ is skew symmetric, where $M(\theta) \in \mathbb{R}^{n \times n}$ is the mass matrix, $\dot{M}(\theta)$ its time derivative, and $C(\theta, \dot{\theta}) \in \mathbb{R}^{n \times n}$ is the Coriolis matrix as defined in Equation (8.19).*

*Proof*   The $(i, j)$th component of $\dot{M} - 2C$ is

$$\dot{m}_{ij}(\theta) - 2c_{ij}(\theta, \dot{\theta}) = \sum_{k=1}^{n} \frac{\partial m_{ij}}{\partial \theta_k} \dot{\theta}_k - \frac{\partial m_{ij}}{\partial \theta_k} \dot{\theta}_k - \frac{\partial m_{ik}}{\partial \theta_j} \dot{\theta}_k + \frac{\partial m_{kj}}{\partial \theta_i} \dot{\theta}_k$$

$$= \sum_{k=1}^{n} \frac{\partial m_{kj}}{\partial \theta_i} \dot{\theta}_k - \frac{\partial m_{ik}}{\partial \theta_j} \dot{\theta}_k.$$

By switching the indices $i$ and $j$, it can be seen that

$$\dot{m}_{ji}(\theta) - 2c_{ji}(\theta, \dot{\theta}) = -(\dot{m}_{ij}(\theta) - 2c_{ij}(\theta, \dot{\theta})),$$

thus proving that $(\dot{M} - 2C)^{\mathrm{T}} = -(\dot{M} - 2C)$ as claimed.   □

### 8.1.3   Understanding the Mass Matrix

The kinetic energy $\frac{1}{2}\dot{\theta}^{\mathrm{T}} M(\theta) \dot{\theta}$ is a generalization of the familiar expression $\frac{1}{2} m v^{\mathrm{T}} v$ for a point mass. The fact that the mass matrix $M(\theta)$ is positive definite, meaning that $\dot{\theta}^{\mathrm{T}} M(\theta) \dot{\theta} > 0$ for all $\dot{\theta} \neq 0$, is a generalization of the fact that the mass of a point mass is always positive, $m > 0$. In both cases, if the velocity is nonzero, the kinetic energy must be positive.

On the one hand, for a point mass with dynamics expressed in Cartesian coordinates as $f = m\ddot{x}$, the mass is independent of the direction of acceleration, and the acceleration $\ddot{x}$ is always "parallel" to the force, in the sense that $\ddot{x}$ is a scalar multiple of $f$. A mass matrix $M(\theta)$, on the other hand, presents a different effective mass in different acceleration directions, and $\ddot{\theta}$ is not generally a scalar multiple of $\tau$ even when $\dot{\theta} = 0$. To visualize the direction dependence of the effective mass, we can map a unit ball of joint accelerations $\{\ddot{\theta} \mid \ddot{\theta}^{\mathrm{T}} \ddot{\theta} = 1\}$ through the mass matrix $M(\theta)$ to generate a joint force–torque ellipsoid when the mechanism is at rest ($\dot{\theta} = 0$). An example is shown in Figure 8.3 for the 2R arm of Figure 8.1, with $L_1 = L_2 = \mathfrak{m}_1 = \mathfrak{m}_2 = 1$, at two different joint configurations: $(\theta_1, \theta_2) = (0°, 90°)$ and $(\theta_1, \theta_2) = (0°, 150°)$. The torque ellipsoid can be interpreted as a direction-dependent mass ellipsoid: the same joint acceleration magnitude $\|\ddot{\theta}\|$ requires different joint torque magnitudes $\|\tau\|$ depending on the acceleration direction. The directions of the principal axes of the mass ellipsoid are given by the eigenvectors $v_i$ of $M(\theta)$ and the lengths of the principal semi-axes are given by the corresponding eigenvalues $\lambda_i$. The acceleration $\ddot{\theta}$ is only a scalar multiple of $\tau$ when $\tau$ is along a principal axis of the ellipsoid.

It is easier to visualize the mass matrix if it is represented as an effective mass of the end-effector, since it is possible to feel this mass directly by grabbing and moving the end-effector. If you grabbed the endpoint of the 2R robot, depending on the direction you applied force to it, how massy would it feel? Let us denote the effective mass matrix at the end-effector as $\Lambda(\theta)$, and the velocity of the

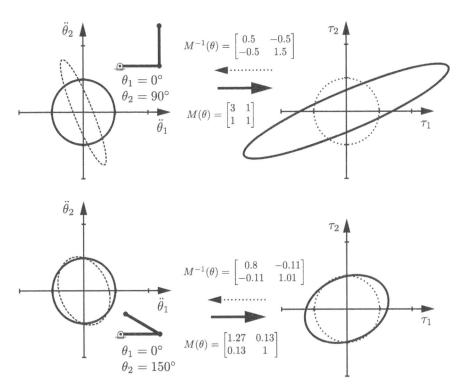

**Figure 8.3** (Bold lines) A unit ball of accelerations in $\ddot{\theta}$ maps through the mass matrix $M(\theta)$ to a torque ellipsoid that depends on the configuration of the 2R arm. These torque ellipsoids may be interpreted as mass ellipsoids. The mapping is shown for two arm configurations: $(0°, 90°)$ and $(0°, 150°)$. (Dotted lines) A unit ball in $\tau$ maps through $M^{-1}(\theta)$ to an acceleration ellipsoid.

end-effector as $V = (\dot{x}, \dot{y})$. We know that the kinetic energy of the robot must be the same regardless of the coordinates we use, so

$$\frac{1}{2}\dot{\theta}^{\mathrm{T}}M(\theta)\dot{\theta} = \frac{1}{2}V^{\mathrm{T}}\Lambda(\theta)V. \tag{8.20}$$

Assuming the Jacobian $J(\theta)$ satisfying $V = J(\theta)\dot{\theta}$ is invertible, Equation (8.20) can be rewritten as follows:

$$V^{\mathrm{T}}\Lambda V = (J^{-1}V)^{\mathrm{T}}M(J^{-1}V)$$
$$= V^{\mathrm{T}}(J^{-\mathrm{T}}MJ^{-1})V.$$

In other words, the end-effector mass matrix is

$$\Lambda(\theta) = J^{-\mathrm{T}}(\theta)M(\theta)J^{-1}(\theta). \tag{8.21}$$

Figure 8.4 shows the end-effector mass ellipsoids, with principal-axis directions given by the eigenvectors of $\Lambda(\theta)$ and principal semi-axis lengths given by its eigenvalues, for the same two 2R robot configurations as in Figure 8.3. The endpoint acceleration $(\ddot{x}, \ddot{y})$ is a scalar multiple of the force $(f_x, f_y)$ applied at

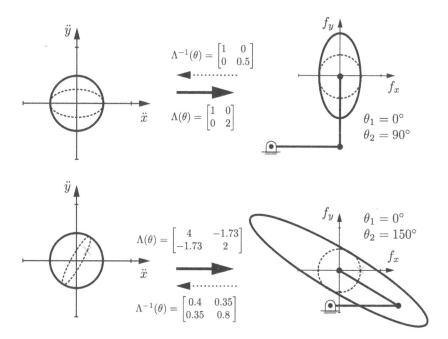

**Figure 8.4** (Bold lines) A unit ball of accelerations in $(\ddot{x}, \ddot{y})$ maps through the end-effector mass matrix $\Lambda(\theta)$ to an end-effector force ellipsoid that depends on the configuration of the 2R arm. For the configuration $(\theta_1, \theta_2) = (0°, 90°)$, a force in the $f_y$-direction exactly feels both masses $m_1$ and $m_2$, while a force in the $f_x$-direction feels only $m_2$. (Dotted lines) A unit ball in $f$ maps through $\Lambda^{-1}(\theta)$ to an acceleration ellipsoid. The $\times$ symbols for $(\theta_1, \theta_2) = (0°, 150°)$ indicate an example endpoint force $(f_x, f_y) = (1, 0)$ and its corresponding acceleration $(\ddot{x}, \ddot{y}) = (0.4, 0.35)$, showing that the force and acceleration at the endpoint are not aligned.

the endpoint only if the force is along a principal axis of the ellipsoid. Unless $\Lambda(\theta)$ is of the form $cI$, where $c > 0$ is a scalar and $I$ is the identity matrix, the mass at the endpoint feels different from a point mass.

The change in apparent endpoint mass as a function of the configuration of the robot is an issue for robots used as haptic displays. One way to reduce the sensation of a changing mass to the user is to make the mass of the links as small as possible.

Note that the ellipsoidal interpretations of the relationship between forces and accelerations defined here are only relevant at zero velocity, where there are no Coriolis or centripetal terms.

### 8.1.4    Lagrangian Dynamics versus Newton–Euler Dynamics

In the rest of this chapter, we focus on the Newton–Euler recursive method for calculating robot dynamics. Using the tools we have developed so far, the Newton–Euler formulation allows computationally efficient computer implementation, particularly for robots with many degrees of freedom, without the need for differentiation. The resulting equations of motion are, and must be, identical with those derived using the energy-based Lagrangian method.

The Newton–Euler method builds on the dynamics of a single rigid body, so we begin there.

## 2    Dynamics of a Single Rigid Body

### 2.1    Classical Formulation

Consider a rigid body consisting of a number of rigidly connected point masses, where point mass $i$ has mass $\mathsf{m}_i$ and the total mass is $\mathsf{m} = \sum_i \mathsf{m}_i$. Let $r_i = (x_i, y_i, z_i)$ be the fixed location of mass $i$ in a body frame $\{\mathrm{b}\}$, where the origin of this frame is the unique point such that

$$\sum_i \mathsf{m}_i r_i = 0.$$

This point is known as the **center of mass**. If some other point happens to be inconveniently chosen as the origin, then the frame $\{\mathrm{b}\}$ should be moved to the center of mass at $(1/\mathsf{m}) \sum_i \mathsf{m}_i r_i$ (in the inconvenient frame) and the $r_i$ recalculated in the center-of-mass frame.

Now assume that the body is moving with a body twist $\mathcal{V}_b = (\omega_b, v_b)$, and let $p_i(t)$ be the time-varying position of $\mathsf{m}_i$, initially located at $r_i$, in the inertial frame $\{\mathrm{b}\}$. Then

$$\dot{p}_i = v_b + \omega_b \times p_i,$$

$$\ddot{p}_i = \dot{v}_b + \frac{d}{dt}\omega_b \times p_i + \omega_b \times \frac{d}{dt}p_i$$

$$= \dot{v}_b + \dot{\omega}_b \times p_i + \omega_b \times (v_b + \omega_b \times p_i).$$

Substituting $r_i$ for $p_i$ on the right-hand side and using our skew-symmetric notation (see Equation (3.30)), we get

$$\ddot{p}_i = \dot{v}_b + [\dot{\omega}_b]r_i + [\omega_b]v_b + [\omega_b]^2 r_i.$$

Taking as a given that $f_i = \mathsf{m}_i \ddot{p}_i$ for a point mass, the force acting on $\mathsf{m}_i$ is

$$f_i = \mathsf{m}_i(\dot{v}_b + [\dot{\omega}_b]r_i + [\omega_b]v_b + [\omega_b]^2 r_i),$$

which implies a moment

$$m_i = [r_i]f_i.$$

The total force and moment acting on the body is expressed as the wrench $\mathcal{F}_b$:

$$\mathcal{F}_b = \begin{bmatrix} m_b \\ f_b \end{bmatrix} = \begin{bmatrix} \sum_i m_i \\ \sum_i f_i \end{bmatrix}.$$

To simplify the expressions for $f_b$ and $m_b$, keep in mind that $\sum_i \mathsf{m}_i r_i = 0$

(and therefore $\sum_i \mathsf{m}_i[r_i] = 0$) and, for $a, b \in \mathbb{R}^3$, $[a] = -[a]^T$, $[a]b = -[b]a$, and $[a][b] = ([b][a])^T$. Focusing on the linear dynamics,

$$f_b = \sum_i \mathsf{m}_i(\dot{v}_b + [\dot{\omega}_b]r_i + [\omega_b]v_b + [\omega_b]^2 r_i)$$

$$= \sum_i \mathsf{m}_i(\dot{v}_b + [\omega_b]v_b) - \sum_i \mathsf{m}_i[r_i]\dot{\omega}_b^{\;0} + \sum_i \mathsf{m}_i[r_i][\omega_b]\omega_b^{\;0}$$

$$= \sum_i \mathsf{m}_i(\dot{v}_b + [\omega_b]v_b)$$

$$= \mathsf{m}(\dot{v}_b + [\omega_b]v_b). \tag{8.22}$$

The velocity product term $\mathsf{m}[\omega_b]v_b$ arises from the fact that, for $\omega_b \neq 0$, a constant $v_b \neq 0$ corresponds to a changing linear velocity in an inertial frame.

Now focusing on the rotational dynamics,

$$m_b = \sum_i \mathsf{m}_i[r_i](\dot{v}_b + [\dot{\omega}_b]r_i + [\omega_b]v_b + [\omega_b]^2 r_i)$$

$$= \sum_i \mathsf{m}_i[r_i]\dot{v}_b^{\;0} + \sum_i \mathsf{m}_i[r_i][\omega_b]v_b^{\;0}$$

$$\quad + \sum_i \mathsf{m}_i[r_i]([\dot{\omega}_b]r_i + [\omega_b]^2 r_i)$$

$$= \sum_i \mathsf{m}_i\left(-[r_i]^2\dot{\omega}_b - [r_i]^T[\omega_b]^T[r_i]\omega_b\right)$$

$$= \sum_i \mathsf{m}_i\left(-[r_i]^2\dot{\omega}_b - [\omega_b][r_i]^2\omega_b\right)$$

$$= \left(-\sum_i \mathsf{m}_i[r_i]^2\right)\dot{\omega}_b + [\omega_b]\left(-\sum_i \mathsf{m}_i[r_i]^2\right)\omega_b$$

$$= \mathcal{I}_b\dot{\omega}_b + [\omega_b]\mathcal{I}_b\omega_b, \tag{8.23}$$

where $\mathcal{I}_b = -\sum_i \mathsf{m}_i[r_i]^2 \in \mathbb{R}^{3\times 3}$ is the body's **rotational inertia matrix**. Equation (8.23) is known as **Euler's equation** for a rotating rigid body.

In Equation (8.23), note the presence of a term linear in the angular acceleration, $\mathcal{I}_b\dot{\omega}_b$, and a term quadratic in the angular velocities, $[\omega_b]\mathcal{I}_b\omega_b$, just as we saw for the mechanisms in Section 8.1. Also, $\mathcal{I}_b$ is symmetric and positive definite, just like the mass matrix for a mechanism, and the rotational kinetic energy is given by the quadratic

$$\mathcal{K} = \frac{1}{2}\omega_b^T\mathcal{I}_b\omega_b.$$

One difference is that $\mathcal{I}_b$ is constant whereas the mass matrix $M(\theta)$ changes with the configuration of the mechanism.

Writing out the individual entries of $\mathcal{I}_b$, we get

$$
\mathcal{I}_b = \begin{bmatrix} \sum \mathfrak{m}_i(y_i^2 + z_i^2) & -\sum \mathfrak{m}_i x_i y_i & -\sum \mathfrak{m}_i x_i z_i \\ -\sum \mathfrak{m}_i x_i y_i & \sum \mathfrak{m}_i(x_i^2 + z_i^2) & -\sum \mathfrak{m}_i y_i z_i \\ -\sum \mathfrak{m}_i x_i z_i & -\sum \mathfrak{m}_i y_i z_i & \sum \mathfrak{m}_i(x_i^2 + y_i^2) \end{bmatrix}
$$

$$
= \begin{bmatrix} \mathcal{I}_{xx} & \mathcal{I}_{xy} & \mathcal{I}_{xz} \\ \mathcal{I}_{xy} & \mathcal{I}_{yy} & \mathcal{I}_{yz} \\ \mathcal{I}_{xz} & \mathcal{I}_{yz} & \mathcal{I}_{zz} \end{bmatrix}.
$$

The summations can be replaced by volume integrals over the body $\mathcal{B}$, using the differential volume element $dV$, with the point masses $\mathfrak{m}_i$ replaced by a mass density function $\rho(x, y, z)$:

$$
\left.\begin{aligned}
\mathcal{I}_{xx} &= \int_{\mathcal{B}} (y^2 + z^2)\rho(x, y, z)\, dV \\
\mathcal{I}_{yy} &= \int_{\mathcal{B}} (x^2 + z^2)\rho(x, y, z)\, dV \\
\mathcal{I}_{zz} &= \int_{\mathcal{B}} (x^2 + y^2)\rho(x, y, z)\, dV \\
\mathcal{I}_{xy} &= -\int_{\mathcal{B}} xy\rho(x, y, z)\, dV \\
\mathcal{I}_{xz} &= -\int_{\mathcal{B}} xz\rho(x, y, z)\, dV \\
\mathcal{I}_{yz} &= -\int_{\mathcal{B}} yz\rho(x, y, z)\, dV.
\end{aligned}\right\} \tag{8.24}
$$

If the body has uniform density, $\mathcal{I}_b$ is determined exclusively by the shape of the rigid body (see Figure 8.5).

Given an inertia matrix $\mathcal{I}_b$, the **principal axes of inertia** are given by the eigenvectors and eigenvalues of $\mathcal{I}_b$. Let $v_1, v_2, v_3$ be the eigenvectors of $\mathcal{I}_b$ and $\lambda_1, \lambda_2, \lambda_3$ be the corresponding eigenvalues. Then the principal axes of inertia are in the directions of $v_1, v_2, v_3$, and the scalar moments of inertia about these axes, the **principal moments of inertia**, are $\lambda_1, \lambda_2, \lambda_3 > 0$. One principal axis maximizes the moment of inertia among all axes passing through the center of mass, and another minimizes the moment of inertia. For bodies with symmetry, often the principal axes of inertia are apparent. They may not be unique; for a uniform-density solid sphere, for example, any three orthogonal axes intersecting at the center of mass constitute a set of principal axes, and the minimum principal moment of inertia is equal to the maximum principal moment of inertia.

If the principal axes of inertia are aligned with the axes of $\{b\}$, the off-diagonal terms of $\mathcal{I}_b$ are all zero, and the eigenvalues are the scalar moments of inertia $\mathcal{I}_{xx}$, $\mathcal{I}_{yy}$, and $\mathcal{I}_{zz}$ about the $\hat{x}$-, $\hat{y}$-, and $\hat{z}$-axes, respectively. In this case, the

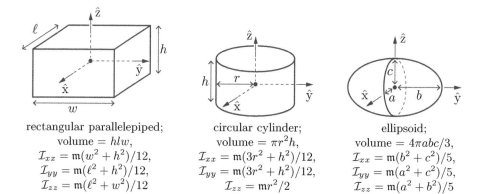

rectangular parallelepiped;
volume $= hlw$,
$\mathcal{I}_{xx} = \mathsf{m}(w^2 + h^2)/12$,
$\mathcal{I}_{yy} = \mathsf{m}(\ell^2 + h^2)/12$,
$\mathcal{I}_{zz} = \mathsf{m}(\ell^2 + w^2)/12$

circular cylinder;
volume $= \pi r^2 h$,
$\mathcal{I}_{xx} = \mathsf{m}(3r^2 + h^2)/12$,
$\mathcal{I}_{yy} = \mathsf{m}(3r^2 + h^2)/12$,
$\mathcal{I}_{zz} = \mathsf{m}r^2/2$

ellipsoid;
volume $= 4\pi abc/3$,
$\mathcal{I}_{xx} = \mathsf{m}(b^2 + c^2)/5$,
$\mathcal{I}_{yy} = \mathsf{m}(a^2 + c^2)/5$,
$\mathcal{I}_{zz} = \mathsf{m}(a^2 + b^2)/5$

**Figure 8.5** The principal axes and the inertia about the principal axes for uniform-density bodies of mass m. Note that the $\hat{x}$- and $\hat{y}$-principal axes of the cylinder are not unique.

equations of motion (8.23) simplify to

$$m_b = \begin{bmatrix} \mathcal{I}_{xx}\dot{\omega}_x + (\mathcal{I}_{zz} - \mathcal{I}_{yy})\omega_y\omega_z \\ \mathcal{I}_{yy}\dot{\omega}_y + (\mathcal{I}_{xx} - \mathcal{I}_{zz})\omega_x\omega_z \\ \mathcal{I}_{zz}\dot{\omega}_z + (\mathcal{I}_{yy} - \mathcal{I}_{xx})\omega_x\omega_y \end{bmatrix}, \tag{8.25}$$

where $\omega_b = (\omega_x, \omega_y, \omega_z)$. When possible, we choose the axes of {b} to be aligned with the principal axes of inertia, in order to reduce the number of nonzero entries in $\mathcal{I}_b$ and to simplify the equations of motion.

Examples of common uniform-density solid bodies, their principal axes of inertia, and the principal moments of inertia obtained by solving the integrals (8.24), are given in Figure 8.5.

An inertia matrix $\mathcal{I}_b$ can be expressed in a rotated frame {c} described by the rotation matrix $R_{bc}$. Denoting this inertia matrix as $\mathcal{I}_c$, and knowing that the kinetic energy of the rotating body is independent of the chosen frame, we have

$$\frac{1}{2}\omega_c^{\mathrm{T}}\mathcal{I}_c\omega_c = \frac{1}{2}\omega_b^{\mathrm{T}}\mathcal{I}_b\omega_b$$

$$= \frac{1}{2}(R_{bc}\omega_c)^{\mathrm{T}}\mathcal{I}_b(R_{bc}\omega_c)$$

$$= \frac{1}{2}\omega_c^{\mathrm{T}}(R_{bc}^{\mathrm{T}}\mathcal{I}_b R_{bc})\omega_c.$$

In other words,

$$\mathcal{I}_c = R_{bc}^{\mathrm{T}}\mathcal{I}_b R_{bc}. \tag{8.26}$$

If the axes of {b} are not aligned with the principal axes of inertia then we can diagonalize the inertia matrix by expressing it instead in the rotated frame {c}, where the columns of $R_{bc}$ correspond to the eigenvectors of $\mathcal{I}_b$.

Sometimes it is convenient to represent the inertia matrix in a frame at a point not at the center of mass of the body, for example at a joint. **Steiner's theorem** can be stated as follows.

**Theorem 8.2**    *The inertia matrix $\mathcal{I}_q$ about a frame aligned with {b}, but at a point $q = (q_x, q_y, q_z)$ in {b}, is related to the inertia matrix $\mathcal{I}_b$ calculated at the center of mass by*

$$\mathcal{I}_q = \mathcal{I}_b + \mathfrak{m}(q^\mathsf{T} q I - q q^\mathsf{T}), \tag{8.27}$$

*where $I$ is the $3 \times 3$ identity matrix and $\mathfrak{m}$ is the mass of the body.*

Steiner's theorem is a more general statement of the parallel-axis theorem, which states that the scalar inertia $\mathcal{I}_d$ about an axis parallel to, but a distance $d$ from, an axis through the center of mass is related to the scalar inertia $\mathcal{I}_{\text{cm}}$ about the axis through the center of mass by

$$\mathcal{I}_d = \mathcal{I}_{\text{cm}} + \mathfrak{m}d^2. \tag{8.28}$$

Equations (8.26) and (8.27) are useful for calculating the inertia of a rigid body consisting of component rigid bodies. First we calculate the inertia matrices of the $n$ component bodies in terms of frames at their individual centers of mass. Then we choose a common frame {common} (e.g., at the center of mass of the composite rigid body) and use Equations (8.26) and (8.27) to express each inertia matrix in this common frame. Once the individual inertia matrices are expressed in {common}, they can be summed to get the inertia matrix $\mathcal{I}_{\text{common}}$ for the composite rigid body.

In the case of motion confined to the $\hat{x}$–$\hat{y}$-plane, where $\omega_b = (0, 0, \omega_z)$ and the inertia of the body about the $\hat{z}$-axis through the center of mass is given by the scalar $\mathcal{I}_{zz}$, the spatial rotational dynamics (8.23) reduces to the planar rotational dynamics

$$m_z = \mathcal{I}_{zz} \dot{\omega}_z,$$

and the rotational kinetic energy is

$$\mathcal{K} = \frac{1}{2} \mathcal{I}_{zz} \omega_z^2.$$

## 2.2    Twist–Wrench Formulation

The linear dynamics (8.22) and the rotational dynamics (8.23) can be written in the following combined form:

$$\begin{bmatrix} m_b \\ f_b \end{bmatrix} = \begin{bmatrix} \mathcal{I}_b & 0 \\ 0 & \mathfrak{m}I \end{bmatrix} \begin{bmatrix} \dot{\omega}_b \\ \dot{v}_b \end{bmatrix} + \begin{bmatrix} [\omega_b] & 0 \\ 0 & [\omega_b] \end{bmatrix} \begin{bmatrix} \mathcal{I}_b & 0 \\ 0 & \mathfrak{m}I \end{bmatrix} \begin{bmatrix} \omega_b \\ v_b \end{bmatrix}, \tag{8.29}$$

where $I$ is the $3 \times 3$ identity matrix. With the benefit of hindsight, and also making use of the fact that $[v]v = v \times v = 0$ and $[v]^\mathsf{T} = -[v]$, we can write

Equation (8.29) in the following equivalent form:

$$
\begin{bmatrix} m_b \\ f_b \end{bmatrix} = \begin{bmatrix} \mathcal{I}_b & 0 \\ 0 & \mathrm{m}I \end{bmatrix} \begin{bmatrix} \dot{\omega}_b \\ \dot{v}_b \end{bmatrix} + \begin{bmatrix} [\omega_b] & [v_b] \\ 0 & [\omega_b] \end{bmatrix} \begin{bmatrix} \mathcal{I}_b & 0 \\ 0 & \mathrm{m}I \end{bmatrix} \begin{bmatrix} \omega_b \\ v_b \end{bmatrix}
$$

$$
= \begin{bmatrix} \mathcal{I}_b & 0 \\ 0 & \mathrm{m}I \end{bmatrix} \begin{bmatrix} \dot{\omega}_b \\ \dot{v}_b \end{bmatrix} - \begin{bmatrix} [\omega_b] & 0 \\ [v_b] & [\omega_b] \end{bmatrix}^{\mathrm{T}} \begin{bmatrix} \mathcal{I}_b & 0 \\ 0 & \mathrm{m}I \end{bmatrix} \begin{bmatrix} \omega_b \\ v_b \end{bmatrix}. \tag{8.30}
$$

Written this way, each term can now be identified with six-dimensional spatial quantities as follows:

(a) The vectors $(\omega_b, v_b)$ and $(m_b, f_b)$ can be respectively identified with the body twist $\mathcal{V}_b$ and body wrench $\mathcal{F}_b$,

$$
\mathcal{V}_b = \begin{bmatrix} \omega_b \\ v_b \end{bmatrix}, \qquad \mathcal{F}_b = \begin{bmatrix} m_b \\ f_b \end{bmatrix}. \tag{8.31}
$$

(b) The **spatial inertia matrix** $\mathcal{G}_b \in \mathbb{R}^{6\times 6}$ is defined as

$$
\mathcal{G}_b = \begin{bmatrix} \mathcal{I}_b & 0 \\ 0 & \mathrm{m}I \end{bmatrix}. \tag{8.32}
$$

As an aside, the kinetic energy of the rigid body can be expressed in terms of the spatial inertia matrix as

$$
\text{kinetic energy} = \frac{1}{2}\omega_b^{\mathrm{T}}\mathcal{I}_b\omega_b + \frac{1}{2}\mathrm{m}v_b^{\mathrm{T}}v_b = \frac{1}{2}\mathcal{V}_b^{\mathrm{T}}\mathcal{G}_b\mathcal{V}_b. \tag{8.33}
$$

(c) The **spatial momentum** $\mathcal{P}_b \in \mathbb{R}^6$ is defined as

$$
\mathcal{P}_b = \begin{bmatrix} \mathcal{I}_b\omega_b \\ \mathrm{m}v_b \end{bmatrix} = \begin{bmatrix} \mathcal{I}_b & 0 \\ 0 & \mathrm{m}I \end{bmatrix} \begin{bmatrix} \omega_b \\ v_b \end{bmatrix} = \mathcal{G}_b\mathcal{V}_b. \tag{8.34}
$$

Observe that the term involving $\mathcal{P}_b$ in Equation (8.30) is left-multiplied by the matrix

$$
- \begin{bmatrix} [\omega_b] & 0 \\ [v_b] & [\omega_b] \end{bmatrix}^{\mathrm{T}}. \tag{8.35}
$$

We now explain the origin and geometric significance of this matrix. First, recall that the cross product of two vectors $\omega_1, \omega_2 \in \mathbb{R}^3$ can be calculated, using the skew-symmetric matrix notation, as follows:

$$
[\omega_1 \times \omega_2] = [\omega_1][\omega_2] - [\omega_2][\omega_1]. \tag{8.36}
$$

The matrix in (8.35) can be thought of as a generalization of the cross-product operation to six-dimensional twists. Specifically, given two twists $\mathcal{V}_1 = (\omega_1, v_1)$

and $\mathcal{V}_2 = (\omega_2, v_2)$, we perform a calculation analogous to (8.36):

$$[\mathcal{V}_1][\mathcal{V}_2] - [\mathcal{V}_2][\mathcal{V}_1] = \left[ \begin{array}{cc} [\omega_1] & v_1 \\ 0 & 0 \end{array} \right] \left[ \begin{array}{cc} [\omega_2] & v_2 \\ 0 & 0 \end{array} \right] - \left[ \begin{array}{cc} [\omega_2] & v_2 \\ 0 & 0 \end{array} \right] \left[ \begin{array}{cc} [\omega_1] & v_1 \\ 0 & 0 \end{array} \right]$$

$$= \left[ \begin{array}{cc} [\omega_1][\omega_2] - [\omega_2][\omega_1] & [\omega_1]v_2 - [\omega_2]v_1 \\ 0 & 0 \end{array} \right]$$

$$= \left[ \begin{array}{cc} [\omega'] & v' \\ 0 & 0 \end{array} \right],$$

which can be written more compactly in vector form as

$$\left[ \begin{array}{c} \omega' \\ v' \end{array} \right] = \left[ \begin{array}{cc} [\omega_1] & 0 \\ [v_1] & [\omega_1] \end{array} \right] \left[ \begin{array}{c} \omega_2 \\ v_2 \end{array} \right].$$

This generalization of the cross product to two twists $\mathcal{V}_1$ and $\mathcal{V}_2$ is called the **Lie bracket** of $\mathcal{V}_1$ and $\mathcal{V}_2$.

**Definition 8.3**    Given two twists $\mathcal{V}_1 = (\omega_1, v_1)$ and $\mathcal{V}_2 = (\omega_2, v_2)$, the **Lie bracket** of $\mathcal{V}_1$ and $\mathcal{V}_2$, written either as $[\mathrm{ad}_{\mathcal{V}_1}]\mathcal{V}_2$ or $\mathrm{ad}_{\mathcal{V}_1}(\mathcal{V}_2)$, is defined as follows:

$$\left[ \begin{array}{cc} [\omega_1] & 0 \\ [v_1] & [\omega_1] \end{array} \right] \left[ \begin{array}{c} \omega_2 \\ v_2 \end{array} \right] = [\mathrm{ad}_{\mathcal{V}_1}]\mathcal{V}_2 = \mathrm{ad}_{\mathcal{V}_1}(\mathcal{V}_2) \in \mathbb{R}^6, \qquad (8.37)$$

where

$$[\mathrm{ad}_{\mathcal{V}}] = \left[ \begin{array}{cc} [\omega] & 0 \\ [v] & [\omega] \end{array} \right] \in \mathbb{R}^{6 \times 6}. \qquad (8.38)$$

**Definition 8.4**    Given a twist $\mathcal{V} = (\omega, v)$ and a wrench $\mathcal{F} = (m, f)$, define the mapping

$$\mathrm{ad}_{\mathcal{V}}^{\mathrm{T}}(\mathcal{F}) = [\mathrm{ad}_{\mathcal{V}}]^{\mathrm{T}}\mathcal{F} = \left[ \begin{array}{cc} [\omega] & 0 \\ [v] & [\omega] \end{array} \right]^{\mathrm{T}} \left[ \begin{array}{c} m \\ f \end{array} \right] = \left[ \begin{array}{c} -[\omega]m - [v]f \\ -[\omega]f \end{array} \right]. \qquad (8.39)$$

Using the notation and definitions above, the dynamic equations for a single rigid body can now be written as

$$\mathcal{F}_b = \mathcal{G}_b \dot{\mathcal{V}}_b - \mathrm{ad}_{\mathcal{V}_b}^{\mathrm{T}}(\mathcal{P}_b)$$
$$= \mathcal{G}_b \dot{\mathcal{V}}_b - [\mathrm{ad}_{\mathcal{V}_b}]^{\mathrm{T}} \mathcal{G}_b \mathcal{V}_b. \qquad (8.40)$$

Note the analogy between Equation (8.40) and the moment equation for a rotating rigid body:

$$m_b = \mathcal{I}_b \dot{\omega}_b - [\omega_b]^{\mathrm{T}} \mathcal{I}_b \omega_b. \qquad (8.41)$$

Equation (8.41) is simply the rotational component of (8.40).

### 8.2.3    Dynamics in Other Frames

The derivation of the dynamic equations (8.40) relies on the use of a center-of-mass frame {b}. It is straightforward to express the dynamics in other frames, however. Let's call one such frame {a}.

Since the kinetic energy of the rigid body must be independent of the frame of representation,

$$
\frac{1}{2}\mathcal{V}_a^{\mathrm{T}}\mathcal{G}_a\mathcal{V}_a = \frac{1}{2}\mathcal{V}_b^{\mathrm{T}}\mathcal{G}_b\mathcal{V}_b
$$

$$
= \frac{1}{2}([\mathrm{Ad}_{T_{ba}}]\mathcal{V}_a)^{\mathrm{T}}\mathcal{G}_b[\mathrm{Ad}_{T_{ba}}]\mathcal{V}_a
$$

$$
= \frac{1}{2}\mathcal{V}_a^{\mathrm{T}}\underbrace{[\mathrm{Ad}_{T_{ba}}]^{\mathrm{T}}\mathcal{G}_b[\mathrm{Ad}_{T_{ba}}]}_{\mathcal{G}_a}\mathcal{V}_a;
$$

for the adjoint representation Ad (see Definition 3.20). In other words, the spatial inertia matrix $\mathcal{G}_a$ in {a} is related to $\mathcal{G}_b$ by

$$
\mathcal{G}_a = [\mathrm{Ad}_{T_{ba}}]^{\mathrm{T}}\mathcal{G}_b[\mathrm{Ad}_{T_{ba}}]. \tag{8.42}
$$

This is a generalization of Steiner's theorem.

Using the spatial inertia matrix $\mathcal{G}_a$, the equations of motion (8.40) in the {b} frame can be expressed equivalently in the {a} frame as

$$
\mathcal{F}_a = \mathcal{G}_a\dot{\mathcal{V}}_a - [\mathrm{ad}_{\mathcal{V}_a}]^{\mathrm{T}}\mathcal{G}_a\mathcal{V}_a, \tag{8.43}
$$

where $\mathcal{F}_a$ and $\mathcal{V}_a$ are the wrench and twist written in the {a} frame. (See Exercise 8.3.) Thus the form of the equations of motion is independent of the frame of representation.

## 8.3    Newton–Euler Inverse Dynamics

We now consider the inverse dynamics problem for an $n$-link open chain connected by one-dof joints. Given the joint positions $\theta \in \mathbb{R}^n$, velocities $\dot{\theta} \in \mathbb{R}^n$, and accelerations $\ddot{\theta} \in \mathbb{R}^n$, the objective is to calculate the right-hand side of the dynamics equation

$$
\tau = M(\theta)\ddot{\theta} + h(\theta, \dot{\theta}).
$$

The main result is a recursive inverse dynamics algorithm consisting of a forward and a backward iteration stage. In the former, the positions, velocities, and accelerations of each link are propagated from the base to the tip while in the backward iterations the forces and moments experienced by each link are propagated from the tip to the base.

## 3.1 Derivation

A body-fixed reference frame $\{i\}$ is attached to the center of mass of each link $i$, $i = 1, \ldots, n$. The base frame is denoted $\{0\}$, and a frame at the end-effector is denoted $\{n+1\}$. This frame is fixed in $\{n\}$.

When the manipulator is at the home position, with all joint variables zero, we denote the configuration of frame $\{j\}$ in $\{i\}$ as $M_{i,j} \in SE(3)$, and the configuration of $\{i\}$ in the base frame $\{0\}$ using the shorthand $M_i = M_{0,i}$. With these definitions, $M_{i-1,i}$ and $M_{i,i-1}$ can be calculated as

$$M_{i-1,i} = M_{i-1}^{-1} M_i \qquad \text{and} \qquad M_{i,i-1} = M_i^{-1} M_{i-1}.$$

The screw axis for joint $i$, expressed in the link frame $\{i\}$, is $\mathcal{A}_i$. This same screw axis is expressed in the space frame $\{0\}$ as $\mathcal{S}_i$, where the two are related by

$$\mathcal{A}_i = \text{Ad}_{M_i^{-1}}(\mathcal{S}_i).$$

Defining $T_{i,j} \in SE(3)$ to be the configuration of frame $\{j\}$ in $\{i\}$ for arbitrary joint variables $\theta$ then $T_{i-1,i}(\theta_i)$, the configuration of $\{i\}$ relative to $\{i-1\}$ given the joint variable $\theta_i$, and $T_{i,i-1}(\theta_i) = T_{i-1,i}^{-1}(\theta_i)$ are calculated as

$$T_{i-1,i}(\theta_i) = M_{i-1,i} e^{[\mathcal{A}_i]\theta_i} \qquad \text{and} \qquad T_{i,i-1}(\theta_i) = e^{-[\mathcal{A}_i]\theta_i} M_{i,i-1}.$$

We further adopt the following notation:

(a) The twist of link frame $\{i\}$, expressed in frame-$\{i\}$ coordinates, is denoted $\mathcal{V}_i = (\omega_i, v_i)$.
(b) The wrench transmitted through joint $i$ to link frame $\{i\}$, expressed in frame-$\{i\}$ coordinates, is denoted $\mathcal{F}_i = (m_i, f_i)$.
(c) Let $\mathcal{G}_i \in \mathbb{R}^{6\times6}$ denote the spatial inertia matrix of link $i$, expressed relative to link frame $\{i\}$. Since we are assuming that all link frames are situated at the link center of mass, $\mathcal{G}_i$ has the block-diagonal form

$$\mathcal{G}_i = \begin{bmatrix} \mathcal{I}_i & 0 \\ 0 & \mathsf{m}_i I \end{bmatrix}, \tag{8.44}$$

where $\mathcal{I}_i$ denotes the $3 \times 3$ rotational inertia matrix of link $i$ and $\mathsf{m}_i$ is the link mass.

With these definitions, we can recursively calculate the twist and acceleration of each link, moving from the base to the tip. The twist $\mathcal{V}_i$ of link $i$ is the sum of the twist of link $i - 1$, but expressed in $\{i\}$, and the added twist due to the joint rate $\dot{\theta}_i$:

$$\mathcal{V}_i = \mathcal{A}_i \dot{\theta}_i + [\text{Ad}_{T_{i,i-1}}] \mathcal{V}_{i-1}. \tag{8.45}$$

The accelerations $\dot{\mathcal{V}}_i$ can also be found recursively. Taking the time derivative of Equation (8.45), we get

$$\dot{\mathcal{V}}_i = \mathcal{A}_i \ddot{\theta}_i + [\text{Ad}_{T_{i,i-1}}] \dot{\mathcal{V}}_{i-1} + \frac{d}{dt}\left([\text{Ad}_{T_{i,i-1}}]\right) \mathcal{V}_{i-1}. \tag{8.46}$$

To calculate the final term in this equation, we express $T_{i,i-1}$ and $\mathcal{A}_i$ as

$$T_{i,i-1} = \begin{bmatrix} R_{i,i-1} & p \\ 0 & 1 \end{bmatrix} \quad \text{and} \quad \mathcal{A}_i = \begin{bmatrix} \omega \\ v \end{bmatrix}.$$

From the fact $\dot{T}_{i,i-1} T_{i,i-1}^{-1} = -[\mathcal{A}_i \dot{\theta}_i]$, we have

$$\dot{R}_{i,i-1} = -[\omega \dot{\theta}_i] R_{i,i-1}, \quad \dot{p} = -[\omega \dot{\theta}_i] p - v \dot{\theta}_i.$$

Then

$$\frac{d}{dt}([\text{Ad}_{T_{i,i-1}}]) \mathcal{V}_{i-1}$$

$$= \frac{d}{dt} \begin{bmatrix} R_{i,i-1} & 0 \\ [p] R_{i,i-1} & R_{i,i-1} \end{bmatrix} \mathcal{V}_{i-1}$$

$$= \begin{bmatrix} -[\omega \dot{\theta}_i] R_{i,i-1} & 0 \\ [-[\omega \dot{\theta}_i] p - v \dot{\theta}_i] R_{i,i-1} - [p][\omega \dot{\theta}_i] R_{i,i-1} & -[\omega \dot{\theta}_i] R_{i,i-1} \end{bmatrix} \mathcal{V}_{i-1}$$

$$= \underbrace{\begin{bmatrix} -[\omega \dot{\theta}_i] & 0 \\ -[v \dot{\theta}_i] & -[\omega \dot{\theta}_i] \end{bmatrix}}_{-[\text{ad}_{\mathcal{A}_i \dot{\theta}_i}]} \underbrace{\begin{bmatrix} R_{i,i-1} & 0 \\ [p] R_{i,i-1} & R_{i,i-1} \end{bmatrix}}_{[\text{Ad}_{T_{i,i-1}}]} \mathcal{V}_{i-1}$$

$$= -[\text{ad}_{\mathcal{A}_i \dot{\theta}_i}] \mathcal{V}_i$$

$$= [\text{ad}_{\mathcal{V}_i}] \mathcal{A}_i \dot{\theta}_i,$$

where the transition from the second equality to the third follows from the Jacobi identity $a \times (b \times c) + b \times (c \times a) + c \times (a \times b) = 0$ for all $a, b, c \in \mathbb{R}^3$, and the transition from the fourth equality to the fifth follows from the identity $[\text{ad}_{\mathcal{V}_1}]\mathcal{V}_2 = -[\text{ad}_{\mathcal{V}_2}]\mathcal{V}_1$. Substituting this result into Equation (8.46), we get

$$\dot{\mathcal{V}}_i = \mathcal{A}_i \ddot{\theta}_i + [\text{Ad}_{T_{i,i-1}}] \dot{\mathcal{V}}_{i-1} + [\text{ad}_{\mathcal{V}_i}] \mathcal{A}_i \dot{\theta}_i, \tag{8.47}$$

i.e., the acceleration of link $i$ is the sum of three components: a component due to the joint acceleration $\ddot{\theta}_i$, a component due to the acceleration of link $i-1$ expressed in $\{i\}$, and a velocity-product component.

Once we have determined all the link twists and accelerations moving outward from the base, we can calculate the joint torques or forces by moving inward from the tip. The rigid-body dynamics (8.40) tells us the total wrench that acts on link $i$ given $\mathcal{V}_i$ and $\dot{\mathcal{V}}_i$. Furthermore, the total wrench acting on link $i$ is the sum of the wrench $\mathcal{F}_i$ transmitted through joint $i$ and the wrench applied to the link through joint $i+1$ (or, for link $n$, the wrench applied to the link by the environment at the end-effector frame $\{n+1\}$), expressed in the frame $i$. Therefore, we have the equality

$$\mathcal{G}_i \dot{\mathcal{V}}_i - \text{ad}_{\mathcal{V}_i}^{\text{T}}(\mathcal{G}_i \mathcal{V}_i) = \mathcal{F}_i - \text{Ad}_{T_{i+1,i}}^{\text{T}}(\mathcal{F}_{i+1}); \tag{8.48}$$

see Figure 8.6. Solving from the tip toward the base, at each link $i$ we solve for the only unknown in Equation (8.48): $\mathcal{F}_i$. Since joint $i$ has only one degree of freedom, five dimensions of the six-vector $\mathcal{F}_i$ are provided "for free" by the

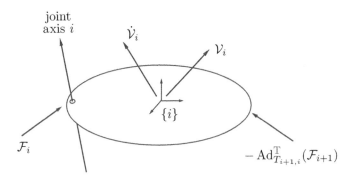

**Figure 8.6** Free-body diagram illustrating the moments and forces exerted on link $i$.

structure of the joint, and the actuator only has to provide the scalar force or torque in the direction of the joint's screw axis:

$$\tau_i = \mathcal{F}_i^{\mathrm{T}} \mathcal{A}_i. \tag{8.49}$$

Equation (8.49) provides the torques required at each joint, solving the inverse dynamics problem.

## .3.2   Newton–Euler Inverse Dynamics Algorithm

**Initialization**

Attach a frame $\{0\}$ to the base, frames $\{1\}$ to $\{n\}$ to the centers of mass of links $\{1\}$ to $\{n\}$, and a frame $\{n+1\}$ at the end-effector, fixed in the frame $\{n\}$. Define $M_{i,i-1}$ to be the configuration of $\{i-1\}$ in $\{i\}$ when $\theta_i = 0$. Let $\mathcal{A}_i$ be the screw axis of joint $i$ expressed in $\{i\}$, and $\mathcal{G}_i$ be the $6 \times 6$ spatial inertia matrix of link $i$. Define $\mathcal{V}_0$ to be the twist of the base frame $\{0\}$ expressed in $\{0\}$ coordinates. (This quantity is typically zero.) Let $\mathfrak{g} \in \mathbb{R}^3$ be the gravity vector expressed in $\{0\}$ coordinates, and define $\dot{\mathcal{V}}_0 = (\dot{\omega}_0, \dot{v}_0) = (0, -\mathfrak{g})$. (Gravity is treated as an acceleration of the base in the opposite direction.) Define $\mathcal{F}_{n+1} = \mathcal{F}_{\mathrm{tip}} = (m_{\mathrm{tip}}, f_{\mathrm{tip}})$ to be the wrench applied to the environment by the end-effector, expressed in the end-effector frame $\{n+1\}$.

**Forward iterations**

Given $\theta, \dot{\theta}, \ddot{\theta}$, for $i = 1$ to $n$ do

$$T_{i,i-1} = e^{-[\mathcal{A}_i]\theta_i} M_{i,i-1}, \tag{8.50}$$
$$\mathcal{V}_i = \mathrm{Ad}_{T_{i,i-1}}(\mathcal{V}_{i-1}) + \mathcal{A}_i \dot{\theta}_i, \tag{8.51}$$
$$\dot{\mathcal{V}}_i = \mathrm{Ad}_{T_{i,i-1}}(\dot{\mathcal{V}}_{i-1}) + \mathrm{ad}_{\mathcal{V}_i}(\mathcal{A}_i)\dot{\theta}_i + \mathcal{A}_i \ddot{\theta}_i. \tag{8.52}$$

**Backward iterations**

For $i = n$ to 1 do

$$\mathcal{F}_i = \mathrm{Ad}_{T_{i+1,i}}^{\mathrm{T}}(\mathcal{F}_{i+1}) + \mathcal{G}_i \dot{\mathcal{V}}_i - \mathrm{ad}_{\mathcal{V}_i}^{\mathrm{T}}(\mathcal{G}_i \mathcal{V}_i), \tag{8.53}$$

$$\tau_i = \mathcal{F}_i^{\mathrm{T}} \mathcal{A}_i. \tag{8.54}$$

## 8.4     Dynamic Equations in Closed Form

In this section we show how the equations in the recursive inverse dynamics algorithm can be organized into a closed-form set of dynamics equations $\tau = M(\theta)\ddot{\theta} + c(\theta, \dot{\theta}) + g(\theta)$.

Before doing so, we prove our earlier assertion that the total kinetic energy $\mathcal{K}$ of the robot can be expressed as $\mathcal{K} = \frac{1}{2}\dot{\theta}^{\mathrm{T}} M(\theta)\dot{\theta}$. We do so by noting that $\mathcal{K}$ can be expressed as the sum of the kinetic energies of each link:

$$\mathcal{K} = \frac{1}{2} \sum_{i=1}^{n} \mathcal{V}_i^{\mathrm{T}} \mathcal{G}_i \mathcal{V}_i, \tag{8.55}$$

where $\mathcal{V}_i$ is the twist of link frame $\{i\}$ and $\mathcal{G}_i$ is the spatial inertia matrix of link $i$ as defined by Equation (8.32) (both are expressed in link-frame-$\{i\}$ coordinates). Let $T_{0i}(\theta_1, \ldots, \theta_i)$ denote the forward kinematics from the base frame $\{0\}$ to link frame $\{i\}$, and let $J_{ib}(\theta)$ denote the body Jacobian obtained from $T_{0i}^{-1}\dot{T}_{0i}$. Note that $J_{ib}$ as defined is a $6 \times i$ matrix; we turn it into a $6 \times n$ matrix by filling in all entries of the last $n - i$ columns with zeros. With this definition of $J_{ib}$, we can write

$$\mathcal{V}_i = J_{ib}(\theta)\dot{\theta}, \qquad i = 1, \ldots, n.$$

The kinetic energy can then be written

$$\mathcal{K} = \frac{1}{2} \dot{\theta}^{\mathrm{T}} \left( \sum_{i=1}^{n} J_{ib}^{\mathrm{T}}(\theta) \mathcal{G}_i J_{ib}(\theta) \right) \dot{\theta}. \tag{8.56}$$

The term inside the parentheses is precisely the mass matrix $M(\theta)$:

$$M(\theta) = \sum_{i=1}^{n} J_{ib}^{\mathrm{T}}(\theta) \mathcal{G}_i J_{ib}(\theta). \tag{8.57}$$

We now return to the original task of deriving a closed-form set of dynamic equations. We start by defining the following stacked vectors:

$$\mathcal{V} = \begin{bmatrix} \mathcal{V}_1 \\ \vdots \\ \mathcal{V}_n \end{bmatrix} \in \mathbb{R}^{6n}, \tag{8.58}$$

$$\mathcal{F} = \begin{bmatrix} \mathcal{F}_1 \\ \vdots \\ \mathcal{F}_n \end{bmatrix} \in \mathbb{R}^{6n}. \tag{8.59}$$

Further, define the following matrices:

$$A = \begin{bmatrix} \mathcal{A}_1 & 0 & \cdots & 0 \\ 0 & \mathcal{A}_2 & \cdots & 0 \\ \vdots & \vdots & \ddots & \vdots \\ 0 & \cdots & \cdots & \mathcal{A}_n \end{bmatrix} \in \mathbb{R}^{6n \times n}, \tag{8.60}$$

$$\mathcal{G} = \begin{bmatrix} \mathcal{G}_1 & 0 & \cdots & 0 \\ 0 & \mathcal{G}_2 & \cdots & 0 \\ \vdots & \vdots & \ddots & \vdots \\ 0 & \cdots & \cdots & \mathcal{G}_n \end{bmatrix} \in \mathbb{R}^{6n \times 6n}, \tag{8.61}$$

$$[\mathrm{ad}_{\mathcal{V}}] = \begin{bmatrix} [\mathrm{ad}_{\mathcal{V}_1}] & 0 & \cdots & 0 \\ 0 & [\mathrm{ad}_{\mathcal{V}_2}] & \cdots & 0 \\ \vdots & \vdots & \ddots & \vdots \\ 0 & \cdots & \cdots & [\mathrm{ad}_{\mathcal{V}_n}] \end{bmatrix} \in \mathbb{R}^{6n \times 6n}, \tag{8.62}$$

$$[\mathrm{ad}_{\mathcal{A}\dot{\theta}}] = \begin{bmatrix} [\mathrm{ad}_{\mathcal{A}_1\dot{\theta}_1}] & 0 & \cdots & 0 \\ 0 & [\mathrm{ad}_{\mathcal{A}_2\dot{\theta}_2}] & \cdots & 0 \\ \vdots & \vdots & \ddots & \vdots \\ 0 & \cdots & \cdots & [\mathrm{ad}_{\mathcal{A}_n\dot{\theta}_n}] \end{bmatrix} \in \mathbb{R}^{6n \times 6n}, \tag{8.63}$$

$$\mathcal{W}(\theta) = \begin{bmatrix} 0 & 0 & \cdots & 0 & 0 \\ [\mathrm{Ad}_{T_{21}}] & 0 & \cdots & 0 & 0 \\ 0 & [\mathrm{Ad}_{T_{32}}] & \cdots & 0 & 0 \\ \vdots & \vdots & \ddots & \vdots & \vdots \\ 0 & 0 & \cdots & [\mathrm{Ad}_{T_{n,n-1}}] & 0 \end{bmatrix} \in \mathbb{R}^{6n \times 6n}. \tag{8.64}$$

We write $\mathcal{W}(\theta)$ to emphasize the dependence of $\mathcal{W}$ on $\theta$. Finally, define the following stacked vectors:

$$\mathcal{V}_{\mathrm{base}} = \begin{bmatrix} \mathrm{Ad}_{T_{10}}(\mathcal{V}_0) \\ 0 \\ \vdots \\ 0 \end{bmatrix} \in \mathbb{R}^{6n}, \tag{8.65}$$

$$\dot{\mathcal{V}}_{\mathrm{base}} = \begin{bmatrix} \mathrm{Ad}_{T_{10}}(\dot{\mathcal{V}}_0) \\ 0 \\ \vdots \\ 0 \end{bmatrix} \in \mathbb{R}^{6n}, \tag{8.66}$$

$$\mathcal{F}_{\mathrm{tip}} = \begin{bmatrix} 0 \\ \vdots \\ 0 \\ \mathrm{Ad}^{\mathrm{T}}_{T_{n+1,n}}(\mathcal{F}_{n+1}) \end{bmatrix} \in \mathbb{R}^{6n}. \tag{8.67}$$

Note that $\mathcal{A} \in \mathbb{R}^{6n \times n}$ and $\mathcal{G} \in \mathbb{R}^{6n \times 6n}$ are constant block-diagonal matrices, in which $\mathcal{A}$ contains only the kinematic parameters while $\mathcal{G}$ contains only the mass and inertial parameters for each link.

With the above definitions, our earlier recursive inverse dynamics algorithm can be assembled into the following set of matrix equations:

$$\mathcal{V} = \mathcal{W}(\theta)\mathcal{V} + \mathcal{A}\dot{\theta} + \mathcal{V}_{\text{base}}, \tag{8.68}$$

$$\dot{\mathcal{V}} = \mathcal{W}(\theta)\dot{\mathcal{V}} + \mathcal{A}\ddot{\theta} - [\text{ad}_{\mathcal{A}\dot{\theta}}](\mathcal{W}(\theta)\mathcal{V} + \mathcal{V}_{\text{base}}) + \dot{\mathcal{V}}_{\text{base}}, \tag{8.69}$$

$$\mathcal{F} = \mathcal{W}^{\text{T}}(\theta)\mathcal{F} + \mathcal{G}\dot{\mathcal{V}} - [\text{ad}_{\mathcal{V}}]^{\text{T}}\mathcal{G}\mathcal{V} + \mathcal{F}_{\text{tip}}, \tag{8.70}$$

$$\tau = \mathcal{A}^{\text{T}}\mathcal{F}. \tag{8.71}$$

The matrix $\mathcal{W}(\theta)$ has the property that $\mathcal{W}^n(\theta) = 0$ (such a matrix is said to be nilpotent of order $n$), and one consequence verifiable through direct calculation is that $(I - \mathcal{W}(\theta))^{-1} = I + \mathcal{W}(\theta) + \cdots + \mathcal{W}^{n-1}(\theta)$. Defining $\mathcal{L}(\theta) = (I - \mathcal{W}(\theta))^{-1}$, it can further be verified via direct calculation that

$$\mathcal{L}(\theta) = \begin{bmatrix} I & 0 & 0 & \cdots & 0 \\ [\text{Ad}_{T_{21}}] & I & 0 & \cdots & 0 \\ [\text{Ad}_{T_{31}}] & [\text{Ad}_{T_{32}}] & I & \cdots & 0 \\ \vdots & \vdots & \vdots & \ddots & \vdots \\ [\text{Ad}_{T_{n1}}] & [\text{Ad}_{T_{n2}}] & [\text{Ad}_{T_{n3}}] & \cdots & I \end{bmatrix} \in \mathbb{R}^{6n \times 6n}. \tag{8.72}$$

We write $\mathcal{L}(\theta)$ to emphasize the dependence of $\mathcal{L}$ on $\theta$. The earlier matrix equations can now be reorganized as follows:

$$\mathcal{V} = \mathcal{L}(\theta)\left(\mathcal{A}\dot{\theta} + \mathcal{V}_{\text{base}}\right), \tag{8.73}$$

$$\dot{\mathcal{V}} = \mathcal{L}(\theta)\left(\mathcal{A}\ddot{\theta} + [\text{ad}_{\mathcal{A}\dot{\theta}}]\mathcal{W}(\theta)\mathcal{V} + [\text{ad}_{\mathcal{A}\dot{\theta}}]\mathcal{V}_{\text{base}} + \dot{\mathcal{V}}_{\text{base}}\right), \tag{8.74}$$

$$\mathcal{F} = \mathcal{L}^{\text{T}}(\theta)\left(\mathcal{G}\dot{\mathcal{V}} - [\text{ad}_{\mathcal{V}}]^{\text{T}}\mathcal{G}\mathcal{V} + \mathcal{F}_{\text{tip}}\right), \tag{8.75}$$

$$\tau = \mathcal{A}^{\text{T}}\mathcal{F}. \tag{8.76}$$

If the robot applies an external wrench $\mathcal{F}_{\text{tip}}$ at the end-effector, this can be included into the dynamics equation

$$\tau = M(\theta)\ddot{\theta} + c(\theta, \dot{\theta}) + g(\theta) + J^{\text{T}}(\theta)\mathcal{F}_{\text{tip}}, \tag{8.77}$$

where $J(\theta)$ denotes the Jacobian of the forward kinematics expressed in the same reference frame as $\mathcal{F}_{\text{tip}}$, and

$$M(\theta) = \mathcal{A}^{\text{T}}\mathcal{L}^{\text{T}}(\theta)\mathcal{G}\mathcal{L}(\theta)\mathcal{A}, \tag{8.78}$$

$$c(\theta, \dot{\theta}) = -\mathcal{A}^{\text{T}}\mathcal{L}^{\text{T}}(\theta)\left(\mathcal{G}\mathcal{L}(\theta)[\text{ad}_{\mathcal{A}\dot{\theta}}]\mathcal{W}(\theta) + [\text{ad}_{\mathcal{V}}]^{\text{T}}\mathcal{G}\right)\mathcal{L}(\theta)\mathcal{A}\dot{\theta}, \tag{8.79}$$

$$g(\theta) = \mathcal{A}^{\text{T}}\mathcal{L}^{\text{T}}(\theta)\mathcal{G}\mathcal{L}(\theta)\dot{\mathcal{V}}_{\text{base}}. \tag{8.80}$$

# 5    Forward Dynamics of Open Chains

The forward dynamics problem involves solving

$$M(\theta)\ddot{\theta} = \tau(t) - h(\theta, \dot{\theta}) - J^{\mathrm{T}}(\theta)\mathcal{F}_{\mathrm{tip}} \tag{8.81}$$

for $\ddot{\theta}$, given $\theta$, $\dot{\theta}$, $\tau$, and the wrench $\mathcal{F}_{\mathrm{tip}}$ applied by the end-effector (if applicable). The term $h(\theta, \dot{\theta})$ can be computed by calling the inverse dynamics algorithm with $\ddot{\theta} = 0$ and $\mathcal{F}_{\mathrm{tip}} = 0$. The inertia matrix $M(\theta)$ can be computed using Equation (8.57). An alternative is to use $n$ calls of the inverse dynamics algorithm to build $M(\theta)$ column by column. In each of the $n$ calls, set $\mathfrak{g} = 0$, $\dot{\theta} = 0$, and $\mathcal{F}_{\mathrm{tip}} = 0$. In the first call, the column vector $\ddot{\theta}$ is all zeros except for a 1 in the first row. In the second call, $\ddot{\theta}$ is all zeros except for a 1 in the second row, and so on. The $\tau$ vector returned by the $i$th call is the $i$th column of $M(\theta)$, and after $n$ calls the $n \times n$ matrix $M(\theta)$ is constructed.

With $M(\theta)$, $h(\theta, \dot{\theta})$, and $\mathcal{F}_{\mathrm{tip}}$ we can use any efficient algorithm for solving Equation (8.81), which is of the form $M\ddot{\theta} = b$, for $\ddot{\theta}$.

The forward dynamics can be used to simulate the motion of the robot given its initial state, the joint forces–torques $\tau(t)$, and an optional external wrench $\mathcal{F}_{\mathrm{tip}}(t)$, for $t \in [0, t_f]$. First define the function *ForwardDynamics* returning the solution to Equation (8.81), i.e.,

$$\ddot{\theta} = ForwardDynamics(\theta, \dot{\theta}, \tau, \mathcal{F}_{\mathrm{tip}}).$$

Defining the variables $q_1 = \theta$, $q_2 = \dot{\theta}$, the second-order dynamics (8.81) can be converted to two first-order differential equations,

$$\dot{q}_1 = q_2,$$
$$\dot{q}_2 = ForwardDynamics(q_1, q_2, \tau, \mathcal{F}_{\mathrm{tip}}).$$

The simplest method for numerically integrating a system of first-order differential equations of the form $\dot{q} = f(q, t)$, $q \in \mathbb{R}^n$, is the first-order Euler iteration

$$q(t + \delta t) = q(t) + \delta t f(q(t), t),$$

where the positive scalar $\delta t$ denotes the timestep. The Euler integration of the robot dynamics is thus

$$q_1(t + \delta t) = q_1(t) + q_2(t)\delta t,$$
$$q_2(t + \delta t) = q_2(t) + ForwardDynamics(q_1, q_2, \tau, \mathcal{F}_{\mathrm{tip}})\delta t.$$

Given a set of initial values for $q_1(0) = \theta(0)$ and $q_2(0) = \dot{\theta}(0)$, the above equations can be iterated forward in time to obtain the motion $\theta(t) = q_1(t)$ numerically.

## Euler Integration Algorithm for Forward Dynamics

- **Inputs:** The initial conditions $\theta(0)$ and $\dot{\theta}(0)$, the input torques $\tau(t)$ and wrenches at the end-effector $\mathcal{F}_{\mathrm{tip}}(t)$ for $t \in [0, t_f]$, and the number of integration steps $N$.

- **Initialization**: Set the timestep $\delta t = t_f / N$, and set $\theta[0] = \theta(0)$, $\dot{\theta}[0] = \dot{\theta}(0)$.
- **Iteration**: For $k = 0$ to $N - 1$ do

$$\ddot{\theta}[k] = ForwardDynamics(\theta[k], \dot{\theta}[k], \tau(k\delta t), \mathcal{F}_{\text{tip}}(k\delta t)),$$
$$\theta[k + 1] = \theta[k] + \dot{\theta}[k]\delta t,$$
$$\dot{\theta}[k + 1] = \dot{\theta}[k] + \ddot{\theta}[k]\delta t.$$

- **Output**: The joint trajectory $\theta(k\delta t) = \theta[k]$, $\dot{\theta}(k\delta t) = \dot{\theta}[k]$, $k = 0, \ldots, N$.

The result of the numerical integration converges to the theoretical result as the number of integration steps $N$ goes to infinity. Higher-order numerical integration schemes, such as fourth-order Runge–Kutta, can yield a closer approximation with fewer computations than the simple first-order Euler method.

## 8.6     Dynamics in the Task Space

In this section we consider how the dynamic equations change under a transformation to coordinates of the end-effector frame (task-space coordinates). To keep things simple we consider a six-degree-of-freedom open chain with joint-space dynamics

$$\tau = M(\theta)\ddot{\theta} + h(\theta, \dot{\theta}), \qquad \theta \in \mathbb{R}^6, \ \tau \in \mathbb{R}^6. \tag{8.82}$$

We also ignore, for the time being, any end-effector forces $\mathcal{F}_{\text{tip}}$. The twist $\mathcal{V} = (\omega, v)$ of the end-effector is related to the joint velocity $\dot{\theta}$ by

$$\mathcal{V} = J(\theta)\dot{\theta}, \tag{8.83}$$

with the understanding that $\mathcal{V}$ and $J(\theta)$ are always expressed in terms of the same reference frame. The time derivative $\dot{\mathcal{V}}$ is then

$$\dot{\mathcal{V}} = \dot{J}(\theta)\dot{\theta} + J(\theta)\ddot{\theta}. \tag{8.84}$$

At configurations $\theta$ where $J(\theta)$ is invertible, we have

$$\dot{\theta} = J^{-1}\mathcal{V}, \tag{8.85}$$
$$\ddot{\theta} = J^{-1}\dot{\mathcal{V}} - J^{-1}\dot{J}J^{-1}\mathcal{V}. \tag{8.86}$$

Substituting for $\dot{\theta}$ and $\ddot{\theta}$ in Equation (8.82) leads to

$$\tau = M(\theta)\left(J^{-1}\dot{\mathcal{V}} - J^{-1}\dot{J}J^{-1}\mathcal{V}\right) + h(\theta, J^{-1}\mathcal{V}). \tag{8.87}$$

Let $J^{-T}$ denote $(J^{-1})^{T} = (J^{T})^{-1}$. Pre-multiply both sides by $J^{-T}$ to get

$$\begin{aligned} J^{-T}\tau &= J^{-T}MJ^{-1}\dot{\mathcal{V}} - J^{-T}MJ^{-1}\dot{J}J^{-1}\mathcal{V} \\ &\quad + J^{-T}h(\theta, J^{-1}\mathcal{V}). \end{aligned} \tag{8.88}$$

Expressing $J^{-T}\tau$ as the wrench $\mathcal{F}$, the above can be written

$$\mathcal{F} = \Lambda(\theta)\dot{\mathcal{V}} + \eta(\theta, \mathcal{V}), \tag{8.89}$$

where

$$\Lambda(\theta) = J^{-T} M(\theta) J^{-1}, \tag{8.90}$$

$$\eta(\theta, \mathcal{V}) = J^{-T} h(\theta, J^{-1}\mathcal{V}) - \Lambda(\theta)\dot{J}J^{-1}\mathcal{V}. \tag{8.91}$$

These are the dynamic equations expressed in end-effector frame coordinates. If an external wrench $\mathcal{F}$ is applied to the end-effector frame then, assuming the actuators provide zero forces and torques, the motion of the end-effector frame is governed by these equations.

Note that $J(\theta)$ must be invertible (i.e., there must be a one-to-one mapping between joint velocities and end-effector twists) in order to derive the task-space dynamics above. Also note the dependence of $\Lambda(\theta)$ and $\eta(\theta, \mathcal{V})$ on $\theta$. In general, we cannot replace the dependence on $\theta$ by a dependence on the end-effector configuration $X$ because there may be multiple solutions to the inverse kinematics, and the dynamics depends on the specific joint configuration $\theta$.

## .7 Constrained Dynamics

Now consider the case where the $n$-joint robot is subject to a set of $k$ holonomic or nonholonomic Pfaffian velocity constraints of the form

$$A(\theta)\dot{\theta} = 0, \qquad A(\theta) \in \mathbb{R}^{k \times n}. \tag{8.92}$$

(See Section 2.4 for an introduction to Pfaffian constraints.) Such constraints can come from loop-closure constraints; for example, the motion of an end-effector rigidly holding a door handle is subject to $k = 5$ constraints due to the hinges of the door. As another example, a robot writing with a pen is subject to a single constraint that keeps the height of the tip of the pen above the paper at zero. In any case, we assume that the constraints do no work on the robot, i.e., the generalized forces $\tau_{con}$ due to the constraints satisfy

$$\tau_{con}^{T} \dot{\theta} = 0.$$

This assumption means that $\tau_{con}$ must be a linear combination of the columns of $A^{T}(\theta)$, i.e., $\tau_{con} = A^{T}(\theta)\lambda$ for some $\lambda \in \mathbb{R}^{k}$, since these are the generalized forces that do no work when $\dot{\theta}$ is subject to the constraints (8.92):

$$(A^{T}(\theta)\lambda)^{T}\dot{\theta} = \lambda^{T}A(\theta)\dot{\theta} = 0 \qquad \text{for all } \lambda \in \mathbb{R}^{k}.$$

For the writing-robot example, the assumption that the constraint is workless means that there can be no friction between the pen and the paper.

Adding the constraint forces $A^{T}(\theta)\lambda$ to the equations of motion, we can write the $n + k$ constrained equations of motion

$$\tau = M(\theta)\ddot{\theta} + h(\theta, \dot{\theta}) + A^{T}(\theta)\lambda, \tag{8.93}$$

$$A(\theta)\dot{\theta} = 0, \tag{8.94}$$

where $\lambda$ is a set of Lagrange multipliers and $A^{\mathrm{T}}(\theta)\lambda$ are the forces applied against the constraints as expressed as joint forces and torques. From these equations, it should be clear that the robot has $n-k$ velocity freedoms and $k$ "force freedoms" – the constraints allow the robot to create any generalized force of the form $A^{\mathrm{T}}(\theta)\lambda$, independent of the robot's motion. (For the writing robot, there is also an inequality constraint: the robot can only apply pushing forces into the paper and table, not pulling forces.)

Since the constraints $A(\theta)\dot{\theta} = 0$ are satisfied at all times, the time rate of change of the constraints satisfies

$$\dot{A}(\theta)\dot{\theta} + A(\theta)\ddot{\theta} = 0. \tag{8.95}$$

Assuming that $M(\theta)$ and $A(\theta)$ are full rank, we can solve Equation (8.93) for $\ddot{\theta}$,

$$\ddot{\theta} = M^{-1}(\theta)(\tau - h(\theta, \dot{\theta}) - A^{\mathrm{T}}(\theta)\lambda), \tag{8.96}$$

substitute into Equation (8.95), and omit the dependences on $\theta$ and $\dot{\theta}$ for conciseness, to get

$$\dot{A}\dot{\theta} + AM^{-1}(\tau - h - A^{\mathrm{T}}\lambda) = 0. \tag{8.97}$$

After some manipulation, we can solve for the Lagrange multipliers:

$$\lambda = (AM^{-1}A^{\mathrm{T}})^{-1}(AM^{-1}(\tau - h) + \dot{A}\dot{\theta}). \tag{8.98}$$

The constraint force depends on both $\tau$ and the state.

Now, to solve the constrained forward dynamics for $\ddot{\theta}$ and $\lambda$ given $\tau$, we can solve Equation (8.98) for $\lambda$ and plug into Equation (8.96).

Equation (8.93) can be used directly to solve the constrained inverse dynamics for $\tau$ given $\lambda$ and a $\ddot{\theta}$ chosen from the $(n-k)$-dimensional space of accelerations satisfying Equation (8.95). If the constraint acts at the end-effector of the robot, $\lambda$ is related to the wrench the end-effector applies to the constraint by

$$J^{\mathrm{T}}(\theta)\mathcal{F}_{\mathrm{tip}} = A^{\mathrm{T}}(\theta)\lambda,$$

where $J(\theta)$ is the Jacobian satisfying $\mathcal{V} = J(\theta)\dot{\theta}$. If $J(\theta)$ is invertible, then $\mathcal{F}_{\mathrm{tip}} = J^{-\mathrm{T}}(\theta)A^{\mathrm{T}}(\theta)\lambda$. In hybrid motion–force control (Section 11.6), where the objective is to control the motion tangent to the constraints and the wrench against the constraints, the requested wrench $\mathcal{F}_d$ must lie in the column space of $J^{-\mathrm{T}}(\theta)A^{\mathrm{T}}(\theta)$, and the Lagrange multipliers are $\lambda = (J^{-\mathrm{T}}(\theta)A^{\mathrm{T}}(\theta))^{\dagger}\mathcal{F}_d$.

**Example 8.5** Consider the 2R robot of Figure 8.1, reproduced in Figure 8.7 with gravity $g$ equal to zero. The lengths of each link are $L_1 = L_2 = 1$, and the point masses at the ends of each link are $\mathfrak{m}_1 = \mathfrak{m}_2 = 1$. The tip of the robot is at $(x, y)$, and the robot's forward kinematics can be written

$$\begin{bmatrix} x \\ y \end{bmatrix} = \begin{bmatrix} c_1 + c_{12} \\ s_1 + s_{12} \end{bmatrix},$$

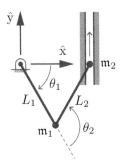

**Figure 8.7** A 2R robot whose tip is constrained to move in a frictionless channel.

where $s_{12}$ and $c_{12}$ are $\sin(\theta_1 + \theta_2)$ and $\cos(\theta_1 + \theta_2)$, respectively. The derivatives of the forward kinematics are

$$\begin{bmatrix} \dot{x} \\ \dot{y} \end{bmatrix} = \underbrace{\begin{bmatrix} -s_1 - s_{12} & -s_{12} \\ c_1 + c_{12} & c_{12} \end{bmatrix}}_{J(\theta)} \begin{bmatrix} \dot{\theta}_1 \\ \dot{\theta}_2 \end{bmatrix},$$

$$\begin{bmatrix} \ddot{x} \\ \ddot{y} \end{bmatrix} = J(\theta)\ddot{\theta} + \underbrace{\begin{bmatrix} -\dot{\theta}_1 c_1 - (\dot{\theta}_1 + \dot{\theta}_2)c_{12} & -(\dot{\theta}_1 + \dot{\theta}_2)c_{12} \\ -\dot{\theta}_1 s_1 - (\dot{\theta}_1 + \dot{\theta}_2)s_{12} & -(\dot{\theta}_1 + \dot{\theta}_2)s_{12} \end{bmatrix}}_{\dot{J}(\theta)} \begin{bmatrix} \dot{\theta}_1 \\ \dot{\theta}_2 \end{bmatrix},$$

where $J(\theta)$ is the Jacobian for velocities expressed as $(\dot{x}, \dot{y})$.

The tip of the robot is constrained to move in a frictionless linear channel at $x = 1$. This holonomic constraint can be expressed in joint coordinates $\theta$ as $c_1 + c_{12} = 1$, and its time derivative can be written $A(\theta)\dot{\theta} = 0$, i.e.,

$$\underbrace{\begin{bmatrix} -s_1 - s_{12} & -s_{12} \end{bmatrix}}_{A(\theta)} \begin{bmatrix} \dot{\theta}_1 \\ \dot{\theta}_2 \end{bmatrix} = \begin{bmatrix} 0 \\ 0 \end{bmatrix}.$$

There are $n = 2$ joint coordinates and $k = 1$ constraint, so $A(\theta) \in \mathbb{R}^{1 \times 2}$. The time derivative of $A(\theta)$ is

$$\dot{A}(\theta) = [-\dot{\theta}_1 c_1 - (\dot{\theta}_1 + \dot{\theta}_2)c_{12} \quad -(\dot{\theta}_1 + \dot{\theta}_2)c_{12}].$$

Consider the case where $\theta_1 = -\pi/3$ and $\theta_2 = 2\pi/3$, as shown in Figure 8.7. The tip is currently moving with the velocity $(\dot{x}, \dot{y}) = (0, 1)$, which implies $\dot{\theta}_1 = 1$ and $\dot{\theta}_2 = 0$. At this state, $A(\theta) = [0 \ -0.866]$ and $\dot{A}(\theta) = [-1 \ -0.5]$. Consulting Equation (8.10) for the mass matrix and velocity-product term, we get

$$M(\theta) = \begin{bmatrix} 2 & 0.5 \\ 0.5 & 1 \end{bmatrix}, \quad h(\theta, \dot{\theta}) = \begin{bmatrix} 0 \\ 0.866 \end{bmatrix}.$$

**Constrained forward dynamics**
Let's solve the constrained forward dynamics for $\ddot{\theta} = (\ddot{\theta}_1, \ddot{\theta}_2)$ and $\lambda$ when the joint torques are $\tau = (\tau_1, \tau_2)$. Solving Equation (8.98) for $\lambda$ and plugging into

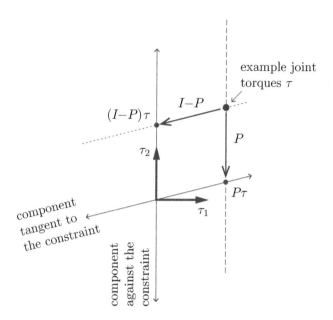

**Figure 8.8** The $(\tau_1, \tau_2)$ joint torque space is partitioned into components against the constraint and components tangent to the constraint, as indicated by the arrowed lines $\tau_1 = 0$ and $0.289\tau_1 - 1.155\tau_2 = 0$, respectively. A set of joint torques $\tau$ can be expressed as $\tau = P\tau + (I - P)\tau$, where the matrix $P$ projects $\tau$ to its component tangent to the constraint, the matrix $I - P$ projects $\tau$ to its component against the constraint, and $I$ is the identity matrix. Joint torques on the dotted line all have the same component against the constraint as $\tau$ but cause different motions of the robot, while joint torques on the dashed line all cause the same motion of the robot as $\tau$ but create different constraint forces.

Equation (8.96), we get

$$\lambda = 0.289\tau_1 - 1.155\tau_2 - 0.167,$$

$$\ddot{\theta}_1 = 0.5\tau_1 + 0.289,$$

$$\ddot{\theta}_2 = -1.155.$$

A few things to notice about the solution:

- At the current state, $\lambda = -0.167$ if $\tau = 0$.
- At the current state, joint torques lying in the one-dimensional subspace satisfying $0.289\tau_1 - 1.155\tau_2 = 0$ do not affect the constraint force.
- At the current state, joint torques lying in the one-dimensional subspace satisfying $\tau_1 = 0$ do not affect the motion of the robot.

The last two observations are illustrated in Figure 8.8. Any set of joint torques $\tau$ can be expressed as the sum of two components: a component that affects the motion of the robot, but not the constraint force, and a component that affects the constraint force, but not the motion.

**Task-space constraint forces**

The force the robot applies against the constraint can be written in terms of a force $f_{\text{tip}} = [f_x \ f_y]^T$ in the task space using the relationship

$$J^T(\theta)f_{\text{tip}} = A^T(\theta)\lambda.$$

Since the Jacobian in this example is invertible, $f_{\text{tip}} = J^{-T}(\theta)A^T(\theta)\lambda$, i.e.,

$$f_{\text{tip}} = \begin{bmatrix} 0.577 & -1.155 \\ 1 & 0 \end{bmatrix} \begin{bmatrix} 0 \\ -0.866 \end{bmatrix} \lambda = \begin{bmatrix} 1 \\ 0 \end{bmatrix} \lambda = \begin{bmatrix} 0.289\tau_1 - 1.155\tau_2 - 0.167 \\ 0 \end{bmatrix},$$

which agrees with our understanding that the robot can only apply forces against the constraint (and vice-versa) in the $f_x$-direction. In this example, if $\tau = 0$, the task-space constraint force is $f_{\text{tip}} = [-0.167 \ 0]^T$, meaning that the robot's tip pushes to the left on the constraint while the constraint pushes back equally to the right to enforce the constraint. In the absence of the constraint, the acceleration of the tip of the robot would have a component to the left.

**Constrained inverse dynamics**

The constrained inverse dynamics involves solving for $\tau$ given a $\ddot\theta$ satisfying Equation (8.95) and $\lambda$. From the results above, we see that any constraint-satisfying $\ddot\theta$ is of the form $(\ddot\theta_1, \ddot\theta_2) = (a, -1.155)$ for any $a \in \mathbb{R}$. Also, assuming we wish to apply a force of $(f_x, f_y) = (f, 0)$ against the channel, since $J(\theta)$ is invertible in this example we have

$$\lambda = (J^{-T}(\theta)A^T(\theta))^\dagger \begin{bmatrix} f \\ 0 \end{bmatrix} = \begin{bmatrix} 1 \\ 0 \end{bmatrix}^\dagger \begin{bmatrix} f \\ 0 \end{bmatrix} = [1 \ 0] \begin{bmatrix} f \\ 0 \end{bmatrix} = f.$$

The solution to the constrained inverse dynamics for $\ddot\theta = (a, -1.155)$ and $\lambda = f$ is given by Equation (8.93):

$$\tau_1 = 2a - 0.578,$$
$$\tau_2 = 0.5a - 0.866f - 0.289.$$

In hybrid motion-force control, $a$ is specifed by a motion controller to track a desired motion along the channel and $f$ is specified by a force controller to achieve a desired force against the channel.

Example 8.5 above considers a particular state of a particular constrained robot with $n = 2$ and $k = 1$. For more general constrained robots, the constraints specify an $(n - k)$-dimensional subspace of actuator forces and torques tangent to the constraints and a $k$-dimensional subpace against the constraints.

Combining Equations (8.98) and (8.93) and manipulating, we can write the dynamics projected to the $(n - k)$-dimensional space tangent to the constraints,

$$P\tau = P(M\ddot\theta + h), \tag{8.99}$$

where

$$P = I - A^T(AM^{-1}A^T)^{-1}AM^{-1} \tag{8.100}$$

and $I$ is the $n \times n$ identity matrix. The $n \times n$ projection matrix $P(\theta)$ has rank $n - k$, and it maps generalized forces $\tau$ to $P(\theta)\tau$, projecting away the generalized force components $(I - P(\theta))\tau$ that act against the constraints while retaining the generalized forces tangent to the constraints. In Example 8.5, illustrated in Figure 8.8, the projections $P$ and $I - P$ are

$$P = \begin{bmatrix} 1 & 0 \\ 0.25 & 0 \end{bmatrix}, \quad I - P = \begin{bmatrix} 0 & 0 \\ -0.25 & 1 \end{bmatrix}.$$

Equation (8.99) can be rearranged into the related form

$$P_{\ddot{\theta}}\ddot{\theta} = P_{\ddot{\theta}}M^{-1}(\tau - h), \tag{8.101}$$

where the rank $n - k$ matrix $P_{\ddot{\theta}}$ is

$$P_{\ddot{\theta}} = M^{-1}PM = I - M^{-1}A^{\mathrm{T}}(AM^{-1}A^{\mathrm{T}})^{-1}A = P^{\mathrm{T}}. \tag{8.102}$$

In Section 11.6 we discuss the related topic of hybrid motion–force control, in which the goal at each instant is to simultaneously achieve a desired motion tangent to the constraints and a desired force against the constraints. In that section we use the task-space dynamics to represent the task-space end-effector motions and wrenches more naturally.

## 8.8     Robot Dynamics in the URDF

As described in Section 4.2 and illustrated in the UR5 Universal Robot Description Format file, the inertial properties of link $i$ are described in the URDF by the link elements `mass`, `origin` (the position and orientation of the center-of-mass frame relative to a frame attached at joint $i$), and `inertia`, which specifies the six elements of the symmetric rotational inertia matrix on or above the diagonal. To fully write the robot's dynamics, for joint $i$ we need in addition the joint element `origin`, specifying the position and orientation of link $i$'s joint frame relative to link $(i - 1)$'s joint frame when $\theta_i = 0$, and the element `axis`, which specifies the axis of motion of joint $i$. We leave to the exercises the translation of these elements into the quantities needed for the Newton–Euler inverse dynamics algorithm.

## 8.9     Actuation, Gearing, and Friction

Until now we have been assuming the existence of actuators that directly provide commanded forces and torques. In practice there are many types of actuators (e.g., electric, hydraulic, and pneumatic) and mechanical power transformers (e.g., gearheads), and the actuators can be located at the joints themselves or remotely, with mechanical power transmitted by cables or timing belts. Each combination of these has its own characteristics that can play a significant role

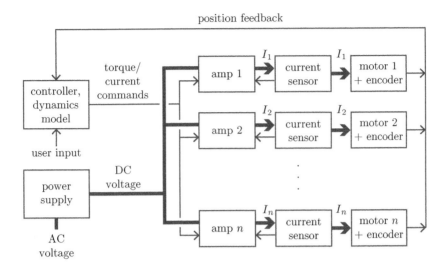

**Figure 8.9** A block diagram of a typical $n$-joint robot. The bold lines correspond to high-power signals while the thin lines correspond to communication signals.

in the "extended dynamics" mapping the actual control inputs (e.g., the current requested of amplifiers connected to electric motors) to the motion of the robot.

In this section we provide an introduction to some of the issues associated with one particular, and common, configuration: geared DC electric motors located at each joint. This is the configuration used in the Universal Robots UR5, for example.

Figure 8.9 shows the electrical block diagram for a typical $n$-joint robot driven by DC electric motors. For concreteness, we assume that each joint is revolute. A power supply converts the wall AC voltage to a DC voltage to power the amplifier associated with each motor. A control box takes user input, for example in the form of a desired trajectory, as well as position feedback from encoders located at each joint. Using the desired trajectory, a model of the robot's dynamics, and the measured error in the current robot state relative to the desired robot state, the controller calculates the torque required of each actuator. Since DC electric motors nominally provide a torque proportional to the current through the motor, this torque command is equivalent to a current command. Each motor amplifier then uses a current sensor (shown as external to the amplifier in Figure 8.9, but in reality internal to the amplifier) to continually adjust the voltage across the motor to try to achieve the requested current.[3] The motion of the motor is sensed by the motor encoder, and the position information is sent back to the controller.

The commanded torque is typically updated at around 1000 times per second

[3] The voltage is typically a time-averaged voltage achieved by the duty cycle of a voltage rapidly switching between a maximum positive voltage and a maximum negative voltage.

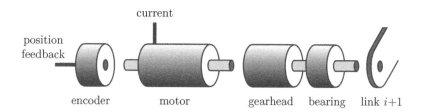

**Figure 8.10** The outer cases of the encoder, motor, gearhead, and bearing are fixed in link $i$, while the gearhead output shaft supported by the bearing is fixed in link $i + 1$.

(1 kHz), and the amplifier's voltage control loop may be updated at a rate ten times that or more.

Figure 8.10 is a conceptual representation of the motor and other components for a single axis. The motor has a single shaft extending from both ends of the motor: one end drives a rotary encoder, which measures the position of the joint, and the other end becomes the input to a gearhead. The gearhead increases the torque while reducing the speed, since most DC electric motors with an appropriate power rating provide torques that are too low to be useful for robotics applications. The purpose of the bearing is to support the gearhead output, freely transmitting torques about the gearhead axis while isolating the gearhead (and motor) from wrench components due to link $i + 1$ in the other five directions. The outer cases of the encoder, motor, gearhead, and bearing are all fixed relative to each other and to link $i$. It is also common for the motor to have some kind of brake, not shown.

### 8.9.1   DC Motors and Gearing

A DC motor consists of a **stator** and a **rotor** that rotates relative to the stator. DC electric motors create torque by sending current through windings in a magnetic field created by permanent magnets, where the magnets are attached to the stator and the windings are attached to the rotor, or vice versa. A DC motor has multiple windings, some of which are energized and some of which are inactive at any given time. The windings that are energized are chosen as a function of the angle of the rotor relative to the stator. This "commutation" of the windings occurs mechanically using brushes (brushed motors) or electrically using control circuitry (brushless motors). Brushless motors have the advantage of no brush wear and higher continuous torque, since the windings are typically attached to the motor housing where the heat due to the resistance of the windings can be more easily dissipated. In our basic introduction to DC motor modeling, we do not distinguish between brushed and brushless motors.

Figure 8.11 shows a brushed DC motor with an encoder and a gearhead.

The torque $\tau$, measured in newton-meters (N m), created by a DC motor is

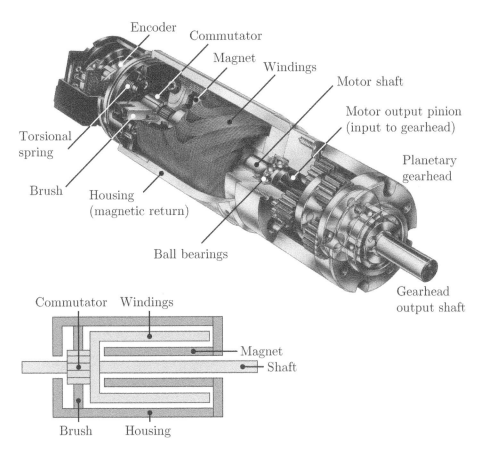

**Figure 8.11** (Top) A cutaway view of a Maxon brushed DC motor with an encoder and gearhead. (Cutaway image courtesy of Maxon Precision Motors, Inc., maxonmotorusa.com.) The motor's rotor consists of the windings, commutator ring, and shaft. Each of the several windings connects different segments of the commutator, and as the motor rotates, the two brushes slide over the commutator ring and make contact with different segments, sending current through one or more windings. One end of the motor shaft turns the encoder, and the other end is input to the gearhead. (Bottom) A simplified cross-section of the motor only, showing the stator (brushes, housing, and magnets) in dark gray and the rotor (windings, commutator, and shaft) in light gray.

governed by the equation

$$\tau = k_t I,$$

where $I$, measured in amps (A), is the current through the windings. The constant $k_t$, measured in newton-meters per amp (N m/A), is called the **torque constant**. The power dissipated as heat by the windings, measured in watts (W), is governed by

$$P_{\text{heat}} = I^2 R,$$

where $R$ is the resistance of the windings in ohms ($\Omega$). To keep the motor windings from overheating, the continuous current flowing through the motor must be limited. Accordingly, in continuous operation, the motor torque must be kept below a continuous-torque limit $\tau_{\text{cont}}$ determined by the thermal properties of the motor.

A simplified model of a DC motor, where all units are in the SI system, can be derived by equating the electrical power consumed by the motor $P_{\text{elec}} = IV$ in watts (W) to the mechanical power $P_{\text{mech}} = \tau w$ (also in W) and other power produced by the motor,

$$IV = \tau w + I^2 R + LI\frac{dI}{dt} + \text{friction and other power-loss terms,}$$

where $V$ is the voltage applied to the motor in volts (V), $w$ is the angular speed of the motor in radians per second (1/s), and $L$ is the inductance due to the windings in henries (H). The terms on the right-hand side are the mechanical power produced by the motor, the power lost to heating the windings due to the resistance of the wires, the power consumed or produced by energizing or de-energizing the inductance of the windings (since the energy stored in an inductor is $\frac{1}{2}LI^2$, and power is the time derivative of energy), and the power lost to friction in bearings, etc. Dropping this last term, replacing $\tau w$ by $k_t I w$, and dividing by $I$, we get the voltage equation

$$V = k_t w + IR + L\frac{dI}{dt}. \tag{8.103}$$

Often Equation (8.103) is written with the **electrical constant** $k_e$ (with units of V s) instead of the torque constant $k_t$, but in SI units (V s or N m/A) the numerical values of the two are identical; they represent the same constant property of the motor. So we prefer to use $k_t$.

The voltage term $k_t w$ in Equation (8.103) is called the **back electromotive force** or **back-emf** for short, and it is what differentiates a motor from being simply a resistor and inductor in series. It also allows a motor, which we usually think of as converting electrical power to mechanical, to be run as a generator, converting mechanical power to electrical. If the motor's electrical inputs are disconnected (so no current can flow) and the shaft is forced to turn by some external torque, you can measure the back-emf voltage $k_t w$ across the motor's inputs.

For simplicity, in the rest of this section we ignore the $L\,dI/dt$ term. This assumption is exactly satisfied when the motor is operating at a constant current. With this assumption, Equation (8.103) can be rearranged to

$$w = \frac{1}{k_t}(V - IR) = \frac{V}{k_t} - \frac{R}{k_t^2}\tau,$$

expressing the speed $w$ as a linear function of $\tau$ (with a slope of $-R/k_t^2$) for a constant $V$. Now assume that the voltage across the motor is limited to the range $[-V_{\max}, +V_{\max}]$ and the current through the motor is limited to $[-I_{\max}, +I_{\max}]$, perhaps by the amplifier or power supply. Then the operating region of the motor

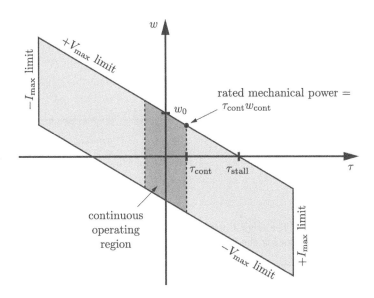

**Figure 8.12** The operating region (light gray) of a current- and voltage-limited DC electric motor, and its continuous operating region (dark gray).

in the torque–speed plane is as shown in Figure 8.12. Note that the signs of $\tau$ and $w$ are opposite in the second and fourth quadrants of this plane, and therefore the product $\tau w$ is negative. When the motor operates in these quadrants, it is actually consuming mechanical power, not producing mechanical power. The motor is acting like a damper.

Focusing on the first quadrant ($\tau \geq 0, w \geq 0, \tau w \geq 0$), the boundary of the operating region is called the **speed–torque curve**. The **no-load speed** $w_0 = V_{\max}/k_t$ at one end of the speed–torque curve is the speed at which the motor spins when it is powered by $V_{\max}$ but is providing no torque. In this operating condition, the back-emf $k_t w$ is equal to the applied voltage, so there is no voltage remaining to create current (or torque). The **stall torque** $\tau_{\text{stall}} = k_t V_{\max}/R$ at the other end of the speed–torque curve is achieved when the shaft is blocked from spinning, so there is no back-emf.

Figure 8.12 also indicates the continuous operating region where $|\tau| \leq \tau_{\text{cont}}$. The motor may be operated intermittently outside the continuous operating region, but extended operation outside the continuous operating region raises the possibility that the motor will overheat.

The motor's rated mechanical power is $P_{\text{rated}} = \tau_{\text{cont}} w_{\text{cont}}$, where $w_{\text{cont}}$ is the speed on the speed–torque curve corresponding to $\tau_{\text{cont}}$. Even if the motor's rated power is sufficient for a particular application, the torque generated by a DC motor is typically too low to be useful. As mentioned earlier, gearing is therefore used to increase the torque while also decreasing the speed. For a gear

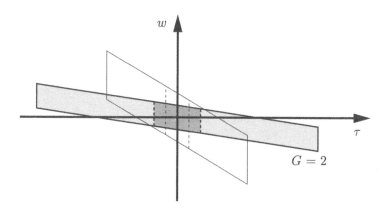

**Figure 8.13** The original motor operating region, and the operating region with a gear ratio $G = 2$ showing the increased torque and decreased speed.

ratio $G$, the output speed of the gearhead is

$$w_{\text{gear}} = \frac{w_{\text{motor}}}{G}.$$

For an ideal gearhead, zero power is lost in the torque conversion, so $\tau_{\text{motor}} w_{\text{motor}} = \tau_{\text{gear}} w_{\text{gear}}$, which implies that

$$\tau_{\text{gear}} = G \tau_{\text{motor}}.$$

In practice, some mechanical power is lost due to friction and impacts between gear teeth, bearings, etc., so

$$\tau_{\text{gear}} = \eta G \tau_{\text{motor}},$$

where $\eta \leq 1$ is the efficiency of the gearhead.

Figure 8.13 shows the operating region of the motor from Figure 8.12 when the motor is geared by $G = 2$ (with $\eta = 1$). The maximum torque doubles, while the maximum speed shrinks by a factor of two. Since many DC motors are capable of no-load speeds of 10 000 rpm or more, robot joints often have gear ratios of 100 or more to achieve an appropriate compromise between speed and torque.

### 8.9.2 Apparent Inertia

The motor's stator is attached to one link and the rotor is attached to another link, possibly through a gearhead. Therefore, when calculating the contribution of a motor to the masses and inertias of the links, the mass and inertia of the stator must be assigned to one link and the mass and inertia of the rotor must be assigned to the other link.

Consider a stationary link 0 with the stator of the joint-1 gearmotor attached to it. The rotational speed of joint 1, the output of the gearhead, is $\dot{\theta}$. Therefore the motor's rotor rotates at $G\dot{\theta}$. The kinetic energy of the rotor is therefore

$$\mathcal{K} = \frac{1}{2}\mathcal{I}_{\text{rotor}}(G\dot{\theta})^2 = \frac{1}{2}\underbrace{G^2\mathcal{I}_{\text{rotor}}}_{\text{apparent inertia}}\dot{\theta}^2,$$

where $\mathcal{I}_{\text{rotor}}$ is the rotor's scalar inertia about the rotation axis and $G^2\mathcal{I}_{\text{rotor}}$ is the **apparent inertia** (often called the **reflected inertia**) of the rotor about the axis. In other words, if you were to grab link 1 and rotate it manually, the inertia contributed by the rotor would feel as if it were a factor $G^2$ larger than its actual inertia, owing to the gearhead.

While the inertia $\mathcal{I}_{\text{rotor}}$ is typically much less than the inertia $\mathcal{I}_{\text{link}}$ of the rest of the link about the rotation axis, the apparent inertia $G^2\mathcal{I}_{\text{rotor}}$ may be on the order of, or even larger than, $\mathcal{I}_{\text{link}}$.

One consequence as the gear ratio becomes large is that the inertia seen by joint $i$ becomes increasingly dominated by the apparent inertia of the rotor. In other words, the torque required of joint $i$ becomes relatively more dependent on $\ddot{\theta}_i$ than on other joint accelerations, i.e., the robot's mass matrix becomes more diagonal. In the limit when the mass matrix has negligible off-diagonal components (and in the absence of gravity), the dynamics of the robot are decoupled – the dynamics at one joint has no dependence on the configuration or motion of the other joints.

As an example, consider the 2R arm of Figure 8.1 with $L_1 = L_2 = \mathsf{m}_1 = \mathsf{m}_2 = 1$. Now assume that each of joint 1 and joint 2 has a motor of mass 1, with a stator of inertia 0.005 and a rotor of inertia 0.00125, and a gear ratio $G$ (with $\eta = 1$). With a gear ratio $G = 10$, the mass matrix is

$$M(\theta) = \begin{bmatrix} 4.13 + 2\cos\theta_2 & 1.01 + \cos\theta_2 \\ 1.01 + \cos\theta_2 & 1.13 \end{bmatrix}.$$

With a gear ratio $G = 100$, the mass matrix is

$$M(\theta) = \begin{bmatrix} 16.5 + 2\cos\theta_2 & 1.13 + \cos\theta_2 \\ 1.13 + \cos\theta_2 & 13.5 \end{bmatrix}.$$

The off-diagonal components are relatively less important for this second robot. The available joint torques of the second robot are ten times that of the first robot so, despite the increases in the mass matrix elements, the second robot is capable of significantly higher accelerations and end-effector payloads. The top speed of each joint of the second robot is ten times less than that of the first robot, however.

If the apparent inertia of the rotor is non-negligible relative to the inertia of the rest of the link, the Newton–Euler inverse dynamics algorithm must be modified to account for it. One approach is to treat the link as consisting of two separate bodies, the geared rotor driving the link and the rest of the link, each with its own center of mass and inertial properties (where the link inertial properties include the inertial properties of the stator of any motor mounted on the link). In the forward iteration, the twist and acceleration of each body is determined while accounting for the gearhead in calculating the rotor's motion. In the backward iteration, the wrench on the link is calculated as the sum of two wrenches: (i) the link wrench as given by Equation (8.53) and (ii) the reaction wrench from the distal rotor. The resultant wrench projected onto the joint axis is then the gear torque $\tau_{\text{gear}}$; dividing $\tau_{\text{gear}}$ by the gear ratio and adding to this the torque resulting from the acceleration of the rotor results in the required motor torque $\tau_{\text{motor}}$. The current command to the DC motor is then $I_{\text{com}} = \tau_{\text{motor}}/(\eta k_t)$.

### 8.9.3    Newton–Euler Inverse Dynamics Algorithm Accounting for Motor Inertias and Gearing

We now reformulate the recursive Newton–Euler inverse dynamics algorithm taking into account the apparent inertias as discussed above. Figure 8.14 illustrates the setup. We assume massless gears and shafts and that the friction between gears as well as the friction between shafts and links is negligible.

**Initialization**
Attach a frame $\{0\}_L$ to the base, frames $\{1\}_L$ to $\{n\}_L$ to the centers of mass of links 1 to $n$, and frames $\{1\}_R$ to $\{n\}_R$ to the centers of mass of rotors 1 to $n$. Frame $\{n+1\}_L$ is attached to the end-effector, which is assumed fixed with respect to frame $\{n\}_L$. Define $M_{i_R,(i-1)_L}$ and $M_{i_L,(i-1)_L}$ to be the configuration of $\{i-1\}_L$ in $\{i\}_R$ and in $\{i\}_L$, respectively, when $\theta_i = 0$. Let $\mathcal{A}_i$ be the screw axis of joint $i$ expressed in $\{i\}_L$. Similarly, let $\mathcal{R}_i$ be the screw axis of rotor $i$ expressed in $\{i\}_R$. Let $\mathcal{G}_{i_L}$ be the $6 \times 6$ spatial inertia matrix of link $i$ that includes the inertia of the attached stator and $\mathcal{G}_{i_R}$ be the $6 \times 6$ spatial inertia matrix of rotor $i$. The gear ratio of motor $i$ is $G_i$. The twists $\mathcal{V}_{0_L}$ and $\dot{\mathcal{V}}_{0_L}$ and the wrench $\mathcal{F}_{(n+1)_L}$ are defined in the same way as $\mathcal{V}_0$, $\dot{\mathcal{V}}_0$, and $\mathcal{F}_{n+1}$ in Section 8.3.2.

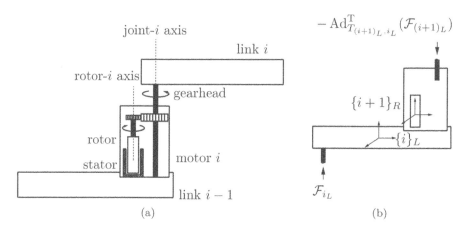

**Figure 8.14** (a) Schematic of a geared motor between links $i-1$ and $i$. (b) The free-body diagram for link $i$, which is analogous to Figure 8.6.

**Forward iterations**

Given $\theta, \dot{\theta}, \ddot{\theta}$, for $i = 1$ to $n$ do

$$T_{i_R,(i-1)_L} = e^{-[\mathcal{R}_i]G_i\theta_i} M_{i_R,(i-1)_L}, \tag{8.104}$$

$$T_{i_L,(i-1)_L} = e^{-[\mathcal{A}_i]\theta_i} M_{i_L,(i-1)_L}, \tag{8.105}$$

$$\mathcal{V}_{i_R} = \text{Ad}_{T_{i_R,(i-1)_L}}(\mathcal{V}_{(i-1)_L}) + \mathcal{R}_i G_i \dot{\theta}_i, \tag{8.106}$$

$$\mathcal{V}_{i_L} = \text{Ad}_{T_{i_L,(i-1)_L}}(\mathcal{V}_{(i-1)_L}) + \mathcal{A}_i \dot{\theta}_i, \tag{8.107}$$

$$\dot{\mathcal{V}}_{i_R} = \text{Ad}_{T_{i_R,(i-1)_L}}(\dot{\mathcal{V}}_{(i-1)_L}) + \text{ad}_{\mathcal{V}_{i_R}}(\mathcal{R}_i)G_i\dot{\theta}_i + \mathcal{R}_i G_i \ddot{\theta}_i, \tag{8.108}$$

$$\dot{\mathcal{V}}_{i_L} = \text{Ad}_{T_{i_L,(i-1)_L}}(\dot{\mathcal{V}}_{(i-1)_L}) + \text{ad}_{\mathcal{V}_{i_L}}(\mathcal{A}_i)\dot{\theta}_i + \mathcal{A}_i \ddot{\theta}_i. \tag{8.109}$$

**Backward iterations**

For $i = n$ to $1$ do

$$\mathcal{F}_{i_L} = \text{Ad}_{T_{(i+1)_L,i_L}}^{\text{T}}(\mathcal{F}_{(i+1)_L}) + \mathcal{G}_{i_L}\dot{\mathcal{V}}_{i_L} - \text{ad}_{\mathcal{V}_{i_L}}^{\text{T}}(\mathcal{G}_{i_L}\mathcal{V}_{i_L})$$
$$+ \text{Ad}_{T_{(i+1)_R,i_L}}^{\text{T}}(\mathcal{G}_{(i+1)_R}\dot{\mathcal{V}}_{(i+1)_R} - \text{ad}_{\mathcal{V}_{(i+1)_R}}^{\text{T}}(\mathcal{G}_{(i+1)_R}\mathcal{V}_{(i+1)_R})), \tag{8.110}$$

$$\tau_{i,\text{gear}} = \mathcal{A}_i^{\text{T}}\mathcal{F}_{i_L}, \tag{8.111}$$

$$\tau_{i,\text{motor}} = \frac{\tau_{i,\text{gear}}}{G_i} + \mathcal{R}_i^{\text{T}}(\mathcal{G}_{i_R}\dot{\mathcal{V}}_{i_R} - \text{ad}_{\mathcal{V}_{i_R}}^{\text{T}}(\mathcal{G}_{i_R}\mathcal{V}_{i_R})). \tag{8.112}$$

In the backward iteration stage, the quantity $\mathcal{F}_{(n+1)_L}$ occurring in the first step of the backward iteration is taken to be the external wrench applied to the end-effector (expressed in the $\{n+1\}_L$ frame), with $\mathcal{G}_{(n+1)_R}$ set to zero; $\mathcal{F}_{i_L}$ denotes the wrench applied to link $i$ via the motor $i$ gearhead (expressed in the $\{i_L\}$ frame); $\tau_{i,\text{gear}}$ is the torque generated at the motor $i$ gearhead; and $\tau_{i,\text{motor}}$ is the torque at rotor $i$.

Note that if there is no gearing then no modification to the original Newton–Euler inverse dynamics algorithm is necessary; the stator is attached to one link and the rotor is attached to another link. Robots constructed with a motor at each axis and no gearheads are sometimes called **direct-drive robots**. Direct-drive robots have low friction, but they see limited use because typically the motors must be large and heavy to generate appropriate torques.

No modification is needed to the Lagrangian approach to the dynamics to handle geared motors, provided that we can correctly represent the kinetic energy of the faster-spinning rotors.

## 8.9.4 Friction

The Lagrangian and Newton–Euler dynamics do not account for friction at the joints, but the friction forces and torques in gearheads and bearings may be significant. Friction is a complex phenomenon that is the subject of considerable current research; any friction model is a gross attempt to capture the average behavior of the micromechanics of contact.

Friction models often include a **static friction** term and a velocity-dependent

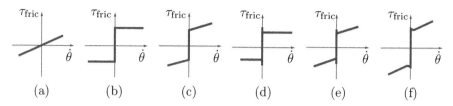

**Figure 8.15** Examples of velocity-dependent friction models. (a) Viscous friction, $\tau_{\text{fric}} = b_{\text{viscous}}\dot{\theta}$. (b) Coulomb friction, $\tau_{\text{fric}} = b_{\text{static}}\,\text{sgn}(\dot{\theta})$; $\tau_{\text{fric}}$ can take any value in $[-b_{\text{static}}, b_{\text{static}}]$ at zero velocity. (c) Static plus viscous friction, $\tau_{\text{fric}} = b_{\text{static}}\,\text{sgn}(\dot{\theta}) + b_{\text{viscous}}\dot{\theta}$. (d) Static and kinetic friction, requiring $\tau_{\text{fric}} \geq |b_{\text{static}}|$ to initiate motion and then $\tau_{\text{fric}} = b_{\text{kinetic}}\,\text{sgn}(\dot{\theta})$ during motion, where $b_{\text{static}} > b_{\text{kinetic}}$. (e) Static, kinetic, and viscous friction. (f) A friction law exhibiting the Stribeck effect – at low velocities, the friction decreases as the velocity increases.

viscous friction term. The presence of a static friction term means that a nonzero torque is required to cause the joint to begin to move. The viscous friction term indicates that the amount of friction torque increases with increasing velocity of the joint. See Figure 8.15 for some examples of velocity-dependent friction models.

Other factors may contribute to the friction at a joint, including the loading of the joint bearings, the time the joint has been at rest, the temperature, etc. The friction in a gearhead often increases as the gear ratio $G$ increases.

### 8.9.5    Joint and Link Flexibility

In practice, a robot's joints and links are likely to exhibit some flexibility. For example, the flexspline element of a harmonic drive gearhead achieves essentially zero backlash by being somewhat flexible. A model of a joint with harmonic drive gearing, then, could include a relatively stiff torsional spring between the motor's rotor and the link to which the gearhead is attached.

Similarly, links themselves are not infinitely stiff. Their finite stiffness is exhibited as vibrations along the link.

Flexible joints and links introduce extra states to the dynamics of the robot, significantly complicating the dynamics and control. While many robots are designed to be stiff in order to minimize these complexities, in some cases this is impractical owing to the extra link mass required to create the stiffness.

### 8.10    Summary

- Given a set of generalized coordinates $\theta$ and generalized forces $\tau$, the Euler–Lagrange equations can be written

$$\tau = \frac{d}{dt}\frac{\partial \mathcal{L}}{\partial \dot{\theta}} - \frac{\partial L}{\partial \theta},$$

where $\mathcal{L}(\theta, \dot{\theta}) = \mathcal{K}(\theta, \dot{\theta}) - \mathcal{P}(\theta)$, $\mathcal{K}$ is the kinetic energy of the robot, and $\mathcal{P}$ is the potential energy of the robot.

- The equations of motion of a robot can be written in the following equivalent forms:

$$
\begin{aligned}
\tau &= M(\theta)\ddot{\theta} + h(\theta, \dot{\theta}) \\
&= M(\theta)\ddot{\theta} + c(\theta, \dot{\theta}) + g(\theta) \\
&= M(\theta)\ddot{\theta} + \dot{\theta}^{\mathrm{T}}\Gamma(\theta)\dot{\theta} + g(\theta) \\
&= M(\theta)\ddot{\theta} + C(\theta, \dot{\theta})\dot{\theta} + g(\theta),
\end{aligned}
$$

where $M(\theta)$ is the $n \times n$ symmetric positive-definite mass matrix, $h(\theta, \dot{\theta})$ is the sum of the generalized forces due to the gravity and quadratic velocity terms, $c(\theta, \dot{\theta})$ are quadratic velocity forces, $g(\theta)$ are gravitational forces, $\Gamma(\theta)$ is an $n \times n \times n$ matrix of Christoffel symbols of the first kind obtained from partial derivatives of $M(\theta)$ with respect to $\theta$, and $C(\theta, \dot{\theta})$ is the $n \times n$ Coriolis matrix whose $(i, j)$th entry is given by

$$
c_{ij}(\theta, \dot{\theta}) = \sum_{k=1}^{n} \Gamma_{ijk}(\theta)\dot{\theta}_k.
$$

If the end-effector of the robot is applying a wrench $\mathcal{F}_{\text{tip}}$ to the environment, the term $J^{\mathrm{T}}(\theta)\mathcal{F}_{\text{tip}}$ should be added to the right-hand side of the robot's dynamic equations.

- The symmetric positive-definite rotational inertia matrix of a rigid body is

$$
\mathcal{I}_b = \begin{bmatrix} \mathcal{I}_{xx} & \mathcal{I}_{xy} & \mathcal{I}_{xz} \\ \mathcal{I}_{xy} & \mathcal{I}_{yy} & \mathcal{I}_{yz} \\ \mathcal{I}_{xz} & \mathcal{I}_{yz} & \mathcal{I}_{zz} \end{bmatrix},
$$

where

$$
\begin{aligned}
\mathcal{I}_{xx} &= \int_{\mathcal{B}}(y^2 + z^2)\rho(x, y, z)dV, & \mathcal{I}_{yy} &= \int_{\mathcal{B}}(x^2 + z^2)\rho(x, y, z)dV, \\
\mathcal{I}_{zz} &= \int_{\mathcal{B}}(x^2 + y^2)\rho(x, y, z)dV, & \mathcal{I}_{xy} &= -\int_{\mathcal{B}} xy\rho(x, y, z)dV, \\
\mathcal{I}_{xz} &= -\int_{\mathcal{B}} xz\rho(x, y, z)dV, & \mathcal{I}_{yz} &= -\int_{\mathcal{B}} yz\rho(x, y, z)dV,
\end{aligned}
$$

$\mathcal{B}$ is the volume of the body, $dV$ is a differential volume element, and $\rho(x, y, z)$ is the density function.

- If $\mathcal{I}_b$ is defined in a frame {b} at the center of mass, with axes aligned with the principal axes of inertia, then $\mathcal{I}_b$ is diagonal.
- If {b} is at the center of mass but its axes are not aligned with the principal axes of inertia, there always exists a rotated frame {c} defined by the rotation matrix $R_{bc}$ such that $\mathcal{I}_c = R_{bc}^{\mathrm{T}}\mathcal{I}_b R_{bc}$ is diagonal.
- If $\mathcal{I}_b$ is defined in a frame {b} at the center of mass then $\mathcal{I}_q$, the inertia in a frame {q} aligned with {b} but displaced from the origin of {b} by $q \in \mathbb{R}^3$ in {b} coordinates, is

$$
\mathcal{I}_q = \mathcal{I}_b + \mathfrak{m}(q^{\mathrm{T}}qI - qq^{\mathrm{T}}).
$$

- The spatial inertia matrix $\mathcal{G}_b$ expressed in a frame {b} at the center of mass is defined as the $6 \times 6$ matrix

$$\mathcal{G}_b = \begin{bmatrix} \mathcal{I}_b & 0 \\ 0 & \mathfrak{m}I \end{bmatrix}.$$

  In a frame {a} at a configuration $T_{ba}$ relative to {b}, the spatial inertia matrix is

$$\mathcal{G}_a = [\mathrm{Ad}_{T_{ba}}]^\mathrm{T} \mathcal{G}_b [\mathrm{Ad}_{T_{ba}}].$$

- The Lie bracket of two twists $\mathcal{V}_1$ and $\mathcal{V}_2$ is

$$\mathrm{ad}_{\mathcal{V}_1}(\mathcal{V}_2) = [\mathrm{ad}_{\mathcal{V}_1}]\mathcal{V}_2,$$

  where

$$[\mathrm{ad}_\mathcal{V}] = \begin{bmatrix} [\omega] & 0 \\ [v] & [\omega] \end{bmatrix} \in \mathbb{R}^{6 \times 6}.$$

- The twist–wrench formulation of the rigid-body dynamics of a single rigid body is

$$\mathcal{F}_b = \mathcal{G}_b \dot{\mathcal{V}}_b - [\mathrm{ad}_{\mathcal{V}_b}]^\mathrm{T} \mathcal{G}_b \mathcal{V}_b.$$

  The equations have the same form if $\mathcal{F}$, $\mathcal{V}$, and $\mathcal{G}$ are all expressed in the same frame, regardless of the frame.

- The kinetic energy of a rigid body is $\frac{1}{2} \mathcal{V}_b^\mathrm{T} \mathcal{G}_b \mathcal{V}_b$, and the kinetic energy of an open-chain robot is $\frac{1}{2} \dot{\theta}^\mathrm{T} M(\theta) \dot{\theta}$.

- The forward–backward Newton–Euler inverse dynamics algorithm is the following:

  **Initialization:** Attach a frame {0} to the base, frames {1} to {n} to the centers of mass of links {1} to {n}, and a frame {n+1} at the end-effector, fixed in the frame {n}. Define $M_{i,i-1}$ to be the configuration of {i − 1} in {i} when $\theta_i = 0$. Let $\mathcal{A}_i$ be the screw axis of joint $i$ expressed in {i}, and $\mathcal{G}_i$ be the $6 \times 6$ spatial inertia matrix of link $i$. Define $\mathcal{V}_0$ to be the twist of the base frame {0} expressed in base-frame coordinates. (This quantity is typically zero.) Let $\mathfrak{g} \in \mathbb{R}^3$ be the gravity vector expressed in base-frame-{0} coordinates, and define $\dot{\mathcal{V}}_0 = (0, -\mathfrak{g})$. (Gravity is treated as an acceleration of the base in the opposite direction.) Define $\mathcal{F}_{n+1} = \mathcal{F}_\mathrm{tip} = (m_\mathrm{tip}, f_\mathrm{tip})$ to be the wrench applied to the environment by the end-effector expressed in the end-effector frame {n + 1}.

  **Forward iterations:** Given $\theta, \dot{\theta}, \ddot{\theta}$, for $i = 1$ to $n$ do

$$T_{i,i-1} = e^{-[\mathcal{A}_i]\theta_i} M_{i,i-1},$$
$$\mathcal{V}_i = \mathrm{Ad}_{T_{i,i-1}}(\mathcal{V}_{i-1}) + \mathcal{A}_i \dot{\theta}_i,$$
$$\dot{\mathcal{V}}_i = \mathrm{Ad}_{T_{i,i-1}}(\dot{\mathcal{V}}_{i-1}) + \mathrm{ad}_{\mathcal{V}_i}(\mathcal{A}_i)\dot{\theta}_i + \mathcal{A}_i \ddot{\theta}_i.$$

**Backward iterations:** For $i = n$ to 1 do

$$\mathcal{F}_i = \text{Ad}^{\text{T}}_{T_{i+1,i}}(\mathcal{F}_{i+1}) + \mathcal{G}_i \dot{\mathcal{V}}_i - \text{ad}^{\text{T}}_{\mathcal{V}_i}(\mathcal{G}_i \mathcal{V}_i),$$

$$\tau_i = \mathcal{F}_i^{\text{T}} \mathcal{A}_i.$$

- Let $J_{ib}(\theta)$ be the Jacobian relating $\dot{\theta}$ to the body twist $\mathcal{V}_i$ in link $i$'s center-of-mass frame $\{i\}$. Then the mass matrix $M(\theta)$ of the manipulator can be expressed as

$$M(\theta) = \sum_{i=1}^{n} J_{ib}^{\text{T}}(\theta) \mathcal{G}_i J_{ib}(\theta).$$

- The forward dynamics problem involves solving

$$M(\theta)\ddot{\theta} = \tau(t) - h(\theta, \dot{\theta}) - J^{\text{T}}(\theta)\mathcal{F}_{\text{tip}}$$

for $\ddot{\theta}$, using any efficient solver of equations of the form $Ax = b$.

- The robot's dynamics $M(\theta)\ddot{\theta} + h(\theta, \dot{\theta})$ can be expressed in the task space as

$$\mathcal{F} = \Lambda(\theta)\dot{\mathcal{V}} + \eta(\theta, \mathcal{V}),$$

where $\mathcal{F}$ is the wrench applied to the end-effector, $\mathcal{V}$ is the twist of the end-effector, and $\mathcal{F}$, $\mathcal{V}$, and the Jacobian $J(\theta)$ are all defined in the same frame. The task-space mass matrix $\Lambda(\theta)$ and gravity and quadratic velocity forces $\eta(\theta, \mathcal{V})$ are

$$\Lambda(\theta) = J^{-\text{T}} M(\theta) J^{-1},$$
$$\eta(\theta, \mathcal{V}) = J^{-\text{T}} h(\theta, J^{-1}\mathcal{V}) - \Lambda(\theta)\dot{J}J^{-1}\mathcal{V}.$$

- Define two $n \times n$ projection matrices of rank $n - k$

$$P(\theta) = I - A^{\text{T}}(AM^{-1}A^{\text{T}})^{-1}AM^{-1},$$
$$P_{\ddot{\theta}}(\theta) = M^{-1}PM = I - M^{-1}A^{\text{T}}(AM^{-1}A^{\text{T}})^{-1}A,$$

corresponding to the $k$ Pfaffian constraints, $A(\theta)\dot{\theta} = 0$, $A \in \mathbb{R}^{k \times n}$, acting on the robot. Then the $n + k$ constrained equations of motion

$$\tau = M(\theta)\ddot{\theta} + h(\theta, \dot{\theta}) + A^{\text{T}}(\theta)\lambda,$$
$$A(\theta)\dot{\theta} = 0$$

can be reduced to the following equivalent forms by eliminating the Lagrange multipliers $\lambda$:

$$P\tau = P(M\ddot{\theta} + h),$$
$$P_{\ddot{\theta}}\ddot{\theta} = P_{\ddot{\theta}}M^{-1}(\tau - h).$$

The matrix $P$ projects away the joint force–torque components that act on the constraints without doing work on the robot, and the matrix $P_{\ddot{\theta}}$ projects away acceleration components that do not satisfy the constraints.

- An ideal gearhead (one that is 100% efficient) with a gear ratio $G$ multiplies the torque at the output of a motor by a factor $G$ and divides the speed by the factor $G$, leaving the mechanical power unchanged. The inertia of the motor's rotor about its axis of rotation, as it appears at the output of the gearhead, is $G^2\mathcal{I}_{\text{rotor}}$.

## 8.11    Software

Software functions associated with this chapter are listed below.

adV = ad(V)
Computes $[\text{ad}_{\mathcal{V}}]$.

taulist = InverseDynamics(thetalist,dthetalist,ddthetalist,g,Ftip,
Mlist,Glist,Slist)
Uses Newton–Euler inverse dynamics to compute the $n$-vector $\tau$ of the required joint forces–torques given $\theta$, $\dot{\theta}$, $\ddot{\theta}$, $\mathfrak{g}$, $\mathcal{F}_{\text{tip}}$, a list of transforms $M_{i-1,i}$ specifying the configuration of the center-of-mass frame of link $\{i\}$ relative to $\{i-1\}$ when the robot is at its home position, a list of link spatial inertia matrices $\mathcal{G}_i$, and a list of joint screw axes $\mathcal{S}_i$ expressed in the base frame.

M = MassMatrix(thetalist,Mlist,Glist,Slist)
Computes the mass matrix $M(\theta)$ given the joint configuration $\theta$, a list of transforms $M_{i-1,i}$, a list of link spatial inertia matrices $\mathcal{G}_i$, and a list of joint screw axes $\mathcal{S}_i$ expressed in the base frame.

c = VelQuadraticForces(thetalist,dthetalist,Mlist,Glist,Slist)
Computes $c(\theta,\dot{\theta})$ given the joint configuration $\theta$, the joint velocities $\dot{\theta}$, a list of transforms $M_{i-1,i}$, a list of link spatial inertia matrices $\mathcal{G}_i$, and a list of joint screw axes $\mathcal{S}_i$ expressed in the base frame.

grav = GravityForces(thetalist,g,Mlist,Glist,Slist)
Computes $g(\theta)$ given the joint configuration $\theta$, the gravity vector $\mathfrak{g}$, a list of transforms $M_{i-1,i}$, a list of link spatial inertia matrices $\mathcal{G}_i$, and a list of joint screw axes $\mathcal{S}_i$ expressed in the base frame.

JTFtip = EndEffectorForces(thetalist,Ftip,Mlist,Glist,Slist)
Computes $J^{\text{T}}(\theta)\mathcal{F}_{\text{tip}}$ given the joint configuration $\theta$, the wrench $\mathcal{F}_{\text{tip}}$ applied by the end-effector, a list of transforms $M_{i-1,i}$, a list of link spatial inertia matrices $\mathcal{G}_i$, and a list of joint screw axes $\mathcal{S}_i$ expressed in the base frame.

ddthetalist = ForwardDynamics(thetalist,dthetalist,taulist,g,Ftip,
Mlist,Glist,Slist)
Computes $\ddot{\theta}$ given the joint configuration $\theta$, the joint velocities $\dot{\theta}$, the joint forces-torques $\tau$, the gravity vector $\mathfrak{g}$, the wrench $\mathcal{F}_{\text{tip}}$ applied by the end-effector, a

list of transforms $M_{i-1,i}$, a list of link spatial inertia matrices $\mathcal{G}_i$, and a list of joint screw axes $\mathcal{S}_i$ expressed in the base frame.

`[thetalistNext,dthetalistNext] = EulerStep(thetalist,dthetalist,`
`ddthetalist,dt)`
Computes a first-order Euler approximation to $\{\theta(t + \delta t), \dot{\theta}(t + \delta t)\}$ given the joint configuration $\theta(t)$, the joint velocities $\dot{\theta}(t)$, the joint accelerations $\ddot{\theta}(t)$, and a timestep $\delta t$.

`taumat = InverseDynamicsTrajectory(thetamat,dthetamat,ddthetamat,`
`g,Ftipmat,Mlist,Glist,Slist)`
The variable `thetamat` is an $N \times n$ matrix of robot joint variables $\theta$, where the $i$th row corresponds to the $n$-vector of joint variables $\theta(t)$ at time $t = (i - 1)\delta t$, where $\delta t$ is the timestep. The variables `dthetamat`, `ddthetamat`, and `Ftipmat` similarly represent $\dot{\theta}$, $\ddot{\theta}$, and $\mathcal{F}_{\text{tip}}$ as a function of time. Other inputs include the gravity vector $\mathfrak{g}$, a list of transforms $M_{i-1,i}$, a list of link spatial inertia matrices $\mathcal{G}_i$, and a list of joint screw axes $\mathcal{S}_i$ expressed in the base frame. This function computes an $N \times n$ matrix `taumat` representing the joint forces-torques $\tau(t)$ required to generate the trajectory specified by $\theta(t)$ and $\mathcal{F}_{\text{tip}}(t)$. Note that it is not necessary to specify $\delta t$. The velocities $\dot{\theta}(t)$ and accelerations $\ddot{\theta}(t)$ should be consistent with $\theta(t)$.

`[thetamat,dthetamat] = ForwardDynamicsTrajectory(thetalist,`
`dthetalist,taumat,g,Ftipmat,Mlist,Glist,Slist,dt,intRes)`
This function numerically integrates the robot's equations of motion using Euler integration. The outputs are $N \times n$ matrices `thetamat` and `dthetamat`, where the $i$th rows correspond respectively to the $n$-vectors $\theta((i-1)\delta t)$ and $\dot{\theta}((i-1)\delta t)$. The inputs are the initial state $\theta(0)$, $\dot{\theta}(0)$, an $N \times n$ matrix of joint forces–torques $\tau(t)$, the gravity vector $\mathfrak{g}$, an $N \times n$ matrix of end-effector wrenches $\mathcal{F}_{\text{tip}}(t)$, a list of transforms $M_{i-1,i}$, a list of link spatial inertia matrices $\mathcal{G}_i$, a list of joint screw axes $\mathcal{S}_i$ expressed in the base frame, the timestep $\delta t$, and the number of integration steps to take during each timestep (a positive integer).

## .12    Notes and References

An accessible general reference on rigid-body dynamics that covers both the Newton–Euler and Lagrangian formulations is Greenwood (2006). A more classical reference that covers a wide range of topics in dynamics is Whittaker (1917).

A recursive inverse dynamics algorithm for open chains using the classical screw-theoretic machinery of twists and wrenches was first formulated by Featherstone (the collection of twists, wrenches, and the corresponding analogues of accelerations, momentum, and inertias, are collectively referred to as spatial vector notation); this formulation, as well as more efficient extensions based on articulated body inertias, are described in Featherstone (1983, 2008).

The recursive inverse dynamics algorithm presented in this chapter was first described in Park et al. (1995) and makes use of standard operators from the theory of Lie groups and Lie algebras. An important practical advantage of this approach is that analytic formulas can be derived for taking the first- and higher-order derivatives of the dynamics. This has important consequences for dynamics-based motion optimization: the availability of analytic gradients can greatly improve the convergence and robustness of motion optimization algorithms. These and other related issues are explored in, e.g., Lee et al. (2005).

The task-space formulation was first initiated by Khatib (1987), who referred to it as the operational-space formulation. Note that the task-space formulation involves taking time derivatives of the forward kinematics Jacobian, i.e., $\dot{J}(\theta)$. Using either the body or space Jacobian, $\dot{J}(\theta)$ can in fact be evaluated analytically; this is explored in an exercise at the end of this chapter.

A brief history of the evolution of robot dynamics algorithms, as well as pointers to references on the more general subject of multibody system dynamics (of which robot dynamics can be considered a subfield) can be found in Featherstone and Orin (2016).

## 8.13     Exercises

**Exercise 8.1**   Derive the formulas given in Figure 8.5 for:

(a) a rectangular parallelpiped;
(b) a circular cylinder;
(c) an ellipsoid.

**Exercise 8.2**   Consider a cast iron dumbbell consisting of a cylinder connecting two solid spheres at either end of the cylinder. The density of the dumbbell is $7500$ kg/m$^3$. The cylinder has a diameter of 4 cm and a length of 20 cm. Each sphere has a diameter of 20 cm.

(a) Find the approximate rotational inertia matrix $\mathcal{I}_b$ in a frame {b} at the center of mass with axes aligned with the principal axes of inertia of the dumbbell.
(b) Write down the spatial inertia matrix $\mathcal{G}_b$.

**Exercise 8.3**   Rigid-body dynamics in an arbitrary frame.

(a) Show that Equation (8.42) is a generalization of Steiner's theorem.
(b) Derive Equation (8.43).

**Exercise 8.4**   The 2R open-chain robot of Figure 8.16, referred to as a rotational inverted pendulum or Furuta pendulum, is shown in its zero position. Assuming that the mass of each link is concentrated at the tip and neglecting

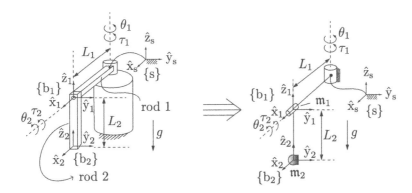

**Figure 8.16** 2R rotational inverted pendulum. (Left) Its construction; (right) the model.

its thickness, the robot can be modeled as shown in the right-hand figure. As-
sume that $m_1 = m_2 = 2$, $L_1 = L_2 = 1$, $g = 10$, and the link inertias $\mathcal{I}_1$ and $\mathcal{I}_2$
(expressed in their respective link frames $\{b_1\}$ and $\{b_2\}$) are

$$\mathcal{I}_1 = \begin{bmatrix} 0 & 0 & 0 \\ 0 & 4 & 0 \\ 0 & 0 & 4 \end{bmatrix}, \qquad \mathcal{I}_2 = \begin{bmatrix} 4 & 0 & 0 \\ 0 & 4 & 0 \\ 0 & 0 & 0 \end{bmatrix}.$$

(a) Derive the dynamic equations and determine the input torques $\tau_1$ and $\tau_2$
when $\theta_1 = \theta_2 = \pi/4$ and the joint velocities and accelerations are all zero.
(b) Draw the torque ellipsoid for the mass matrix $M(\theta)$ when $\theta_1 = \theta_2 = \pi/4$.

**Exercise 8.5**  Prove the following Lie bracket identity (called the Jacobi identity) for arbitrary twists $\mathcal{V}_1$, $\mathcal{V}_2$, $\mathcal{V}_3$:

$$\mathrm{ad}_{\mathcal{V}_1}(\mathrm{ad}_{\mathcal{V}_2}(\mathcal{V}_3)) + \mathrm{ad}_{\mathcal{V}_3}(\mathrm{ad}_{\mathcal{V}_1}(\mathcal{V}_2)) + \mathrm{ad}_{\mathcal{V}_2}(\mathrm{ad}_{\mathcal{V}_3}(\mathcal{V}_1)) = 0.$$

**Exercise 8.6**  The evaluation of $\dot{J}(\theta)$, the time derivative of the forward kinematics Jacobian, is needed in the calculation of the frame accelerations $\dot{\mathcal{V}}_i$ in
the Newton–Euler inverse dynamics algorithm and also in the formulation of the
task-space dynamics. Letting $J_i(\theta)$ denote the $i$th column of $J(\theta)$, we have

$$\frac{d}{dt}J_i(\theta) = \sum_{j=1}^{n} \frac{\partial J_i}{\partial \theta_j}\dot{\theta}_j.$$

(a) Suppose that $J(\theta)$ is the space Jacobian. Show that

$$\frac{\partial J_i}{\partial \theta_j} = \begin{cases} \mathrm{ad}_{J_j}(J_i) & \text{for } i > j \\ 0 & \text{for } i \leq j. \end{cases}$$

(b) Now suppose that $J(\theta)$ is the body Jacobian. Show that

$$\frac{\partial J_i}{\partial \theta_j} = \begin{cases} \mathrm{ad}_{J_i}(J_j) & \text{for } i < j \\ 0 & \text{for } i \geq j. \end{cases}$$

**Exercise 8.7**   Show that the time derivative of the mass matrix $\dot{M}(\theta)$ can be written explicitly as

$$\dot{M} = -\mathcal{A}^{\mathrm{T}}\mathcal{L}^{\mathrm{T}}\mathcal{W}^{\mathrm{T}}[\mathrm{ad}_{\mathcal{A}\dot{\theta}}]^{\mathrm{T}}\mathcal{L}^{\mathrm{T}}\mathcal{G}\mathcal{L}\mathcal{A} - \mathcal{A}^{\mathrm{T}}\mathcal{L}^{\mathrm{T}}\mathcal{G}\mathcal{L}[\mathrm{ad}_{\mathcal{A}\dot{\theta}}]\mathcal{W}\mathcal{L}\mathcal{A},$$

with the matrices as defined in the closed-form dynamics formulation.

**Exercise 8.8**   Explain intuitively the shapes of the end-effector force ellipsoids in Figure 8.4 on the basis of the point masses and the Jacobians.

**Exercise 8.9**   Consider a motor with rotor inertia $\mathcal{I}_{\mathrm{rotor}}$ connected through a gearhead of gear ratio $G$ to a load with scalar inertia $\mathcal{I}_{\mathrm{link}}$ about the rotation axis. The load and motor are said to be **inertia matched** if, for any given torque $\tau_m$ at the motor, the acceleration of the load is maximized. The acceleration of the load can be written

$$\ddot{\theta} = \frac{G\tau_m}{\mathcal{I}_{\mathrm{link}} + G^2\mathcal{I}_{\mathrm{rotor}}}.$$

Solve for the inertia-matching gear ratio $\sqrt{\mathcal{I}_{\mathrm{link}}/\mathcal{I}_{\mathrm{rotor}}}$ by solving $d\ddot{\theta}/dG = 0$.

**Exercise 8.10**   Give the steps that rearrange Equation (8.99) to get Equation (8.101). Remember that $P(\theta)$ is not full rank and cannot be inverted.

**Exercise 8.11**   Program a function to calculate $h(\theta, \dot{\theta}) = c(\theta, \dot{\theta}) + g(\theta)$ efficiently using Newton–Euler inverse dynamics.

**Exercise 8.12**   Give the equations that would convert the joint and link descriptions in a robot's URDF file to the data Mlist, Glist, and Slist, suitable for using with the Newton–Euler algorithm InverseDynamicsTrajectory.

**Exercise 8.13**   The efficient evaluation of $M(\theta)$.

(a) Develop conceptually a computationally efficient algorithm for determining the mass matrix $M(\theta)$ using Equation (8.57).

(b) Implement this algorithm.

**Exercise 8.14**   The function InverseDynamicsTrajectory requires the user to enter not only a time sequence of joint variables thetamat but also a time sequence of joint velocities dthetamat and accelerations ddthetamat. Instead, the function could use numerical differencing to find approximately the joint velocities and accelerations at each timestep, using only thetamat. Write an alternative InverseDynamicsTrajectory function that does not require the user to enter dthetamat and ddthetamat. Verify that it yields similar results.

**Exercise 8.15**   Dynamics of the UR5 robot.

(a) Write the spatial inertia matrices $\mathcal{G}_i$ of the six links of the UR5, given the center-of-mass frames and mass and inertial properties defined in the URDF in Section 4.2.

(b) Simulate the UR5 falling under gravity with acceleration $g = 9.81$ m/s$^2$ in the $-\hat{z}_\mathrm{s}$-direction. The robot starts at its zero configuration and zero joint torques are applied. Simulate the motion for three seconds, with at least 100 integration steps per second. (Ignore the effects of friction and the geared rotors.)

# 9 Trajectory Generation

During robot motion, the robot controller is provided with a steady stream of goal positions and velocities to track. This specification of the robot position as a function of time is called a **trajectory**. In some cases, the trajectory is completely specified by the task – for example, the end-effector may be required to track a known moving object. In other cases, as when the task is simply to move from one position to another in a given time, we have freedom to design the trajectory to meet these constraints. This is the domain of **trajectory planning**. The trajectory should be a sufficiently smooth function of time, and it should respect any given limits on joint velocities, accelerations, or torques.

In this chapter we consider a trajectory as the combination of a **path**, a purely geometric description of the sequence of configurations achieved by the robot, and a **time scaling**, which specifies the times when those configurations are reached. We consider three cases: point-to-point straight-line trajectories in both joint space and task space; trajectories passing through a sequence of timed **via points**; and minimum-time trajectories along specified paths taking actuator limits into consideration. Finding paths that avoid obstacles is left to Chapter 10.

## 9.1 Definitions

A **path** $\theta(s)$ maps a scalar path parameter $s$, assumed to be 0 at the start of the path and 1 at the end, to a point in the robot's configuration space $\Theta$, $\theta : [0, 1] \rightarrow \Theta$. As $s$ increases from 0 to 1, the robot moves along the path. Sometimes $s$ is taken to be time and is allowed to vary from time $s = 0$ to the total motion time $s = T$, but it is often useful to separate the role of the geometric path parameter $s$ from the time parameter $t$. A **time scaling** $s(t)$ assigns a value $s$ to each time $t \in [0, T]$, $s : [0, T] \rightarrow [0, 1]$.

Together, a path and a time scaling define a **trajectory** $\theta(s(t))$, or $\theta(t)$ for short. Using the chain rule, the velocity and acceleration along the trajectory can be written as

$$\dot{\theta} = \frac{d\theta}{ds} \dot{s}, \tag{9.1}$$

$$\ddot{\theta} = \frac{d\theta}{ds} \ddot{s} + \frac{d^2\theta}{ds^2} \dot{s}^2. \tag{9.2}$$

To ensure that the robot's acceleration (and therefore dynamics) is well defined, each of $\theta(s)$ and $s(t)$ must be twice differentiable.

## .2 Point-to-Point Trajectories

The simplest type of motion is from rest at one configuration to rest at another. We call this a point-to-point motion. The simplest type of path for point-to-point motion is a straight line. Straight-line paths and their time scalings are discussed below.

### .2.1 Straight-Line Paths

A "straight line" from a start configuration $\theta_{start}$ to an end configuration $\theta_{end}$ could be defined in joint space or in task space. The advantage of a straight-line path from $\theta_{start}$ to $\theta_{end}$ in joint space is simplicity: since joint limits typically take the form $\theta_{i,min} \leq \theta_i \leq \theta_{i,max}$ for each joint $i$, the allowable joint configurations form a convex set $\Theta_{free}$ in joint space, so the straight line between any two endpoints in $\Theta_{free}$ also lies in $\Theta_{free}$. The straight line can be written

$$\theta(s) = \theta_{start} + s(\theta_{end} - \theta_{start}), \qquad s \in [0, 1] \qquad (9.3)$$

with derivatives

$$\frac{d\theta}{ds} = \theta_{end} - \theta_{start}, \qquad (9.4)$$

$$\frac{d^2\theta}{ds^2} = 0. \qquad (9.5)$$

Straight lines in joint space generally do not yield straight-line motion of the end-effector in task space. If task-space straight-line motions are desired, the start and end configurations can be specified by $X_{start}$ and $X_{end}$ in task space. If $X_{start}$ and $X_{end}$ are represented by a minimum set of coordinates then a straight line is defined as $X(s) = X_{start} + s(X_{end} - X_{start}), s \in [0, 1]$. Compared with the case when joint coordinates are used, the following issues must be addressed:

- If the path passes near a kinematic singularity, the joint velocities may become unreasonably large for almost all time scalings of the path.
- Since the robot's reachable task space may not be convex in $X$ coordinates, some points on a straight line between two reachable endpoints may not be reachable (Figure 9.1).

In addition to the issues above, if $X_{start}$ and $X_{end}$ are represented as elements of $SE(3)$ instead of as a minimum set of coordinates, then there is the question of how to define a "straight" line in $SE(3)$. A configuration of the form $X_{start} + s(X_{end} - X_{start})$ does not generally lie in $SE(3)$.

One option is to use the screw motion (simultaneous rotation about and

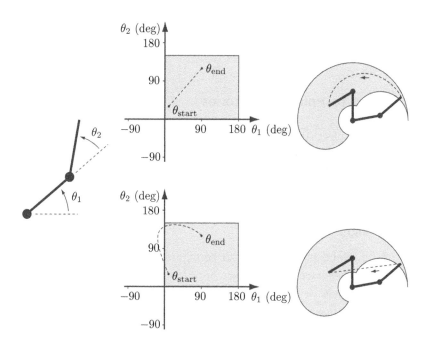

**Figure 9.1** (Left) A 2R robot with joint limits $0° \leq \theta_1 \leq 180°$, $0° \leq \theta_2 \leq 150°$. (Top center) A straight-line path in joint space and (top right) the corresponding motion of the end-effector in task space (dashed line). The reachable endpoint configurations, subject to joint limits, are indicated in gray. (Bottom center) This curved line in joint space and (bottom right) the corresponding straight-line path in task space (dashed line) would violate the joint limits.

translation along a fixed screw axis) that moves the robot's end-effector from $X_{\text{start}} = X(0)$ to $X_{\text{end}} = X(1)$. To derive this $X(s)$, we can write the start and end configurations explicitly in the $\{s\}$ frame as $X_{s,\text{start}}$ and $X_{s,\text{end}}$ and use our subscript cancellation rule to express the end configuration in the start frame:

$$X_{\text{start,end}} = X_{\text{start},s} X_{s,\text{end}} = X_{s,\text{start}}^{-1} X_{s,\text{end}}.$$

Then $\log(X_{s,\text{start}}^{-1} X_{s,\text{end}})$ is the matrix representation of the twist, expressed in the $\{\text{start}\}$ frame, that takes $X_{\text{start}}$ to $X_{\text{end}}$ in unit time. The path can therefore be written as

$$X(s) = X_{\text{start}} \exp(\log(X_{\text{start}}^{-1} X_{\text{end}})s), \tag{9.6}$$

where $X_{\text{start}}$ is post-multiplied by the matrix exponential since the twist is represented in the $\{\text{start}\}$ frame, not the fixed world frame $\{s\}$.

This screw motion provides a "straight-line" motion in the sense that the screw axis is constant. The origin of the end-effector does not generally follow a straight line in Cartesian space, since it is following a screw motion. It may be preferable to decouple the rotational motion from the translational motion.

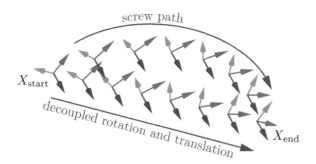

$X_{\text{start}}$

screw path

decoupled rotation and translation

$X_{\text{end}}$

**Figure 9.2** A path following a constant screw motion versus a decoupled path where the frame origin follows a straight line and the angular velocity is constant.

Writing $X = (R, p)$, we can define the path

$$p(s) = p_{\text{start}} + s(p_{\text{end}} - p_{\text{start}}), \tag{9.7}$$
$$R(s) = R_{\text{start}} \exp(\log(R_{\text{start}}^{\text{T}} R_{\text{end}})s) \tag{9.8}$$

where the frame origin follows a straight line while the axis of rotation is constant in the body frame. Figure 9.2 illustrates a screw path and a decoupled path for the same $X_{\text{start}}$ and $X_{\text{end}}$.

## .2.2 Time Scaling a Straight-Line Path

A time scaling $s(t)$ of a path should ensure that the motion is appropriately smooth and that any constraints on robot velocity and acceleration are satisfied. For a straight-line path in joint space of the form (9.3), the time-scaled joint velocities and accelerations are $\dot{\theta} = \dot{s}(\theta_{\text{end}} - \theta_{\text{start}})$ and $\ddot{\theta} = \ddot{s}(\theta_{\text{end}} - \theta_{\text{start}})$, respectively. For a straight-line path in task space parametrized by a minimum set of coordinates $X \in \mathbb{R}^m$, simply replace $\theta$, $\dot{\theta}$, and $\ddot{\theta}$ by $X$, $\dot{X}$, and $\ddot{X}$.

### .2.2.1 Polynomial Time Scaling
### Third-Order Polynomials
A convenient form for the time scaling $s(t)$ is a cubic polynomial of time,

$$s(t) = a_0 + a_1 t + a_2 t^2 + a_3 t^3. \tag{9.9}$$

A point-to-point motion in time $T$ imposes the initial constraints $s(0) = \dot{s}(0) = 0$ and the terminal constraints $s(T) = 1$ and $\dot{s}(T) = 0$. Evaluating Equation (9.9) and its derivative

$$\dot{s}(t) = a_1 + 2a_2 t + 3a_3 t^2 \tag{9.10}$$

at $t = 0$ and $t = T$ and solving the four constraints for $a_0, \ldots, a_3$, we find

$$a_0 = 0, \quad a_1 = 0, \quad a_2 = \frac{3}{T^2}, \quad a_3 = -\frac{2}{T^3}.$$

Plots of $s(t)$, $\dot{s}(t)$, and $\ddot{s}(t)$ are shown in Figure 9.3.

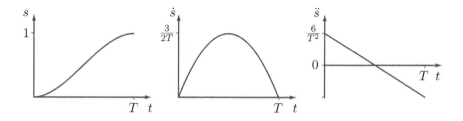

**Figure 9.3** Plots of $s(t)$, $\dot{s}(t)$, and $\ddot{s}(t)$ for a third-order polynomial time scaling.

Substituting $s = a_2 t^2 + a_3 t^3$ into Equation (9.3) yields

$$\theta(t) = \theta_{\text{start}} + \left( \frac{3t^2}{T^2} - \frac{2t^3}{T^3} \right) (\theta_{\text{end}} - \theta_{\text{start}}), \tag{9.11}$$

$$\dot{\theta}(t) = \left( \frac{6t}{T^2} - \frac{6t^2}{T^3} \right) (\theta_{\text{end}} - \theta_{\text{start}}), \tag{9.12}$$

$$\ddot{\theta}(t) = \left( \frac{6}{T^2} - \frac{12t}{T^3} \right) (\theta_{\text{end}} - \theta_{\text{start}}). \tag{9.13}$$

The maximum joint velocities are achieved at the halfway point of the motion, $t = T/2$:

$$\dot{\theta}_{\text{max}} = \frac{3}{2T} (\theta_{\text{end}} - \theta_{\text{start}}).$$

The maximum joint accelerations and decelerations are achieved at $t = 0$ and $t = T$:

$$\ddot{\theta}_{\text{max}} = \left| \frac{6}{T^2} (\theta_{\text{end}} - \theta_{\text{start}}) \right|, \qquad \ddot{\theta}_{\text{min}} = - \left| \frac{6}{T^2} (\theta_{\text{end}} - \theta_{\text{start}}) \right|.$$

If there are known limits on the maximum joint velocities $|\dot{\theta}| \leq \dot{\theta}_{\text{limit}}$ and maximum joint accelerations $|\ddot{\theta}| \leq \ddot{\theta}_{\text{limit}}$, these bounds can be checked to see whether the requested motion time $T$ is feasible. Alternatively, one could solve for $T$ to find the minimum possible motion time that satisfies the most restrictive velocity or acceleration constraint.

### Fifth-Order Polynomials

Because third-order time scaling does not constrain the endpoint path accelerations $\ddot{s}(0)$ and $\ddot{s}(T)$ to be zero, the robot is asked to achieve a discontinuous jump in acceleration at both $t = 0$ and $t = T$. This implies an infinite *jerk*, the derivative of acceleration, which may cause vibration of the robot.

One solution is to constrain the endpoint accelerations to $\ddot{s}(0) = \ddot{s}(T) = 0$. Adding these two constraints to the problem formulation requires the addition of two more design freedoms in the polynomial, yielding a quintic polynomial of time, $s(t) = a_0 + \cdots + a_5 t^5$. We can use the six terminal position, velocity, and acceleration constraints to solve uniquely for $a_0, \ldots, a_5$ (Exercise 9.5), which

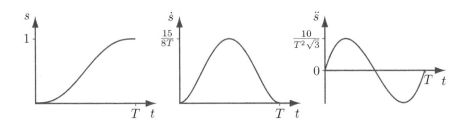

**Figure 9.4** Plots of $s(t)$, $\dot{s}(t)$, and $\ddot{s}(t)$ for a fifth-order polynomial time scaling.

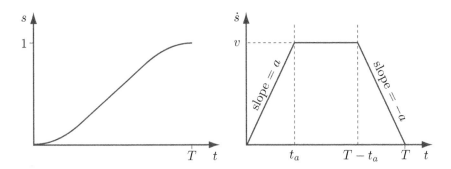

**Figure 9.5** Plots of $s(t)$ and $\dot{s}(t)$ for a trapezoidal motion profile.

yields a smoother motion with a higher maximum velocity than a cubic time scaling. A plot of the time scaling is shown in Figure 9.4.

### 9.2.2.2 Trapezoidal Motion Profiles

Trapezoidal time scalings are quite common in motor control, particularly for the motion of a single joint, and they get their name from their velocity profiles. The point-to-point motion consists of a constant acceleration phase $\ddot{s} = a$ of time $t_a$, followed by a constant velocity phase $\dot{s} = v$ of time $t_v = T - 2t_a$, followed by a constant deceleration phase $\ddot{s} = -a$ of time $t_a$. The resulting $\dot{s}$ profile is a trapezoid and the $s$ profile is the concatenation of a parabola, linear segment, and parabola as a function of time (Figure 9.5).

The trapezoidal time scaling is not as smooth as the cubic time scaling, but it has the advantage that if there are known constant limits on the joint velocities $\dot{\theta}_{\text{limit}} \in \mathbb{R}^n$ and on the joint accelerations $\ddot{\theta}_{\text{limit}} \in \mathbb{R}^n$ then the trapezoidal motion using the largest $v$ and $a$ satisfying

$$|(\theta_{\text{end}} - \theta_{\text{start}})v| \leq \dot{\theta}_{\text{limit}}, \tag{9.14}$$

$$|(\theta_{\text{end}} - \theta_{\text{start}})a| \leq \ddot{\theta}_{\text{limit}} \tag{9.15}$$

is the fastest straight-line motion possible. (See Exercise 9.8.)

If $v^2/a > 1$, the robot never reaches the velocity $v$ during the motion (Exercise 9.10). The three-phase accelerate–coast–decelerate motion becomes a two-phase accelerate–decelerate "bang-bang" motion, and the trapezoidal profile $\dot{s}(t)$ in Figure 9.5 becomes a triangle.

Assuming that $v^2/a \leq 1$, the trapezoidal motion is fully specified by $v$, $a$, $t_a$, and $T$, but only two of these can be specified independently since they must satisfy $s(T) = 1$ and $v = at_a$. It is unlikely that we would specify $t_a$ independently, so we can eliminate it from the equations of motion by the substitution $t_a = v/a$. The motion profile during the three stages (acceleration, coast, deceleration) can then be written in terms of $v$, $a$, and $T$ as follows:

$$\text{for } 0 \leq t \leq \frac{v}{a}, \qquad \ddot{s}(t) = a, \tag{9.16}$$

$$\dot{s}(t) = at, \tag{9.17}$$

$$s(t) = \frac{1}{2}at^2; \tag{9.18}$$

$$\text{for } \frac{v}{a} < t \leq T - \frac{v}{a}, \qquad \ddot{s}(t) = 0, \tag{9.19}$$

$$\dot{s}(t) = v, \tag{9.20}$$

$$s(t) = vt - \frac{v^2}{2a}; \tag{9.21}$$

$$\text{for } T - \frac{v}{a} < t \leq T, \qquad \ddot{s}(t) = -a, \tag{9.22}$$

$$\dot{s}(t) = a(T - t), \tag{9.23}$$

$$s(t) = \frac{2avT - 2v^2 - a^2(t-T)^2}{2a}. \tag{9.24}$$

Since only two of $v$, $a$, and $T$ can be chosen independently, we have three options:

- Choose $v$ and $a$ such that $v^2/a \leq 1$, ensuring a three-stage trapezoidal profile, and solve $s(T) = 1$ (using Equation (9.24)) for $T$:

$$T = \frac{a + v^2}{va}.$$

  If $v$ and $a$ correspond to the highest possible joint velocities and accelerations, this is the minimum possible time for the motion.

- Choose $v$ and $T$ such that $2 \geq vT > 1$, ensuring a three-stage trapezoidal profile and that the top speed $v$ is sufficient to reach $s = 1$ in time $T$, and solve $s(T) = 1$ for $a$:

$$a = \frac{v^2}{vT - 1}.$$

- Choose $a$ and $T$ such that $aT^2 \geq 4$, ensuring that the motion is completed in time, and solve $s(T) = 1$ for $v$:

$$v = \frac{1}{2}\left(aT - \sqrt{a}\sqrt{aT^2 - 4}\right).$$

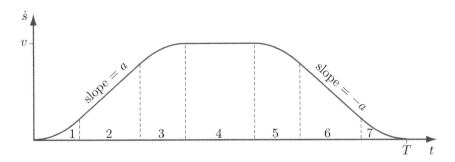

**Figure 9.6** Plot of $\dot{s}(t)$ for an S-curve motion profile consisting of seven stages: (1) constant positive jerk; (2) constant acceleration; (3) constant negative jerk; (4) constant velocity; (5) constant negative jerk; (6) constant deceleration; and (7) constant positive jerk.

### 2.2.3 S-Curve Time Scalings

Just as cubic polynomial time scalings lead to infinite jerk at the beginning and end of the motion, trapezoidal motions cause discontinuous jumps in acceleration at $t \in \{0, t_a, T - t_a, T\}$. A solution is a smoother **S-curve** time scaling, a popular motion profile in motor control because it avoids vibrations or oscillations induced by step changes in acceleration. An S-curve time scaling consists of seven stages: (1) constant jerk $d^3 s/dt^3 = J$ until a desired acceleration $\ddot{s} = a$ is achieved; (2) constant acceleration until the desired $\dot{s} = v$ is being approached; (3) constant negative jerk $-J$ until $\ddot{s}$ equals zero exactly at the time $\dot{s}$ reaches $v$; (4) coasting at constant $v$; (5) constant negative jerk $-J$; (6) constant deceleration $-a$; and (7) constant positive jerk $J$ until $\ddot{s}$ and $\dot{s}$ reach zero exactly at the time $s$ reaches 1.

The $\dot{s}(t)$ profile for an S-curve is shown in Figure 9.6.

Given some subset of $v$, $a$, $J$, and the total motion time $T$, algebraic manipulation reveals the switching time between stages and conditions that ensure that all seven stages are actually achieved, similarly to the case of the trapezoidal motion profile.

## .3 Polynomial Via Point Trajectories

If the goal is to have the robot joints pass through a series of **via points** at specified times, without a strict specification on the shape of path between consecutive points, a simple solution is to use polynomial interpolation to find joint histories $\theta(t)$ directly without first specifying a path $\theta(s)$ and then a time scaling $s(t)$ (Figure 9.7).

Let the trajectory be specified by $k$ via points, with the start point occurring at $T_1 = 0$ and the final point at $T_k = T$. Since each joint history is interpolated individually, we focus on a single joint variable and call it $\beta$ to avoid a prolifera-

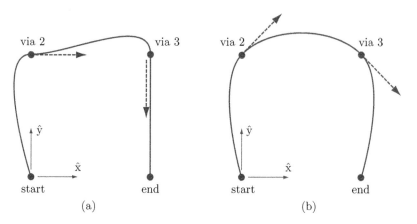

**Figure 9.7** Two paths in an $(x, y)$ space corresponding to piecewise-cubic trajectories interpolating four via points, including a start point and an end point. The velocities at the start and end are zero, and the velocities at vias 2 and 3 are indicated by the dashed tangent vectors. The shape of the path depends on the velocities specified at the via points.

tion of subscripts. At each via point $i \in \{1, \ldots, k\}$, the user specifies the desired position $\beta(T_i) = \beta_i$ and velocity $\dot{\beta}(T_i) = \dot{\beta}_i$. The trajectory has $k - 1$ segments, and the duration of segment $j \in \{1, \ldots, k - 1\}$ is $\Delta T_j = T_{j+1} - T_j$. The joint trajectory during segment $j$ is expressed as the third-order polynomial

$$\beta(T_j + \Delta t) = a_{j0} + a_{j1}\Delta t + a_{j2}\Delta t^2 + a_{j3}\Delta t^3 \qquad (9.25)$$

in terms of the time $\Delta t$ elapsed in segment $j$, where $0 \le \Delta t \le \Delta T_j$. Segment $j$ is subject to the four constraints

$$\beta(T_j) = \beta_j, \qquad\qquad \dot{\beta}(T_j) = \dot{\beta}_j,$$
$$\beta(T_j + \Delta T_j) = \beta_{j+1}, \qquad \dot{\beta}(T_j + \Delta T_j) = \dot{\beta}_{j+1}.$$

Solving these constraints for $a_{j0}, \ldots, a_{j3}$ yields

$$a_{j0} = \beta_j, \qquad\qquad\qquad\qquad\qquad (9.26)$$

$$a_{j1} = \dot{\beta}_j, \qquad\qquad\qquad\qquad\qquad (9.27)$$

$$a_{j2} = \frac{3\beta_{j+1} - 3\beta_j - 2\dot{\beta}_j\Delta T_j - \dot{\beta}_{j+1}\Delta T_j}{\Delta T_j^2}, \qquad (9.28)$$

$$a_{j3} = \frac{2\beta_j + (\dot{\beta}_j + \dot{\beta}_{j+1})\Delta T_j - 2\beta_{j+1}}{\Delta T_j^3}. \qquad (9.29)$$

Figure 9.8 shows the time histories for the interpolation of Figure 9.7(a). In this two-dimensional $(x, y)$ coordinate space the via points $1, \ldots, 4$ occur at times $T_1 = 0$, $T_2 = 1$, $T_3 = 2$, and $T_4 = 3$. The via points are at $(0, 0)$, $(0, 1)$, $(1, 1)$, and $(1, 0)$ with velocities $(0, 0)$, $(1, 0)$, $(0, -1)$, and $(0, 0)$.

Two issues are worth mentioning:

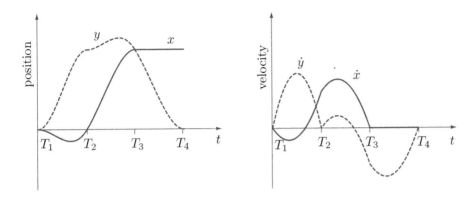

**Figure 9.8** The coordinate time histories for the cubic via-point interpolation of Figure 9.7(a).

- The quality of the interpolated trajectories is improved by "reasonable" combinations of via-point times and via-point velocities. For example, if the user wants to specify a via-point location and time, but not the velocity, a heuristic could be used to choose a via velocity on the basis of the times and coordinate vectors to the via points before and after the via in question. As an example, the trajectory of Figure 9.7(b) is smoother than the trajectory of Figure 9.7(a).
- Cubic via-point interpolation ensures that velocities are continuous at via points, but not accelerations. The approach is easily generalized to the use of fifth-order polynomials and specification of the accelerations at the via points, at the cost of increased complexity of the solution.

If only two points are used (the start and end point), and the velocities at each are zero, the resulting trajectory is identical to the straight-line cubic polynomial time-scaled trajectory discussed in Section 9.2.2.1.

There are many other methods for interpolating a set of via points. For example, B-spline interpolation is popular. In B-spline interpolation, the path may not pass exactly through the via points, but the path is guaranteed to be confined to the convex hull of the via points, unlike the paths in Figure 9.7. This can be important to ensure that joint limits or workspace obstacles are respected.

## 9.4 Time-Optimal Time Scaling

In the case where the path $\theta(s)$ is fully specified by the task or an obstacle-avoiding path planner (e.g., Figure 9.9), the trajectory planning problem reduces to finding a time scaling $s(t)$. One could choose the time scaling to minimize the energy consumed while meeting a time constraint, or to prevent spilling a glass of water that the robot is carrying. One of the most useful time scalings minimizes the time of motion along the path, subject to the robot's actuator limits. Such time-optimal trajectories maximize the robot's productivity.

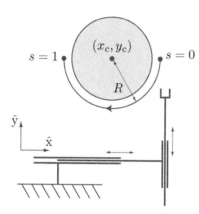

**Figure 9.9** A path planner has returned a semicircular path of radius $R$ around an obstacle in $(x, y)$ space for a robot with two prismatic joints. The path can be represented in terms of a path parameter $s$, as $x(s) = x_c + R \cos s\pi$ and $y(s) = y_c - R \sin s\pi$ for $s \in [0, 1]$. For a 2R robot, inverse kinematics would be used to express the path as a function of $s$ in joint coordinates.

While the trapezoidal time scalings of Section 9.2.2.2 can yield time-optimal trajectories, this is only under the assumption of straight-line motions, constant maximum acceleration $a$, and constant maximum coasting velocity $v$. For most robots, because of state-dependent joint actuator limits and the state-dependent dynamics

$$M(\theta)\ddot{\theta} + c(\theta, \dot{\theta}) + g(\theta) = \tau, \qquad (9.30)$$

the maximum available velocities and accelerations change along the path.

In this section we consider the problem of finding the fastest possible time scaling $s(t)$ that respects the robot's actuator limits. We write the limits on the $i$th actuator as

$$\tau_i^{\min}(\theta, \dot{\theta}) \leq \tau_i \leq \tau_i^{\max}(\theta, \dot{\theta}). \qquad (9.31)$$

The available actuator torque is typically a function of the current joint speed (see Section 8.9.1). For example, for a given maximum voltage of a DC motor, the maximum torque available from the motor drops linearly with the motor's speed.

Before proceeding we recall that the quadratic velocity terms $c(\theta, \dot{\theta})$ in Equation (9.30) can be written equivalently as

$$c(\theta, \dot{\theta}) = \dot{\theta}^{\mathrm{T}} \Gamma(\theta) \dot{\theta},$$

where $\Gamma(\theta)$ is the three-dimensional tensor of Christoffel symbols constructed from partial derivatives of components of the mass matrix $M(\theta)$ with respect to $\theta$. This form shows more clearly the quadratic dependence on velocities. Now, beginning with Equation (9.30), replacing $\dot{\theta}$ by $(d\theta/ds)\dot{s}$ and $\ddot{\theta}$ by $(d\theta/ds)\ddot{s} +$

$(d^2\theta/ds^2)\dot{s}^2$, and rearranging, we get

$$\underbrace{\left(M(\theta(s))\frac{d\theta}{ds}\right)}_{m(s)\in\mathbb{R}^n}\ddot{s} + \underbrace{\left(M(\theta(s))\frac{d^2\theta}{ds^2} + \left(\frac{d\theta}{ds}\right)^{\mathrm{T}}\Gamma(\theta(s))\frac{d\theta}{ds}\right)}_{c(s)\in\mathbb{R}^n}\dot{s}^2 + \underbrace{g(\theta(s))}_{g(s)\in\mathbb{R}^n} = \tau, \quad (9.32)$$

expressed more compactly as the vector equation

$$m(s)\ddot{s} + c(s)\dot{s}^2 + g(s) = \tau, \quad (9.33)$$

where $m(s)$ is the effective inertia of the robot when it is confined to the path $\theta(s)$, $c(s)\dot{s}^2$ comprises the quadratic velocity terms, and $g(s)$ is the gravitational torque.

Similarly, the actuation constraints (9.31) can be expressed as a function of $s$:

$$\tau_i^{\min}(s, \dot{s}) \le \tau_i \le \tau_i^{\max}(s, \dot{s}). \quad (9.34)$$

Substituting the $i$th component of Equation (9.33), we get

$$\tau_i^{\min}(s, \dot{s}) \le m_i(s)\ddot{s} + c_i(s)\dot{s}^2 + g_i(s) \le \tau_i^{\max}(s, \dot{s}). \quad (9.35)$$

Let $L_i(s, \dot{s})$ and $U_i(s, \dot{s})$ be the minimum and maximum accelerations $\ddot{s}$ satisfying the $i$th component of Equation (9.35). Depending on the sign of $m_i(s)$, we have three possibilities:

$$\left.\begin{array}{ll}
\text{if } m_i(s) > 0, & L_i(s, \dot{s}) = \dfrac{\tau_i^{\min}(s, \dot{s}) - c(s)\dot{s}^2 - g(s)}{m_i(s)}, \\[2ex]
& U_i(s, \dot{s}) = \dfrac{\tau_i^{\max}(s, \dot{s}) - c(s)\dot{s}^2 - g(s)}{m_i(s)} \\[2ex]
\text{if } m_i(s) < 0, & L_i(s, \dot{s}) = \dfrac{\tau_i^{\max}(s, \dot{s}) - c(s)\dot{s}^2 - g(s)}{m_i(s)} \\[2ex]
& U_i(s, \dot{s}) = \dfrac{\tau_i^{\min}(s, \dot{s}) - c(s)\dot{s}^2 - g(s)}{m_i(s)} \\[2ex]
\text{if } m_i(s) = 0, & \text{we have a \textit{zero-inertia point}, discussed in Section 9.4.4.}
\end{array}\right\} \quad (9.36)$$

Defining

$$L(s, \dot{s}) = \max_i L_i(s, \dot{s}) \quad \text{and} \quad U(s, \dot{s}) = \min_i U_i(s, \dot{s}),$$

the actuator limits (9.35) can be written as the state-dependent time-scaling constraints

$$L(s, \dot{s}) \le \ddot{s} \le U(s, \dot{s}). \quad (9.37)$$

The time-optimal time-scaling problem can now be stated:

*Given a path $\theta(s), s \in [0, 1]$, an initial state $(s_0, \dot{s}_0) = (0, 0)$, and a final state $(s_f, \dot{s}_f) = (1, 0)$, find a monotonically increasing twice-differentiable time scaling $s : [0, T] \to [0, 1]$ that*

(a) *satisfies $s(0) = \dot{s}(0) = \dot{s}(T) = 0$ and $s(T) = 1$, and*

(b) *minimizes the total travel time $T$ along the path while respecting the actuator constraints* (9.37).

The problem formulation is easily generalized to the case of nonzero initial and final velocities along the path, $\dot{s}(0) > 0$ and $\dot{s}(T) > 0$.

### 9.4.1     The $(s, \dot{s})$ Phase Plane

The problem is easily visualized in the $(s, \dot{s})$ phase plane of the path-constrained robot, with $s$ running from 0 to 1 on a horizontal axis and $\dot{s}$ on a vertical axis. Since $s(t)$ is monotonically increasing, $\dot{s}(t) \geq 0$ for all times $t$ and for all $s \in [0, 1]$. A time scaling of the path is any curve in the phase plane that moves monotonically to the right from $(0, 0)$ to $(1, 0)$ (Figure 9.10). Not all such curves satisfy the acceleration constraints (9.37), however.

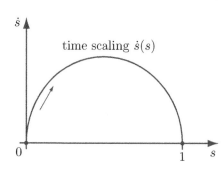

**Figure 9.10** A time scaling in the $(s, \dot{s})$ phase plane is a curve, with $\dot{s} \geq 0$ at all times, connecting the initial path position and velocity $(0, 0)$ to the final position and velocity $(1, 0)$.

To see the effect of the acceleration constraints, at each $(s, \dot{s})$ in the phase plane, we can plot the limits $L(s, \dot{s}) \leq \ddot{s} \leq U(s, \dot{s})$ as a cone constructed from $\dot{s}$, $L$, and $U$, as illustrated in Figure 9.11(a). If $L(s, \dot{s}) \geq U(s, \dot{s})$, the cone disappears – there are no actuator commands that can keep the robot on the path at this state. These **inadmissible** states are indicated in gray in Figure 9.11(a). For any $s$, typically there is a single limit velocity $\dot{s}_{\lim}(s)$ above which all velocities are inadmissible. The function $\dot{s}_{\lim}(s)$ is called the **velocity limit curve**. On the velocity limit curve, $L(s, \dot{s}) = U(s, \dot{s})$, and the cone reduces to a single vector.

For a time scaling to satisfy the acceleration constraints, the tangent of the time-scaling curve must lie inside the feasible cone at all points on the curve. Figure 9.11(b) shows an example of an infeasible time scaling, which demands more deceleration than the actuators can provide at the state indicated.

For a minimum-time motion, the "speed" $\dot{s}$ must be as high as possible at every $s$ while still satisfying the acceleration constraints and the endpoint constraints. To see this, write the total time of motion $T$ as

$$T = \int_0^T 1 \, dt. \tag{9.38}$$

Making the substitution $ds/ds = 1$, and changing the limits of integration from

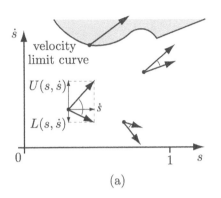
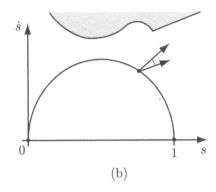

(a)                                              (b)

**Figure 9.11** (a) Acceleration-limited motion cones at four different states. The upper ray of the cone is the sum of $U(s, \dot{s})$ plotted in the vertical direction (the change in velocity) and $\dot{s}$ plotted in the horizontal direction (the change in position). The lower ray of the cone is constructed from $L(s, \dot{s})$ and $\dot{s}$. The points in gray, bounded by the velocity limit curve, have $L(s, \dot{s}) \geq U(s, \dot{s})$: the state is inadmissible and there is no motion cone. On the velocity limit curve the cone is reduced to a single tangent vector. (b) The proposed time scaling is infeasible because the tangent to the curve is outside the motion cone at the state indicated.

0 to $T$ (time) to 0 to 1 ($s$), we get

$$T = \int_0^T 1 \, dt = \int_0^T \frac{ds}{ds} \, dt = \int_0^1 \frac{dt}{ds} \, ds = \int_0^1 \dot{s}^{-1}(s) \, ds. \qquad (9.39)$$

Thus for time to be minimized, $\dot{s}^{-1}(s)$ should be as small as possible, and therefore $\dot{s}(s)$ must be as large as possible, at all $s$, while still satisfying the acceleration constraints (9.37) and the boundary constraints.

This implies that the time scaling must always operate either at the limit $U(s, \dot{s})$ or at the limit $L(s, \dot{s})$, and our only choice is when to switch between these limits. A common solution is a *bang-bang* trajectory: maximum acceleration $U(s, \dot{s})$ followed by a switch to maximum deceleration $L(s, \dot{s})$. (This is similar to the trapezoidal motion profile that never reaches the coasting velocity $v$ in Section 9.2.2.2.) In this case the time scaling is calculated by numerically integrating $U(s, \dot{s})$ forward in $s$ from $(0, 0)$, integrating $L(s, \dot{s})$ backward in $s$ from $(1, 0)$, and finding the intersection of these curves (Figure 9.12(a)). The switch between maximum acceleration and maximum deceleration occurs at the intersection.

In some cases, the velocity limit curve prevents a single-switch solution (Figure 9.12(b)). These cases require an algorithm to find multiple switching points.

### .4.2    The Time-Scaling Algorithm

Finding the optimal time scaling is reduced to finding the switches between maximum acceleration $U(s, \dot{s})$ and maximum deceleration $L(s, \dot{s})$, maximizing the "height" of the curve in the $(s, \dot{s})$ phase plane.

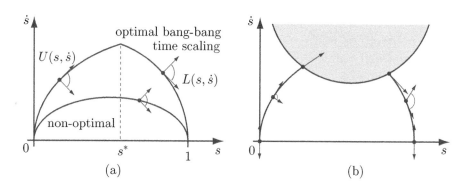

**Figure 9.12** (a) A time-optimal bang-bang time scaling integrates $U(s, \dot{s})$ from $(0, 0)$ and switches to $L(s, \dot{s})$ at a switching point $s^*$. Also shown is a non-optimal time scaling with a tangent inside a motion cone. (b) Sometimes the velocity limit curve prevents a single-switch solution.

### Time-scaling algorithm

1. Initialize an empty list of switches $\mathcal{S} = \{\}$ and a switch counter $i = 0$. Set $(s_i, \dot{s}_i) = (0, 0)$.
2. Integrate the equation $\ddot{s} = L(s, \dot{s})$ backward in time from $(1, 0)$ until $L(s, \dot{s}) > U(s, \dot{s})$ (the velocity limit curve is penetrated) or $s = 0$. Call this phase plane curve $F$.
3. Integrate the equation $\ddot{s} = U(s, \dot{s})$ forward in time from $(s_i, \dot{s}_i)$ until it crosses $F$ or until $U(s, \dot{s}) < L(s, \dot{s})$ (the velocity limit curve is penetrated). Call this curve $A_i$. If $A_i$ crosses $F$ then increment $i$, set $(s_i, \dot{s}_i)$ to the $(s, \dot{s})$ value at which the crossing occurs, and append $s_i$ to the list of switches $\mathcal{S}$. This is a switch from maximum acceleration to maximum deceleration. The problem is solved and $\mathcal{S}$ is the set of switches expressed in the path parameter. If instead the velocity limit curve is penetrated, let $(s_{\text{lim}}, \dot{s}_{\text{lim}})$ be the point of penetration and proceed to the next step.
4. Perform a binary search on the velocity in the range $[0, \dot{s}_{\text{lim}}]$ to find the velocity $\dot{s}'$ such that the curve integrating $\ddot{s} = L(s, \dot{s})$ forward from $(s_{\text{lim}}, \dot{s}')$ touches the velocity limit curve without penetrating it. The binary search is initiated with $\dot{s}_{\text{high}} = \dot{s}_{\text{lim}}$ and $\dot{s}_{\text{low}} = 0$.
   (a) Set the test velocity halfway between $\dot{s}_{\text{low}}$ and $\dot{s}_{\text{high}}$: $\dot{s}_{\text{test}} = (\dot{s}_{\text{high}} + \dot{s}_{\text{low}})/2$. The test point is $(s_{\text{lim}}, \dot{s}_{\text{test}})$.
   (b) If the curve from the test point penetrates the velocity limit curve, set $\dot{s}_{\text{high}}$ equal to $\dot{s}_{\text{test}}$. If instead the curve from the test point hits $\dot{s} = 0$, set $\dot{s}_{\text{low}}$ equal to $\dot{s}_{\text{test}}$. Return to Step 4(a).
   Continue the binary search until a specified tolerance. Let $(s_{\text{tan}}, \dot{s}_{\text{tan}})$ be the point where the resulting curve just touches the velocity limit curve tangentially (or comes closest to the curve without hitting it). The motion cone at

this point is reduced to a single vector $(L(s, \dot{s}) = U(s, \dot{s}))$, tangent to the velocity limit curve.

5. Integrate $\ddot{s} = L(s, \dot{s})$ backwards from $(s_{\text{tan}}, \dot{s}_{\text{tan}})$ until it intersects $A_i$. Increment $i$, set $(s_i, \dot{s}_i)$ to the $(s, \dot{s})$ value at the intersection, and label as $A_i$ the curve segment from $(s_i, \dot{s}_i)$ to $(s_{\text{tan}}, \dot{s}_{\text{tan}})$. Append $s_i$ to the list of switches $\mathcal{S}$. This is a switch from maximum acceleration to maximum deceleration.

6. Increment $i$ and set $(s_i, \dot{s}_i)$ to $(s_{\text{tan}}, \dot{s}_{\text{tan}})$. Append $s_i$ to the list of switches $\mathcal{S}$. This is a switch from maximum deceleration to maximum acceleration. Go to Step 3.

Figure 9.13 shows Steps 2–6 of the time-scaling algorithm. (Step 2) Integration of $\ddot{s} = L(s, \dot{s})$ backward from $(1, 0)$ until the velocity limit curve is reached. (Step 3) Integration of $\ddot{s} = U(s, \dot{s})$ forward from $(0, 0)$ to the intersection $(s_{\text{lim}}, \dot{s}_{\text{lim}})$ with the velocity limit curve. (Step 4) Binary search to find $(s_{\text{lim}}, \dot{s}')$ from which $\ddot{s} = L(s, \dot{s})$, integrated forward from $(s_{\text{lim}}, \dot{s}')$, touches the velocity limit curve tangentially. (Step 5) Integration backward along $L(s, \dot{s})$ from $(s_{\text{tan}}, \dot{s}_{\text{tan}})$ to find the first switch from acceleration to deceleration. (Step 6) The second switch, from deceleration to acceleration, is at $(s_2, \dot{s}_2) = (s_{\text{tan}}, \dot{s}_{\text{tan}})$. (Step 3) Integration forward along $U(s, \dot{s})$ from $(s_2, \dot{s}_2)$ results in intersection with $F$ at $(s_3, \dot{s}_3)$, where a switch occurs from acceleration to deceleration. The optimal time scaling consists of switches at $\mathcal{S} = \{s_1, s_2, s_3\}$.

## 4.3    A Variation on the Time-Scaling Algorithm

Remember that each point $(s, \dot{s})$ below the velocity limit curve has a cone of feasible motions, while each point on the velocity limit curve has a single feasible vector. The only points on the velocity limit curve that can be part of an optimal solution are those where the feasible motion vector is tangent to the velocity limit curve; these are the points $(s_{\text{tan}}, \dot{s}_{\text{tan}})$ referred to above. Recognizing this, the binary search in Step 4 of the time-scaling algorithm, which is essentially searching for a point $(s_{\text{tan}}, \dot{s}_{\text{tan}})$, can be replaced by explicit construction of the velocity limit curve and a search for points on this curve satisfying the tangency condition. See Figure 9.14.

## 4.4    Assumptions and Caveats

The description above covers the major points of the optimal time-scaling algorithm. A few assumptions were glossed over; they are made explicit now.

- **Static posture maintenance.** The algorithm, as described, assumes that the robot can maintain its configuration against gravity at any state $(s, \dot{s} = 0)$. This ensures the existence of valid time scalings, namely, time scalings that move the robot along the path arbitrarily slowly. For some robots and paths this assumption may be violated owing to weakness in the actuators.

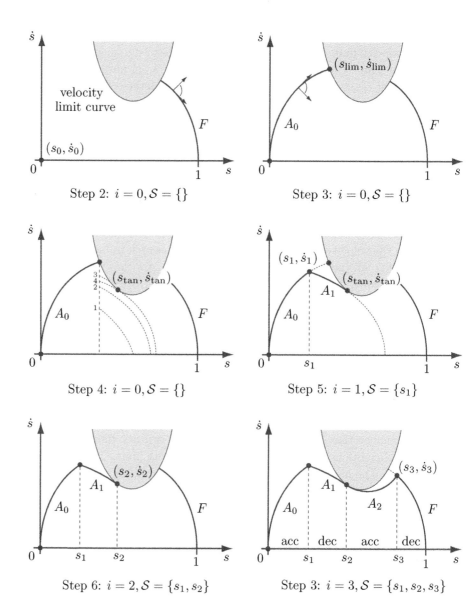

**Figure 9.13** The time-scaling algorithm.

**Figure 9.14** A point on the velocity limit curve can only be a part of a time-optimal time scaling if the feasible motion vector at that point is tangential to the curve. A search along the velocity limit curve illustrated reveals that there are only two points on this particular curve (marked by dots) that can belong to a time-optimal time scaling.

For example, some paths may require some momentum to carry the motion through configurations that the robot cannot maintain statically. The algorithm can be modified to handle such cases.

- **Inadmissible states.** The algorithm assumes that at every $s$ there is a unique velocity limit $\dot{s}_{\lim}(s) > 0$ such that all velocities $\dot{s} \leq \dot{s}_{\lim}(s)$ are admissible and all velocities $\dot{s} > \dot{s}_{\lim}(s)$ are inadmissible. For some models of actuator dynamics or friction this assumption may be violated – there may be isolated "islands" of inadmissible states. The algorithm can be modified to handle this case.
- **Zero-inertia points.** The algorithm assumes that there are no zero-inertia points (Equation (9.36)). If $m_i(s) = 0$ in (9.36) then the torque provided by actuator $i$ has no dependence on the acceleration $\ddot{s}$, and the $i$th actuator constraint in (9.35) directly defines a *velocity* constraint on $\dot{s}$. At a point $s$ with one or more zero components in $m(s)$, the velocity limit curve is defined by the minimum of (a) the velocity constraints defined by the zero-inertia components and (b) the $\dot{s}$ values satisfying $L_i(s, \dot{s}) = U_i(s, \dot{s})$ for the other components. For the algorithm as described, singular arcs of zero-inertia points on the velocity limit curve may lead to rapid switching between $\ddot{s} = U(s, \dot{s})$ and $\ddot{s} = L(s, \dot{s})$. In such cases, choosing an acceleration tangent to the velocity limit curve and lying between $U(s, \dot{s})$ and $L(s, \dot{s})$, preserves time optimality without causing chattering of the controls.

It is worth noting that the time-scaling algorithm generates trajectories with discontinuous acceleration, which could lead to vibrations. Beyond this, inaccuracies in models of robot inertial properties and friction make direct application of the time-scaling algorithm impractical. Finally, since a minimum-time time scaling always saturates at least one actuator, if the robot moves off the planned trajectory, there may be no torque left for corrective action by a feedback controller.

Despite these drawbacks, the time-scaling algorithm provides a deep understanding of the true maximum capabilities of a robot following a path.

## .5    Summary

- A trajectory $\theta(t)$, $\theta : [0, T] \to \Theta$, can be written as $\theta(s(t))$, i.e., as the composition of a path $\theta(s)$, $\theta : [0, 1] \to \Theta$, and a time scaling $s(t)$, $s : [0, T] \to [0, 1]$.
- A straight-line path in joint space can be written $\theta(s) = \theta_{\text{start}} + s(\theta_{\text{end}} - \theta_{\text{start}})$, $s \in [0, 1]$. A similar form holds for straight-line paths in a minimum set of task-space coordinates. A "straight-line" path in $SE(3)$, where $X = (R, p)$, can be decoupled to a Cartesian path and a rotation path:

$$p(s) = p_{\text{start}} + s(p_{\text{end}} - p_{\text{start}}), \tag{9.40}$$
$$R(s) = R_{\text{start}} \exp(\log(R_{\text{start}}^{\text{T}} R_{\text{end}})s). \tag{9.41}$$

- A cubic polynomial $s(t) = a_0 + a_1 t + a_2 t^2 + a_3 t^3$ can be used to time scale a point-to-point motion with zero initial and final velocities. The acceleration undergoes a step change (infinite jerk) at $t = 0$ and $t = T$. A Such an impulse in jerk can cause vibration of the robot.
- A quintic polynomial $s(t) = a_0 + a_1 t + a_2 t^2 + a_3 t^3 + a_4 t^4 + a_5 t^5$ can be used to time-scale a point-to-point motion with zero initial and final velocities and accelerations. The jerk is finite at all times.
- The trapezoidal motion profile is a popular time scaling in point-to-point control, particularly the control of a single motor. The motion consists of three phases: constant acceleration, constant velocity, and constant deceleration, resulting in a trapezoid in $\dot{s}(t)$. Trapezoidal motion involves step changes in acceleration.
- The S-curve motion profile is also popular in point-to-point control of a motor. It consists of seven phases: (1) constant positive jerk; (2) constant acceleration; (3) constant negative jerk; (4) constant velocity; (5) constant negative jerk; (6) constant deceleration; and (7) constant positive jerk.
- Given a set of via points including a start state, a goal state, and other via states through which the robot's motion must pass, as well as the times $T_i$ at which these states should be reached, a series of cubic-polynomial time scalings can be used to generate a trajectory $\theta(t)$ interpolating the via points. To prevent step changes in acceleration at the via points, a series of quintic polynomials can be used instead.
- Given a robot path $\theta(s)$, the dynamics of the robot, and limits on the actuator torques, the actuator constraints can be expressed in terms of $(s, \dot{s})$ as the vector inequalities

$$L(s, \dot{s}) \leq \ddot{s} \leq U(s, \dot{s}).$$

The time-optimal time scaling $s(t)$ is such that the "height" of the curve in the $(s, \dot{s})$ phase plane is maximized while satisfying $s(0) = \dot{s}(0) = \dot{s}(T) = 0$, $s(T) = 1$, and the actuator constraints. The optimal solution always operates at maximum acceleration $U(s, \dot{s})$ or maximum deceleration $L(s, \dot{s})$.

## 9.6     Software

Software functions associated with this chapter are listed below.

`s = CubicTimeScaling(Tf,t)`
Computes $s(t)$ for a cubic time scaling, given $t$ and the total time of motion $T_f$.

`s = QuinticTimeScaling(Tf,t)`
Computes $s(t)$ for a quintic time scaling, given $t$ and the total time of motion $T_f$.

`traj = JointTrajectory(thetastart,thetaend,Tf,N,method)`

Computes a straight-line trajectory in joint space as an $N \times n$ matrix, where each of the $N$ rows is an $n$-vector of the joint variables at an instant in time. The first row is $\theta_{\text{start}}$ and the $N$th row is $\theta_{\text{end}}$. The elapsed time between each row is $T_f/(N-1)$. The parameter `method` equals either 3 for a cubic time scaling or 5 for a quintic time scaling.

`traj = ScrewTrajectory(Xstart,Xend,Tf,N,method)`
Computes a trajectory as a list of $N$ $SE(3)$ matrices, where each matrix represents the configuration of the end-effector at an instant in time. The first matrix is $X_{\text{start}}$, the $N$th matrix is $X_{\text{end}}$, and the motion is along a constant screw axis. The elapsed time between each matrix is $T_f/(N-1)$. The parameter `method` equals either 3 for a cubic time scaling or 5 for a quintic time scaling.

`traj = CartesianTrajectory(Xstart,Xend,Tf,N,method)`
Computes a trajectory as a list of $N$ $SE(3)$ matrices, where each matrix represents the configuration of the end-effector at an instant in time. The first matrix is $X_{\text{start}}$, the $N$th matrix is $X_{\text{end}}$, and the origin of the end-effector frame follows a straight line, decoupled from the rotation. The elapsed time between each matrix is $T_f/(N-1)$. The parameter `method` equals either 3 for a cubic time scaling or 5 for a quintic time scaling.

## .7    Notes and References

Bobrow et al. (1985) and Shin and McKay (1985) published papers nearly simultaneously that independently derived the essence of the time-optimal time-scaling algorithm outlined in Section 9.4. A year earlier, Hollerbach had addressed the restricted problem of finding dynamically feasible time-scaled trajectories for uniform time scalings where the time variable $t$ is replaced by $ct$ for $c > 0$ (Hollerbach, 1984).

The original papers of Bobrow et al. and Shin and McKay were followed by a number of papers refining the methods by addressing zero-inertia points, singularities, algorithm efficiency, and even the presence of constraints and obstacles (Pfeiffer and Johanni, 1987; Slotine and Yang, 1989; Shiller and Dubowsky, 1985, 1988, 1989, 1991; Shiller and Lu, 1992; Pham, 2014; Pham and Stasse, 2015). In particular, a computationally efficient method for finding the points $(s_{\text{tan}}, \dot{s}_{\text{tan}})$, where the optimal time scaling touches the velocity limit curve, is described in Pfeiffer and Johanni (1987) and in Slotine and Yang (1989). This algorithm is used to improve the computational efficiency of the time-scaling algorithm; see, for example, the description and supporting open-source code in Pham (2014) and Pham and Stasse (2015). In this chapter the binary search approach in Step 4 of the time-scaling algorithm follows the presentation in Bobrow et al. (1985) because of its conceptual simplicity.

Other research has focused on numerical methods such as dynamic programming or nonlinear optimization to minimize cost functions such as the actuator energy. One early example of work in this area is by Vukobratović and Kirćanski (1982).

## 9.8    Exercises

**Exercise 9.1**   Consider an elliptical path in the $(x, y)$-plane. The path starts at $(0, 0)$ and proceeds clockwise to $(2, 1)$, $(4, 0)$, $(2, -1)$, and back to $(0, 0)$ (Figure 9.15). Write the path as a function of $s \in [0, 1]$.

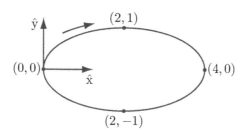

**Figure 9.15** An elliptical path.

**Exercise 9.2**   A cylindrical path in $X = (x, y, z)$ is given by $x = \cos 2\pi s$, $y = \sin 2\pi s$, $z = 2s$, $s \in [0, 1]$, and its time scaling is $s(t) = \frac{1}{4}t + \frac{1}{8}t^2, t \in [0, 2]$. Write down $\dot{X}$ and $\ddot{X}$.

**Exercise 9.3**   Consider a path from $X(0) = X_{\text{start}} \in SE(3)$ to $X(1) = X_{\text{end}} \in SE(3)$ consisting of motion along a constant screw axis. The path is time scaled by some $s(t)$. Write down the twist $\mathcal{V}$ and acceleration $\dot{\mathcal{V}}$ at any point on the path given $\dot{s}$ and $\ddot{s}$.

**Exercise 9.4**   Consider a straight-line path $\theta(s) = \theta_{\text{start}} + s(\theta_{\text{end}} - \theta_{\text{start}}), s \in [0, 1]$ from $\theta_{\text{start}} = (0, 0)$ to $\theta_{\text{end}} = (\pi, \pi/3)$. The motion starts and ends at rest. The feasible joint velocities are $|\dot{\theta}_1|, |\dot{\theta}_2| \leq 2$ rad/s and the feasible joint accelerations are $|\ddot{\theta}_1|, |\ddot{\theta}_2| \leq 0.5$ rad/s$^2$. Find the fastest motion time $T$ using a cubic time scaling that satisfies the joint velocity and acceleration limits.

**Exercise 9.5**   Find the fifth-order polynomial time scaling that satisfies $s(T) = 1$ and $s(0) = \dot{s}(0) = \ddot{s}(0) = \dot{s}(T) = \ddot{s}(T) = 0$.

**Exercise 9.6**   As a function of the total time of motion $T$, find the times at which the acceleration $\ddot{s}$ of the fifth-order polynomial point-to-point time scaling is a maximum or a minimum.

**Exercise 9.7**   If you want to use a polynomial time scaling for point-to-point motion with zero initial and final velocities, accelerations, and jerks, what would be the minimum order of the polynomial?

**Exercise 9.8**   Prove that the trapezoidal time scaling, using the maximum allowable acceleration $a$ and velocity $v$, minimizes the time of motion $T$.

**Exercise 9.9**   Plot by hand the acceleration profile $\ddot{s}(t)$ for a trapezoidal time scaling.

**Exercise 9.10**   If $v$ and $a$ are specified for a trapezoidal time scaling of a robot, prove that $v^2/a \leq 1$ is a necessary condition for the robot to reach the maximum velocity $v$ during the path.

**Exercise 9.11**   If $v$ and $T$ are specified for a trapezoidal time scaling, prove that $vT > 1$ is a necessary condition for the motion to be able to complete in time $T$. Prove that $vT \leq 2$ is a necessary condition for a three-stage trapezoidal motion.

**Exercise 9.12**   If $a$ and $T$ are specified for a trapezoidal time scaling, prove that $aT^2 \geq 4$ is a necessary condition to ensure that the motion completes in time.

**Exercise 9.13**   Consider the case where the maximum velocity $v$ is never reached in a trapezoidal time scaling. The motion becomes a bang-bang motion: constant acceleration $a$ for time $T/2$ followed by constant deceleration $-a$ for time $T/2$. Write down the position $s(t)$, velocity $\dot{s}(t)$, and acceleration $\ddot{s}(t)$ for both phases, in analogy to Equations (9.16)–(9.24).

**Exercise 9.14**   Plot by hand the acceleration profile $\ddot{s}(t)$ for an S-curve time scaling.

**Exercise 9.15**   A seven-stage S-curve is fully specified by the time $t_J$ (the duration of a constant positive or negative jerk), the time $t_a$ (the duration of constant positive or negative acceleration), the time $t_v$ (the duration of constant velocity), the total time $T$, the jerk $J$, the acceleration $a$, and the velocity $v$. Of these seven quantities, how many can be specified independently?

**Exercise 9.16**   A nominal S-curve has seven stages, but it can have fewer if certain inequality constraints are not satisfied. Indicate which cases are possible with fewer than seven stages. Sketch by hand the $\dot{s}(t)$ velocity profiles for these cases.

**Exercise 9.17**   If the S-curve achieves all seven stages and uses a jerk $J$, an acceleration $a$, and a velocity $v$, what is the constant-velocity coasting time $t_v$ in terms of $v$, $a$, $J$, and the total motion time $T$?

**Exercise 9.18**   Write your own via-point cubic-polynomial interpolation trajectory generator program for a two-dof robot. A new position and velocity specification is required for each joint at 1000 Hz. The user specifies a sequence of via-point positions, velocities, and times, and the program generates an array consisting of the joint angles and velocities at every millisecond from time $t = 0$ to time $t = T$, the total duration of the movement. For a test case with at least three via points (one at the start and and one at the end, both with zero velocity, and at least one more via point), plot

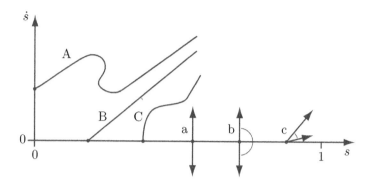

**Figure 9.16** A, B, and C are candidate integral curves, originating from the dots indicated, while a, b, and c are candidate motion cones at $\dot{s} = 0$. Two of the integral curves and two of the motion cones are incorrect.

(a) the path in the joint angle space (similar to Figure 9.7), and
(b) the position and velocity of each joint as a function of time (these plots should look similar to Figure 9.8).

**Exercise 9.19**  Via points with specified positions, velocities, and accelerations can be interpolated using fifth-order polynomials of time. For a fifth-order polynomial segment between via points $j$ and $j + 1$, of duration $\Delta T_j$, with $\beta_j$, $\beta_{j+1}$, $\dot{\beta}_j$, $\dot{\beta}_{j+1}$, $\ddot{\beta}_j$, and $\ddot{\beta}_{j+1}$ specified, solve for the coefficients of the fifth-order polynomial (which is similar to Equations (9.26)–(9.29)). A symbolic math solver will simplify the problem.

**Exercise 9.20**  By hand or by computer, plot a trapezoidal motion profile in the $(s, \dot{s})$-plane.

**Exercise 9.21**  Figure 9.16 shows three candidate motion curves in the $(s, \dot{s})$-plane (A, B, and C) and three candidate motion cones at $\dot{s} = 0$ (a, b, and c). Two of the three curves and two of the three motion cones cannot be correct for any robot dynamics. Indicate which are incorrect and explain your reasoning. Explain why the remaining curve and motion cone are possibilities.

**Exercise 9.22**  Under the assumptions of Section 9.4.4, explain why the time-scaling algorithm of Section 9.4.2 (see Figure 9.13) is correct. In particular,

(a) explain why, in the binary search of Step 4, the curve integrated forward from $(s_{\text{lim}}, \dot{s}_{\text{test}})$ must either hit (or run tangent to) the velocity limit curve or hit the $\dot{s} = 0$-axis (and does not hit the curve $F$, for example);
(b) explain why the final time scaling can only touch the velocity limit curve tangentially; and
(c) explain why the acceleration switches from minimum to maximum at points where the time scaling touches the velocity limit curve.

**Exercise 9.23** Explain how the time-scaling algorithm should be modified to handle the case where the initial and final velocities, at $s = 0$ and $s = 1$, are nonzero.

**Exercise 9.24** Explain how the time-scaling algorithm should be modified if the robot's actuators are too weak to hold it statically at some configurations of the path (the static-posture-maintenance assumption is violated), but the assumptions on inadmissible states and zero-inertia points are satisfied. Valid time scalings may no longer exist. Under what condition(s) should the algorithm terminate and indicate that no valid time scaling exists? (Under the assumptions of Section 9.4.4 the original algorithm always finds a solution and therefore does not check for failure cases.) What do the motion cones look like at states $(s, \dot{s} = 0)$ where the robot cannot hold itself statically?

**Exercise 9.25** Create a computer program that plots the motion cones in the $(s, \dot{s})$-plane for a 2R robot in a horizontal plane. The path is a straight line in joint space from $(\theta_1, \theta_2) = (0, 0)$ to $(\pi/2, \pi/2)$. Use the dynamics from Equation (8.9) (with $g = 0$), then rewrite the dynamics in terms of $s, \dot{s}, \ddot{s}$ instead of $\theta, \dot{\theta}, \ddot{\theta}$. The actuators can provide torques in the range $-\tau_{i,\text{limit}} - b\dot{\theta}_i \leq \tau_i \leq \tau_{i,\text{limit}} - b\dot{\theta}_i$, where $b > 0$ indicates the velocity dependence of the torque. The cones should be drawn at a grid of points in $(s, \dot{s})$. To keep the figure manageable, normalize each cone ray to the same length.

**Exercise 9.26** We have been assuming forward motion on a path, $\dot{s} > 0$. What if we allowed backward motion on a path, $\dot{s} < 0$? This exercise involves drawing motion cones and an integral curve in the $(s, \dot{s})$-plane, including both positive and negative values of $\dot{s}$. Assume that the maximum acceleration is $U(s, \dot{s}) = U > 0$ (constant over the $(s, \dot{s})$-plane) and the maximum deceleration is $L(s, \dot{s}) = L = -U$. You can assume, for example, that $U = 1$ and $L = -1$.

(a) For any constant $s$, draw the motion cones at the five points where $\dot{s}$ takes the values $\{-2, -1, 0, 1, 2\}$.

(b) Assume the motion starts at $(s, \dot{s}) = (0, 0)$ and follows the maximum acceleration $U$ for time $t$. Then it follows the maximum deceleration $L$ for time $2t$. Then it follows $U$ for time $t$. Sketch by hand the integral curve. (The exact shape does not matter, but the curve should have the correct features.)

# 10 Motion Planning

Motion planning is the problem of finding a robot motion from a start state to a goal state that avoids obstacles in the environment and satisfies other constraints, such as joint limits or torque limits. Motion planning is one of the most active subfields of robotics, and it is the subject of entire books. The purpose of this chapter is to provide a practical overview of a few common techniques, using robot arms and mobile robots as the primary example systems (Figure 10.1).

The chapter begins with a brief overview of motion planning. This is followed by foundational material including configuration space obstacles and graph search. We conclude with summaries of several different planning methods.

## 10.1 Overview of Motion Planning

A key concept in motion planning is configuration space, or **C-space** for short. Every point in the C-space $\mathcal{C}$ corresponds to a unique configuration $q$ of the robot, and every configuration of the robot can be represented as a point in C-space. For example, the configuration of a robot arm with $n$ joints can be represented as a list of $n$ joint positions, $q = (\theta_1, \ldots, \theta_n)$. The **free C-space** $\mathcal{C}_{\text{free}}$ consists of the configurations where the robot neither penetrates an obstacle nor violates a joint limit.

In this chapter, unless otherwise stated, we assume that $q$ is an $n$-vector and

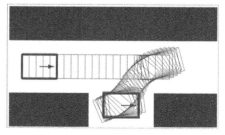

**Figure 10.1** (Left) A robot arm executing an obstacle-avoiding motion plan. The motion plan was generated using Movelt! (Şucan and Chitta, 2016) and visualized using rviz in ROS (the Robot Operating System). (Right) A car-like mobile robot executing parallel parking.

that $\mathcal{C} \subset \mathbb{R}^n$. With some generalization, the concepts of this chapter apply to non-Euclidean C-spaces such as $\mathcal{C} = SE(3)$.

The control inputs available to drive the robot are written as an $m$-vector $u \in \mathcal{U} \subset \mathbb{R}^m$, where $m = n$ for a typical robot arm. If the robot has second-order dynamics, such as that for a robot arm, and the control inputs are forces (equivalently, accelerations), the *state* of the robot is defined by its configuration and velocity, $x = (q, v) \in \mathcal{X}$. For $q \in \mathbb{R}^n$, typically we write $v = \dot{q}$. If we can treat the control inputs as velocities, the state $x$ is simply the configuration $q$. The notation $q(x)$ indicates the configuration $q$ corresponding to the state $x$, and $\mathcal{X}_{\text{free}} = \{x \mid q(x) \in \mathcal{C}_{\text{free}}\}$.

The equations of motion of the robot are written

$$\dot{x} = f(x, u) \tag{10.1}$$

or, in integral form,

$$x(T) = x(0) + \int_0^T f(x(t), u(t))dt. \tag{10.2}$$

### 0.1.1   Types of Motion Planning Problems

With the definitions above, a fairly broad specification of the motion planning problem is the following:

*Given an initial state $x(0) = x_{\text{start}}$ and a desired final state $x_{\text{goal}}$, find a time $T$ and a set of controls $u : [0, T] \to \mathcal{U}$ such that the motion (10.2) satisfies $x(T) = x_{\text{goal}}$ and $q(x(t)) \in \mathcal{C}_{\text{free}}$ for all $t \in [0, T]$.*

It is assumed that a feedback controller (Chapter 11) is available to ensure that the planned motion $x(t)$, $t \in [0, T]$, is followed closely. It is also assumed that an accurate geometric model of the robot and environment is available to evaluate $\mathcal{C}_{\text{free}}$ during motion planning.

There are many variations of the basic problem; some are discussed below.

**Path planning versus motion planning.** The path planning problem is a subproblem of the general motion planning problem. Path planning is the purely geometric problem of finding a collision-free path $q(s), s \in [0, 1]$, from a start configuration $q(0) = q_{\text{start}}$ to a goal configuration $q(1) = q_{\text{goal}}$, without concern for the dynamics, the duration of motion, or constraints on the motion or on the control inputs. It is assumed that the path returned by the path planner can be time scaled to create a feasible trajectory (Chapter 9). This problem is sometimes called the **piano mover's problem**, emphasizing the focus on the geometry of cluttered spaces.

**Control inputs: $m = n$ versus $m < n$.** If there are fewer control inputs $m$ than degrees of freedom $n$, then the robot is incapable of following many paths, even if they are collision-free. For example, a car has $n = 3$ (the position and orientation of the chassis in the plane) but $m = 2$ (forward–backward motion and steering); it cannot slide directly sideways into a parking space.

**Online versus offline.** A motion planning problem requiring an immediate result, perhaps because obstacles appear, disappear, or move unpredictably, calls for a fast, online, planner. If the environment is static then a slower offline planner may suffice.

**Optimal versus satisficing.** In addition to reaching the goal state, we might want the motion plan to minimize (or approximately minimize) a cost $J$, e.g.,

$$J = \int_0^T L(x(t), u(t)) dt.$$

For example, minimizing with $L = 1$ yields a time-optimal motion while minimizing with $L = u^T(t)u(t)$ yields a "minimum-effort" motion.

**Exact versus approximate.** We may be satisfied with a final state $x(T)$ that is sufficiently close to $x_{\text{goal}}$, e.g., $\|x(T) - x_{\text{goal}}\| < \epsilon$.

**With or without obstacles.** The motion planning problem can be challenging even in the absence of obstacles, particularly if $m < n$ or optimality is desired.

### 10.1.2   Properties of Motion Planners

Planners must conform to the properties of the motion planning problem as outlined above. In addition, planners can be distinguished by the following properties.

**Multiple-query versus single-query planning.** If the robot is being asked to solve a number of motion planning problems in an unchanging environment, it may be worth spending the time building a data structure that accurately represents $\mathcal{C}_{\text{free}}$. This data structure can then be searched to solve multiple planning queries efficiently. Single-query planners solve each new problem from scratch.

**"Anytime" planning.** An anytime planner is one that continues to look for better solutions after a first solution is found. The planner can be stopped at any time, for example when a specified time limit has passed, and the best solution returned.

**Completeness.** A motion planner is said to be **complete** if it is guaranteed to find a solution in finite time if one exists, and to report failure if there is no feasible motion plan. A weaker concept is **resolution completeness**. A planner is resolution complete if it is guaranteed to find a solution if one exists at the resolution of a discretized representation of the problem, such as the resolution of a grid representation of $\mathcal{C}_{\text{free}}$. Finally, a planner is **probabilistically complete** if the probability of finding a solution, if one exists, tends to 1 as the planning time goes to infinity.

**Computational complexity.** The computational complexity refers to charac-
terizations of the amount of time the planner takes to run or the amount
of memory it requires. These are measured in terms of the description
of the planning problem, such as the dimension of the C-space or the
number of vertices in the representation of the robot and obstacles. For
example, the time for a planner to run may be exponential in $n$, the
dimension of the C-space. The computational complexity may be expressed in terms of the average case or the worst case. Some planning
algorithms lend themselves easily to computational complexity analysis,
while others do not.

## 10.1.3   Motion Planning Methods

There is no single planner applicable to all motion planning problems. Below is
a broad overview of some of the many motion planners available. Details are left
to the sections indicated.

**Complete methods (Section 10.3).** These methods focus on exact representations of the geometry or topology of $\mathcal{C}_{\text{free}}$, ensuring completeness. For
all but simple or low-degree-of-freedom problems, these representations
are mathematically or computationally prohibitive to derive.
**Grid methods (Section 10.4).** These methods discretize $\mathcal{C}_{\text{free}}$ into a grid and
search the grid for a motion from $q_{\text{start}}$ to a grid point in the goal region.
Modifications of the approach may discretize the state space or control
space or they may use multi-scale grids to refine the representation of
$\mathcal{C}_{\text{free}}$ near obstacles. These methods are relatively easy to implement
and can return optimal solutions but, for a fixed resolution, the memory
required to store the grid, and the time to search it, grow exponentially
with the number of dimensions of the space. This limits the approach to
low-dimensional problems.
**Sampling methods (Section 10.5).** A generic sampling method relies on a
random or deterministic function to choose a sample from the C-space
or state space; a function to evaluate whether the sample is in $\mathcal{X}_{\text{free}}$; a
function to determine the "closest" previous free-space sample; and a
local planner to try to connect to, or move toward, the new sample from
the previous sample. This process builds up a graph or tree representing
feasible motions of the robot. Sampling methods are easy to implement,
tend to be probabilistically complete, and can even solve high-degree-of-
freedom motion planning problems. The solutions tend to be satisficing,
not optimal, and it can be difficult to characterize the computational
complexity.
**Virtual potential fields (Section 10.6).** Virtual potential fields create forces
on the robot that pull it toward the goal and push it away from obstacles.

The approach is relatively easy to implement, even for high-degree-of-freedom systems, and fast to evaluate, often allowing online implementation. The drawback is local minima in the potential function: the robot may get stuck in configurations where the attractive and repulsive forces cancel but the robot is not at the goal state.

**Nonlinear optimization (Section 10.7).** The motion planning problem can be converted to a nonlinear optimization problem by representing the path or controls by a finite number of design parameters, such as the coefficients of a polynomial or a Fourier series. The problem is to solve for the design parameters that minimize a cost function while satisfying constraints on the controls, obstacles, and goal. While these methods can produce near-optimal solutions, they require an initial guess at the solution. Because the objective function and feasible solution space are generally not convex, the optimization process can get stuck far away from a feasible solution, let alone an optimal solution.

**Smoothing (Section 10.8).** Often the motions found by a planner are jerky. A smoothing algorithm can be run on the result of the motion planner to improve the smoothness.

A major trend in recent years has been toward sampling methods, which are easy to implement and can handle high-dimensional problems.

## 10.2     Foundations

Before discussing motion planning algorithms, we establish the concepts used in many of them: configuration space obstacles, collision detection, graphs, and graph search.

### 10.2.1     Configuration Space Obstacles (C-Obstacles)

Determining whether a robot at a configuration $q$ is in collision with a known environment generally requires a complex operation involving a CAD model of the environment and robot. There are a number of free and commercial software packages that can perform this operation, and we will not delve into them here. For our purposes, it is enough to know that the workspace obstacles partition the configuration space $\mathcal{C}$ into two sets, the **free space** $\mathcal{C}_{\text{free}}$ and the **obstacle space** $\mathcal{C}_{\text{obs}}$, where $\mathcal{C} = \mathcal{C}_{\text{free}} \cup \mathcal{C}_{\text{obs}}$. Joint limits are treated as obstacles in the configuration space.

With the concepts of $\mathcal{C}_{\text{free}}$ and $\mathcal{C}_{\text{obs}}$, the path planning problem reduces to the problem of finding a path for a point robot among the obstacles $\mathcal{C}_{\text{obs}}$. If the obstacles break $\mathcal{C}_{\text{free}}$ into separate **connected components**, and $q_{\text{start}}$ and $q_{\text{goal}}$ do not lie in the same connected component, then there is no collision-free path.

The explicit mathematical representation of a C-obstacle can be exceedingly

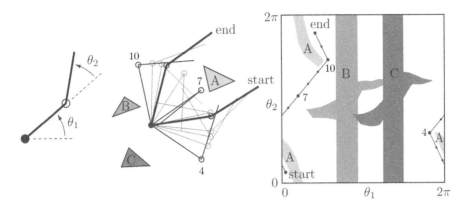

**Figure 10.2** (Left) The joint angles of a 2R robot arm. (Middle) The arm navigating among obstacles A, B, and C. (Right) The same motion in C-space. Three intermediate points, 4, 7, and 10, along the path are labeled.

complex, and for that reason C-obstacles are rarely represented exactly. Despite this, the concept of C-obstacles is very important for understanding motion planning algorithms. The ideas are best illustrated by examples.

### 10.2.1.1 A 2R Planar Arm

Figure 10.2 shows a 2R planar robot arm, with configuration $q = (\theta_1, \theta_2)$, among obstacles A, B, and C in the workspace. The C-space of the robot is represented by a portion of the plane with $0 \leq \theta_1 < 2\pi$, $0 \leq \theta_2 < 2\pi$. Remember from Chapter 2, however, that the topology of the C-space is a torus (or doughnut) since the edge of the square at $\theta_1 = 2\pi$ is connected to the edge $\theta_1 = 0$; similarly, $\theta_2 = 2\pi$ is connected to $\theta_2 = 0$. The square region of $\mathbb{R}^2$ is obtained by slicing the surface of the doughnut twice, at $\theta_1 = 0$ and $\theta_2 = 0$, and laying it flat on the plane.

The C-space on the right in Figure 10.2 shows the workspace obstacles A, B, and C represented as C-obstacles. Any configuration lying inside a C-obstacle corresponds to penetration of the obstacle by the robot arm in the workspace. A free path for the robot arm from one configuration to another is shown in both the workspace and C-space. The path and obstacles illustrate the topology of the C-space. Note that the obstacles break $\mathcal{C}_{\text{free}}$ into three connected components.

### 10.2.1.2 A Circular Planar Mobile Robot

Figure 10.3 shows a top view of a circular mobile robot whose configuration is given by the location of its center, $(x, y) \in \mathbb{R}^2$. The robot translates (moves without rotating) in a plane with a single obstacle. The corresponding C-obstacle is obtained by "growing" (enlarging) the workspace obstacle by the radius of the mobile robot. Any point outside this C-obstacle represents a free configuration of the robot. Figure 10.4 shows the workspace and C-space for two obstacles,

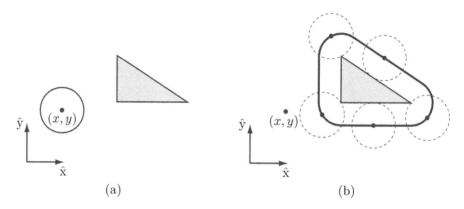

(a)                    (b)

**Figure 10.3** (a) A circular mobile robot (open circle) and a workspace obstacle (gray triangle). The configuration of the robot is represented by $(x, y)$, the center of the robot. (b) In the C-space, the obstacle is "grown" by the radius of the robot and the robot is treated as a point. Any $(x, y)$ configuration outside the bold line is collision-free.

indicating that in this case the mobile robot cannot pass between the two obstacles.

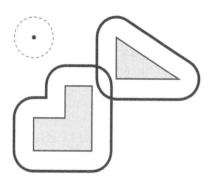

**Figure 10.4** The "grown" C-space obstacles corresponding to two workspace obstacles and a circular mobile robot. The overlapping boundaries mean that the robot cannot move between the two obstacles.

### 10.2.1.3   A Polygonal Planar Mobile Robot That Translates

Figure 10.5 shows the C-obstacle for a polygonal mobile robot translating in the presence of a polygonal obstacle. The C-obstacle is obtained by sliding the robot along the boundary of the obstacle and tracing the position of the robot's reference point.

### 10.2.1.4   A Polygonal Planar Mobile Robot that Translates and Rotates

Figure 10.6 illustrates the C-obstacle for the workspace obstacle and triangular mobile robot of Figure 10.5 if the robot is now allowed to rotate. The C-space is now three dimensional, given by $(x, y, \theta) \in \mathbb{R}^2 \times S^1$. The three-dimensional C-obstacle is the union of two-dimensional C-obstacle slices at angles $\theta \in [0, 2\pi)$. Even for this relatively low-dimensional C-space, an exact representation of the

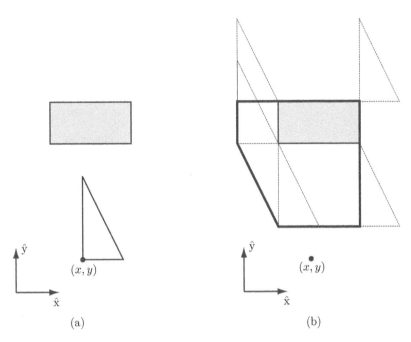

**Figure 10.5** (a) The configuration of a triangular mobile robot, which can translate but not rotate, is represented by the $(x, y)$ location of a reference point. Also shown is a workspace obstacle in gray. (b) The corresponding C-space obstacle (bold outline) is obtained by sliding the robot around the boundary of the obstacle and tracing the position of the reference point.

C-obstacle is quite complex. For this reason, C-obstacles are rarely described exactly.

## 10.2.2  Distance to Obstacles and Collision Detection

Given a C-obstacle $\mathcal{B}$ and a configuration $q$, let $d(q, \mathcal{B})$ be the distance between the robot and the obstacle, where

$$d(q, \mathcal{B}) > 0 \quad \text{(no contact with the obstacle)},$$
$$d(q, \mathcal{B}) = 0 \quad \text{(contact)},$$
$$d(q, \mathcal{B}) < 0 \quad \text{(penetration)}.$$

The distance could be defined as the Euclidean distance between the two closest points of the robot and the obstacle, respectively.

A **distance-measurement algorithm** is one that determines $d(q, \mathcal{B})$. A **collision–detection routine** determines whether $d(q, \mathcal{B}_i) \leq 0$ for any C-obstacle $\mathcal{B}_i$. A collision-detection routine returns a binary result and may or may not utilize a distance-measurement algorithm at its core.

One popular distance-measurement algorithm is the Gilbert–Johnson–Keerthi

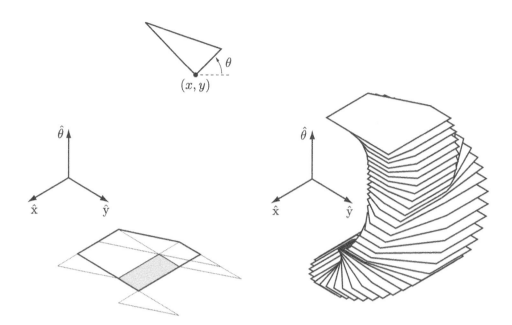

**Figure 10.6** (Top) A triangular mobile robot that can both rotate and translate, represented by the configuration $(x, y, \theta)$. (Left) The C-space obstacle from Figure 10.5(b) when the robot is restricted to $\theta = 0$. (Right) The full three-dimensional C-space obstacle shown in slices at $10°$ increments.

(GJK) algorithm, which efficiently computes the distance between two convex bodies, possibly represented by triangular meshes. Any robot or obstacle can be treated as the union of multiple convex bodies. Extensions of this algorithm are used in many distance-measurement algorithms and collision-detection routines for robotics, graphics, and game-physics engines.

A simpler approach is to approximate the robot and obstacles as unions of overlapping spheres. Approximations must always be **conservative** – the approximation must cover all points of the object – so that if a collision-detection routine indicates a free configuration $q$, then we are guaranteed that the actual geometry is collision-free. As the number of spheres in the representation of the robot and obstacles increases, the closer the approximations come to the actual geometry. An example is shown in Figure 10.7.

Given a robot at $q$ represented by $k$ spheres of radius $R_i$ centered at $r_i(q)$, $i = 1, \ldots, k$, and an obstacle $\mathcal{B}$ represented by $\ell$ spheres of radius $B_j$ centered at $b_j, j = 1, \ldots, \ell$, the distance between the robot and the obstacle can be calculated as

$$d(q, \mathcal{B}) = \min_{i,j} \|r_i(q) - b_j\| - R_i - B_j.$$

Apart from determining whether a particular configuration of the robot is in collision, another useful operation is determining whether the robot collides

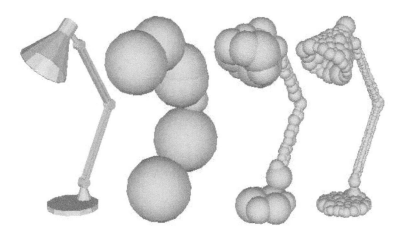

**Figure 10.7** A lamp represented by spheres. The approximation improves as the number of spheres used to represent the lamp increases. Figure from Hubbard (1996) used with permission.

during a particular motion segment. While exact solutions have been developed for particular object geometries and motion types, the general approach is to sample the path at finely spaced points and to "grow" the robot to ensure that if two consecutive configurations are collision-free for the grown robot then the volume swept out by the actual robot between the two configurations is also collision-free.

### 0.2.3   Graphs and Trees

Many motion planners explicitly or implicitly represent the C-space or state space as a **graph**. A graph consists of a collection of nodes $\mathcal{N}$ and a collection of edges $\mathcal{E}$, where each edge $e$ connects two nodes. In motion planning, a node typically represents a configuration or state while an edge between nodes $n_1$ and $n_2$ indicates the ability to move from $n_1$ to $n_2$ without penetrating an obstacle or violating other constraints.

A graph can be either **directed** or **undirected**. In an undirected graph, each edge is bidirectional: if the robot can travel from $n_1$ to $n_2$ then it can also travel from $n_2$ to $n_1$. In a directed graph, or **digraph** for short, each edge allows travel in only one direction. The same two nodes can have two edges between them, allowing travel in opposite directions.

Graphs can also be **weighted** or **unweighted**. In a weighted graph, each edge has a positive cost associated with traversing it. In an unweighted graph each edge has the same cost (e.g., 1). Thus the most general type of graph we consider is a weighted digraph.

A **tree** is a digraph in which (1) there are no cycles and (2) each node has at

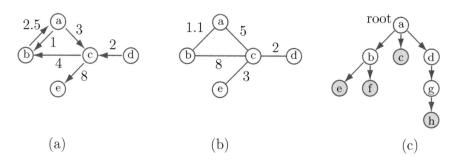

**Figure 10.8** (a) A weighted digraph. (b) A weighted undirected graph. (c) A tree. The leaves are shaded gray.

most one **parent** node (i.e., at most one edge leading to the node). A tree has one **root** node with no parents and a number of **leaf** nodes with no **child**.

A digraph, undirected graph, and tree are illustrated in Figure 10.8.

Given $N$ nodes, any graph can be represented by a matrix $A \in \mathbb{R}^{N \times N}$, where element $a_{ij}$ of the matrix represents the cost of the edge from node $i$ to node $j$; a zero or negative value indicates no edge between the nodes. Graphs and trees can be represented more compactly as a list of nodes, each with links to its neighbors.

### 10.2.4    Graph Search

Once the free space is represented as a graph, a motion plan can be found by searching the graph for a path from the start to the goal. One of the most powerful and popular graph search algorithms is $A^*$ (pronounced "A star") search.

#### 10.2.4.1    $A^*$ Search

The $A^*$ search algorithm efficiently finds a minimum-cost path on a graph when the cost of the path is simply the sum of the positive edge costs along the path.

Given a graph described by a set of nodes $\mathcal{N} = \{1, \ldots, N\}$, where node 1 is the start node, and a set of edges $\mathcal{E}$, the $A^*$ algorithm makes use of the following data structures:

- a sorted list OPEN of the nodes from which exploration is still to be done, and a list CLOSED of nodes for which this exploration has already taken place;
- a matrix cost[node1,node2] encoding the set of edges, where a positive value corresponds to the cost of moving from node1 to node2 (a negative value indicates that no edge exists);
- an array past_cost[node] of the minimum cost found so far to reach node node from the start node; and
- a search tree defined by an array parent[node], which contains for each node a link to the node preceding it in the shortest path found so far from the start node to that node.

To initialize the search, the matrix `cost` is constructed to encode the edges, the list `OPEN` is initialized to the start node 1, the cost to reach the start node (`past_cost[1]`) is initialized to 0, and `past_cost[node]` for node $\in \{2, \ldots, N\}$ is initialized to infinity (or a large number), indicating that currently we have no idea of the cost of reaching those nodes.

At each step of the algorithm, the first node in `OPEN` is removed from `OPEN` and called `current`. The node `current` is also added to `CLOSED`. The first node in `OPEN` is one that minimizes the total estimated cost of the best path to the goal that passes through that node. The estimated cost is calculated as

```
est_total_cost[node] = past_cost[node]
                     + heuristic_cost_to_go(node)
```

where `heuristic_cost_to_go(node)` $\geq 0$ is an optimistic (underestimating) estimate of the actual cost-to-go to the goal from `node`. For many path planning problems, an appropriate choice for the heuristic is the straight-line distance to the goal, ignoring any obstacles.

Because `OPEN` is a list sorted according to the estimated total cost, inserting a new node at the correct location in `OPEN` entails a small computational price.

If the node `current` is in the goal set then the search is finished and the path is reconstructed from the `parent` links. If not, for each neighbor `nbr` of `current` in the graph which is not also in `CLOSED`, the `tentative_past_cost` for `nbr` is calculated as `past_cost[current] + cost[current,nbr]`. If

$$\text{tentative\_past\_cost} < \text{past\_cost[nbr]},$$

then `nbr` can be reached with less cost than was previously thought, so `past_cost[nbr]` is set to `tentative_past_cost` and `parent[nbr]` is set to `current`. The node `nbr` is then added (or moved) in `OPEN` according to its estimated total cost.

The algorithm then returns to the beginning of the main loop, removing the first node from `OPEN` and calling it `current`. If `OPEN` is empty then there is no solution.

The $A^*$ algorithm is guaranteed to return a minimum-cost path, as nodes are only checked for inclusion in the goal set when they have the minimum total estimated cost of all nodes. If the node `current` is in the goal set then `heuristic_cost_to_go(current)` is zero and, since all edge costs are positive, we know that any path found in the future must have a cost greater than or equal to `past_cost[current]`. Therefore the path to `current` must be a shortest path. (There may be other paths of the same cost.)

If the heuristic "cost-to-go" is calculated exactly, considering obstacles, then $A^*$ will explore from the minimum number of nodes necessary to solve the problem. Of course, calculating the cost-to-go exactly is equivalent to solving the path planning problem, so this is impractical. Instead, the heuristic cost-to-go should be calculated quickly and should be as close as possible to the actual cost-to-go to ensure that the algorithm runs efficiently. Using an optimistic cost-to-go ensures an optimal solution.

The algorithm $A^*$ is an example of the general class of **best-first** searches, which always explore from the node currently deemed "best" by some measure.

The $A^*$ search algorithm is described in pseudocode in Algorithm 10.1.

---

**Algorithm 10.1** $A^*$ search.

---

1: OPEN $\leftarrow \{1\}$
2: past_cost[1] $\leftarrow 0$, past_cost[node] $\leftarrow$ infinity for node $\in \{2, \ldots, N\}$
3: **while** OPEN is not empty **do**
4:     current $\leftarrow$ first node in OPEN, remove from OPEN
5:     add current to CLOSED
6:     **if** current is in the goal set **then**
7:         **return** SUCCESS and the path to current
8:     **end if**
9:     **for** each nbr of current not in CLOSED **do**
10:         tentative_past_cost $\leftarrow$ past_cost[current]+cost[current,nbr]
11:         **if** tentative_past_cost < past_cost[nbr] **then**
12:             past_cost[nbr] $\leftarrow$ tentative_past_cost
13:             parent[nbr] $\leftarrow$ current
14:             put (or move) nbr in sorted list OPEN according to
                    est_total_cost[nbr] $\leftarrow$ past_cost[nbr] +
                        heuristic_cost_to_go(nbr)
15:         **end if**
16:     **end for**
17: **end while**
18: **return** FAILURE

---

### 10.2.4.2    Other Search Methods

- **Dijkstra's method.** If the heuristic cost-to-go is always estimated as zero then $A^*$ always explores from the OPEN node that has been reached with minimum past cost. This variant is called Dijkstra's algorithm, which preceded $A^*$ historically. Dijkstra's algorithm is also guaranteed to find a minimum-cost path but on many problems it runs more slowly than $A^*$ owing to the lack of a heuristic look-ahead function to help guide the search.

- **Breadth-first search.** If each edge in $\mathcal{E}$ has the same cost, Dijkstra's algorithm reduces to breadth-first search. All nodes one edge away from the start node are considered first, then all nodes two edges away, etc. The first solution found is therefore a minimum-cost path.

- **Suboptimal $A^*$ search.** If the heuristic cost-to-go is overestimated by multiplying the optimistic heuristic by a constant factor $\eta > 1$, the $A^*$ search will be biased to explore from nodes closer to the goal rather than nodes with a low past cost. This may cause a solution to be found more quickly but, unlike the case of an optimistic cost-to-go heuristic, the solution will not be guaranteed to be optimal. One possibility is to run $A^*$ with an inflated cost-to-go to find an initial solution, then rerun the search with progressively smaller values of $\eta$ until the time allotted for the search has expired or a solution is found with $\eta = 1$.

## 0.3    Complete Path Planners

Complete path planners rely on an exact representation of the free C-space $\mathcal{C}_{\text{free}}$. These techniques tend to be mathematically and algorithmically sophisticated, and impractical for many real systems, so we do not delve into them in detail.

One approach to complete path planning, which we will see in modified form in Section 10.5, is based on representing the complex high-dimensional space $\mathcal{C}_{\text{free}}$ by a one-dimensional **roadmap** $R$ with the following properties:

(a) **Reachability.** From every point $q \in \mathcal{C}_{\text{free}}$, a free path to a point $q' \in R$ can be found trivially (e.g., a straight-line path).
(b) **Connectivity.** For each connected component of $\mathcal{C}_{\text{free}}$, there is one connected component of $R$.

With such a roadmap, the planner can find a path between any two points $q_{\text{start}}$ and $q_{\text{goal}}$ in the same connected component of $\mathcal{C}_{\text{free}}$ by simply finding paths from $q_{\text{start}}$ to a point $q'_{\text{start}} \in R$, from a point $q'_{\text{goal}} \in R$ to $q_{\text{goal}}$, and from $q'_{\text{start}}$ to $q'_{\text{goal}}$ on the roadmap $R$. If a path can be found trivially between $q_{\text{start}}$ and $q_{\text{goal}}$, the roadmap may not even be used.

While constructing a roadmap of $\mathcal{C}_{\text{free}}$ is complex in general, some problems admit simple roadmaps. For example, consider a polygonal robot translating among polygonal obstacles in the plane. As can be seen in Figure 10.5, the C-obstacles in this case are also polygons. A suitable roadmap is the weighted undirected **visibility graph**, with nodes at the vertices of the C-obstacles and edges between the nodes that can "see" each other (i.e., the line segment between the vertices does not intersect an obstacle). The weight associated with each edge is the Euclidean distance between the nodes.

Not only is this a suitable roadmap $R$, but it allows us to use an $A^*$ search to find a shortest path between any two configurations in the same connected component of $\mathcal{C}_{\text{free}}$, as the shortest path is guaranteed either to be a straight line from $q_{\text{start}}$ to $q_{\text{goal}}$ or to consist of a straight line from $q_{\text{start}}$ to a node $q'_{\text{start}} \in R$, a straight line from a node $q'_{\text{goal}} \in R$ to $q_{\text{goal}}$, and a path along the straight edges of $R$ from $q'_{\text{start}}$ to $q'_{\text{goal}}$ (Figure 10.9). Note that the shortest path requires the robot to graze the obstacles, so we implicitly treat $\mathcal{C}_{\text{free}}$ as including its boundary.

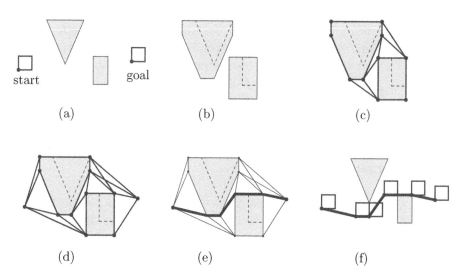

**Figure 10.9** (a) The start and goal configurations for a square mobile robot (reference point shown) in an environment with a triangular and a rectangular obstacle. (b) The grown C-obstacles. (c) The visibility graph roadmap $R$ of $\mathcal{C}_{\text{free}}$. (d) The full graph consists of $R$ plus nodes at $q_{\text{start}}$ and $q_{\text{goal}}$, along with the links connecting these nodes to the visible nodes of $R$. (e) Searching the graph results in the shortest path, shown in bold. (f) The robot is shown traversing the path.

## 10.4    Grid Methods

A search algorithm like $A^*$ requires a discretization of the search space. The simplest discretization of C-space is a grid. For example, if the configuration space is $n$-dimensional and we desire $k$ grid points along each dimension, the C-space is represented by $k^n$ grid points.

The $A^*$ algorithm can be used as a path planner for a C-space grid, with the following minor modifications:

- The definition of a "neighbor" of a grid point must be chosen: is the robot constrained to move in axis-aligned directions in configuration space or can it move in multiple dimensions simultaneously? For example, for a two-dimensional C-space, neighbors could be 4-connected (on the cardinal points of a compass: north, south, east, and west) or 8-connected (diagonals allowed), as shown in Figure 10.10(a). If diagonal motions are allowed, the cost to diagonal neighbors should be penalized appropriately. For example, the cost to a north, south, east or west neighbor could be 1, while the cost to a diagonal neighbor could be $\sqrt{2}$. If integers are desired, for efficiency of the implementation, the approximate costs 5 and 7 could be used.

- If only axis-aligned motions are used, the heuristic cost-to-go should be based on the **Manhattan distance**, not the Euclidean distance. The Manhattan

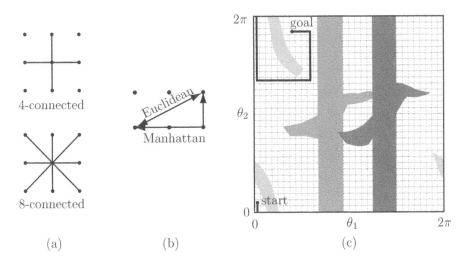

**Figure 10.10** (a) A 4-connected grid point and an 8-connected grid point for a space $n = 2$. (b) Grid points spaced at unit intervals. The Euclidean distance between the two points indicated is $\sqrt{5}$ while the Manhattan distance is 3. (c) A grid representation of the C-space and a minimum-length Manhattan-distance path for the problem of Figure 10.2.

distance counts the number of "city blocks" that must be traveled, with the rule that diagonals through a block are not possible (Figure 10.10(b)).
- A node `nbr` is added to `OPEN` only if the step from `current` to `nbr` is collision-free. (The step may be considered collision-free if a grown version of the robot at `nbr` does not intersect any obstacles.)
- Other optimizations are possible, owing to the known regular structure of the grid.

An $A^*$ grid-based path planner is resolution-complete: it will find a solution if one exists at the level of discretization of the C-space. The path will be a shortest path subject to the allowed motions.

Figure 10.10(c) illustrates grid-based path planning for the 2R robot example of Figure 10.2. The C-space is represented as a grid with $k = 32$, i.e., there is a resolution of $360°/32 = 11.25°$ for each joint. This yields a total of $32^2 = 1024$ grid points.

The grid-based planner, as described, is a single-query planner: it solves each path planning query from scratch. However, if the same $q_{\text{goal}}$ will be used in the same environment for multiple path planning queries, it may be worth pre-processing the entire grid to enable fast path planning. This is the **wavefront** planner, illustrated in Figure 10.11.

Although grid-based path planning is easy to implement, it is only appropriate for low-dimensional C-spaces. The reason is that the number of grid points, and hence the computational complexity of the path planner, increases exponentially with the number of dimensions $n$. For instance, a resolution $k = 100$ in a C-space

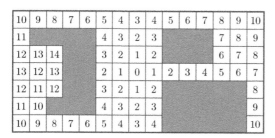

**Figure 10.11** A wavefront planner on a two-dimensional grid. The goal configuration is given a score of 0. Then all collision-free 4-neighbors are given a score of 1. The process continues, breadth-first, with each free neighbor (that does not have a score already) assigned the score of its parent plus 1. Once every grid cell in the connected component of the goal configuration is assigned a score, planning from any location in the connected component is trivial: at every step, the robot simply moves "downhill" to a neighbor with a lower score. Grid points in collision receive a high score.

with $n = 3$ dimensions leads to $k^n = 1$ million grid nodes, while $n = 5$ leads to 10 billion grid nodes and $n = 7$ leads to 100 trillion nodes. An alternative is to reduce the resolution $k$ along each dimension, but this leads to a coarse representation of C-space that may miss free paths.

### 10.4.1    Multi-Resolution Grid Representation

One way to reduce the computational complexity of a grid-based planner is to use a multi-resolution grid representation of $\mathcal{C}_{\text{free}}$. Conceptually, a grid point is considered an obstacle if any part of the rectilinear cell centered on the grid point touches a C-obstacle. To refine the representation of the obstacle, an obstacle cell can be subdivided into smaller cells. Each dimension of the original cell is split in half, resulting in $2^n$ subcells for an $n$-dimensional space. Any cells that are still in contact with a C-obstacle are then subdivided further, up to a specified maximum resolution.

The advantage of this representation is that only the portions of C-space near obstacles are refined to high resolution, while those away from obstacles are represented by a coarse resolution. This allows the planner to find paths using short steps through cluttered spaces while taking large steps through wide open space. The idea is illustrated in Figure 10.12, which uses only 10 cells to represent an obstacle at the same resolution as a fixed grid that uses 64 cells.

For $n = 2$, this multi-resolution representation is called a **quadtree**, as each obstacle cell subdivides into $2^n = 4$ cells. For $n = 3$, each obstacle cell subdivides into $2^n = 8$ cells, and the representation is called an **octree**.

The multi-resolution representation of $\mathcal{C}_{\text{free}}$ can be built in advance of the search or incrementally as the search is being performed. In the latter case, if the step from `current` to `nbr` is found to be in collision, the step size can be halved until the step is free or the minimum step size is reached.

original cell     subdivision 1     subdivision 2     subdivision 3

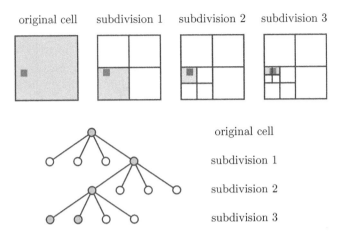

original cell

subdivision 1

subdivision 2

subdivision 3

**Figure 10.12** At the original C-space cell resolution, a small obstacle (indicated by the dark square) causes the whole cell to be labeled an obstacle. Subdividing the cell once shows that at least three quarters of the cell is actually free. Three levels of subdivision results in a representation using ten total cells: four at subdivision level 3, three at subdivision level 2, and three at subdivision level 1. The cells shaded light gray are the obstacle cells in the final representation. The subdivision of the original cell is shown in the lower panel as a tree, specifically a quadtree, where the leaves of the tree are the final cells in the representation.

## 0.4.2   Grid Methods with Motion Constraints

The above grid-based planners operate under the assumption that the robot can go from one cell to any neighboring cell in a regular C-space grid. This may not be possible for some robots. For example, a car cannot reach, in one step, a "neighbor" cell that is to the side of it. Also, motions for a fast-moving robot arm should be planned in the state space, not just C-space, to take the arm dynamics into account. In the state space, the robot arm cannot move in certain directions (Figure 10.13).

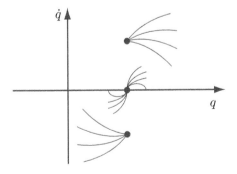

**Figure 10.13** Sample trajectories emanating from three initial states in the phase space of a dynamic system with $q \in \mathbb{R}$. If the initial state has $\dot{q} > 0$, the trajectory cannot move to the left (corresponding to negative motion in $q$) instantaneously. Similarly, if the initial state has $\dot{q} < 0$, the trajectory cannot move to the right instantaneously.

Grid-based planners must be adapted to account for the motion constraints of the particular robot. In particular, the constraints may result in a directed

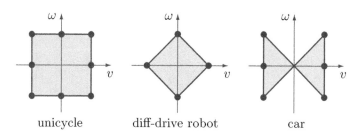

**Figure 10.14** Discretizations of the control sets for unicycle, diff-drive, and car-like robots.

graph. One approach is to discretize the robot controls while still making use of a grid on the C-space or state space, as appropriate. Details for a wheeled mobile robot and a dynamic robot arm are described next.

### 10.4.2.1  Grid-Based Path Planning for a Wheeled Mobile Robot

As described in Section 13.3, the controls for simplified models of unicycle, diff-drive, and car-like robots are $(v, \omega)$, i.e., the forward–backward linear velocity and the angular velocity. The control sets for these mobile robots are shown in Figure 10.14. Also shown are proposed discretizations of the controls, as dots. Other discretizations could be chosen.

Using the control discretization, we can use a variant of Dijkstra's algorithm to find short paths (Algorithm 10.2).

The search expands from $q_{\text{start}}$ by integrating forward each control for a time $\Delta t$, creating new nodes for the paths that are collision-free. Each node keeps track of the control used to reach the node as well as the cost of the path to the node. The cost of the path to a new node is the sum of the cost of the previous node, **current**, plus the cost of the action.

Integration of the controls does not move the mobile robot to exact grid points. Instead, the C-space grid comes into play in lines 9 and 10. When a node is expanded, the grid cell it sits in is marked "occupied." Subsequently, any node in this occupied cell will be pruned from the search. This prevents the search from expanding nodes that are close by nodes reached with a lower cost.

No more than **MAXCOUNT** nodes, where **MAXCOUNT** is a value chosen by the user, are considered during the search.

The time $\Delta t$ should be chosen so that each motion step is "small." The size of the grid cells should be chosen as large as possible while ensuring that integration of any control for a time $\Delta t$ will move the mobile robot outside its current grid cell.

The planner terminates when **current** lies inside the goal region, or when there are no more nodes left to expand (perhaps because of obstacles), or when **MAXCOUNT** nodes have been considered. Any path found is optimal for the choice of cost function and other parameters to the problem. The planner actually

---

**Algorithm 10.2** Grid-based Dijkstra planner for a wheeled mobile robot.

---

1: OPEN ← {$q_{\text{start}}$}
2: past_cost [$q_{\text{start}}$] ← 0
3: counter ← 1
4: **while** OPEN is not empty and counter < MAXCOUNT **do**
5:    current ← first node in OPEN, remove from OPEN
6:    **if** current is in the goal set **then**
7:       **return** SUCCESS and the path to current
8:    **end if**
9:    **if** current is not in a previously occupied C-space grid cell **then**
10:       mark grid cell occupied
11:       counter ← counter + 1
12:       **for** each control in the discrete control set **do**
13:          integrate control forward a short time $\Delta t$ from current to $q_{\text{new}}$
14:          **if** the path to $q_{\text{new}}$ is collision-free **then**
15:             compute cost of the path to $q_{\text{new}}$
16:             place $q_{\text{new}}$ in OPEN, sorted by cost
17:             parent [$q_{\text{new}}$] ← current
18:          **end if**
19:       **end for**
20:    **end if**
21: **end while**
22: **return** FAILURE

---

runs faster in somewhat cluttered spaces, as the obstacles help to guide the exploration.

Some examples of motion plans for a car are shown in Figure 10.15.

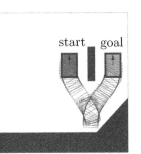 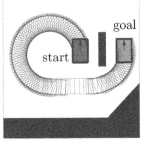

**Figure 10.15** (Left) A minimum-cost path for a car-like robot where each action has identical cost, favoring a short path. (Right) A minimum-cost path where reversals are penalized. Penalizing reversals requires a modification to Algorithm 10.2.

### 0.4.2.2 Grid-Based Motion Planning for a Robot Arm

One method for planning the motion for a robot arm is to decouple the problem into a path planning problem followed by a time scaling of the path:

(a) Apply a grid-based or other path planner to find an obstacle-free path in C-space.

(b) Time scale the path to find the fastest trajectory that respects the robot's dynamics, as described in Section 9.4, or use any less aggressive time scaling.

Since the motion planning problem is broken into two steps (path planning followed by time scaling), the resultant motion will not be time-optimal in general.

Another approach is to plan directly in the state space. Given a state $(q, \dot{q})$ of the robot arm, let $\mathcal{A}(q, \dot{q})$ represent the set of accelerations that are feasible on the basis of the limited joint torques. To discretize the controls, the set $\mathcal{A}(q, \dot{q})$ is intersected with a grid of points of the form

$$\sum_{i=1}^{n} ca_i \hat{e}_i,$$

where $c$ is any integer, $a_i > 0$ is the acceleration step size in the $\ddot{q}_i$-direction, and $\hat{e}_i$ is a unit vector in the $i$th direction (Figure 10.16).

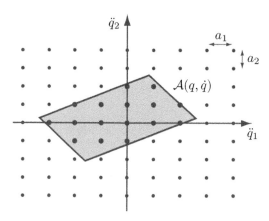

Figure 10.16 The instantaneously available acceleration set $\mathcal{A}(q, \dot{q})$ for a two-joint robot, intersected with a grid spaced at $a_1$ in $\ddot{q}_1$ and $a_2$ in $\ddot{q}_2$, gives the discretized control actions (shown as larger dots).

As the robot moves, the acceleration set $\mathcal{A}(q, \dot{q})$ changes but the grid remains fixed. Because of this, and assuming a fixed integration time $\Delta t$ at each "step" in a motion plan, the reachable states of the robot (after any integral number of steps) are confined to a grid in state space. To see this, consider a single joint angle of the robot, $q_1$, and assume for simplicity zero initial velocity, $\dot{q}_1(0) = 0$. The velocity at timestep $k$ takes the form

$$\dot{q}_1(k) = \dot{q}_1(k-1) + c(k)a_1\Delta t,$$

where $c(k)$ is any value in a finite set of integers. By induction, the velocity at any timestep must be of the form $a_1 k_v \Delta t$, where $k_v$ is an integer. The position at timestep $k$ takes the form

$$q_1(k) = q_1(k-1) + \dot{q}_1(k-1)\Delta t + \frac{1}{2}c(k)a_1(\Delta t)^2.$$

Substituting the velocity from the previous equation, we find that the position at any timestep must be of the form $a_1 k_p (\Delta t)^2/2 + q_1(0)$, where $k_p$ is an integer.

To find a trajectory from a start node to a goal set, a breadth-first search can be employed to create a search tree on the state space nodes. When exploration is made from a node $(q, \dot{q})$ in the state space, the set $\mathcal{A}(q, \dot{q})$ is evaluated to find the discrete set of control actions. New nodes are created by integrating the control actions for time $\Delta t$. A node is discarded if the path to it is in collision or if it has been reached previously (i.e., by a trajectory taking the same or less time).

Because the joint angles and angular velocities are bounded, the state space grid is finite and therefore it can be searched in finite time. The planner is resolution-complete and returns a time-optimal trajectory, subject to the resolution specified in the control grid and timestep $\Delta t$.

The control-grid step sizes $a_i$ must be chosen small enough that $\mathcal{A}(q, \dot{q})$, for any feasible state $(q, \dot{q})$, contains a representative set of points of the control grid. Choosing a finer grid for the controls, or a smaller timestep $\Delta t$, creates a finer grid in the state space and a higher likelihood of finding a solution amidst obstacles. It also allows the choice of a smaller goal set while keeping points of the state space grid inside the set.

Finer discretization comes at a computational cost. If the resolution of the control discretization is increased by a factor $r$ in each dimension (i.e., each $a_i$ is reduced to $a_i/r$), and the timestep size is divided by a factor $\tau$, the computation time spent growing the search tree for a given total motion time $T$ increases by a factor $r^{n\tau}$, where $n$ is the number of joints. For example, increasing the control-grid resolution by a factor $r = 2$ and decreasing the timestep by a factor $\tau = 4$ for a three-joint robot results in a search that is likely to take $2^{3 \times 4} = 4096$ times longer to complete. The high computational complexity of the planner makes it impractical beyond a few degrees of freedom.

The description above ignores one important issue: the feasible control set $\mathcal{A}(q, \dot{q})$ changes during a timestep, so the control chosen at the beginning of the timestep may no longer be feasible by the end of the timestep. For that reason, a conservative approximation $\tilde{\mathcal{A}}(q, \dot{q}) \subset \mathcal{A}(q, \dot{q})$ should be used instead. This set should remain feasible over the duration of a timestep regardless of which control action is chosen. How to determine such a conservative approximation $\tilde{\mathcal{A}}(q, \dot{q})$ is beyond the scope of this chapter, but it has to do with bounds on how rapidly the arm's mass matrix $M(q)$ changes with $q$ and how fast the robot is moving. At low speeds $\dot{q}$ and short durations $\Delta t$, the conservative set $\tilde{\mathcal{A}}(q, \dot{q})$ is very close to $\mathcal{A}(q, \dot{q})$.

## 10.5   Sampling Methods

Each grid-based method discussed above delivers optimal solutions subject to the chosen discretization. A drawback of these approaches, however, is their high computational complexity, making them unsuitable for systems having more than a few degrees of freedom.

A different class of planners, known as sampling methods, relies on a random or deterministic function to choose a sample from the C-space or state space; a function to evaluate whether a sample or motion is in $\mathcal{X}_{\text{free}}$; a function to determine nearby previous free-space samples; and a simple local planner to try to connect to, or move toward, the new sample. These functions are used to build up a graph or tree representing feasible motions of the robot.

Sampling methods generally give up on the resolution-optimal solutions of a grid search in exchange for the ability to find satisficing solutions quickly in high-dimensional state spaces. The samples are chosen to form a roadmap or search tree that quickly approximates the free space $\mathcal{X}_{\text{free}}$ using fewer samples than would typically be required by a fixed high-resolution grid, where the number of grid points increases exponentially with the dimension of the search space. Most sampling methods are probabilistically complete: the probability of finding a solution, when one exists, approaches 100% as the number of samples goes to infinity.

Two major classes of sampling methods are rapidly exploring random trees (RRTs) and probabilistic roadmaps (PRMs). The former use a tree representation for single-query planning in either C-space or state space, while PRMs are primarily C-space planners that create a roadmap graph for multiple-query planning.

### 10.5.1  The RRT Algorithm

The RRT algorithm searches for a collision-free motion from an initial state $x_{\text{start}}$ to a goal set $\mathcal{X}_{\text{goal}}$. It is applied to kinematic problems, where the state $x$ is simply the configuration $q$, as well as to dynamic problems, where the state includes the velocity. The basic RRT grows a single tree from $x_{\text{start}}$ as outlined in Algorithm 10.3.

In a typical implementation for a kinematic problem (where $x$ is simply $q$), the sampler in line 3 chooses $x_{\text{samp}}$ randomly from an almost-uniform distribution over $\mathcal{X}$, with a slight bias toward states in $\mathcal{X}_{\text{goal}}$. The closest node $x_{\text{nearest}}$ in the search tree $T$ (line 4) is the node minimizing the Euclidean distance to $x_{\text{samp}}$. The state $x_{\text{new}}$ (line 5) is chosen as the state a small distance $d$ from $x_{\text{nearest}}$ on the straight line to $x_{\text{samp}}$. Because $d$ is small, a very simple local planner, e.g., one that returns a straight-line motion, will often find a motion connecting $x_{\text{nearest}}$ to $x_{\text{new}}$. If the motion is collision-free, the new state $x_{\text{new}}$ is added to the search tree $T$.

The net effect is that the nearly uniformly distributed samples "pull" the tree toward them, causing the tree to rapidly explore $\mathcal{X}_{\text{free}}$. An example of the effect of this pulling action on exploration is shown in Figure 10.17.

The basic algorithm leaves the programmer with many choices: how to sample from $\mathcal{X}$ (line 3), how to define the "nearest" node in $T$ (line 4), and how to plan the motion to make progress toward $x_{\text{samp}}$ (line 5). Even a small change to the sampling method, for example, can yield a dramatic change in the running time of

---

**Algorithm 10.3** RRT algorithm.

---

1: initialize search tree $T$ with $x_{\text{start}}$
2: **while** $T$ is less than the maximum tree size **do**
3:     $x_{\text{samp}} \leftarrow$ sample from $\mathcal{X}$
4:     $x_{\text{nearest}} \leftarrow$ nearest node in $T$ to $x_{\text{samp}}$
5:     employ a local planner to find a motion from $x_{\text{nearest}}$ to $x_{\text{new}}$ in
           the direction of $x_{\text{samp}}$
6:     **if** the motion is collision-free **then**
7:         add $x_{\text{new}}$ to $T$ with an edge from $x_{\text{nearest}}$ to $x_{\text{new}}$
8:         **if** $x_{\text{new}}$ is in $\mathcal{X}_{\text{goal}}$ **then**
9:             **return** SUCCESS and the motion to $x_{\text{new}}$
10:        **end if**
11:    **end if**
12: **end while**
13: **return** FAILURE

---

 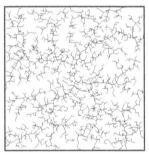

**Figure 10.17** (Left) A tree generated by applying a uniformly-distributed random motion from a randomly chosen tree node does not explore very far. (Right) A tree generated by the RRT algorithm using samples drawn randomly from a uniform distribution. Both trees have 2000 nodes. Figure from LaValle and Kuffner (1999) used with permission.

the planner. A wide variety of planners have been proposed in the literature based on these choices and other variations. Some of these variations are described below.

### 0.5.1.1 Line 3: The Sampler

The most obvious sampler is one that samples randomly from a uniform distribution over $\mathcal{X}$. This is straightforward for Euclidean C-spaces $\mathbb{R}^n$, as well as for $n$-joint robot C-spaces $T^n = S^1 \times \cdots \times S^1$ ($n$ times), where we can choose a uniform distribution over each joint angle, and for the C-space $\mathbb{R}^2 \times S^1$ for a mobile robot in the plane, where we can choose a uniform distribution over $\mathbb{R}^2$ and $S^1$ individually. The notion of a uniform distribution on some other curved C-spaces, for example $SO(3)$, is less straightforward.

For dynamic systems, a uniform distribution over the state space can be defined as the cross product of a uniform distribution over C-space and a uniform distribution over a bounded velocity set.

Although the name "rapidly-exploring random trees" is derived from the idea of a random sampling strategy, in fact the samples need not be generated randomly. For example, a deterministic sampling scheme that generates a progressively finer (multi-resolution) grid on $\mathcal{X}$ could be employed instead. To reflect this more general view, the approach has been called *rapidly-exploring dense trees* (RDTs), emphasizing the key point that the samples should eventually become dense in the state space (i.e., as the number of samples goes to infinity, the samples become arbitrarily close to every point in $\mathcal{X}$).

### 10.5.1.2  Line 4: Defining the Nearest Node

Finding the "nearest" node depends on a definition of distance on $\mathcal{X}$. For an unconstrained kinematic robot on $\mathcal{C} = \mathbb{R}^n$, a natural choice for the distance between two points is simply the Euclidean distance. For other spaces, the choice is less obvious.

As an example, for a car-like robot with a C-space $\mathbb{R}^2 \times S^1$, which configuration is closest to the configuration $x_{\mathrm{samp}}$: one that is rotated 20 degrees relative to $x_{\mathrm{samp}}$, one that is 2 meters straight behind it, or one that is 1 meter straight to the side of it (Figure 10.18)? Since the motion constraints prevent spinning in place or moving directly sideways, the configuration that is 2 meters straight behind is best positioned to make progress toward $x_{\mathrm{samp}}$. Thus defining a notion of distance requires

- combining components of different units (e.g., degrees, meters, degrees/s, meters/s) into a single distance measure; and
- taking into account the motion constraints of the robot.

The closest node $x_{\mathrm{nearest}}$ should perhaps be defined as the one that can reach $x_{\mathrm{samp}}$ the fastest, but computing this is as hard as solving the motion planning problem.

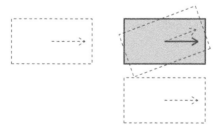

**Figure 10.18** Which of the three dashed configurations of the car is "closest" to the configuration in gray?

A simple choice of a distance measure from $x$ to $x_{\mathrm{samp}}$ is the weighted sum of the distances along the different components of $x_{\mathrm{samp}} - x$. The weights express the relative importance of the different components. If more is known about the set of states that the robot can reach from a state $x$ in limited time, this information can be used in determining the nearest node. In any case, the nearest node should be computed quickly. Finding a nearest neighbor is a common problem in computational geometry, and various algorithms, such as $k$d trees and hashing, can be used to solve it efficiently.

### 0.5.1.3 Line 5: The Local Planner

The job of the local planner is to find a motion from $x_{\text{nearest}}$ to some point $x_{\text{new}}$ which is closer to $x_{\text{samp}}$. The planner should be simple and it should run quickly. Three examples are as follows.

**A straight-line planner.** The plan is a straight line to $x_{\text{new}}$, which may be chosen at $x_{\text{samp}}$ or at a fixed distance $d$ from $x_{\text{nearest}}$ on the straight line to $x_{\text{samp}}$. This is suitable for kinematic systems with no motion constraints.

**Discretized controls planner.** For systems with motion constraints, such as wheeled mobile robots or dynamic systems, the controls can be discretized into a discrete set $\{u_1, u_2, \ldots\}$, as in the grid methods with motion constraints (Section 10.4.2 and Figures 10.14 and 10.16). Each control is integrated from $x_{\text{nearest}}$ for a fixed time $\Delta t$ using $\dot{x} = f(x, u)$. Among the new states reached without collision, the state that is closest to $x_{\text{samp}}$ is chosen as $x_{\text{new}}$.

**Wheeled robot planners.** For a wheeled mobile robot, local plans can be found using Reeds–Shepp curves, as described in Section 13.3.3.

Other robot-specific local planners can be designed.

### 0.5.1.4 Other RRT Variants

The performance of the basic RRT algorithm depends heavily on the choice of sampling method, distance measure, and local planner. Beyond these choices, two other variants of the basic RRT are outlined below.

**Bidirectional RRT**

The bidirectional RRT grows two trees: one "forward" from $x_{\text{start}}$ and one "backward" from $x_{\text{goal}}$. The algorithm alternates between growing the forward tree and growing the backward tree, and every so often it attempts to connect the two trees by choosing $x_{\text{samp}}$ from the other tree. The advantage of this approach is that a single goal state $x_{\text{goal}}$ can be reached exactly, rather than just a goal set $\mathcal{X}_{\text{goal}}$. Another advantage is that, in many environments, the two trees are likely to find each other much more quickly than a single "forward" tree will find a goal set.

The major problem is that the local planner might not be able to connect the two trees exactly. For example, the discretized controls planner of Section 10.5.1.3 is highly unlikely to create a motion exactly to a node in the other tree. In this case, the two trees may be considered more or less connected when points on each tree are sufficiently close. The "broken" discontinuous trajectory can be returned and patched by a smoothing method (Section 10.8).

**RRT***

The basic RRT algorithm returns SUCCESS once a motion to $\mathcal{X}_{\text{goal}}$ is found. An alternative is to continue running the algorithm and to terminate the search only when another termination condition is reached (e.g., a maximum running time or a maximum tree size). Then the motion with the minimum cost can be returned. In this way, the RRT solution may continue to improve as time goes by. Because edges in the tree are never deleted or changed, however, the RRT generally does not converge to an optimal solution.

The RRT* algorithm is a variation on the single-tree RRT that continually rewires the search tree to ensure that it always encodes the shortest path from $x_{\text{start}}$ to each node in the tree. The basic approach works for C-space path planning with no motion constraints, allowing exact paths from any node to any other node.

To modify the RRT to the RRT*, line 7 of the RRT algorithm, which inserts $x_{\text{new}}$ in $T$ with an edge from $x_{\text{nearest}}$ to $x_{\text{new}}$, is replaced by a test of all the nodes $x \in \mathcal{X}_{\text{near}}$ in $T$ that are sufficiently near to $x_{\text{new}}$. An edge to $x_{\text{new}}$ is created from the $x \in \mathcal{X}_{\text{near}}$ by the local planner that (1) has a collision-free motion and (2) minimizes the total cost of the path from $x_{\text{start}}$ to $x_{\text{new}}$, not just the cost of the added edge. The total cost is the cost to reach the candidate $x \in \mathcal{X}_{\text{near}}$ plus the cost of the new edge.

The next step is to consider each $x \in \mathcal{X}_{\text{near}}$ to see whether it could be reached at lower cost by a motion through $x_{\text{new}}$. If so, the parent of $x$ is changed to $x_{\text{new}}$. In this way, the tree is incrementally rewired to eliminate high-cost motions in favor of the minimum-cost motions available so far.

The definition of $\mathcal{X}_{\text{near}}$ depends on the number of samples in the tree, details of the sampling method, the dimension of the search space, and other factors.

Unlike the RRT, the solution provided by RRT* approaches the optimal solution as the number of sample nodes increases. Like the RRT, the RRT* algorithm is probabilistically complete. Figure 10.19 demonstrates the rewiring behavior of RRT* compared to that of RRT for a simple example in $\mathcal{C} = \mathbb{R}^2$.

## 10.5.2    The PRM Algorithm

The PRM uses sampling to build a roadmap representation of $\mathcal{C}_{\text{free}}$ (Section 10.3) before answering any specific queries. The roadmap is an undirected graph: the robot can move in either direction along any edge exactly from one node to the next. For this reason, PRMs primarily apply to kinematic problems for which an exact local planner exists that can find a path (ignoring obstacles) from any $q_1$ to any other $q_2$. The simplest example is a straight-line planner for a robot with no kinematic constraints.

Once the roadmap is built, a particular start node $q_{\text{start}}$ can be added to the

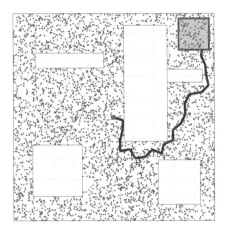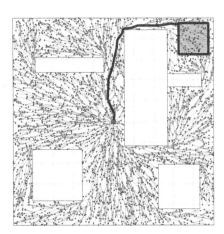

**Figure 10.19** (Left) The tree generated by an RRT after 5,000 nodes. The goal region is the square at the top right corner, and the shortest path is indicated. (Right) The tree generated by RRT* after 5,000 nodes. Figure from Karaman and Frazzoli (2010) used with permission.

graph by attempting to connect it to the roadmap, starting with the closest node. The same is done for the goal node $q_{goal}$. The graph is then searched for a path, typically using $A^*$. Thus the query can be answered efficiently once the roadmap has been built.

The use of PRMs allows the possibility of building a roadmap quickly and efficiently relative to constructing a roadmap using a high-resolution grid representation. The reason is that the volume fraction of the C-space that is "visible" by the local planner from a given configuration does not typically decrease exponentially with increasing dimension of the C-space.

The algorithm for constructing a roadmap $R$ with $N$ nodes is outlined in Algorithm 10.4 and illustrated in Figure 10.20.

A key choice in the PRM roadmap-construction algorithm is how to sample from $C_{free}$. While the default might be sampling randomly from a uniform distribution on $C$ and eliminating configurations in collision, it has been shown that sampling more densely near obstacles can improve the likelihood of finding narrow passages, thus significantly reducing the number of samples needed to properly represent the connectivity of $C_{free}$. Another option is deterministic multi-resolution sampling.

## 10.6    Virtual Potential Fields

Virtual potential field methods are inspired by potential energy fields in nature, such as gravitational and magnetic fields. From physics we know that a potential field $\mathcal{P}(q)$ defined over $C$ induces a force $F = -\partial\mathcal{P}/\partial q$ that drives an object from high to low potential. For example, if $h$ is the height above the Earth's surface in

---

**Algorithm 10.4** PRM roadmap construction algorithm (undirected graph).

1: **for** $i = 1, \ldots, N$ **do**
2:    $q_i \leftarrow$ sample from $\mathcal{C}_{\text{free}}$
3:    add $q_i$ to $R$
4: **end for**
5: **for** $i = 1, \ldots, N$ **do**
6:    $\mathcal{N}(q_i) \leftarrow k$ closest neighbors of $q_i$
7:    **for each** $q \in \mathcal{N}(q_i)$ **do**
8:       **if** there is a collision-free local path from $q$ to $q_i$ and
         there is not already an edge from $q$ to $q_i$ **then**
9:          add an edge from $q$ to $q_i$ to the roadmap $R$
10:       **end if**
11:    **end for**
12: **end for**
13: **return** $R$

---

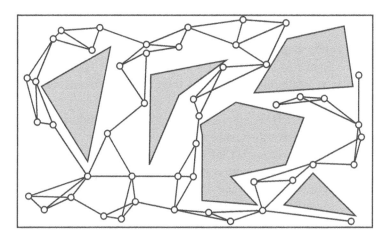

**Figure 10.20** An example PRM roadmap for a point robot in $\mathcal{C} = \mathbb{R}^2$. The $k = 3$ closest neighbors are taken into consideration for connection to a sample node $q$. The degree of a node can be greater than three since it may be a close neighbor of many nodes.

a uniform gravitational potential field ($g = 9.81$ m/s$^2$) then the potential energy of a mass $m$ is $\mathcal{P}(h) = mgh$ and the force acting on it is $F = -\partial \mathcal{P}/\partial h = -mg$. The force will cause the mass to fall to the Earth's surface.

In robot motion control, the goal configuration $q_{\text{goal}}$ is assigned a low virtual potential and obstacles are assigned a high virtual potential. Applying a force to the robot proportional to the negative gradient of the virtual potential naturally pushes the robot toward the goal and away from the obstacles.

A virtual potential field is very different from the planners we have seen so far. Typically the gradient of the field can be calculated quickly, so the motion can be calculated in real time (reactive control) instead of planned in advance.

With appropriate sensors, the method can even handle obstacles that move or appear unexpectedly. The drawback of the basic method is that the robot can get stuck in local minima of the potential field, away from the goal, even when a feasible motion to the goal exists. In certain cases it is possible to design the potential to guarantee that the only local minimum is at the goal, eliminating this problem.

## 10.6.1  A Point in C-space

Let's begin by assuming a point robot in its C-space. A goal configuration $q_{goal}$ is typically encoded by a quadratic potential energy "bowl" with zero energy at the goal,

$$\mathcal{P}_{goal}(q) = \frac{1}{2}(q - q_{goal})^{\mathrm{T}} K (q - q_{goal}),$$

where $K$ is a symmetric positive-definite weighting matrix (for example, the identity matrix). The force induced by this potential is

$$F_{goal}(q) = -\frac{\partial \mathcal{P}_{goal}}{\partial q} = K(q_{goal} - q),$$

an attractive force proportional to the distance from the goal.

The repulsive potential induced by a C-obstacle $\mathcal{B}$ can be calculated from the distance $d(q, \mathcal{B})$ to the obstacle (Section 10.2.2):

$$\mathcal{P}_{\mathcal{B}}(q) = \frac{k}{2d^2(q, \mathcal{B})}, \tag{10.3}$$

where $k > 0$ is a scaling factor. The potential is only properly defined for points outside the obstacle, $d(q, \mathcal{B}) > 0$. The force induced by the obstacle potential is

$$F_{\mathcal{B}}(q) = -\frac{\partial \mathcal{P}_{\mathcal{B}}}{\partial q} = \frac{k}{d^3(q, \mathcal{B})} \frac{\partial d}{\partial q}.$$

The total potential is obtained by summing the attractive goal potential and the repulsive obstacle potentials,

$$\mathcal{P}(q) = \mathcal{P}_{goal}(q) + \sum_i \mathcal{P}_{\mathcal{B}_i}(q),$$

yielding a total force

$$F(q) = F_{goal}(q) + \sum_i F_{\mathcal{B}_i}(q).$$

Note that the sum of the attractive and repulsive potentials may not give a minimum (zero force) exactly at $q_{goal}$. Also, it is common to put a bound on the maximum potential and force, as the simple obstacle potential (10.3) would otherwise yield unbounded potentials and forces near the boundaries of obstacles.

Figure 10.21 shows a potential field for a point in $\mathbb{R}^2$ with three circular obstacles. The contour plot of the potential field clearly shows the global minimum near the center of the space (near the goal marked with a +), a local minimum

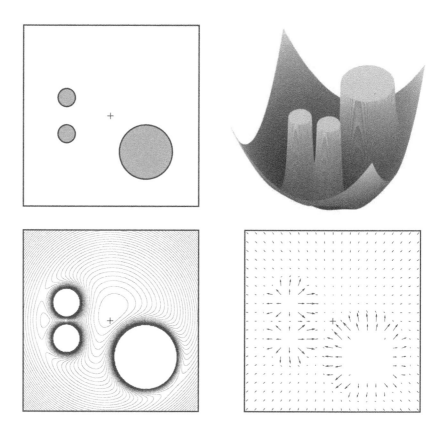

**Figure 10.21** (Top left) Three obstacles and a goal point, marked with a $+$, in $\mathbb{R}^2$. (Top right) The potential function summing the bowl-shaped potential pulling the robot to the goal with the repulsive potentials of the three obstacles. The potential function saturates at a specified maximum value. (Bottom left) A contour plot of the potential function, showing the global minimum, a local minimum, and four saddles: between each obstacle and the boundary of the workspace, and between the two small obstacles. (Bottom right) Forces induced by the potential function.

near the two obstacles on the left, as well as saddles (critical points that are a maximum in one direction and a minimum in the other direction) near the obstacles. Saddles are generally not a problem, as a small perturbation allows continued progress toward the goal. Local minima away from the goal are a problem, however, as they attract nearby states.

To actually control the robot using the calculated $F(q)$, we have several options, two of which are:

- Apply the calculated force plus damping,

$$u = F(q) - B\dot{q}. \tag{10.4}$$

If $B$ is positive definite then it dissipates energy for all $\dot{q} \neq 0$, reducing oscillation and guaranteeing that the robot will come to rest. If $B = 0$, the robot continues to move while maintaining constant total energy, which is the sum of the initial kinetic energy $\frac{1}{2}\dot{q}^T(0)M(q(0))\dot{q}(0)$ and the initial virtual potential energy $\mathcal{P}(q(0))$.

The motion of the robot under the control law (10.4) can be visualized as a ball rolling in gravity on the potential surface of Figure 10.21, where the dissipative force is rolling friction.

- Treat the calculated force as a commanded velocity instead:

$$\dot{q} = F(q). \tag{10.5}$$

This automatically eliminates oscillations.

Using the simple obstacle potential (10.3), even distant obstacles have a nonzero effect on the motion of the robot. To speed up evaluation of the repulsive terms, distant obstacles could be ignored. We can define a range of influence of the obstacles $d_{\text{range}} > 0$ so that the potential is zero for all $d(q, \mathcal{B}) \geq d_{\text{range}}$:

$$U_\mathcal{B}(q) = \begin{cases} \dfrac{k}{2}\left(\dfrac{d_{\text{range}} - d(q,\mathcal{B})}{d_{\text{range}}d(q,\mathcal{B})}\right)^2 & \text{if } d(q,\mathcal{B}) < d_{\text{range}} \\ 0 & \text{otherwise.} \end{cases}$$

Another issue is that $d(q, \mathcal{B})$ and its gradient are generally difficult to calculate. An approach to dealing with this is described in Section 10.6.3.

## 0.6.2  Navigation Functions

A significant problem with the potential field method is local minima. While potential fields may be appropriate for relatively uncluttered spaces or for rapid response to unexpected obstacles, they are likely to get the robot stuck in local minima for many practical applications.

One method that avoids this issue is the wavefront planner of Figure 10.11. The wavefront algorithm creates a local-minimum-free potential function by a breadth-first traversal of every cell reachable from the goal cell in a grid representation of the free space. Therefore, if a solution exists to the motion planning problem then simply moving "downhill" at every step is guaranteed to bring the robot to the goal.

Another approach to local-minimum-free gradient following is based on replacing the virtual potential function with a **navigation function**. A navigation function $\varphi(q)$ is a type of virtual potential function that

1. is smooth (or at least twice differentiable) on $q$;
2. has a bounded maximum value (e.g., 1) on the boundaries of all obstacles;
3. has a single minimum at $q_{\text{goal}}$; and
4. has a full-rank Hessian $\partial^2\varphi/\partial q^2$ at all critical points $q$ where $\partial\varphi/\partial q = 0$ (i.e., $\varphi(q)$ is a **Morse** function).

Condition 1 ensures that the Hessian $\partial^2 \varphi / \partial q^2$ exists. Condition 2 puts an upper bound on the virtual potential energy of the robot. The key conditions are 3 and 4. Condition 3 ensures that of the critical points of $\varphi(q)$ (including minima, maxima, and saddles), there is only one minimum, at $q_{\text{goal}}$. This ensures that $q_{\text{goal}}$ is at least locally attractive. There may be saddle points that are minima along a subset of directions, but condition 4 ensures that the set of initial states that are attracted to any saddle point has empty interior (zero measure), and therefore almost every initial state converges to the unique minimum $q_{\text{goal}}$.

While constructing navigation potential functions with only a single minimum is nontrivial, Rimon and Koditschek (1991) showed how to construct them for the particular case of an $n$-dimensional $\mathcal{C}_{\text{free}}$ consisting of all points inside an $n$-sphere of radius $R$ and outside smaller spherical obstacles $\mathcal{B}_i$ of radius $r_i$ centered at $q_i$, i.e., $\{q \in \mathbb{R}^n \mid \|q\| \leq R$ and $\|q - q_i\| > r_i$ for all $i\}$. This is called a **sphere world**. While a real C-space is unlikely to be a sphere world, Rimon and Koditschek showed that the boundaries of the obstacles, and the associated navigation function, can be deformed to a much broader class of **star-shaped** obstacles. A star-shaped obstacle is one that has a center point from which the line segment to any point on the obstacle boundary is contained completely within the obstacle. A **star world** is a star-shaped C-space which has star-shaped obstacles. Thus finding a navigation function for an arbitrary star world reduces to finding a navigation function for a "model" sphere world that has centers at the centers of the star-shaped obstacles, then stretching and deforming that navigation function to one that fits the star world. Rimon and Koditschek gave a systematic procedure to accomplish this.

Figure 10.22 shows a deformation of a navigation function on a model sphere world to a star world for the case $\mathcal{C} \subset \mathbb{R}^2$.

### 10.6.3   Workspace Potential

A difficulty in calculating the repulsive force from an obstacle is obtaining the distance to the obstacle, $d(q, \mathcal{B})$. One approach that avoids an exact calculation is to represent the boundary of an obstacle as a set of point obstacles, and to represent the robot by a small set of control points. Let the Cartesian location of control point $i$ on the robot be written $f_i(q) \in \mathbb{R}^3$ and boundary point $j$ of the obstacle be $c_j \in \mathbb{R}^3$. Then the distance between the two points is $\|f_i(q) - c_j\|$, and the potential at the control point $i$ due to the obstacle point $j$ is

$$P'_{ij}(q) = \frac{k}{2\|f_i(q) - c_j\|^2},$$

yielding the repulsive force at the control point

$$F'_{ij}(q) = -\frac{\partial P'_{ij}}{\partial q} = \frac{k}{\|f_i(q) - c_j\|^4} \left(\frac{\partial f_i}{\partial q}\right)^{\text{T}} (f_i(q) - c_j) \in \mathbb{R}^3.$$

To turn the linear force $F'_{ij}(q) \in \mathbb{R}^3$ into a generalized force $F_{ij}(q) \in \mathbb{R}^n$

acting on the robot arm or mobile robot, we first find the Jacobian $J_i(q) \in \mathbb{R}^{3 \times n}$ relating $\dot{q}$ to the linear velocity of the control point $\dot{f}_i$:

$$\dot{f}_i = \frac{\partial f_i}{\partial q} \dot{q} = J_i(q)\dot{q}.$$

By the principle of virtual work, the generalized force $F_{ij}(q) \in \mathbb{R}^n$ due to the repulsive linear force $F'_{ij}(q) \in \mathbb{R}^3$ is simply

$$F_{ij}(q) = J_i^{\mathrm{T}}(q)F'_{ij}(q).$$

Now the total force $F(q)$ acting on the robot is the sum of the easily calculated attractive force $F_{\mathrm{goal}}(q)$ and the repulsive forces $F_{ij}(q)$ for all $i$ and $j$.

### 0.6.4 Wheeled Mobile Robots

The preceding analysis assumes that a control force $u = F(q) - B\dot{q}$ (control law (10.4)) or a velocity $\dot{q} = F(q)$ (control law (10.5)) can be applied in any direction. If the robot is a wheeled mobile robot subject to rolling constraints $A(q)\dot{q} = 0$, however, the calculated $F(q)$ must be projected to controls $F_{\mathrm{proj}}(q)$ that move the robot tangentially to the constraints. For a kinematic robot employing the control law $\dot{q} = F_{\mathrm{proj}}(q)$, a suitable projection is

$$F_{\mathrm{proj}}(q) = \left( I - A^{\mathrm{T}}(q)\big(A(q)A^{\mathrm{T}}(q)\big)^{-1}A(q)\right)F(q).$$

For a dynamic robot employing the control law $u = F_{\mathrm{proj}}(q) - B\dot{q}$, the projection was discussed in Section 8.7.

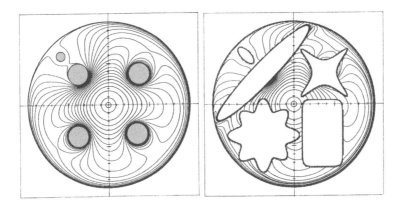

**Figure 10.22** (Left) A model "sphere world" with five circular obstacles. The contour plot of a navigation function is shown. The goal is at $(0,0)$. Note that the obstacles induce saddle points near the obstacles, but no local minima. (Right) A "star world" obtained by deforming the obstacles and the potential while retaining a navigation function. Figure from Rimon and Koditschek (1991) used with permission from the American Mathematical Society.

### 10.6.5    Use of Potential Fields in Planners

A potential field can be used in conjunction with a path planner. For example, a best-first search such as $A^*$ can use the potential as an estimate of the cost-to-go. Incorporating a search function prevents the planner from getting permanently stuck in local minima.

## 10.7    Nonlinear Optimization

The motion planning problem can be expressed as a general nonlinear optimization, with equality and inequality constraints, taking advantage of a number of software packages to solve such problems. Nonlinear optimization problems can be solved by gradient-based methods, such as sequential quadratic programming (SQP), or non-gradient methods, such as simulated annealing, Nelder–Mead optimization, and genetic programming. Like many nonlinear optimization problems, these methods are not generally guaranteed to find a feasible solution when one exists, let alone an optimal one. For methods that use gradients of the objective function and constraints, however, we can expect a locally optimal solution if we start the process with a guess that is "close" to a solution.

The general problem can be written as follows:

$$
\begin{array}{llll}
\text{find} & u(t), q(t), T & & (10.6) \\
\text{minimizing} & J(u(t), q(t), T) & & (10.7) \\
\text{subject to} & \dot{x}(t) = f(x(t), u(t)), & \forall t \in [0, T], & (10.8) \\
& u(t) \in \mathcal{U}, & \forall t \in [0, T], & (10.9) \\
& q(t) \in \mathcal{C}_{\text{free}}, & \forall t \in [0, T], & (10.10) \\
& x(0) = x_{\text{start}}, & & (10.11) \\
& x(T) = x_{\text{goal}}. & & (10.12)
\end{array}
$$

To solve this problem approximately by nonlinear optimization, the control $u(t)$, trajectory $q(t)$, and equality and inequality constraints (10.8)–(10.12) must be discretized. This is typically done by ensuring that the constraints are satisfied at a fixed number of points distributed evenly over the interval $[0, T]$ and choosing a finite-parameter representation of the position and/or control histories. We have at least three choices of how to parametrize the position and controls:

(a) **Parametrize the trajectory** $q(t)$. In this case, we solve for the parametrized trajectory directly. At any time the controls $u(t)$ are calculated using the equations of motion. This approach does not apply to systems with fewer controls than configuration variables, $m < n$.

(b) **Parametrize the control** $u(t)$. We solve for $u(t)$ directly. Calculating the state $x(t)$ requires integrating the equations of motion.

(c) **Parametrize both** $q(t)$ **and** $u(t)$. We have a larger number of variables since we parametrize both $q(t)$ and $u(t)$. Also, we have a larger number of constraints, as $q(t)$ and $u(t)$ must satisfy the dynamic equations $\dot{x} = f(x, u)$ explicitly, typically at a fixed number of points distributed evenly over the interval $[0, T]$. We must be careful to choose the parametrizations of $q(t)$ and $u(t)$ to be consistent with each other, so that the dynamic equations can be satisfied at these points.

A trajectory or control history can be parametrized in any number of ways. The parameters can be the coefficients of a polynomial in time, the coefficients of a truncated Fourier series, spline coefficients, wavelet coefficients, piecewise constant acceleration or force segments, etc. For example, the control $u_i(t)$ could be represented by $p + 1$ coefficients $a_j$ of a polynomial in time:

$$u_i(t) = \sum_{j=0}^{p} a_j t^j.$$

In addition to the parameters for the state or control history, the total time $T$ may be another control parameter. The choice of parametrization has implications for the efficiency of the calculation of $q(t)$ and $u(t)$ at a given time $t$. It also determines the sensitivity of the state and control to the parameters and whether each parameter affects the profiles at all times $[0, T]$ or just on a finite-time support base. These are important factors in the stability and efficiency of the numerical optimization.

## 10.8 Smoothing

The axis-aligned motions of a grid planner and the randomized motions of sampling planners may lead to jerky motion of a robot. One approach to dealing with this issue is to let the planner handle the work of searching globally for a solution, then post-process the resulting motion to make it smoother.

There are many ways to do this; two possibilities are outlined below.

### Nonlinear Optimization

While gradient-based nonlinear optimization may fail to find a solution if initialized with a random initial trajectory, it can make an effective post-processing step, since the plan initializes the optimization with a "reasonable" solution. The initial motion must be converted to a parametrized representation of the controls, and the cost $J(u(t), q(t), T)$ can be expressed as a function of $u(t)$ or $q(t)$. For example, the cost function

$$J = \frac{1}{2} \int_0^T \dot{u}^T(t) \dot{u}(t) dt$$

penalizes rapidly changing controls. This has an analogy in human motor control, where the smoothness of human arm motions has been attributed to minimization of the rate of change of torques at the joints (Uno et al., 1989).

**Subdivide and Reconnect**

A local planner can be used to attempt a connection between two distant points on a path. If this new connection is collision-free, it replaces the original path segment. Since the local planner is designed to produce short, smooth, paths, the new path is likely to be shorter and smoother than the original. This test-and-replace procedure can be applied iteratively to randomly chosen points on the path. Another possibility is to use a recursive procedure that subdivides the path first into two pieces and attempts to replace each piece with a shorter path; then, if either portion cannot be replaced by a shorter path, it subdivides again; and so on.

## 10.9　Summary

- A fairly general statement of the motion planning problem is as follows. Given an initial state $x(0) = x_{start}$ and a desired final state $x_{goal}$, find a time $T$ and a set of controls $u : [0, T] \rightarrow \mathcal{U}$ such that the motion satisfies $x(T) \in \mathcal{X}_{goal}$ and $q(x(t)) \in \mathcal{C}_{free}$ for all $t \in [0, T]$.

- Motion planning problems can be classified in the following categories: path planning versus motion planning; fully actuated versus constrained or underactuated; online versus offline; optimal versus satisficing; exact versus approximate; with or without obstacles.

- Motion planners can be characterized by the following properties: multiple-query versus single-query; anytime planning or not; complete, resolution complete, probabilistically complete, or none of the above; and their degree of computational complexity.

- Obstacles partition the C-space into free C-space, $\mathcal{C}_{free}$, and obstacle space, $\mathcal{C}_{obs}$, where $\mathcal{C} = \mathcal{C}_{free} \cup \mathcal{C}_{obs}$. Obstacles may split $\mathcal{C}_{free}$ into separate connected components. There is no feasible path between configurations in different connected components.

- A conservative check of whether a configuration $q$ is in collision uses a simplified "grown" representation of the robot and obstacles. If there is no collision between the grown bodies, then the configuration is guaranteed collision-free. Checking whether a path is collision-free usually involves sampling the path at finely spaced points and ensuring that if the individual configurations are collision-free then the swept volume of the robot path is collision-free.

- The C-space geometry is often represented by a graph consisting of nodes and edges between the nodes, where edges represent feasible paths. The graph can be undirected (edges flow in both directions) or directed (edges flow in only one direction). Edges can be unweighted or weighted according to their cost of traversal. A tree is a directed graph with no cycles in which each node has at most one parent.

- A roadmap path planner uses a graph representation of $\mathcal{C}_{\text{free}}$, and path planning problems can be solved using a simple path from $q_{\text{start}}$ onto the roadmap, a path along the roadmap, and a simple path from the roadmap to $q_{\text{goal}}$.

- The $A^*$ algorithm is a popular search method that finds minimum-cost paths on a graph. It operates by always exploring from a node that is (1) unexplored and (2) on a path with minimum estimated total cost. The estimated total cost is the sum of the weights for the edges encountered in reaching the node from the start node plus an estimate of the cost-to-go to the goal. To ensure that the search returns an optimal solution, the cost-to-go estimate should be optimistic.

- A grid-based path planner discretizes the C-space into a graph consisting of neighboring points on a regular grid. A multi-resolution grid can be used to allow large steps in wide open spaces and smaller steps near obstacle boundaries.

- Discretizing the control set allows robots with motion constraints to take advantage of grid-based methods. If integrating a control does not land the robot exactly on a grid point, the new state may still be pruned if a state in the same grid cell has already been achieved with a lower cost.

- The basic RRT algorithm grows a single search tree from $x_{\text{start}}$ to find a motion to $\mathcal{X}_{\text{goal}}$. It relies on a sampler to find a sample $x_{\text{samp}}$ in $\mathcal{X}$, an algorithm to find the closest node $x_{\text{nearest}}$ in the search tree, and a local planner to find a motion from $x_{\text{nearest}}$ to a point closer to $x_{\text{samp}}$. The sampling is chosen to cause the tree to explore $\mathcal{X}_{\text{free}}$ quickly.

- The bidirectional RRT grows a search tree from both $x_{\text{start}}$ and $x_{\text{goal}}$ and attempts to join them up. The RRT* algorithm returns solutions that tend toward the optimal as the planning time goes to infinity.

- The PRM builds a roadmap of $\mathcal{C}_{\text{free}}$ for multiple-query planning. The roadmap is built by sampling $\mathcal{C}_{\text{free}}$ $N$ times, then using a local planner to attempt to connect each sample with several of its nearest neighbors. The roadmap is searched using $A^*$.

- Virtual potential fields are inspired by potential energy fields such as gravitational and electromagnetic fields. The goal point creates an attractive potential while obstacles create repulsive potentials. The total potential $\mathcal{P}(q)$ is the sum of these, and the virtual force applied to the robot is $F(q) = -\partial \mathcal{P}/\partial q$. The robot is controlled by applying this force plus damping or by simulating first-order dynamics and driving the robot with $F(q)$ as a velocity. Potential field methods are conceptually simple but may get the robot stuck in local minima away from the goal.

- A navigation function is a potential function with no local minima. Navigation functions result in near-global convergence to $q_{\text{goal}}$. While they are difficult to design in general, they can be designed systematically for certain environments.

- Motion planning problems can be converted to general nonlinear optimization problems with equality and inequality constraints. While optimization methods can be used to find smooth near-optimal motions, they tend to get stuck in local minima in cluttered C-spaces. Optimization methods typically require a good initial guess at a solution.
- Motions returned by grid-based and sampling-based planners tend to be jerky. Smoothing the plan using nonlinear optimization or subdivide-and-reconnect can improve the quality of the motion.

## 10.10    Notes and References

Excellent books covering motion planning include the original text by Latombe (1991) and the more recent ones by Choset et al. (2005) and LaValle (2006). Other summaries of the state of the art in motion planning can be found in the *Handbook of Robotics* (Kavraki and LaValle, 2016) and, particularly for robots subject to nonholonomic and actuation constraints, in the *Control Handbook* (Lynch et al., 2011), the *Encyclopedia of Systems and Control* (Lynch, 2015), and the textbook by Murray et al. (1994). Search algorithms and other algorithms for artificial intelligence are covered in detail in Russell and Norvig (2009).

Landmark early work on motion planning for Shakey the Robot at SRI led to the development of $A^*$ search in 1968 by Hart et al. (1968). This work built on the then newly established approach to dynamic programming for optimal decision-making, as described by Bellman and Dreyfus (1962), and it improved on the performance of Dijkstra's algorithm (Dijkstra, 1959). A suboptimal "anytime" variant of $A^*$ was proposed in Likhachev et al. (2003). Early work on multi-resolution path planning is described in Kambhampati and Davis (1986), Lozano-Perez (2001), Faverjon (1984), and Herman (1986). This work was based on hierarchical decompositions of C-space (Samet, 1984).

One early line of work focused on exact characterization of the free C-space in the presence of obstacles. The visibility-graph approach for polygons moving among polygons was developed by Lozano-Pérez and Wesley (1979). In more general settings, researchers used sophisticated algorithms and mathematical methods to develop cellular decompositions and exact roadmaps of the free C-space. Important examples of this work are found in a series of papers by Schwartz and Sharir on the piano movers' problem (Schwartz and Sharir, 1983a,b,c) and in Canny's PhD thesis (Canny, 1988).

As a result of the mathematical sophistication and high computational complexity needed to exactly represent the topology of C-spaces, a movement formed in the 1990s to approximately represent C-spaces using samples, and that movement carries on strongly today. This line of work has followed two main branches, probabilistic roadmaps (PRMs) (Kavraki et al., 1996) and rapidly exploring random trees (RRTs) (LaValle and Kuffner, 1999, 2001b,a). Owing to their ability to handle complex high-dimensional C-spaces relatively efficiently, research in sampling-based planners has exploded, and some subsequent work is summarized in Choset et al. (2005) and LaValle (2006). The bidirectional RRT and RRT*, highlighted in this chapter, are described in LaValle (2006) and Karaman and Frazzoli (2011), respectively.

The grid-based approach to motion planning for a wheeled mobile robot was introduced by Barraquand and Latombe (1993), and the grid-based approach to time-optimal motion planning for a robot arm with dynamic constraints was introduced in Canny et al. (1988) and Donald and Xavier (1995b,a).

The GJK algorithm for collision detection was derived in Gilbert et al. (1988). Open-source collision-detection packages are implemented in the Open Motion Planning Library (Şucan et al., 2012) and in the Robot Operating System (ROS). An approach to approximating polyhedra with spheres for fast collision detection is described in Hubbard (1996).

The potential field approach to motion planning and real-time obstacle avoidance was first introduced by Khatib and is summarized in Khatib (1986). A search-based planner using a potential field to guide the search was described by Barraquand et al. (1992). The construction of navigation functions, i.e., potential functions with a unique local minimum, is described in a series of papers by Koditschek and Rimon (Koditschek and Rimon, 1990; Koditschek, 1991a,b; Rimon and Koditschek, 1991, 1992).

Nonlinear optimization-based motion planning has been formulated in a number of publications, including: the classic computer graphics paper by Witkin and Kass (1988) using optimization to generate the motions of an animated jumping lamp; work on generating motion plans for dynamic nonprehensile manipulation (Lynch and Mason, 1999); Newton algorithms for optimal motions of mechanisms (Lee et al., 2005); and more recent developments in short-burst sequential action control, which solves both the motion planning and feedback control problems (Ansari and Murphey, 2016; Tzorakoleftherakis et al., 2016). Path smoothing for mobile robot paths by subdivide and reconnect is described in Laumond et al. (1994).

## 0.11 Exercises

**Exercise 10.1** One path is **homotopic** to another if it can be continuously deformed into the other without moving the endpoints. In other words, it can be stretched and pulled like a rubber band, but it cannot be cut and pasted back together. For the C-space in Figure 10.2, draw a path from the start to the goal that is not homotopic to the one shown.

**Exercise 10.2** Label the connected components in Figure 10.2. For each connected component, draw a picture of the robot for one configuration in the connected component.

**Exercise 10.3** Assume that $\theta_2$ joint angles in the range $[175°, 185°]$ result in self-collision for the robot of Figure 10.2. Draw the new joint-limit C-obstacle on top of the existing C-obstacles and label the resulting connected components of $\mathcal{C}_{\text{free}}$. For each connected component, draw a picture of the robot for one configuration in the connected component.

**Exercise 10.4**  Draw the C-obstacle corresponding to the obstacle and translating planar robot in Figure 10.23.

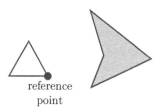

reference
point

**Figure 10.23** Exercise 10.4.

**Exercise 10.5**  Write a program that accepts as input the coordinates of a polygonal robot (relative to a reference point on the robot) and the coordinates of a polygonal obstacle and produces as output a drawing of the corresponding C-space obstacle. In Mathematica, you may find the function `ConvexHull` useful. In MATLAB, try `convhull`.

**Exercise 10.6**  Calculating a square root can be computationally expensive. For a robot and an obstacle represented as collections of spheres (Section 10.2.2), provide a method for calculating the distance between the robot and obstacle that minimizes the use of square roots.

**Exercise 10.7**  Draw the visibility roadmap for the C-obstacles and $q_{\text{start}}$ and $q_{\text{goal}}$ in Figure 10.24. Indicate the shortest path.

● goal

start

**Figure 10.24** Planning problem for Exercise 10.7.

**Exercise 10.8**  Not all edges of the visibility roadmap described in Section 10.3 are needed. Prove that an edge between two vertices of C-obstacles need not be included in the roadmap if either end of the edge does not hit the obstacle tangentially (i.e., it hits at a concave vertex). In other words, if the edge ends by "colliding" with an obstacle, it will never be used in a shortest path.

**Exercise 10.9** Implement an $A^*$ path planner for a point robot in a plane with obstacles. The planar region is a $100 \times 100$ area. The program generates a graph consisting of $N$ nodes and $E$ edges, where $N$ and $E$ are chosen by the user. After generating $N$ randomly chosen nodes, the program should connect randomly chosen nodes by edges until $E$ unique edges have been generated. The cost associated with each edge is the Euclidean distance between the nodes. Finally, the program should display the graph, search the graph using $A^*$ for the shortest path between nodes 1 and $N$, and display the shortest path or indicate FAILURE if no path exists. The heuristic cost-to-go is the Euclidean distance to the goal.

**Exercise 10.10** Modify the $A^*$ planner in Exercise 10.9 to use a heuristic cost-to-go equal to ten times the distance to the goal node. Compare the running time with the original $A^*$ when they are run on the same graphs. (You may need to use large graphs to notice any effect.) Are the solutions found with the new heuristic optimal?

**Exercise 10.11** Modify the $A^*$ algorithm from Exercise 10.9 to use Dijkstra's algorithm instead. Comment on the relative running times of $A^*$ and Dijkstra's algorithm when each is run on the same graphs.

**Exercise 10.12** Write a program that accepts the vertices of polygonal obstacles from a user, as well as the specification of a 2R robot arm, rooted at $(x, y) = (0, 0)$, with link lengths $L_1$ and $L_2$. Each link is simply a line segment. Generate the C-space obstacles for the robot by sampling the two joint angles at $k$-degree intervals (e.g., $k = 5$) and checking for intersections between the line segments and the polygon. Plot the obstacles in the workspace, and in the C-space grid use a black square or dot at each configuration colliding with an obstacle. (Hint: At the core of this program is a subroutine to see whether two line segments intersect. If the segments' corresponding infinite lines intersect, you can check whether this intersection is within the line segments.)

**Exercise 10.13** Write an $A^*$ grid path planner for a 2R robot with obstacles and display on the C-space the paths you find. (See Exercise 10.12 and Figure 10.10.)

**Exercise 10.14** Implement the grid-based path planner for a wheeled mobile robot (Algorithm 10.2), given the control discretization. Choose a simple method to represent obstacles and check for collisions. Your program should plot the obstacles and show the path that is found from the start to the goal.

**Exercise 10.15** Write an RRT planner for a point robot moving in a plane with obstacles. Free space and obstacles are represented by a two-dimensional array, where each element corresponds to a grid cell in the two-dimensional space. The occurrence of a 1 in an element of the array means that there is an obstacle there, and a 0 indicates that the cell is in free space. Your program should plot the obstacles, the tree that is formed, and show the path that is found from the start to the goal.

**Exercise 10.16**   Do the same as for the previous exercise, except that obstacles are now represented by line segments. The line segments can be thought of as the boundaries of obstacles.

**Exercise 10.17**   Write a PRM planner to solve the same problem as in Exercise 10.15.

**Exercise 10.18**   Write a program to implement a virtual potential field for a 2R robot in an environment with point obstacles. The two links of the robot are line segments, and the user specifies the goal configuration of the robot, the start configuration of the robot, and the location of the point obstacles in the workspace. Put two control points on each link of the robot and transform the workspace potential forces to configuration space potential forces. In one workspace figure, draw an example environment consisting of a few point obstacles and the robot at its start and goal configurations. In a second C-space figure, plot the potential function as a contour plot over $(\theta_1, \theta_2)$, and overlay a planned path from a start configuration to a goal configuration. The robot uses the kinematic control law $\dot{q} = F(q)$.

See whether you can create a planning problem that results in convergence to an undesired local minimum for some initial arm configurations but succeeds in finding a path to the goal for other initial arm configurations.

# 11  Robot Control

A robot arm can exhibit a number of different behaviors, depending on the task and its environment. It can act as a source of programmed motions for tasks such as moving an object from one place to another or tracing a trajectory for a spray paint gun. It can act as a source of forces, as when applying a polishing wheel to a workpiece. In tasks such as writing on a chalkboard, it must control forces in some directions (the force must press the chalk against the board) and motions in others (the motion must be in the plane of the board). When the purpose of the robot is to act as a haptic display, rendering a virtual environment, we may want it to act like a spring, damper, or mass, yielding in response to forces applied to it.

In each of these cases, it is the job of the robot controller to convert the task specification to forces and torques at the actuators. Control strategies that achieve the behaviors described above are known as **motion control, force control, hybrid motion–force control**, or **impedance control**. Which of these behaviors is appropriate depends on both the task and the environment. For example, a force-control goal makes sense when the end-effector is in contact with something but not when it is moving in free space. We also have a fundamental constraint imposed by the mechanics, irrespective of the environment: the robot cannot independently control the motion and force in the same direction. If the robot imposes a motion then the environment will determine the force, and if the robot imposes a force then the environment will determine the motion.

Once we have chosen a control goal consistent with the task and environment, we can use feedback control to achieve it. Feedback control uses position, velocity, and force sensors to measure the actual behavior of the robot, compares it with the desired behavior, and modulates the control signals sent to the actuators. Feedback is used in nearly all robot systems.

In this chapter we focus on: feedback control for motion control, both in the joint space and in the task space; force control; hybrid motion–force control; and impedance control.

## 11.1     Control System Overview

A typical control block diagram is shown in Figure 11.1(a). The sensors are typically: potentiometers, encoders, or resolvers for joint position and angle sensing; tachometers for joint velocity sensing; joint force–torque sensors; and/or multi-axis force–torque sensors at the "wrist" between the end of the arm and the end-effector. The controller samples the sensors and updates its control signals to the actuators at a rate of hundreds to a few thousands of Hz. In most robotic applications control update rates higher than this are of limited benefit, given the time constants associated with the dynamics of the robot and environment. In our analysis we will ignore the fact that the sampling time is nonzero and treat controllers as if they were implemented in continuous time.

While tachometers can be used for direct velocity sensing, a common approach is to use a digital filter to numerically-difference the position signals at successive timesteps. A low-pass filter is often used in combination with the differencing filter to reduce the high-frequency signal content due to quantization of the differenced position signals.

As discussed in Section 8.9, there are a number of different technologies for creating mechanical power, transforming the speeds and forces, and transmitting to the robot joints. In this chapter we lump each joint's amplifier, actuator, and transmission together and treat them as a transformer from low-power control signals to forces and torques. This assumption, along with the assumption of perfect sensors, allows us to simplify the block diagram of Figure 11.1(a) to the one shown in Figure 11.1(b), where the controller produces the forces and torques directly. The rest of this chapter deals with the control algorithms that go inside the "controller" box in Figure 11.1(b).

Real robot systems are subject to flexibility and vibrations in the joints and links, backlash at the gears and transmissions, actuator saturation limits, and limited resolution of the sensors. These raise significant issues in design and control but they are beyond the scope of this chapter.

## 11.2     Error Dynamics

In this section we focus on the controlled dynamics of a single joint, as the concepts generalize easily to the case of a multi-joint robot.

If the desired joint position is $\theta_d(t)$ and the actual joint position is $\theta(t)$ then we define the joint error to be

$$\theta_e(t) = \theta_d(t) - \theta(t).$$

The differential equation governing the evolution of the joint error $\theta_e(t)$ of the controlled system is called the **error dynamics**. The purpose of the feedback controller is to create an error dynamics such that $\theta_e(t)$ tends to zero, or a small value, as $t$ increases.

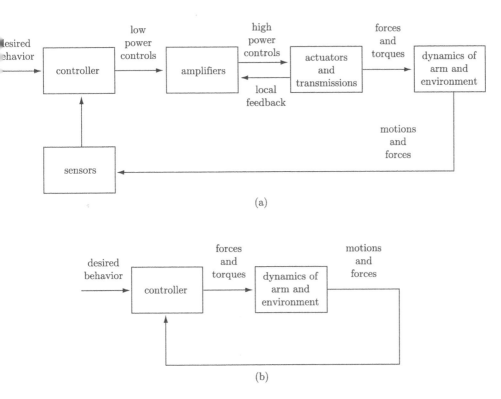

**Figure 11.1** (a) A typical robot control system. An inner control loop is used to help the amplifier and actuator to achieve the desired force or torque. For example, a DC motor amplifier in torque control mode may sense the current actually flowing through the motor and implement a local controller to better match the desired current, since the current is proportional to the torque produced by the motor. Alternatively the motor controller may directly sense the torque by using a strain gauge on the motor's output gearing, and close a local torque-control loop using that feedback. (b) A simplified model with ideal sensors and a controller block that directly produces forces and torques. This assumes ideal behavior of the amplifier and actuator blocks in part (a). Not shown are the disturbance forces that can be injected before the dynamics block, or disturbance forces or motions injected after the dynamics block.

## 1.2.1    Error Response

A common way to test how well a controller works is to specify a nonzero initial error $\theta_e(0)$ and see how quickly, and how completely, the controller reduces the initial error. We define the (unit) **error response** to be the response $\theta_e(t), t > 0$, of the controlled system for the initial conditions $\theta_e(0) = 1$ and $\dot{\theta}_e(0) = \ddot{\theta}_e(0) = \cdots = 0$.

An ideal controller would drive the error to zero instantly and keep the error at zero for all time. In practice it takes time to reduce the error, and the error may never be completely eliminated. As illustrated in Figure 11.2, a typical error response $\theta_e(t)$ can be described by a **transient response** and a **steady-**

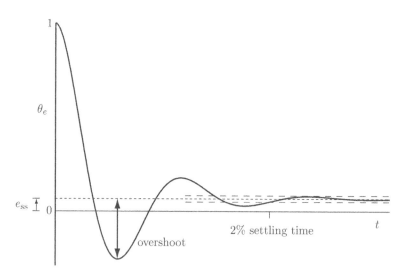

**Figure 11.2** An example error response showing the steady-state error $e_{\mathrm{ss}}$, the overshoot, and the 2% settling time.

state response. The steady-state response is characterized by the **steady-state error** $e_{\mathrm{ss}}$, which is the asymptotic error $\theta_e(t)$ as $t \to \infty$. The transient response is characterized by the **overshoot** and (2%) **settling time**. The 2% settling time is the first time $T$ such that $|\theta_e(t) - e_{\mathrm{ss}}| \le 0.02(\theta_e(0) - e_{\mathrm{ss}})$ for all $t \ge T$ (see the pair of long-dashed lines). Overshoot occurs if the error response initially overshoots the final steady-state error, and in this case the overshoot is defined as

$$\text{overshoot} = \left| \frac{\theta_{e,\min} - e_{\mathrm{ss}}}{\theta_e(0) - e_{\mathrm{ss}}} \right| \times 100\%,$$

where $\theta_{e,\min}$ is the least positive value achieved by the error.

A good error response is characterized by

- little or no steady-state error,
- little or no overshoot, and
- a short 2% settling time.

### 11.2.2   Linear Error Dynamics

In this chapter we work primarily with **linear** systems with error dynamics described by linear ordinary differential equations of the form

$$a_p \theta_e^{(p)} + a_{p-1}\theta_e^{(p-1)} + \cdots + a_2\ddot{\theta}_e + a_1\dot{\theta}_e + a_0\theta_e = c. \tag{11.1}$$

This is a $p$th-order differential equation, because $p$ time derivatives of $\theta_e$ are present. The differential equation (11.1) is **homogeneous** if the constant $c$ is zero and **nonhomogeneous** if $c \neq 0$.

For homogeneous ($c = 0$) linear error dynamics, the $p$th-order differential equation (11.1) can be rewritten as

$$\theta_e^{(p)} = -\frac{1}{a_p}(a_{p-1}\theta_e^{(p-1)} + \cdots + a_2\ddot{\theta}_e + a_1\dot{\theta}_e + a_0\theta_e)$$
$$= -a'_{p-1}\theta_e^{(p-1)} - \cdots - a'_2\ddot{\theta}_e - a'_1\dot{\theta}_e - a'_0\theta_e. \tag{11.2}$$

This $p$th-order differential equation can be expressed as $p$ coupled first-order differential equations by defining the vector $x = (x_1, \ldots, x_p)$, where

$$x_1 = \theta_e,$$
$$x_2 = \dot{x}_1 = \dot{\theta}_e,$$
$$\vdots \quad \vdots$$
$$x_p = \dot{x}_{p-1} = \theta_e^{(p-1)},$$

and writing Equation (11.2) as

$$\dot{x}_p = -a'_0 x_1 - a'_1 x_2 - \cdots - a'_{p-1} x_p.$$

Then $\dot{x}(t) = Ax(t)$, where

$$A = \begin{bmatrix} 0 & 1 & 0 & \cdots & 0 & 0 \\ 0 & 0 & 1 & \cdots & 0 & 0 \\ \vdots & \vdots & \vdots & \ddots & \vdots & \vdots \\ 0 & 0 & 0 & \cdots & 1 & 0 \\ 0 & 0 & 0 & \cdots & 0 & 1 \\ -a'_0 & -a'_1 & -a'_2 & \cdots & -a'_{p-2} & -a'_{p-1} \end{bmatrix} \in \mathbb{R}^{p \times p}.$$

By analogy with the scalar first-order differential equation $\dot{x}(t) = ax(t)$, which has solution $x(t) = e^{at}x(0)$, the vector differential equation $\dot{x}(t) = Ax(t)$ has solution $x(t) = e^{At}x(0)$ using the matrix exponential, as we saw in Section 3.2.3.1. Also analogous to the scalar differential equation, whose solution converges to the equilibrium $x = 0$ from any initial condition if $a$ is negative, the differential equation $\dot{x}(t) = Ax(t)$ converges to $x = 0$ if the matrix $A$ is negative definite, i.e., all eigenvalues of $A$ (which may be complex) have negative real components.

The eigenvalues of $A$ are given by the roots of the characteristic polynomial of $A$, i.e., the complex values $s$ satisfying

$$\det(sI - A) = s^p + a'_{p-1}s^{p-1} + \cdots + a'_2 s^2 + a'_1 s + a'_0 = 0. \tag{11.3}$$

Equation (11.3) is also the characteristic equation associated with the $p$th-order differential equation (11.1).

A necessary condition for each root of Equation (11.3) to have a negative real component is that all coefficients $a'_0, \ldots, a'_{p-1}$ must be positive. This condition is also sufficient for $p = 1$ and 2. For $p = 3$, the condition $a'_2 a'_1 > a'_0$ must also hold. For higher-order systems, other conditions must hold.

If each root of Equation (11.3) has a negative real component, we call the error dynamics **stable**. If any of the roots has a positive real component, the error dynamics are **unstable**, and the error $\|\theta_e(t)\|$ can grow without bound as $t \to \infty$.

For second-order error dynamics, a good mechanical analogy to keep in mind is the linear mass–spring–damper (Figure 11.3). The position of the mass $m$ is $\theta_e$ and an external force $f$ is applied to the mass. The damper applies a force $-b\dot{\theta}_e$ to the mass, where $b$ is the damping constant, and the spring applies a force $-k\theta_e$ to the mass, where $k$ is the spring constant. Therefore the equation of motion of the mass can be written as

$$m\ddot{\theta}_e + b\dot{\theta}_e + k\theta_e = f. \tag{11.4}$$

In the limit as the mass $m$ approaches zero, the second-order dynamics (11.4) reduces to the first-order dynamics

$$b\dot{\theta}_e + k\theta_e = f. \tag{11.5}$$

By the first-order dynamics, an external force generates a velocity rather than an acceleration.

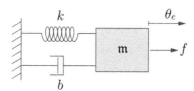

**Figure 11.3** A linear mass–spring–damper.

In the following subsections we consider the first- and second-order error responses for the homogeneous case ($f = 0$) with $b, k > 0$, ensuring that the error dynamics are stable and that the error converges to zero ($e_{\mathrm{ss}} = 0$).

### 11.2.2.1  First-Order Error Dynamics

The first-order error dynamics (11.5) with $f = 0$ can be written in the form

$$\dot{\theta}_e(t) + \frac{k}{b}\theta_e(t) = 0$$

or

$$\dot{\theta}_e(t) + \frac{1}{t}\theta_e(t) = 0, \tag{11.6}$$

where $t = b/k$ is called the **time constant** of the first-order differential equation. The solution to the differential equation (11.6) is

$$\theta_e(t) = e^{-t/t}\theta_e(0). \tag{11.7}$$

The time constant $t$ is the time at which the first-order exponential decay has decayed to approximately 37% of its initial value. The error response $\theta_e(t)$ is defined by the initial condition $\theta_e(0) = 1$. Plots of the error response are shown in Figure 11.4 for different time constants. The steady-state error is zero, there is no overshoot in the decaying exponential error response, and the 2% settling time is determined by solving

$$\frac{\theta_e(t)}{\theta_e(0)} = 0.02 = e^{-t/\mathrm{t}}$$

for $t$. Solving, we get

$$\ln 0.02 = -t/\mathrm{t} \quad \rightarrow \quad t = 3.91\mathrm{t},$$

a 2% settling time of approximately 4t. The response gets faster as the spring constant $k$ increases or the damping constant $b$ decreases.

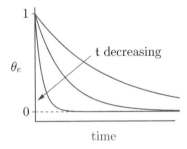

Figure 11.4 The first-order error response for three different time constants t.

## 1.2.2.2  Second-Order Error Dynamics

The second-order error dynamics

$$\ddot{\theta}_e(t) + \frac{b}{\mathrm{m}}\dot{\theta}_e(t) + \frac{k}{\mathrm{m}}\theta_e(t) = 0$$

can be written in the **standard second-order form**

$$\ddot{\theta}_e(t) + 2\zeta\omega_n\dot{\theta}_e(t) + \omega_n^2\theta_e(t) = 0, \tag{11.8}$$

where $\omega_n$ is called the **natural frequency** and $\zeta$ is called the **damping ratio**. For the mass–spring–damper, $\omega_n = \sqrt{k/m}$ and $\zeta = b/(2\sqrt{km})$. The two roots of the characteristic polynomial

$$s^2 + 2\zeta\omega_n s + \omega_n^2 = 0 \tag{11.9}$$

are

$$s_1 = -\zeta\omega_n + \omega_n\sqrt{\zeta^2 - 1} \quad \text{and} \quad s_2 = -\zeta\omega_n - \omega_n\sqrt{\zeta^2 - 1}. \tag{11.10}$$

The second-order error dynamics (11.8) is stable if and only if $\zeta\omega_n > 0$ and $\omega_n^2 > 0$.

If the error dynamics is stable, then there are three types of solutions $\theta_e(t)$ to the differential equation, depending on whether the roots $s_{1,2}$ are real and unequal ($\zeta > 1$), real and equal ($\zeta = 1$), or complex conjugates ($\zeta < 1$).

- **Overdamped:** $\zeta > 1$. The roots $s_{1,2}$ are real and distinct, and the solution to the differential equation (11.8) is

$$\theta_e(t) = c_1 e^{s_1 t} + c_2 e^{s_2 t},$$

where $c_1$ and $c_2$ can be calculated from the initial conditions. The response is the sum of two decaying exponentials, with time constants $t_1 = -1/s_1$ and $t_2 = -1/s_2$. The "slower" time constant in the solution is given by the less negative root, $s_1 = -\zeta \omega_n + \omega_n \sqrt{\zeta^2 - 1}$.

  The initial conditions for the (unit) error response are $\theta_e(0) = 1$ and $\dot{\theta}_e(0) = 0$, and the constants $c_1$ and $c_2$ can be calculated as

$$c_1 = \frac{1}{2} + \frac{\zeta}{2\sqrt{\zeta^2 - 1}} \qquad \text{and} \qquad c_2 = \frac{1}{2} - \frac{\zeta}{2\sqrt{\zeta^2 - 1}}.$$

- **Critically damped:** $\zeta = 1$. The roots $s_{1,2} = -\omega_n$ are equal and real, and the solution is

$$\theta_e(t) = (c_1 + c_2 t)e^{-\omega_n t},$$

  i.e., a decaying exponential multiplied by a linear function of time. The time constant of the decaying exponential is $t = 1/\omega_n$. For the error response with $\theta_e(0) = 1$ and $\dot{\theta}_e(0) = 0$,

$$c_1 = 1 \qquad \text{and} \qquad c_2 = \omega_n.$$

- **Underdamped:** $\zeta < 1$. The roots $s_{1,2}$ are complex conjugates at $s_{1,2} = -\zeta \omega_n \pm j\omega_d$, where $\omega_d = \omega_n \sqrt{1 - \zeta^2}$ is the **damped natural frequency**. The solution is

$$\theta_e(t) = (c_1 \cos \omega_d t + c_2 \sin \omega_d t) e^{-\zeta \omega_n t},$$

  i.e., a decaying exponential (time constant $t = 1/(\zeta \omega_n)$) multiplied by a sinusoid. For the error response with $\theta_e(0) = 1$ and $\dot{\theta}_e(0) = 0$,

$$c_1 = 1 \qquad \text{and} \qquad c_2 = \frac{\zeta}{\sqrt{1 - \zeta^2}}.$$

Example root locations for the overdamped, critically damped, and underdamped cases, as well as their error responses $\theta_e(t)$, are shown in Figure 11.5. This figure also shows the relationship between the root locations and properties of the transient response: roots further to the left in the complex plane correspond to shorter settling times, and roots further away from the real axis correspond to greater overshoot and oscillation. These general relationships between root locations and transient response properties also hold for higher-order systems with more than two roots.

If the second-order error dynamics (11.8) is stable, the steady-state error $e_{ss}$ is zero regardless of whether the error dynamics is overdamped, underdamped, or critically damped. The 2% settling time is approximately 4t, where t corresponds to the "slower" root $s_1$ if the error dynamics is overdamped. The overshoot is

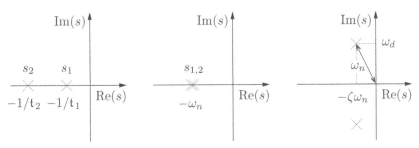

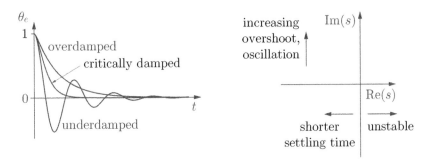

**Figure 11.5** (Top) Example root locations for overdamped, critically damped, and underdamped second-order systems. (Bottom left) Error responses for overdamped, critically damped, and underdamped second-order systems. (Bottom right) Relationship of the root locations to properties of the transient response.

zero for overdamped and critically damped error dynamics and, for underdamped error dynamics, the overshoot can be calculated by finding the first time (after $t = 0$) where the error response satisfies $\dot{\theta}_e = 0$. This is the peak of the overshoot, and it occurs at

$$t_p = \pi/\omega_d.$$

Substituting this into the underdamped error response, we get

$$\theta_e(t_p) = \theta_e\left(\frac{\pi}{\omega_d}\right) = \left(\cos\left(\omega_d\frac{\pi}{\omega_d}\right) + \frac{\zeta}{\sqrt{1-\zeta^2}}\sin\left(\omega_d\frac{\pi}{\omega_d}\right)\right)e^{-\zeta\omega_n\pi/\omega_d}$$

$$= -e^{-\pi\zeta/\sqrt{1-\zeta^2}}.$$

Therefore, by our definition of overshoot, the overshoot is $e^{-\pi\zeta/\sqrt{1-\zeta^2}} \times 100\%$. Thus $\zeta = 0.1$ gives an overshoot of 73%, $\zeta = 0.5$ gives an overshoot of 16%, and $\zeta = 0.8$ gives an overshoot of 1.5%.

## 11.3    Motion Control with Velocity Inputs

As discussed in Chapter 8, we typically assume that there is direct control of the forces or torques at robot joints, and the robot's dynamics transforms those controls to joint accelerations. In some cases, however, we can assume that there is direct control of the joint velocities, for example when the actuators are stepper motors. In this case the velocity of a joint is determined directly by the frequency of the pulse train sent to the stepper.[1] Another example occurs when the amplifier for an electric motor is placed in velocity control mode – the amplifier attempts to achieve the joint velocity requested by the user, rather than a joint force or torque.

In this section we will assume that the control inputs are joint velocities. In Section 11.4, and indeed in the rest of the chapter, the control inputs are assumed to be joint forces and torques.

The motion control task can be expressed in joint space or task space. When the trajectory is expressed in task space, the controller is fed a steady stream of end-effector configurations $X_d(t)$, and the goal is to command joint velocities that cause the robot to track this trajectory. In joint space, the controller is fed a steady stream of desired joint positions $\theta_d(t)$.

The main ideas are well illustrated by a robot with a single joint, so we begin there and then generalize to a multi-joint robot.

### 11.3.1    Motion Control of a Single Joint

#### 11.3.1.1    Feedforward Control
Given a desired joint trajectory $\theta_d(t)$, the simplest type of control would be to choose the commanded velocity $\dot{\theta}(t)$ as

$$\dot{\theta}(t) = \dot{\theta}_d(t), \tag{11.11}$$

where $\dot{\theta}_d(t)$ comes from the desired trajectory. This is called a **feedforward** or **open-loop controller**, since no feedback (sensor data) is needed to implement it.

#### 11.3.1.2    Feedback Control
In practice, position errors will accumulate over time under the feedforward control law (11.11). An alternative strategy is to measure the actual position of each joint continually and implement a **feedback controller**.

**P Control and First-Order Error Dynamics**
The simplest feedback controller is

$$\dot{\theta}(t) = K_p(\theta_d(t) - \theta(t)) = K_p\theta_e(t), \tag{11.12}$$

---

[1] This assumes that the torque requirements are low enough that the stepper motor can keep up with the pulse train.

where $K_p > 0$. This controller is called a proportional controller, or **P controller**, because it creates a corrective control proportional to the position error $\theta_e(t) = \theta_d(t) - \theta(t)$. In other words, the constant **control gain** $K_p$ acts somewhat like a virtual spring that tries to pull the actual joint position to the desired joint position.

The P controller is an example of a **linear controller**, as it creates a control signal that is a linear combination of the error $\theta_e(t)$ and possibly its time derivatives and time integrals.

The case where $\theta_d(t)$ is constant, i.e., $\dot{\theta}_d(t) = 0$, is called **setpoint control**. In setpoint control, the error dynamics

$$\dot{\theta}_e(t) = \overset{0}{\cancel{\dot{\theta}_d(t)}} - \dot{\theta}(t)$$

is written as follows after substituting in the P controller $\dot{\theta}(t) = K_p\theta_e(t)$:

$$\dot{\theta}_e(t) = -K_p\theta_e(t) \quad \rightarrow \quad \dot{\theta}_e(t) + K_p\theta_e(t) = 0.$$

This is a first-order error dynamic equation (11.6) with time constant $t = 1/K_p$. The decaying exponential error response is illustrated in Figure 11.4. The steady-state error is zero, there is no overshoot, and the 2% settling time is $4/K_p$. A larger $K_p$ means a faster response.

Now consider the case where $\theta_d(t)$ is not constant but $\dot{\theta}_d(t)$ is constant, i.e., $\dot{\theta}_d(t) = c$. Then the error dynamics under the P controller can be written

$$\dot{\theta}_e(t) = \dot{\theta}_d(t) - \dot{\theta}(t) = c - K_p\theta_e(t),$$

which we rewrite as

$$\dot{\theta}_e(t) + K_p\theta_e(t) = c.$$

This is a first-order nonhomogeneous linear differential equation with solution

$$\theta_e(t) = \frac{c}{K_p} + \left(\theta_e(0) - \frac{c}{K_p}\right)e^{-K_pt},$$

which converges to the nonzero value $c/K_p$ as time goes to infinity. Unlike the case of setpoint control, the steady-state error $e_{ss}$ is nonzero; the joint position always lags behind the moving reference. The steady-state error $c/K_p$ can be made small by choosing the control gain $K_p$ large, but there are practical limits on how large $K_p$ can be. For one thing, real joints have velocity limits that may prevent the realization of the large commanded velocities associated with a large $K_p$. For another, large values of $K_p$ may cause instability when implemented by a discrete-time digital controller – the large gain may result in a large change in $\theta_e$ during a single servo cycle, meaning that the control action late in the servo cycle is in response to sensor data that is no longer relevant.

### PI Control and Second-Order Error Dynamics
An alternative to using a large gain $K_p$ is to introduce another term in the control law. A proportional-integral controller, or **PI controller**, adds a term

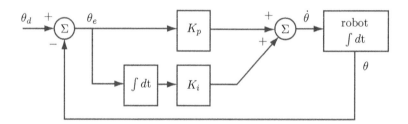

**Figure 11.6** The block diagram of a PI controller that produces a commanded velocity $\dot{\theta}$ as input to the robot.

that is proportional to the time-integral of the error:

$$\dot{\theta}(t) = K_p \theta_e(t) + K_i \int_0^t \theta_e(\mathrm{t})\, dt, \qquad (11.13)$$

where $t$ is the current time and t is the variable of integration. The PI controller block diagram is illustrated in Figure 11.6.

With this controller, the error dynamics for a constant $\dot{\theta}_d(t)$ becomes

$$\dot{\theta}_e(t) + K_p \theta_e(t) + K_i \int_0^t \theta_e(\mathrm{t})\, dt = c.$$

Taking the time derivative of this dynamics, we get

$$\ddot{\theta}_e(t) + K_p \dot{\theta}_e(t) + K_i \theta_e(t) = 0. \qquad (11.14)$$

We can rewrite this equation in the standard second-order form (11.8), with natural frequency $\omega_n = \sqrt{K_i}$ and damping ratio $\zeta = K_p/(2\sqrt{K_i})$.

Relating the PI controller of Equation (11.14) to the mass–spring–damper of Figure 11.3, the gain $K_p$ plays the role of $b/\mathrm{m}$ for the mass–spring–damper (a larger $K_p$ means a larger damping constant $b$), and the gain $K_i$ plays the role of $k/\mathrm{m}$ (a larger $K_i$ means a larger spring constant $k$).

The PI-controlled error dynamics equation is stable if $K_i > 0$ and $K_p > 0$, and the roots of the characteristic equation are

$$s_{1,2} = -\frac{K_p}{2} \pm \sqrt{\frac{K_p^2}{4} - K_i}.$$

Let's hold $K_p$ equal to 20 and plot the roots in the complex plane as $K_i$ grows from zero (Figure 11.7). This plot, or any plot of the roots as one parameter is varied, is called a **root locus**.

For $K_i = 0$, the characteristic equation $s^2 + K_p s + K_i = s^2 + 20s = s(s + 20) = 0$ has roots at $s_1 = 0$ and $s_2 = -20$. As $K_i$ increases, the roots move toward each other on the real axis of the $s$-plane as shown in the left-hand panel in Figure 11.7. Because the roots are real and unequal, the error dynamics equation is overdamped ($\zeta = K_p/(2\sqrt{K_i}) > 1$, case I) and the error response is sluggish due to the time constant $\mathrm{t}_1 = -1/s_1$ of the exponential corresponding to the "slow" root. As $K_i$ increases, the damping ratio decreases, the "slow"

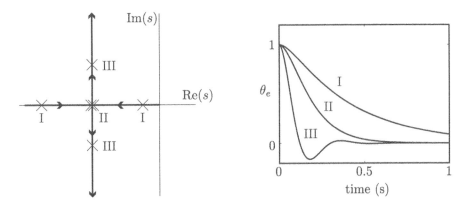

**Figure 11.7** (Left) The complex roots of the characteristic equation of the error dynamics of the PI velocity-controlled joint for a fixed $K_p = 20$ as $K_i$ increases from zero. This is known as a root locus plot. (Right) The error response to an initial error $\theta_e = 1$, $\dot{\theta}_e = 0$, is shown for overdamped ($\zeta = 1.5$, $K_i = 44.4$, case I), critically damped ($\zeta = 1$, $K_i = 100$, case II), and underdamped ($\zeta = 0.5$, $K_i = 400$, case III) cases.

root moves left (while the "fast" root moves right), and the response gets faster. When $K_i$ reaches 100, the two roots meet at $s_{1,2} = -10 = -\omega_n = K_p/2$, and the error dynamics equation is critically damped ($\zeta = 1$, case II). The error response has a short 2% settling time of $4t = 4/(\zeta\omega_n) = 0.4$ s and no overshoot or oscillation. As $K_i$ continues to grow, the damping ratio $\zeta$ falls below 1 and the roots move vertically off the real axis, becoming complex conjugates at $s_{1,2} = -10 \pm j\sqrt{K_i - 100}$ (case III). The error dynamics is underdamped, and the response begins to exhibit overshoot and oscillation as $K_i$ increases. The settling time is unaffected as the time constant $t = 1/(\zeta\omega_n)$ remains constant.

According to our simple model of the PI controller, we could always choose $K_p$ and $K_i$ for critical damping ($K_i = K_p^2/4$) and increase $K_p$ and $K_i$ without bound to make the error response arbitrarily fast. As described above, however, there are practical limits. Within these practical limits, $K_p$ and $K_i$ should be chosen to yield critical damping.

Figure 11.8 shows for comparison the performances of a P controller and a PI controller attempting to track a constant-velocity trajectory. The proportional gain $K_p$ is the same in both cases, while $K_i = 0$ for the P controller. From the shape of the response, it appears that $K_i$ in the PI controller was chosen to be a bit too large, making the system underdamped. It is also clear that $e_{ss} = 0$ for the PI controller but $e_{ss} \neq 0$ for the P controller, agreeing with our analysis above.

If the desired velocity $\dot{\theta}_d(t)$ is anything other than constant, the PI controller cannot be expected to eliminate steady-state error completely. If it changes slowly, however, then a well-designed PI controller can be expected to provide better tracking performance than a P controller.

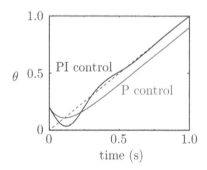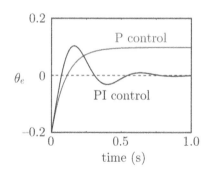

**Figure 11.8** The motion of P-controlled and PI-controlled joints, with initial position error, tracking a reference trajectory (dashed) where $\dot{\theta}_d(t)$ is constant. (Left) The responses $\theta(t)$. (Right) The error responses $\theta_e(t) = \theta_d(t) - \theta(t)$.

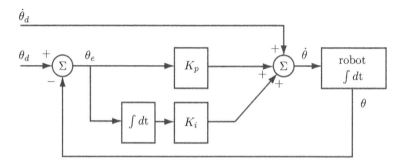

**Figure 11.9** The block diagram of feedforward plus PI feedback control that produces a commanded velocity $\dot{\theta}$ as input to the robot.

### 11.3.1.3 Feedforward Plus Feedback Control

A drawback of feedback control is that an error is required before the joint begins to move. It would be preferable to use our knowledge of the desired trajectory $\theta_d(t)$ to initiate motion before any error accumulates.

We can combine the advantages of feedforward control, which commands motion even when there is no error, with the advantages of feedback control, which limits the accumulation of error, as follows:

$$\dot{\theta}(t) = \dot{\theta}_d(t) + K_p \theta_e(t) + K_i \int_0^t \theta_e(\mathrm{t}) \, dt. \tag{11.15}$$

This feedforward–feedback controller, illustrated in Figure 11.9, is our preferred control law for producing a commanded velocity to the joint.

### 11.3.2 Motion Control of a Multi-Joint Robot

The single-joint PI feedback plus feedforward controller (11.15) generalizes immediately to robots with $n$ joints. The reference position $\theta_d(t)$ and actual position

$\theta(t)$ are now $n$-vectors, and the gains $K_p$ and $K_i$ are diagonal $n \times n$ matrices of the form $k_p I$ and $k_i I$, where the scalars $k_p$ and $k_i$ are positive and $I$ is the $n \times n$ identity matrix. Each joint is subject to the same stability and performance analysis as the single joint in Section 11.3.1.

### 1.3.3 Task-Space Motion Control

We can express the feedforward plus feedback control law in task space. Let $X(t) \in SE(3)$ be the configuration of the end-effector as a function of time and $\mathcal{V}_b(t)$ be the end-effector twist expressed in the end-effector frame {b}, i.e., $[\mathcal{V}_b] = X^{-1}\dot{X}$. The desired motion is given by $X_d(t)$ and $[\mathcal{V}_d] = X_d^{-1}\dot{X}_d$. A task-space version of the control law (11.15) is

$$\mathcal{V}_b(t) = [\mathrm{Ad}_{X^{-1}X_d}]\mathcal{V}_d(t) + K_p X_e(t) + K_i \int_0^t X_e(\mathrm{t})\, d\mathrm{t}. \tag{11.16}$$

The term $[\mathrm{Ad}_{X^{-1}X_d}]\mathcal{V}_d$ expresses the feedforward twist $\mathcal{V}_d$ in the actual end-effector frame at $X$ (which could also be written $X_{sb}$) rather than the desired end-effector frame $X_d$ (which could also be written $X_{sd}$). When the end-effector is at the desired configuration ($X = X_d$), this term reduces to $\mathcal{V}_d$. Also, the configuration error $X_e(t)$ is not simply $X_d(t) - X(t)$, since it does not make sense to subtract elements of $SE(3)$. Instead, as we saw in Section 6.2, $X_e$ should refer to the twist which, if followed for unit time, takes $X$ to $X_d$. The $se(3)$ representation of this twist, expressed in the end-effector frame, is $[X_e] = \log(X^{-1}X_d)$.

As in Section 11.3.2, the diagonal gain matrices $K_p, K_i \in \mathbb{R}^{6 \times 6}$ take the form $k_p I$ and $k_i I$, respectively, where $k_p, k_i > 0$.

The commanded joint velocities $\dot{\theta}$ realizing $\mathcal{V}_b$ from the control law (11.16) can be calculated using the inverse velocity kinematics from Section 6.3,

$$\dot{\theta} = J_b^\dagger(\theta)\mathcal{V}_b,$$

where $J_b^\dagger(\theta)$ is the pseudoinverse of the body Jacobian.

Motion control in task space can be defined using other representations of the end-effector configuration and velocity. For example, for a minimal coordinate representation of the end-effector configuration $x \in \mathbb{R}^m$, the control law can be written

$$\dot{x}(t) = \dot{x}_d(t) + K_p(x_d(t) - x(t)) + K_i \int_0^t (x_d(\mathrm{t}) - x(\mathrm{t}))\, d\mathrm{t}. \tag{11.17}$$

For a hybrid configuration representation $X = (R, p)$, with velocities represented by $(\omega_b, \dot{p})$:

$$\begin{bmatrix} \omega_b(t) \\ \dot{p}(t) \end{bmatrix} = \begin{bmatrix} R^{\mathrm{T}}(t)R_d(t) & 0 \\ 0 & I \end{bmatrix} \begin{bmatrix} \omega_d(t) \\ \dot{p}_d(t) \end{bmatrix} + K_p X_e(t) + K_i \int_0^t X_e(\mathrm{t})\, d\mathrm{t}, \tag{11.18}$$

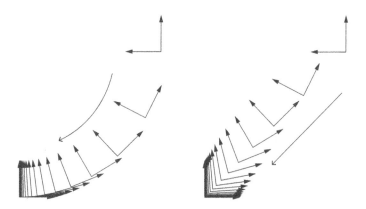

**Figure 11.10** (Left) The end-effector configuration converging to the origin under the control law (11.16), where the end-effector velocity is represented as the body twist $\mathcal{V}_b$. (Right) The end-effector configuration converging to the origin under the control law (11.18), where the end-effector velocity is represented as $(\omega_b, \dot{p})$.

where

$$X_e(t) = \left[ \begin{array}{c} \omega_e(t) \\ p_d(t) - p(t) \end{array} \right],$$

where $[\omega_e(t)] = \log(R^\mathrm{T}(t)R_d(t))$.

Figure 11.10 shows the performance of the control law (11.16), where the end-effector velocity is the body twist $\mathcal{V}_b$, and the performance of the control law (11.18), where the end-effector velocity is $(\omega_b, \dot{p})$. The control task is to stabilize $X_d$ at the origin from the initial configuration

$$R_0 = \left[ \begin{array}{ccc} 0 & -1 & 0 \\ 1 & 0 & 0 \\ 0 & 0 & 1 \end{array} \right], \qquad p_0 = \left[ \begin{array}{c} 1 \\ 1 \\ 0 \end{array} \right].$$

The feedforward velocity is zero and $K_i = 0$. Figure 11.10 shows the different paths followed by the end-effector. The decoupling of linear and angular control in the control law (11.18) is visible in the straight-line motion of the origin of the end-effector frame.

An application of the control law (11.16) to mobile manipulation can be found in Section 13.5.

## 11.4     Motion Control with Torque or Force Inputs

Stepper-motor-controlled robots are generally limited to applications with low or predictable force–torque requirements. Also, robot-control engineers do not rely on the velocity-control modes of off-the-shelf amplifiers for electric motors, because these velocity-control algorithms do not make use of a dynamic model of the robot. Instead, robot-control engineers use amplifiers in torque-control mode: the input to the amplifier is the desired torque (or force). This allows the robot-control engineer to use a dynamic model of the robot in the design of the control law.

In this section, the controller generates joint torques and forces to try to track a desired trajectory in joint space or task space. Once again, the main ideas are well illustrated by a robot with a single joint, so we begin there and then generalize to a multi-joint robot.

## 11.4.1 Motion Control of a Single Joint

Consider a single motor attached to a single link, as shown in Figure 11.11. Let $\tau$ be the motor's torque and $\theta$ be the angle of the link. The dynamics can be written as

$$\tau = M\ddot{\theta} + \mathfrak{m}gr\cos\theta, \qquad (11.19)$$

where $M$ is the scalar inertia of the link about the axis of rotation, $\mathfrak{m}$ is the mass of the link, $r$ is the distance from the axis to the center of mass of the link, and $g \geq 0$ is the gravitational acceleration.

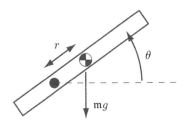

**Figure 11.11** A single-joint robot rotating under gravity. The center of mass is indicated by the checkered disk.

According to the model (11.19) there is no dissipation: if the link were made to move and $\tau$ were then set to zero, the link would move forever. This is unrealistic, of course; there is bound to be friction at the various bearings, gears, and transmissions. Friction modeling is an active research area, but in a simple model, rotational friction is due to viscous friction forces, so that

$$\tau_{\text{fric}} = b\dot{\theta}, \qquad (11.20)$$

where $b > 0$. Adding the friction torque, our final model is

$$\tau = M\ddot{\theta} + \mathfrak{m}gr\cos\theta + b\dot{\theta}, \qquad (11.21)$$

which we may write more compactly as

$$\tau = M\ddot{\theta} + h(\theta, \dot{\theta}), \qquad (11.22)$$

where $h$ contains all terms that depend only on the state, not the acceleration.

For concreteness in the following simulations, we set $M = 0.5 \text{ kg m}^2$, $\mathfrak{m} = 1$ kg, $r = 0.1$ m, and $b = 0.1$ N m s/rad. In some examples the link moves in a horizontal plane, so $g = 0$. In other examples, the link moves in a vertical plane, so $g = 9.81$ m/s$^2$.

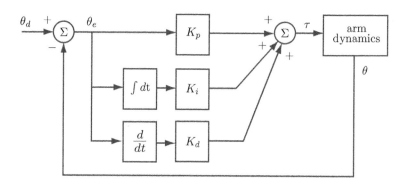

**Figure 11.12** Block diagram of a PID controller.

### 11.4.1.1   Feedback Control: PID Control

A common feedback controller is linear proportional-integral-derivative control, or **PID control**. The PID controller is simply the PI controller (Equation (11.13)) with an added term proportional to the time derivative of the error,

$$\tau = K_p\theta_e + K_i \int \theta_e(t)dt + K_d\dot{\theta}_e, \qquad (11.23)$$

where the control gains $K_p$, $K_i$, and $K_d$ are positive. The proportional gain $K_p$ acts as a virtual spring that tries to reduce the position error $\theta_e = \theta_d - \theta$. The derivative gain $K_d$ acts as a virtual damper that tries to reduce the velocity error $\dot{\theta}_e = \dot{\theta}_d - \dot{\theta}$. The integral gain can be used to reduce or eliminate steady-state errors. The PID controller block diagram is given in Figure 11.12.

#### PD Control and Second-Order Error Dynamics

For now let's consider the case where $K_i = 0$. This is known as PD control. Let's also assume the robot moves in a horizontal plane ($g = 0$). Substituting the PD control law into the dynamics (11.21), we get

$$M\ddot{\theta} + b\dot{\theta} = K_p(\theta_d - \theta) + K_d(\dot{\theta}_d - \dot{\theta}). \qquad (11.24)$$

If the control objective is setpoint control at a constant $\theta_d$ with $\dot{\theta}_d = \ddot{\theta}_d = 0$, then $\theta_e = \theta_d - \theta$, $\dot{\theta}_e = -\dot{\theta}$, and $\ddot{\theta}_e = -\ddot{\theta}$. Equation (11.24) can be rewritten as

$$M\ddot{\theta}_e + (b + K_d)\dot{\theta}_e + K_p\theta_e = 0, \qquad (11.25)$$

or, in the standard second-order form (11.8), as

$$\ddot{\theta}_e + \frac{b + K_d}{M}\dot{\theta}_e + \frac{K_p}{M}\theta_e = 0 \quad \rightarrow \quad \ddot{\theta}_e + 2\zeta\omega_n\dot{\theta}_e + \omega_n^2\theta_e = 0, \qquad (11.26)$$

where the damping ratio $\zeta$ and the natural frequency $\omega_n$ are

$$\zeta = \frac{b + K_d}{2\sqrt{K_p M}} \quad \text{and} \quad \omega_n = \sqrt{\frac{K_p}{M}}.$$

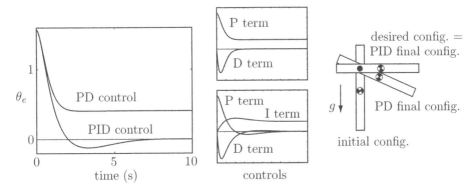

**Figure 11.13** (Left) The tracking errors for a PD controller with $K_d = 2$ N m s/rad and $K_p = 2.205$ N m/rad for critical damping, and a PID controller with the same PD gains and $K_i = 1$ N m/(rad s). The arm starts at $\theta(0) = -\pi/2, \dot{\theta}(0) = 0$, with a goal state $\theta_d = 0, \dot{\theta}_d = 0$. (Middle) The individual contributions of the terms in the PD and PID control laws. Note that the nonzero I (integral) term for the PID controller allows the P (proportional) term to drop to zero. (Right) The initial and final configurations, with the center of mass indicated by checkered disks.

For stability, $b + K_d$ and $K_p$ must be positive. If the error dynamics equation is stable then the steady-state error is zero. For no overshoot and a fast response, the gains $K_d$ and $K_p$ should be chosen to satisfy critical damping ($\zeta = 1$). For a fast response, $K_p$ should be chosen to be as high as possible, subject to practical issues such as actuator saturation, undesired rapid torque changes (chattering), vibrations of the structure due to unmodeled flexibility in the joints and links, and possibly even instability due to the finite servo rate frequency.

**PID Control and Third-Order Error Dynamics**

Now consider the case of setpoint control where the link moves in a vertical plane ($g > 0$). With the PD control law above, the error dynamics can now be written

$$M\ddot{\theta}_e + (b + K_d)\dot{\theta}_e + K_p\theta_e = \mathfrak{m}gr\cos\theta. \tag{11.27}$$

This implies that the joint comes to rest at a configuration $\theta$ satisfying $K_p\theta_e = \mathfrak{m}gr\cos\theta$, i.e., the final error $\theta_e$ is nonzero when $\theta_d \neq \pm\pi/2$. The reason is that the robot must provide a nonzero torque to hold the link at rest at $\theta \neq \pm\pi/2$, but the PD control law creates a nonzero torque at rest only if $\theta_e \neq 0$. We can make this steady-state error small by increasing the gain $K_p$ but, as discussed above, there are practical limits.

To eliminate the steady-state error, we return to the PID controller by setting $K_i > 0$. This allows a nonzero steady-state torque even with zero position error; only the *integrated* error must be nonzero. Figure 11.13 demonstrates the effect of adding the integral term to the controller.

To see how this works, write down the setpoint error dynamics

$$M\ddot{\theta}_e + (b + K_d)\dot{\theta}_e + K_p\theta_e + K_i \int \theta_e(\mathrm{t})dt = \tau_{\text{dist}}, \tag{11.28}$$

where $\tau_{\text{dist}}$ is a disturbance torque substituted for the gravity term $\mathfrak{m}gr\cos\theta$. Taking derivatives of both sides, we get the third-order error dynamics

$$M\theta_e^{(3)} + (b + K_d)\ddot{\theta}_e + K_p\dot{\theta}_e + K_i\theta_e = \dot{\tau}_{\text{dist}}. \tag{11.29}$$

If $\tau_{\text{dist}}$ is constant then the right-hand side of Equation (11.29) is zero, and its characteristic equation is

$$s^3 + \frac{b + K_d}{M}s^2 + \frac{K_p}{M}s + \frac{K_i}{M} = 0. \tag{11.30}$$

If all roots of Equation (11.30) have a negative real part then the error dynamics is stable, and $\theta_e$ converges to zero. (While the disturbance torque due to gravity is not constant as the link rotates, it approaches a constant as $\dot{\theta}$ goes to zero, and therefore similar reasoning holds close to the equilibrium $\theta_e = 0$.)

For all the roots of Equation (11.30) to have a negative real component, the following conditions on the control gains must be satisfied for stability (Section 11.2.2.2):

$$K_d > -b$$
$$K_p > 0$$
$$\frac{(b + K_d)K_p}{M} > K_i > 0.$$

Thus the new gain $K_i$ must satisfy both a lower *and* an upper bound (Figure 11.14). A reasonable design strategy is to choose $K_p$ and $K_d$ for a good transient response and then choose $K_i$ large enough that it is helpful in reducing or eliminating steady-state errors but small enough that it does not significantly impact stability. In the example of Figure 11.13, the relatively large $K_i$ worsens the transient response, giving significant overshoot, but steady-state error is eliminated.

In practice, $K_i = 0$ for many robot controllers, since stability is paramount. Other techniques can be employed to limit the adverse stability effects of integral control, such as **integrator anti-windup**, which places a limit on how large the error integral is allowed to grow.

Pseudocode for the PID control algorithm is given in Figure 11.15.

While our analysis has focused on setpoint control, the PID controller applies perfectly well to trajectory following, where $\dot{\theta}_d(t) \neq 0$. Integral control will not eliminate tracking error along arbitrary trajectories, however.

## 11.4.1.2   Feedforward Control

Another strategy for trajectory following is to use a model of the robot's dynamics to proactively generate torques instead of waiting for errors. Let the

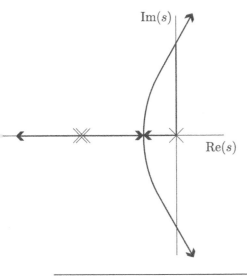

Im(s)

Re(s)

**Figure 11.14** The movement of the three roots of Equation (11.30) as $K_i$ increases from zero. First a PD controller is chosen with $K_p$ and $K_d$ yielding critical damping, giving rise to two collocated roots on the negative real axis. Adding an infinitesimal gain $K_i > 0$ creates a third root at the origin. As we increase $K_i$, one of the two collocated roots moves to the left on the negative real axis while the other two roots move toward each other, meet, break away from the real axis, begin curving to the right, and finally move into the right half-plane when $K_i = (b + K_d)K_p/M$. The system is unstable for larger values of $K_i$.

```
time = 0                    // dt = servo cycle time
eint = 0                    // error integral
qprev = senseAngle          // initial joint angle q
loop
  [qd,qdotd] = trajectory(time) // from trajectory generator

  q = senseAngle            // sense actual joint angle
  qdot = (q - qprev)/dt     // simple velocity calculation
  qprev = q

  e = qd - q
  edot = qdotd - qdot
  eint = eint + e*dt

  tau = Kp*e + Kd*edot + Ki*eint
  commandTorque(tau)

  time = time + dt
end loop
```

**Figure 11.15** Pseudocode for PID control.

controller's model of the dynamics be

$$\tau = \tilde{M}(\theta)\ddot{\theta} + \tilde{h}(\theta,\dot{\theta}), \tag{11.31}$$

where the model is perfect if $\tilde{M}(\theta) = M(\theta)$ and $\tilde{h}(\theta,\dot{\theta}) = h(\theta,\dot{\theta})$. Note that the inertia model $\tilde{M}(\theta)$ is written as a function of the configuration $\theta$. While the inertia of our simple one-joint robot is not a function of configuration, writing the equations in this way allows us to re-use Equation (11.31) for multi-joint systems in Section 11.4.2.

```
time = 0                                    // dt = servo cycle time
loop
  [qd,qdotd,qdotdotd] = trajectory(time)    // trajectory generator
  tau = Mtilde(qd)*qdotdotd + htilde(qd,qdotd) // calculate dynamics
  commandTorque(tau)
  time = time + dt
end loop
```

**Figure 11.16** Pseudocode for feedforward control.

Given $\theta_d$, $\dot{\theta}_d$, and $\ddot{\theta}_d$ from a trajectory generator, the feedforward torque is calculated as

$$\tau(t) = \tilde{M}(\theta_d(t))\ddot{\theta}_d(t) + \tilde{h}(\theta_d(t), \dot{\theta}_d(t)). \tag{11.32}$$

If the model of the robot dynamics is exact, and there are no initial state errors, then the robot follows the desired trajectory exactly.

A pseudocode implementation of feedforward control is given in Figure 11.16.

Figure 11.17 shows two examples of feedforward-trajectory following for the link under gravity. Here, the controller's dynamic model is correct except that it has $\tilde{r} = 0.08$ m, when actually $r = 0.1$ m. In Task 1 the error stays small, as unmodeled gravity effects provide a spring-like force to $\theta = -\pi/2$, accelerating the robot at the beginning and decelerating it at the end. In Task 2, unmodeled gravity effects act against the desired motion, resulting in a larger tracking error.

Because there are always modeling errors, feedforward control is always used in conjunction with feedback, as discussed next.

### 11.4.1.3    Feedforward Plus Feedback Linearization

All practical controllers use feedback, as no model of robot and environment dynamics will be perfect. Nonetheless, a good model can be used to improve performance and simplify analysis.

Let's combine PID control with a model of the robot dynamics $\{\tilde{M}, \tilde{h}\}$ to achieve the error dynamics

$$\ddot{\theta}_e + K_d\dot{\theta}_e + K_p\theta_e + K_i \int \theta_e(t)dt = 0 \tag{11.33}$$

along arbitrary trajectories, not just to a setpoint. The error dynamics (11.33) and a proper choice of PID gains ensure exponential decay of the trajectory error.

Since $\ddot{\theta}_e = \ddot{\theta}_d - \ddot{\theta}$, to achieve the error dynamics (11.33) we choose the robot's commanded acceleration to be

$$\ddot{\theta} = \ddot{\theta}_d - \ddot{\theta}_e,$$

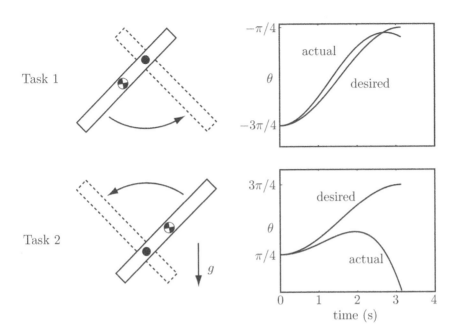

**Figure 11.17** Results of feedforward control with an incorrect model: $\tilde{r} = 0.08$ m but $r = 0.1$ m. The checkered disks indicate the center of mass. The desired trajectory in Task 1 is $\theta_d(t) = -\pi/2 - (\pi/4)\cos t$ for $0 \le t \le \pi$. The desired trajectory for Task 2 is $\theta_d(t) = \pi/2 - (\pi/4)\cos t$ for $0 \le t \le \pi$.

then combine this with Equation (11.33) to get

$$\ddot{\theta} = \ddot{\theta}_d + K_d \dot{\theta}_e + K_p \theta_e + K_i \int \theta_e(t)dt. \qquad (11.34)$$

Substituting $\ddot{\theta}$ from Equation (11.34) into a model of the robot dynamics $\{\tilde{M}, \tilde{h}\}$, we get the **feedforward plus feedback linearizing controller**, also called the **inverse dynamics controller** or the **computed torque controller**:

$$\boxed{\tau = \tilde{M}(\theta)\left(\ddot{\theta}_d + K_p\theta_e + K_i \int \theta_e(t)dt + K_d\dot{\theta}_e\right) + \tilde{h}(\theta, \dot{\theta}).} \qquad (11.35)$$

This controller includes a feedforward component due to the use of the planned acceleration $\ddot{\theta}_d$ and is called feedback linearizing because feedback of $\theta$ and $\dot{\theta}$ is used to generate the linear error dynamics. The $\tilde{h}(\theta, \dot{\theta})$ term cancels the dynamics that depends nonlinearly on the state, and the inertia model $\tilde{M}(\theta)$ converts the desired joint accelerations into joint torques, realizing the simple linear error dynamics (11.33).

A block diagram of the computed torque controller is shown in Figure 11.18. The gains $K_p, K_i$, and $K_d$ are chosen to place the roots of the characteristic

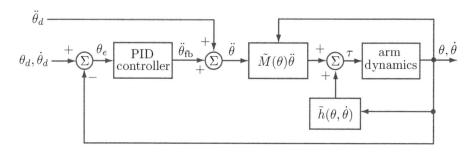

**Figure 11.18** Computed torque control. The feedforward acceleration $\ddot{\theta}_d$ is added to the acceleration $\ddot{\theta}_{\text{fb}}$ computed by the PID feedback controller to create the commanded acceleration $\ddot{\theta}$.

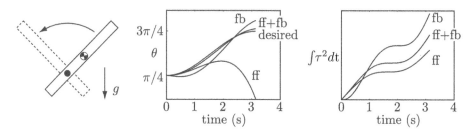

**Figure 11.19** Performance of feedforward only (ff), feedback only (fb), and computed torque control (ff+fb). The PID gains are taken from Figure 11.13, and the feedforward modeling error is taken from Figure 11.17. The desired motion is Task 2 from Figure 11.17 (left-hand plot). The center plot shows the tracking performance of the three controllers. The right-hand plot shows $\int \tau^2(t)dt$, a standard measure of the control effort, for each of the three controllers. These plots show typical behavior: the computed torque controller yields better tracking than either feedforward or feedback alone, with less control effort than feedback alone.

equation so as to achieve good transient response. In practice $K_i$ is often chosen to be zero.

Figure 11.19 shows the typical behavior of computed torque control relative to feedforward and feedback only. Pseudocode is given in Figure 11.20.

## 11.4.2    Motion Control of a Multi-Joint Robot

The methods applied above for a single-joint robot carry over directly to $n$-joint robots. The difference is that the dynamics (11.22) now takes the more general, vector-valued, form

$$\tau = M(\theta)\ddot{\theta} + h(\theta, \dot{\theta}), \tag{11.36}$$

where the $n \times n$ positive-definite mass matrix $M$ is now a function of the configuration $\theta$. In general, the components of the dynamics (11.36) are coupled – the acceleration of a joint is a function of the positions, velocities, and torques at other joints.

```
time = 0                              // dt = cycle time
eint = 0                              // error integral
qprev = senseAngle                    // initial joint angle q
loop
  [qd,qdotd,qdotdotd] = trajectory(time) // from trajectory generator

  q = senseAngle                      // sense actual joint angle
  qdot = (q - qprev)/dt               // simple velocity calculation
  qprev = q

  e = qd - q
  edot = qdotd - qdot
  eint = eint + e*dt

  tau = Mtilde(q)*(qdotdotd+Kp*e+Kd*edot+Ki*eint) + htilde(q,qdot)
  commandTorque(tau)

  time = time + dt
end loop
```

**Figure 11.20** Pseudocode for the computed torque controller.

We distinguish between two types of control for multi-joint robots: **decentralized** control, where each joint is controlled separately with no sharing of information between joints, and **centralized** control, where full state information for each of the $n$ joints is available to calculate the controls for each joint.

### 11.4.2.1 Decentralized Multi-Joint Control
The simplest method for controlling a multi-joint robot is to apply at each joint an independent controller, such as the single-joint controllers discussed in Section 11.4.1. Decentralized control is appropriate when the dynamics are decoupled, at least approximately. The dynamics are decoupled when the acceleration of each joint depends only on the torque, position, and velocity of that joint. This requires that the mass matrix be diagonal, as in Cartesian or **gantry** robots, where the first three axes are prismatic and orthogonal. This kind of robot is equivalent to three single-joint systems.

Approximate decoupling is also achieved in highly geared robots in the absence of gravity. The mass matrix $M(\theta)$ is nearly diagonal, as it is dominated by the apparent inertias of the motors themselves (see Section 8.9.2). Significant friction at the individual joints also contributes to the decoupling of the dynamics.

### 11.4.2.2 Centralized Multi-Joint Control
When gravity forces and torques are significant and coupled, or when the mass matrix $M(\theta)$ is not well approximated by a diagonal matrix, decentralized control may not yield acceptable performance. In this case the computed torque controller (11.35) of Figure 11.18 can be generalized to a multi-joint robot. The

configurations $\theta$ and $\theta_d$ and the error $\theta_e = \theta_d - \theta$ are now $n$-vectors, and the positive scalar gains become positive-definite matrices $K_p, K_i, K_d$:

$$\tau = \tilde{M}(\theta)\left(\ddot{\theta}_d + K_p\theta_e + K_i \int \theta_e(\mathrm{t})dt + K_d\dot{\theta}_e\right) + \tilde{h}(\theta, \dot{\theta}). \qquad (11.37)$$

Typically, we choose the gain matrices as $k_pI$, $k_iI$, and $k_dI$, where $k_p$, $k_i$, and $k_d$ are nonnegative scalars. Commonly, $k_i$ is chosen to be zero. In the case of an exact dynamics model for $\tilde{M}$ and $\tilde{h}$, the error dynamics of each joint reduces to the linear dynamics (11.33). The block diagram and pseudocode for this control algorithm are found in Figures 11.18 and 11.20, respectively.

Implementing the control law (11.37) requires calculating potentially complex dynamics. We may not have a good model of these dynamics, or the equations may be too computationally expensive to calculate at servo rate. In this case, if the desired velocities and accelerations are small, an approximation to (11.37) can be obtained using only PID control and gravity compensation:

$$\tau = K_p\theta_e + K_i \int \theta_e(\mathrm{t})dt + K_d\dot{\theta}_e + \tilde{g}(\theta). \qquad (11.38)$$

With zero friction, perfect gravity compensation, and PD setpoint control ($K_i = 0$ and $\dot{\theta}_d = \ddot{\theta}_d = 0$), the controlled dynamics can be written as

$$M(\theta)\ddot{\theta} + C(\theta, \dot{\theta})\dot{\theta} = K_p\theta_e - K_d\dot{\theta}, \qquad (11.39)$$

where the Coriolis and centripetal terms are written $C(\theta, \dot{\theta})\dot{\theta}$. We can now define a virtual "error energy," which is the sum of an "error potential energy" stored in the virtual spring $K_p$ and an "error kinetic energy":

$$V(\theta_e, \dot{\theta}_e) = \frac{1}{2}\theta_e^{\mathrm{T}}K_p\theta_e + \frac{1}{2}\dot{\theta}_e^{\mathrm{T}}M(\theta)\dot{\theta}_e. \qquad (11.40)$$

Since $\dot{\theta}_d = 0$, this reduces to

$$V(\theta_e, \dot{\theta}) = \frac{1}{2}\theta_e^{\mathrm{T}}K_p\theta_e + \frac{1}{2}\dot{\theta}^{\mathrm{T}}M(\theta)\dot{\theta}. \qquad (11.41)$$

Taking the time derivative and substituting (11.39) into it, we get

$$\dot{V} = -\dot{\theta}^{\mathrm{T}}K_p\theta_e + \dot{\theta}^{\mathrm{T}}M(\theta)\ddot{\theta} + \frac{1}{2}\dot{\theta}^{\mathrm{T}}\dot{M}(\theta)\dot{\theta}$$

$$= -\dot{\theta}^{\mathrm{T}}K_p\theta_e + \dot{\theta}^{\mathrm{T}}\left(K_p\theta_e - K_d\dot{\theta} - C(\theta, \dot{\theta})\dot{\theta}\right) + \frac{1}{2}\dot{\theta}^{\mathrm{T}}\dot{M}(\theta)\dot{\theta}. \qquad (11.42)$$

Rearranging, and using the fact that $\dot{M} - 2C$ is skew symmetric (Proposition 8.1.2), we get

$$\dot{V} = -\dot{\theta}^{\mathrm{T}}K_p\theta_e + \dot{\theta}^{\mathrm{T}}\left(K_p\theta_e - K_d\dot{\theta}\right) + \frac{1}{2}\dot{\theta}^{\mathrm{T}}\underbrace{\left(\dot{M}(\theta) - 2C(\theta, \dot{\theta})\right)}_{0}\dot{\theta}$$

$$= -\dot{\theta}^{\mathrm{T}}K_d\dot{\theta} \leq 0. \qquad (11.43)$$

This shows that the error energy is decreasing when $\dot{\theta} \neq 0$. If $\dot{\theta} = 0$ and $\theta \neq \theta_d$, the virtual spring ensures that $\ddot{\theta} \neq 0$, so $\dot{\theta}_e$ will again become nonzero and more error energy will be dissipated. Thus, by the Krasovskii–LaSalle invariance principle (Exercise 11.12), the total error energy decreases monotonically and the robot converges to rest at $\theta_d$ ($\theta_e = 0$) from any initial state.

### 1.4.3  Task-Space Motion Control

In Section 11.4.2 we focused on motion control in joint space. On the one hand, this is convenient because joint limits are easily expressed in this space, and the robot should be able to execute any joint-space path respecting these limits. Trajectories are naturally described by the joint variables, and there are no issues of singularities or redundancy.

On the other hand, since the robot interacts with the external environment and objects in it, it may be more convenient to express the motion as a trajectory of the end-effector in task space. Let the end-effector trajectory be specified by $(X(t), \mathcal{V}_b(t))$, where $X \in SE(3)$ and $[\mathcal{V}_b] = X^{-1}\dot{X}$, i.e., the twist $\mathcal{V}_b$ is expressed in the end-effector frame {b}. Provided that the corresponding trajectory in joint space is feasible, we now have two options for control: (1) convert to a joint-space trajectory and proceed with controls as in Section 11.4.2; or (2) express the robot dynamics and control law in the task space.

The first option is to convert the trajectory to joint space. The forward kinematics are $X = T(\theta)$ and $\mathcal{V}_b = J_b(\theta)\dot{\theta}$. Then the joint-space trajectory is obtained from the task-space trajectory using inverse kinematics (Chapter 6):

$$\text{(inverse kinematics)} \quad \theta(t) = T^{-1}(X(t)), \tag{11.44}$$

$$\dot{\theta}(t) = J_b^\dagger(\theta(t))\mathcal{V}_b(t), \tag{11.45}$$

$$\ddot{\theta}(t) = J_b^\dagger(\theta(t))\left(\dot{\mathcal{V}}_b(t) - \dot{J}_b(\theta(t))\dot{\theta}(t)\right). \tag{11.46}$$

A drawback of this approach is that we must calculate the inverse kinematics, $J_b^\dagger$, and $\dot{J}_b$, which may require significant computing power.

The second option is to express the robot's dynamics in task-space coordinates, as discussed in Section 8.6. Recall the task-space dynamics

$$\mathcal{F}_b = \Lambda(\theta)\dot{\mathcal{V}}_b + \eta(\theta, \mathcal{V}_b).$$

The joint forces and torques $\tau$ are related to the wrenches $\mathcal{F}_b$ expressed in the end-effector frame by $\tau = J_b^{\mathrm{T}}(\theta)\mathcal{F}_b$.

We can now write a control law in task space inspired by the computed torque control law in joint coordinates (11.37),

$$\tau =$$

$$J_b^{\mathrm{T}}(\theta)\left(\tilde{\Lambda}(\theta)\left(\frac{d}{dt}([\mathrm{Ad}_{X^{-1}X_d}]\mathcal{V}_d) + K_p X_e + K_i \int X_e(t)dt + K_d \mathcal{V}_e\right) + \tilde{\eta}(\theta, \mathcal{V}_b)\right),$$

$$\tag{11.47}$$

where $\{\tilde{\Lambda}, \tilde{\eta}\}$ represents the controller's dynamics model and $\frac{d}{dt}([\mathrm{Ad}_{X^{-1}X_d}]\mathcal{V}_d)$ is the feedforward acceleration expressed in the actual end-effector frame at $X$ (this term can be approximated as $\dot{\mathcal{V}}_d$ at states close to the reference state). The configuration error $X_e$ satisfies $[X_e] = \log(X^{-1}X_d)$: $X_e$ is the twist, expressed in the end-effector frame, which, if followed for unit time, would move the current configuration $X$ to the desired configuration $X_d$. The velocity error is calculated as

$$\mathcal{V}_e = [\mathrm{Ad}_{X^{-1}X_d}]\mathcal{V}_d - \mathcal{V}.$$

The transform $[\mathrm{Ad}_{X^{-1}X_d}]$ expresses the reference twist $\mathcal{V}_d$, which is expressed in the frame $X_d$, as a twist in the end-effector frame at $X$, in which the actual velocity $\mathcal{V}$ is represented, so the two expressions can be differenced.

## 11.5     Force Control

When the task is not to create motions at the end-effector but to apply forces and torques to the environment, **force control** is needed. Pure force control is only possible if the environment provides resistance forces in every direction (e.g., if the end-effector is embedded in concrete or attached to a spring providing resistance in every motion direction). Pure force control is something of an abstraction, as robots are usually able to move freely in at least *some* direction. It is a useful abstraction, however, that leads to hybrid motion-force control as discussed in Section 11.6.

In ideal force control, the force applied by the end-effector is unaffected by disturbance motions applied to the end-effector. This is dual to the case of ideal motion control, where the motion is unaffected by disturbance forces.

Let $\mathcal{F}_{\mathrm{tip}}$ be the wrench applied by the manipulator to the environment. The manipulator dynamics can be written as

$$M(\theta)\ddot{\theta} + c(\theta, \dot{\theta}) + g(\theta) + b(\dot{\theta}) + J^{\mathrm{T}}(\theta)\mathcal{F}_{\mathrm{tip}} = \tau, \tag{11.48}$$

where $\mathcal{F}_{\mathrm{tip}}$ and $J(\theta)$ are defined in the same frame (the space frame or the end-effector frame). Since the robot typically moves slowly (or not at all) during a force control task, we can ignore the acceleration and velocity terms to get

$$g(\theta) + J^{\mathrm{T}}(\theta)\mathcal{F}_{\mathrm{tip}} = \tau. \tag{11.49}$$

In the absence of any direct measurements of the force–torque at the robot end-effector, joint-angle feedback alone can be used to implement the force-control law

$$\tau = \tilde{g}(\theta) + J^{\mathrm{T}}(\theta)\mathcal{F}_d, \tag{11.50}$$

where $\tilde{g}(\theta)$ is a model of the gravitational torques and $\mathcal{F}_d$ is the desired wrench. This control law requires a good model for gravity compensation as well as precise control of the torques produced at the robot joints. In the case of a DC electric

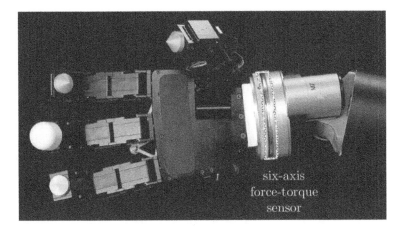

**Figure 11.21** A six-axis force–torque sensor mounted between a robot arm and its end-effector.

motor without gearing, torque control can be achieved by current control of the motor. In the case of a highly geared actuator, a large friction torque in the gearing can degrade the quality of torque control achieved using only current control. In this case, the output of the gearing can be instrumented with strain gauges to measure the joint torque directly; this information is fed back to a local controller that modulates the motor current to achieve the desired output torque.

Another solution is to equip the robot arm with a six-axis force-torque sensor between the arm and the end-effector to directly measure the end-effector wrench $\mathcal{F}_{\text{tip}}$ (Figure 11.21). Consider a PI force controller[2] with a feedforward term and gravity compensation,

$$\tau = \tilde{g}(\theta) + J^{\mathrm{T}}(\theta)\left(\mathcal{F}_d + K_{fp}\mathcal{F}_e + K_{fi}\int \mathcal{F}_e(t)dt\right), \tag{11.51}$$

where $\mathcal{F}_e = \mathcal{F}_d - \mathcal{F}_{\text{tip}}$ and $K_{fp}$ and $K_{fi}$ are positive-definite proportional and integral gain matrices, respectively. In the case of perfect gravity modeling, plugging the force controller (11.51) into the dynamics (11.49), we get the error dynamics

$$K_{fp}\mathcal{F}_e + K_{fi}\int \mathcal{F}_e(t)dt = 0. \tag{11.52}$$

In the case of a nonzero but constant force disturbance on the right-hand side

---

[2] Derivative control is not typically relevant for two reasons: (1) force measurements are often noisy, so their computed time derivatives are nearly meaningless; and (2) we are assuming the direct control of the joint torques and forces, and our simple rigid-body dynamics models imply direct transmission to the end-effector forces – there is no dynamics that integrates our control commands to produce the desired behavior, unlike with motion control.

of (11.52), arising from an incorrect model of $\tilde{g}(\theta)$, for example, we take the derivative to get

$$K_{fp}\dot{\mathcal{F}}_e + K_{fi}\mathcal{F}_e = 0, \tag{11.53}$$

showing that $\mathcal{F}_e$ converges to zero for positive-definite $K_{fp}$ and $K_{fi}$.

The control law (11.51) is simple and appealing but potentially dangerous if incorrectly applied. If there is nothing for the robot to push against, it will accelerate in a failing attempt to create end-effector forces. Since a typical force-control task requires little motion, we can limit this acceleration by adding velocity damping. This gives the modified control law

$$\tau = \tilde{g}(\theta) + J^{\mathrm{T}}(\theta) \left( \mathcal{F}_d + K_{fp}\mathcal{F}_e + K_{fi} \int \mathcal{F}_e(t)dt - K_{\mathrm{damp}}\mathcal{V} \right), \tag{11.54}$$

where $K_{\mathrm{damp}}$ is positive definite.

## 11.6     Hybrid Motion–Force Control

Most tasks requiring the application of controlled forces also require the generation of controlled motions. **Hybrid motion-force control** is used to achieve this. If the the task space is $n$-dimensional then we are free to specify $n$ of the $2n$ forces and motions at any time $t$; the other $n$ are determined by the environment. Apart from this constraint, we also should not specify forces and motions in the "same direction," as then they are not independent.

For example, consider a two-dimensional environment modeled by a damper, $f = B_{\mathrm{env}}v$, where

$$B_{\mathrm{env}} = \begin{bmatrix} 2 & 1 \\ 1 & 1 \end{bmatrix}.$$

Defining the components of $v$ and $f$ as $(v_1, v_2)$ and $(f_1, f_2)$, we have $f_1 = 2v_1 + v_2$ and $f_2 = v_1 + v_2$. We have $n = 2$ freedoms to choose among the $2n = 4$ velocities and forces at any time. As an example, we can specify both $f_1$ and $v_1$ independently, because $B_{\mathrm{env}}$ is not diagonal. Then $v_2$ and $f_2$ are determined by $B_{\mathrm{env}}$. We cannot independently control both $f_1$ and $2v_1 + v_2$, as these are in the "same direction" according to the environment.

### 11.6.1     Natural and Artificial Constraints

A particularly interesting case occurs when the environment is infinitely stiff (rigid constraints) in $k$ directions and unconstrained in $n - k$ directions. In this case, we cannot choose *which* of the $2n$ motions and forces to specify – the contact with the environment chooses the $k$ directions in which the robot can freely apply forces and the $n - k$ directions of free motion. For example, consider a task space with the $n = 6$ dimensions of $SE(3)$. Then a robot firmly grasping a cabinet door has $6 - k = 1$ motion freedom of its end-effector, i.e., rotation about the cabinet hinges, and therefore $k = 5$ force freedoms; the robot can apply any wrench that has zero moment about the axis of the hinges without moving the door.

As another example, a robot writing on a chalkboard may freely control the force into the board ($k = 1$), but it cannot penetrate the board; it may freely move with $6 - k = 5$ degrees of freedom (two specifying the motion of the tip of the chalk in the plane of the board and three describing the orientation of the chalk), but it cannot independently control the forces in these directions.

The chalk example comes with two caveats. The first is due to friction – the chalk-wielding robot can actually control forces tangent to the plane of the board provided that the requested motion in the plane of the board is zero and the requested tangential forces do not exceed the static friction limit determined by the friction coefficient and the normal force into the board (see the discussion of friction modeling in Section 12.2). Within this regime, the robot has three motion freedoms (rotations about three axes intersecting at the contact between the chalk and the board) and three linear force freedoms. Second, the robot could decide to pull away from the board. In this regime, the robot would have six motion freedoms and no force freedoms. Thus the configuration of the robot is not the only factor determining the directions of the motion and force freedoms. Nonetheless, in this section we consider the simplified case where the motion and force freedoms are determined solely by the robot's configuration, and all constraints are equality constraints. For example, the inequality velocity constraint of the board (the chalk cannot penetrate the board) is treated as an equality constraint (the robot also does not pull the chalk away from the board).

As a final example, consider a robot erasing a frictionless chalkboard using an eraser modeled as a rigid block (Figure 11.22). Let $X(t) \in SE(3)$ be the configuration of the block's frame {b} relative to a space frame {s}. The body-frame twist and wrench are written $\mathcal{V}_b = (\omega_x, \omega_y, \omega_z, v_x, v_y, v_z)$ and $\mathcal{F}_b = (m_x, m_y, m_z, f_x, f_y, f_z)$, respectively. Maintaining contact with the board puts $k = 3$ constraints on the twist:

$$\omega_x = 0,$$
$$\omega_y = 0,$$
$$v_z = 0.$$

In the language of Chapter 2, these velocity constraints are holonomic – the differential constraints can be integrated to give configuration constraints.

These constraints are called **natural constraints**, specified by the environment. There are $6 - k = 3$ natural constraints on the wrench, too: $m_z = f_x = f_y = 0$. In light of the natural constraints, we can freely specify any twist of the eraser satisfying the $k = 3$ velocity constraints and any wrench satisfying the $6 - k = 3$ wrench constraints (provided that $f_z < 0$, to maintain contact with the

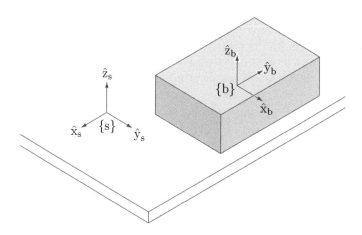

**Figure 11.22** The fixed space frame {s} is attached to the chalkboard and the body frame {b} is attached to the center of the eraser.

board). These motion and force specifications are called **artificial constraints.**
Below is an example set of artificial constraints with the corresponding natural
constraints:

| natural constraint | artificial constraint |
|:---:|:---:|
| $\omega_x = 0$ | $m_x = 0$ |
| $\omega_y = 0$ | $m_y = 0$ |
| $m_z = 0$ | $\omega_z = 0$ |
| $f_x = 0$ | $v_x = k_1$ |
| $f_y = 0$ | $v_y = 0$ |
| $v_z = 0$ | $f_z = k_2 < 0$ |

The artificial constraints cause the eraser to move with $v_x = k_1$ while applying
a constant force $k_2$ against the board.

### 11.6.2    A Hybrid Motion–Force Controller

We now return to the problem of designing a hybrid motion–force controller. If
the environment is rigid, then we can express the $k$ natural constraints on the
velocity in task space as the Pfaffian constraints

$$A(\theta)\mathcal{V} = 0, \tag{11.55}$$

where $A(\theta) \in \mathbb{R}^{k \times 6}$ for twists $\mathcal{V} \in \mathbb{R}^6$. This formulation includes holonomic and
nonholonomic constraints.

If the task-space dynamics of the robot (Section 8.6), in the absence of constraints, is given by

$$\mathcal{F} = \Lambda(\theta)\dot{\mathcal{V}} + \eta(\theta, \mathcal{V}),$$

where $\tau = J^{\mathrm{T}}(\theta)\mathcal{F}$ are the joint torques and forces created by the actuators, then the constrained dynamics, following Section 8.7, is

$$\mathcal{F} = \Lambda(\theta)\dot{\mathcal{V}} + \eta(\theta, \mathcal{V}) + \underbrace{A^{\mathrm{T}}(\theta)\lambda}_{\mathcal{F}_{\text{tip}}}, \tag{11.56}$$

where $\lambda \in \mathbb{R}^k$ are Lagrange multipliers and $\mathcal{F}_{\text{tip}}$ is the wrench that the robot applies against the constraints. The requested wrench $\mathcal{F}_d$ must lie in the column space of $A^{\mathrm{T}}(\theta)$.

Since Equation (11.55) must be satisfied at all times, we can replace it by the time derivative

$$A(\theta)\dot{\mathcal{V}} + \dot{A}(\theta)\mathcal{V} = 0. \tag{11.57}$$

To ensure that Equation (11.57) is satisfied when the system state already satisfies $A(\theta)\mathcal{V} = 0$, any requested acceleration $\dot{\mathcal{V}}_d$ should satisfy $A(\theta)\dot{\mathcal{V}}_d = 0$.

Now solving Equation (11.56) for $\dot{\mathcal{V}}$, substituting the result into (11.57), and solving for $\lambda$, we get

$$\lambda = (A\Lambda^{-1}A^{\mathrm{T}})^{-1}(A\Lambda^{-1}(\mathcal{F} - \eta) - \dot{A}\mathcal{V}), \tag{11.58}$$

where we have used $-A\dot{\mathcal{V}} = \dot{A}\mathcal{V}$ from Equation (11.57). Using Equation (11.58), we can calculate the wrench $\mathcal{F}_{\text{tip}} = A^{\mathrm{T}}(\theta)\lambda$ that the robot applies against the constraints.

Substituting Equation (11.58) into Equation (11.56) and manipulating, the $n$ equations of the constrained dynamics (11.56) can be expressed as the $n - k$ independent motion equations

$$P(\theta)\mathcal{F} = P(\theta)(\Lambda(\theta)\dot{\mathcal{V}} + \eta(\theta, \mathcal{V})), \tag{11.59}$$

where

$$P = I - A^{\mathrm{T}}(A\Lambda^{-1}A^{\mathrm{T}})^{-1}A\Lambda^{-1} \tag{11.60}$$

and $I$ is the identity matrix. The $n \times n$ matrix $P(\theta)$ has rank $n - k$ and projects an arbitrary manipulator wrench $\mathcal{F}$ onto the subspace of wrenches that move the end-effector tangent to the constraints. The rank-$k$ matrix $I - P(\theta)$ projects an arbitrary wrench $\mathcal{F}$ onto the subspace of wrenches that act against the constraints. Thus $P$ partitions the $n$-dimensional force space into wrenches that address the motion control task and wrenches that address the force control task.

Our hybrid motion–force controller is simply the sum of a task-space motion controller, derived from the computed torque control law (11.47), and a task-space force controller (11.51), each projected to generate forces in its appropriate subspace. Assuming wrenches and twists expressed in the end-effector frame {b},

$$
\tau = J_b^{\mathrm{T}}(\theta) \left( \underbrace{P(\theta) \left( \tilde{\Lambda}(\theta) \left( \frac{d}{dt}([\mathrm{Ad}_{X^{-1}X_d}]\mathcal{V}_d) + K_p X_e + K_i \int X_e(t)dt + K_d \mathcal{V}_e \right) \right)}_{\text{motion control}} \right.
$$

$$
+ \underbrace{(I - P(\theta)) \left( \mathcal{F}_d + K_{fp}\mathcal{F}_e + K_{fi} \int \mathcal{F}_e(t)dt \right)}_{\text{force control}}
$$

$$
\left. + \underbrace{\tilde{\eta}(\theta, \mathcal{V}_b)}_{\text{Coriolis and gravity}} \right). \tag{11.61}
$$

Because the dynamics of the two controllers are decoupled by the orthogonal projections $P$ and $I - P$, the controller inherits the error dynamics and stability analyses of the individual force and motion controllers on their respective subspaces.

A difficulty in implementing the hybrid control law (11.61) in rigid environments is knowing the form of the constraints $A(\theta)\mathcal{V} = 0$ active at any time. This is necessary to specify the desired motion and force and to calculate the projections, but any model of the environment will have some uncertainty. One approach to dealing with this issue is to use a real-time estimation algorithm to identify the constraint directions on the basis of force feedback. Another is to sacrifice some performance by choosing low feedback gains, which makes the motion controller "soft" and the force controller more tolerant of force error. We can also build passive compliance into the structure of the robot itself to achieve a similar effect. In any case, some passive compliance is unavoidable, owing to flexibility in the joints and links.

## 11.7    Impedance Control

Ideal hybrid motion–force control in rigid environments demands extremes in robot **impedance**, which characterizes the change in endpoint motion as a function of disturbance forces. Ideal motion control corresponds to high impedance (little change in motion due to force disturbances) while ideal force control corresponds to low impedance (little change in force due to motion disturbances). In practice, there are limits to a robot's achievable impedance range.

In this section we consider the problem of impedance control, where the robot end-effector is asked to render particular mass, spring, and damper properties.[3] For example, a robot used as a haptic surgical simulator could be tasked with mimicking the mass, stiffness, and damping properties of a virtual surgical instrument in contact with virtual tissue.

---

[3] A popular subcategory of impedance control is **stiffness control** or **compliance control**, where the robot renders a virtual spring only.

The dynamics for a one-degree-of-freedom robot rendering an impedance can be written

$$\mathfrak{m}\ddot{x} + b\dot{x} + kx = f, \tag{11.62}$$

where $x$ is the position, $\mathfrak{m}$ is the mass, $b$ is the damping, $k$ is the stiffness, and $f$ is the force applied by the user (Figure 11.23). Loosely, we say that the robot renders high impedance if one or more of the $\{\mathfrak{m}, b, k\}$ parameters, usually including $b$ or $k$, is large. Similarly, we say that the impedance is low if all these parameters are small.

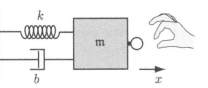

**Figure 11.23** A robot creating a one-dof mass–spring–damper virtual environment. A human hand applies a force $f$ to the haptic interface.

More formally, taking the Laplace transform[4] of Equation (11.62), we get

$$(\mathfrak{m}s^2 + bs + k)X(s) = F(s), \tag{11.63}$$

and the impedance is defined by the transfer function from position perturbations to forces, $Z(s) = F(s)/X(s)$. Thus impedance is frequency dependent, with a low-frequency response dominated by the spring and a high-frequency response dominated by the mass. The **admittance**, $Y(s)$, is the inverse of the impedance: $Y(s) = Z^{-1}(s) = X(s)/F(s)$.

A good motion controller is characterized by high impedance (low admittance), since $\Delta X = Y\Delta F$. If the admittance $Y$ is small then force perturbations $\Delta F$ produce only small position perturbations $\Delta X$. Similarly, a good force controller is characterized by low impedance (high admittance), since $\Delta F = Z\Delta X$ and a small $Z$ implies that motion perturbations produce only small force perturbations.

The goal of impedance control is to implement the task-space behavior

$$M\ddot{x} + B\dot{x} + Kx = f_{\text{ext}}, \tag{11.64}$$

where $x \in \mathbb{R}^n$ is the task-space configuration in a minimum set of coordinates, e.g., $x \in \mathbb{R}^3$; $M, B$, and $K$ are the positive-definite virtual mass, damping, and stiffness matrices to be simulated by the robot, and $f_{\text{ext}}$ is a force applied to the robot, perhaps by a user. The values of $M, B$, and $K$ may change, depending on the location in the virtual environment, in order to represent distinct objects for instance, but we focus on the case of constant values. We could also replace $\ddot{x}$, $\dot{x}$, and $x$ with small displacements $\Delta\ddot{x}$, $\Delta\dot{x}$, and $\Delta x$ from reference values in a controlled motion of the robot, but we will dispense with any such extra notation here.

The behavior (11.64) could be implemented in terms of twists and wrenches

---

[4] If you are unfamiliar with the Laplace transform and transfer functions, do not panic! We do not need the details here.

instead, replacing $f_{\text{ext}}$ by the (body or spatial) wrench $\mathcal{F}_{\text{ext}}$, $\dot{x}$ by the twist $\mathcal{V}$, $\ddot{x}$ by $\dot{\mathcal{V}}$, and $x$ by the exponential coordinates $\mathcal{S}\theta$. Alternatively, the linear and rotational behaviors can be decoupled, as discussed in Section 11.4.3.

There are two common ways to achieve the behavior (11.64).

- The robot senses the endpoint motion $x(t)$ and commands joint torques and forces to create $-f_{\text{ext}}$, the force to display to the user. Such a robot is called **impedance controlled**, as it implements a transfer function $Z(s)$ from motions to forces. Theoretically, an impedance-controlled robot should only be coupled to an admittance-type environment.
- The robot senses $f_{\text{ext}}$ using a wrist force–torque sensor and controls its motions in response. Such a robot is called **admittance controlled**, as it implements a transfer function $Y(s)$ from forces to motions. Theoretically, an admittance-controlled robot should only be coupled to an impedance-type environment.

### 11.7.1    Impedance-Control Algorithm

In an impedance-control algorithm, encoders, tachometers, and possibly accelerometers are used to estimate the joint and endpoint positions, velocities, and possibly accelerations. Often impedance-controlled robots are not equipped with a wrist force–torque sensor and instead rely on their ability to precisely control joint torques to render the appropriate end-effector force $-f_{\text{ext}}$ (Equation (11.64)). A good control law might be

$$
\tau = J^{\text{T}}(\theta) \left( \underbrace{\tilde{\Lambda}(\theta)\ddot{x} + \tilde{\eta}(\theta,\dot{x})}_{\text{arm dynamics compensation}} \;-\; \underbrace{(M\ddot{x} + B\dot{x} + Kx)}_{f_{\text{ext}}} \right), \tag{11.65}
$$

where the task-space dynamics model $\{\tilde{\Lambda}, \tilde{\eta}\}$ is expressed in terms of the coordinates $x$. Addition of an end-effector force–torque sensor allows the use of feedback terms to achieve more closely the desired interaction force $-f_{\text{ext}}$.

In the control law (11.65), it is assumed that $\ddot{x}$, $\dot{x}$, and $x$ are measured directly. Measurement of the acceleration $\ddot{x}$ is likely to be noisy, and there is the problem of attempting to compensate for the robot's mass after the acceleration has been sensed. Therefore, it is not uncommon to eliminate the mass compensation term $\tilde{\Lambda}(\theta)\ddot{x}$ and to set $M = 0$. The mass of the arm will be apparent to the user, but impedance-controlled manipulators are often designed to be lightweight. It is also not uncommon to assume small velocities and replace the nonlinear dynamics compensation with a simpler gravity-compensation model.

Problems can arise when (11.65) is used to simulate stiff environments (the case of large $K$). On the one hand, small changes in position, measured by encoders for example, lead to large changes in motor torques. This effective high gain, coupled with delays, sensor quantization, and sensor errors, can lead to oscillatory behavior or instability. On the other hand, the effective gains are low when emulating low-impedance environments. A lightweight backdrivable manipulator can excel at emulating such environments.

## .7.2  Admittance-Control Algorithm

In an admittance-control algorithm the force $f_{ext}$ applied by the user is sensed by the wrist load cell, and the robot responds with an end-effector acceleration satisfying Equation (11.64). A simple approach is to calculate the desired end-effector acceleration $\ddot{x}_d$ according to

$$M\ddot{x}_d + B\dot{x} + Kx = f_{ext},$$

where $(x, \dot{x})$ is the current state. Solving, we get

$$\ddot{x}_d = M^{-1}(f_{ext} - B\dot{x} - Kx). \tag{11.66}$$

For the Jacobian $J(\theta)$ defined by $\dot{x} = J(\theta)\dot{\theta}$, the desired joint accelerations $\ddot{\theta}_d$ can be solved as

$$\ddot{\theta}_d = J^\dagger(\theta)(\ddot{x}_d - \dot{J}(\theta)\dot{\theta}),$$

and inverse dynamics used to calculate the commanded joint forces and torques $\tau$. Simplified versions of this control law can be obtained when the goal is to simulate only a spring or a damper. To make the response smoother in the face of noisy force measurements, the force readings can be low-pass filtered.

Simulating a low-mass environment is challenging for admittance-controlled robots, as small forces produce large accelerations. The effective large gains can produce instability. Admittance control by a highly geared robot can excel at emulating stiff environments, however.

## 1.8  Low-Level Joint Force–Torque Control

Throughout this chapter we have been assuming that each joint produces the torque or force requested of it. In practice this ideal is not exactly achieved, and there are different approaches to approximating it. Some of the most common approaches using electric motors (Section 8.9.1) are listed below, along with their advantages and disadvantages relative to the previously listed approach. Here we assume a revolute joint and a rotary motor.

### Current Control of a Direct-Drive Motor

In this configuration, each joint has a motor amplifier and an electric motor with no gearhead. The torque of the motor approximately obeys the relationship $\tau = k_t I$, i.e., the torque is proportional to the current through the motor. The amplifier takes the requested torque, divides by the torque constant $k_t$, and generates the motor current $I$. To create the desired current, a current sensor integrated with the amplifier continuously measures the actual current through the motor, and the amplifier uses a local feedback control loop to adjust the time-averaged voltage across the motor to achieve the desired current. This local feedback loop runs at a higher rate than the control loop that generates the requested torques. A typical example is 10 kHz for the local current control loop and 1 kHz for the outer control loop requesting joint torques.

An issue with this configuration is that often an ungeared motor must be quite large to create sufficient torque for the application. The configuration can work if the motors are fixed to the ground and connected to the end-effector through cables or a closed-chain linkage. If the motors are moving, as do the motors at the joints of a serial chain, for example, large ungeared motors are often impractical.

### Current Control of a Geared Motor

This configuration is similar to the previous one, except that the motor has a gearhead (Section 8.9.1). A gear ratio $G > 1$ increases the torque available to the joint.

**Advantage:** A smaller motor can provide the necessary torques. The motor also operates at higher speeds, where it is more efficient at converting electrical power to mechanical power.

**Disadvantage:** The gearhead introduces backlash (the output of the gearhead can move without the input moving, making motion control near zero velocity challenging) and friction. Backlash can be nearly eliminated by using particular types of gearing, such as harmonic drive gears. Friction, however, cannot be eliminated. The nominal torque at the gearhead output is $Gk_tI$, but friction in the gearhead reduces the torque available and creates significant uncertainty in the torque actually produced.

### Current Control of a Geared Motor with Local Strain Gauge Feedback

This configuration is similar to the previous one, except that the harmonic drive gearing is instrumented with strain gauges that sense how much torque is actually being delivered at the output of the gearhead. This torque information is used by the amplifier in a local feedback controller to adjust the current in the motor so as to achieve the requested torque.

**Advantage:** Putting the sensor at the output of the gearing allows compensation of frictional uncertainties.

**Disadvantage:** There is additional complexity of the joint configuration. Also, harmonic drive gearing achieves near-zero backlash by introducing some torsional compliance in the gearset, and the added dynamics due to the presence of this torsional spring can complicate high-speed motion control.

### Series Elastic Actuator

A series elastic actuator (SEA) consists of an electric motor with a gearhead (often a harmonic drive gearhead) and a torsional spring attaching the output of the gearhead to the output of the actuator. It is similar to the previous configuration, except that the torsional spring constant of the added spring is much lower than the spring constant of the harmonic drive gearing. The angular deflection $\Delta\phi$ of the spring is often measured by optical, magnetic, or capacitive encoders. The torque delivered to the output of the actuator is $k\Delta\phi$, where $k$ is the torsional spring constant. The spring's deflection is fed to a local feedback controller that controls the current to the motor so as to achieve the desired spring deflection, and therefore the desired torque.

**Advantage:** The addition of the torsional spring makes the joint naturally "soft," and therefore well suited for human-robot interaction tasks. It also protects the gearing and motor from shocks at the output, such as when the output link hits something hard in the environment.

**Disadvantage:** There is additional complexity of the joint configuration. Also, the added dynamics due to the softer spring make it more challenging to control high-speed or high-frequency motions at the output.

In 2011, NASA's Robonaut 2 (R2) became the first humanoid robot in space, performing operations on the International Space Station. Robonaut 2 incorporates a number of SEAs, including the hip actuator shown in Figure 11.24.

## 11.9 Other Topics

### Robust Control
While all stable feedback controllers confer some amount of robustness of operation to uncertainty, the field of robust control deals with designing controllers that explicitly guarantee the performance of a robot subject to bounded parametric uncertainties such as those in its inertial properties.

### Adaptive Control
The adaptive control of robots involves estimating the robot's inertial or other parameters during execution and updating the control law in real time to incorporate those estimates.

### Iterative Learning Control
Iterative learning control (ILC) generally focuses on repetitive tasks. If a robot performs the same pick-and-place operation over and over again, the trajectory errors from the previous execution can be used to modify the feedforward control for the next execution. In this way, the robot improves its performance over time, driving the execution error toward zero. This type of learning control differs from adaptive control in that the "learned" information is generally nonparametric and useful only for a single trajectory. However, ILC can account for effects that have not been parametrized in a particular model.

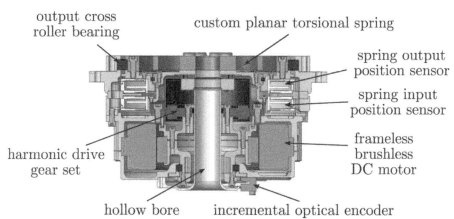
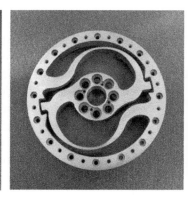

output cross roller bearing

custom planar torsional spring

spring output position sensor

spring input position sensor

frameless brushless DC motor

harmonic drive gear set

hollow bore    incremental optical encoder

**Figure 11.24** (Top left) The Robonaut 2 on the International Space Station. (Top middle) R2's hip joint SEA. (Top right) The custom torsional spring. The inner ring of hole mounts connects to the harmonic gearhead output, and the outer ring of hole mounts is the output of the SEA, connecting to the next link. The spring is designed with hard stops after approximately 0.07 rad of deflection. (Bottom) A cross-section of the SEA. The deflection $\Delta\phi$ of the torsional spring is determined by differencing the deflection readings at the spring input and the spring output. The optical encoder and spring deflection sensors provide an estimate of the joint angle. The motor controller–amplifier is located at the SEA, and it communicates with the centralized controller using a serial communication protocol. The hollow bore allows cables to pass through the interior of the SEA. All images courtesy of NASA.

### Passive Compliance and Flexible Manipulators

All robots unavoidably have some passive compliance. Modeling this compliance can be as simple as assuming torsional springs at each revolute joint (e.g., to account for finite stiffness in the flexsplines of harmonic drive gearing) or as complicated as treating links as flexible beams. Two significant effects of flexibility are (1) a mismatch between the motor angle reading, the true joint angle, and the endpoint location of the attached link, and (2) increased order of the dynamics of the robot. These issues raise challenging problems in control, particularly when the vibration modes are at low frequencies.

Some robots are specifically designed for passive compliance, particularly those meant for contact interactions with humans or the environment. Such robots may sacrifice motion-control performance in favor of safety. One passively compliant actuator is the series elastic actuator, described above.

### Variable-Impedance Actuators

The impedance of a joint is typically controlled using a feedback control law, as described in Section 11.7. There are limits to the bandwidth of this control, however; a joint that is actively controlled to behave as a spring will only achieve spring-like behavior in respect of low-frequency perturbations.

A new class of actuators, called **variable-impedance actuators** or **variable-stiffness actuators**, is intended to give actuators the desired passive mechanical impedance without the bandwidth limitations of an active control law. As an example, a variable-stiffness actuator may consist of two motors, with one motor independently controlling the mechanical stiffness of the joint (e.g., using the setpoint of an internal nonlinear spring) while the other motor produces a torque.

## 1.10   Summary

- The performance of a feedback controller is often tested by specifying a nonzero initial error $\theta_e(0)$. The error response is typically characterized by the overshoot, the 2% settling time, and the steady-state error.
- The linear error dynamics

$$a_p\theta_e^{(p)} + a_{p-1}\theta_e^{(p-1)} + \cdots + a_2\ddot{\theta}_e + a_1\dot{\theta}_e + a_0\theta_e = 0$$

  is stable, and all initial errors converge to zero, if and only if all the complex roots $s_1, \ldots, s_p$ of the characteristic equation

$$a_p s^p + a_{p-1}s^{p-1} + \cdots + a_2 s^2 + a_1 s + a_0 = 0$$

  have real components less than zero, i.e., $\mathrm{Re}(s_i) < 0$ for all $i = 1, \ldots, p$.
- Stable second-order linear error dynamics can be written in the standard form

$$\ddot{\theta}_e + 2\zeta\omega_n\dot{\theta}_e + \omega_n^2\theta_e = 0,$$

  where $\zeta$ is the damping ratio and $\omega_n$ is the natural frequency. The roots of the characteristic equation are

$$s_{1,2} = -\zeta\omega_n \pm \omega_n\sqrt{\zeta^2 - 1}.$$

  The error dynamics are overdamped if $\zeta > 1$, critically damped if $\zeta = 1$, and underdamped if $\zeta < 1$.

- The feedforward plus PI feedback controller generating joint velocity commands for a multi-joint robot is

$$\dot{\theta}(t) = \dot{\theta}_d(t) + K_p \theta_e(t) + K_i \int_0^t \theta_e(\mathrm{t})\, dt,$$

  where $K_p = k_p I$ and $K_i = k_i I$. The joint error $\theta_e(t)$ converges to zero as $t$ goes to infinity, for setpoint control or constant reference velocities, provided that $k_p > 0$ and $k_i > 0$.

- A task-space version of the feedforward plus PI feedback controller generating twists expressed in the end-effector frame is written

$$\mathcal{V}_b(t) = [\mathrm{Ad}_{X^{-1}X_d}]\mathcal{V}_d(t) + K_p X_e(t) + K_i \int_0^t X_e(\mathrm{t})\, dt,$$

  where $[X_e] = \log(X^{-1}X_d)$.

- The PID joint-space feedback controller generating joint forces and torques is

$$\tau = K_p \theta_e + K_i \int \theta_e(\mathrm{t})dt + K_d \dot{\theta}_e,$$

  where $\theta_e = \theta_d - \theta$ and $\theta_d$ is the vector of the desired joint angles.

- The joint-space computed torque controller is

$$\tau = \tilde{M}(\theta)\left(\ddot{\theta}_d + K_p\theta_e + K_i \int \theta_e(\mathrm{t})dt + K_d \dot{\theta}_e\right) + \tilde{h}(\theta,\dot{\theta}).$$

  This controller cancels nonlinear terms, uses feedforward control to proactively generate the desired acceleration $\ddot{\theta}_d$, and uses linear feedback control for stabilization.

- For robots without joint friction and a perfect model of gravity forces, joint-space PD setpoint control plus gravity compensation,

$$\tau = K_p\theta_e + K_d\dot{\theta} + \tilde{g}(\theta),$$

  yields global convergence to $\theta_e = 0$ by the Krasovskii–LaSalle invariance principle.

- Task-space force control can be achieved by the controller

$$\tau = \tilde{g}(\theta) + J^{\mathrm{T}}(\theta)\left(\mathcal{F}_d + K_{fp}\mathcal{F}_e + K_{fi}\int \mathcal{F}_e(\mathrm{t})dt - K_{\mathrm{damp}}\mathcal{V}\right),$$

  consisting of gravity compensation, feedforward force control, PI force feedback, and damping to prevent fast motion.

- Rigid constraints in the environment specify $6 - k$ free motion directions and $k$ constraint directions in which forces can be applied. These constraints can be represented as $A(\theta)\mathcal{V} = 0$. A wrench $\mathcal{F}$ can be partitioned as $\mathcal{F} = P(\theta)\mathcal{F} + (I - P(\theta))\mathcal{F}$, where $P(\theta)$ projects onto wrenches that move the end-effector tangent to the constraints and $I - P(\theta)$ projects onto wrenches that act against the constraints. The projection matrix $P(\theta)$ is written in terms of the task-space mass matrix $\Lambda(\theta)$ and constraints $A(\theta)$ as

$$P = I - A^{\mathrm{T}}(A\Lambda^{-1}A^{\mathrm{T}})^{-1}A\Lambda^{-1}.$$

- An impedance controller measures end-effector motions and creates endpoint forces to mimic a mass–spring–damper system. An admittance controller measures end-effector forces and creates endpoint motions to achieve the same purpose.

## 1.11    Software

Software functions associated with this chapter are listed below.

taulist = ComputedTorque(thetalist,dthetalist,eint,g, Mlist,Glist,Slist,thetalistd,dthetalistd,ddthetalistd,Kp,Ki,Kd)
This function computes the joint controls $\tau$ for the computed torque control law (11.35) at a particular time instant. The inputs are the $n$-vectors of joint variables, joint velocities, and joint error integrals; the gravity vector $\mathfrak{g}$; a list of transforms $M_{i-1,i}$ describing the link center of mass locations; a list of link spatial inertia matrices $\mathcal{G}_i$; a list of joint screw axes $\mathcal{S}_i$ expressed in the base frame; the $n$-vectors $\theta_d$, $\dot{\theta}_d$, and $\ddot{\theta}_d$ describing the desired motion; and the scalar PID gains $k_p$, $k_i$, and $k_d$, where the gain matrices are just $K_p = k_pI$, $K_i = k_iI$, and $K_d = k_dI$.

[taumat,thetamat] = SimulateControl(thetalist,dthetalist,g, Ftipmat,Mlist,Glist,Slist,thetamatd,dthetamatd,ddthetamatd, gtilde,Mtildelist,Gtildelist,Kp,Ki,Kd,dt,intRes)
This function simulates the computed torque controller (11.35) over a given desired trajectory. The inputs include the initial state of the robot, given by $\theta(0)$ and $\dot{\theta}(0)$; the gravity vector $\mathfrak{g}$; an $N \times 6$ matrix of wrenches applied by the end-effector, where each of the $N$ rows correspond to an instant in time in the trajectory; a list of transforms $M_{i-1,i}$ describing the link center-of-mass locations; a list of link spatial-inertia matrices $\mathcal{G}_i$; a list of joint screw axes $\mathcal{S}_i$ expressed in the base frame; the $N \times n$ matrices of the desired joint positions, velocities, and accelerations, where each of the $N$ rows corresponds to an instant in time; a (possibly incorrect) model of the gravity vector; a (possibly incorrect) model of the transforms $M_{i-1,i}$; a (possibly incorrect) model of the link inertia matrices; the scalar PID gains $k_p$, $k_i$, and $k_d$, where the gain matrices are $K_p = k_pI$, $K_i = k_iI$, and $K_d = k_dI$; the timestep between each of the $N$ rows in the matrices defining the desired trajectory; and the number of integration steps to take during each timestep.

## 11.12     Notes and References

The computed torque controller originates from research in the 1970s (Paul, 1972; Markiewicz, 1973; Bejczy, 1974; Raibert and Horn, 1978), and issues with its practical implementation (e.g., its computational complexity and modeling errors) have driven much of the subsequent research in nonlinear control, robust control, iterative learning control, and adaptive control. Proportional-derivative (PD) control plus gravity compensation was suggested and analyzed in Takegaki and Arimoto (1981), and subsequent analysis and modification of the basic controller was reviewed in Kelly (1997).

The task-space approach to motion control, also called operational-space control, was originally outlined in Luh et al. (1980) and Khatib (1987). A geometric approach to tracking control for mechanical systems is presented in Bullo and Murray (1999), where the configuration space for the system can be a generic manifold, including $SO(3)$ and $SE(3)$.

The notion of natural and artificial constraints in hybrid motion–force control was first described by Mason (1981), and an early hybrid motion–force controller based on these concepts was reported in Raibert and Craig (1981). As pointed out by Duffy (1990), we must take care in specifying the subspaces in which motions and forces can be controlled. The approach to hybrid motion–force control in this chapter mirrors the geometric approach of Liu and Li (2002). Impedance control was first described by Hogan in a series of papers (Hogan, 1985a,b,c). The stiffness matrix for a rigid body whose configuration is represented by $X \in SE(3)$ is discussed in Howard et al. (1998) and in Lončarić (1987).

Robot control builds on the well-established field of linear control (e.g., Franklin et al., 2014; Åström and Murray, 2008) and the growing field of nonlinear control (Isidori, 1995; Jurdjevic, 1997; Khalil, 2014; Nijmeijer and van der Schaft, 1990; Sastry, 1999). General references on robot control include: the edited volume by de Wit et al. (2012); the textbooks by Spong et al. (2005), Siciliano et al. (2009), Craig (2004), and Murray et al. (1994); chapters in the *Handbook of Robotics* on motion control (Chung et al., 2016) and force control (Villani and Schutter, 2016); the chapter on robot motion control in the *Encyclopedia of Systems and Control* (Spong, 2015); and, for underactuated and nonholonomic robots, chapters in the *Control Handbook* (Lynch et al., 2011) and in the *Encyclopedia of Systems and Control* (Lynch, 2015).

The principles governing SEAs are laid out in Pratt and Williamson (1995), and NASA's Robonaut 2 and its SEAs are described on the Robonaut 2 website (Robonaut 2, 2016), in Diftler et al. (2011), and in Mehling (2015). Variable-impedance actuators were reviewed in Vanderbrought et al. (2013).

## .13   Exercises

**Exercise 11.1**   Classify the following robot tasks as motion control, force control, hybrid motion–force control, impedance control, or some combination. Justify your answer.

(a) Tightening a screw with a screwdriver.
(b) Pushing a box along the floor.
(c) Pouring a glass of water.
(d) Shaking hands with a human.
(e) Throwing a baseball to hit a target.
(f) Shoveling snow.
(g) Digging a hole.
(h) Giving a back massage.
(i) Vacuuming the floor.
(j) Carrying a tray of glasses.

**Exercise 11.2**   The 2% settling time of an underdamped second-order system is approximately $t = 4/(\zeta \omega_n)$, for $e^{-\zeta \omega_n t} = 0.02$. What is the 5% settling time?

**Exercise 11.3**   Solve for any constants and give the specific equation for an underdamped second-order system with $\omega_n = 4$, $\zeta = 0.2$, $\theta_e(0) = 1$, and $\dot{\theta}_e(0) = 0$. Calculate the damped natural frequency, approximate overshoot, and 2% settling time. Plot the solution on a computer and measure the exact overshoot and settling time.

**Exercise 11.4**   Solve for any constants and give the specific equation for an underdamped second-order system with $\omega_n = 10$, $\zeta = 0.1$, $\theta_e(0) = 0$, and $\dot{\theta}_e(0) = 1$. Calculate the damped natural frequency. Plot the solution on a computer.

**Exercise 11.5**   Consider a pendulum in a gravitational field with $g = 10$ m/s$^2$. The pendulum consists of a 2 kg mass at the end of a 1 m massless rod. The pendulum joint has a viscous-friction coefficient of $b = 0.1$ N m s/rad.

(a) Write the equation of motion of the pendulum in terms of $\theta$, where $\theta = 0$ corresponds to the "hanging down" configuration.
(b) Linearize the equation of motion about the stable "hanging down" equilibrium. To do this, replace any trigonometric terms in $\theta$ with the linear term in the Taylor expansion. Give the effective mass and spring constants $m$ and $k$ in the linearized dynamics $m\ddot{\theta} + b\dot{\theta} + k\theta = 0$. At the stable equilibrium, what is the damping ratio? Is the system underdamped, critically damped, or overdamped? If it is underdamped, what is the damped natural frequency? What is the time constant of convergence to the equilibrium and the 2% settling time?
(c) Now write the linearized equations of motion for $\theta = 0$ at the balanced upright configuration. What is the effective spring constant $k$?

(d) You add a motor at the joint of the pendulum to stabilize the upright position, and you choose a P controller $\tau = K_p\theta$. For what values of $K_p$ is the upright position stable?

**Exercise 11.6**   Develop a controller for a one-dof mass–spring–damper system of the form $m\ddot{x} + b\dot{x} + kx = f$, where $f$ is the control force and $m = 4$ kg, $b = 2$ Ns/m, and $k = 0.1$ N/m.

(a) What is the damping ratio of the uncontrolled system? Is the uncontrolled system overdamped, underdamped, or critically damped? If it is underdamped, what is the damped natural frequency? What is the time constant of convergence to the origin?

(b) Choose a P controller $f = K_p x_e$, where $x_e = x_d - x$ is the position error and $x_d = 0$. What value of $K_p$ yields critical damping?

(c) Choose a D controller $f = K_d \dot{x}_e$, where $\dot{x}_d = 0$. What value of $K_d$ yields critical damping?

(d) Choose a PD controller that yields critical damping and a 2% settling time of 0.01 s.

(e) For the PD controller above, if $x_d = 1$ and $\dot{x}_d = \ddot{x}_d = 0$, what is the steady-state error $x_e(t)$ as $t$ goes to infinity? What is the steady-state control force?

(f) Now insert a PID controller for $f$. Assume $x_d \neq 0$ and $\dot{x}_d = \ddot{x}_d = 0$. Write the error dynamics in terms of $\ddot{x}_e$, $\dot{x}_e$, $x_e$, and $\int x_e(t)dt$ on the left-hand side and a constant forcing term on the right-hand side. (Hint: You can write $kx$ as $-k(x_d - x) + kx_d$.) Take the time derivative of this equation and give the conditions on $K_p$, $K_i$, and $K_d$ for stability. Show that zero steady-state error is possible with a PID controller.

**Exercise 11.7**   Simulation of a one-dof robot and robot controller.

(a) Write a simulator for a one-joint robot consisting of a motor rotating a link in gravity using the model parameters given in Section 11.4.1. The simulator should consist of: (1) a dynamics function that takes as input the current state of the robot and the torque applied by the motor and gives as output the acceleration of the robot; and (2) a numerical integrator that uses the dynamics function to calculate the new state of the system over a series of timesteps $\Delta t$. A first-order Euler integration method suffices for this problem (e.g., $\theta(k+1) = \theta(k) + \dot{\theta}(k)\Delta t$, $\dot{\theta}(k+1) = \dot{\theta}(k) + \ddot{\theta}(k)\Delta t$). Test the simulator in two ways: (1) starting the robot at rest at $\theta = -\pi/2$ and applying a constant torque of 0.5 N m; and (2) starting the robot at rest at $\theta = -\pi/4$ and applying zero torque. For both examples, plot the position as a function of time for sufficient duration to see the basic behavior. Ensure that the behavior makes sense. A reasonable choice of $\Delta t$ is 1 ms.

(b) Add two more functions to your simulator: (1) a trajectory generator function that takes the current time and returns the desired state and acceleration of the robot; and (2) a control function that takes the current state of the robot and information from the trajectory generator and returns a control torque. The simplest trajectory generator would return $\theta = \theta_{d1}$ and $\dot{\theta} = \ddot{\theta} = 0$ for all time $t < T$, and $\theta = \theta_{d2} \neq \theta_{d1}$ and $\dot{\theta} = \ddot{\theta} = 0$ for all time $t \geq T$. This trajectory is a step function in position. Use a PD feedback controller for the control function and set $K_p = 10$ N m/rad. For a well-tuned choice of $K_d$, give $K_d$ (including units) and plot the position as a function of time over 2 seconds for an initial state at rest at $\theta = -\pi/2$ and a step trajectory with $\theta_{d1} = -\pi/2$ and $\theta_{d2} = 0$. The step occurs at $T = 1$ s.

(c) Demonstrate two different choices of $K_d$ that yield (1) overshoot and (2) sluggish response with no overshoot. Give the gains and the position plots.

(d) Add a nonzero $K_i$ to your original well-tuned PD controller to eliminate steady-state error. Give the PID gains and plot the results of the step test.

**Exercise 11.8**   Modify the simulation of the one-joint robot in Exercise 11.7 to model a flexible transmission from the motor to the link with a stiffness of 500 Nm/rad. Tune a PID controller to give a good response for a desired trajectory with a step transition from $\theta = -\pi/2$ to $\theta = 0$. Give the gains and plot the response.

**Exercise 11.9**   Simulation of a two-dof robot and robot controller (Figure 11.25).

(a) *Dynamics.* Derive the dynamics of a 2R robot under gravity (Figure 11.25). The mass of link $i$ is $\mathfrak{m}_i$, the center of mass is a distance $r_i$ from the joint, the scalar inertia of link $i$ about the joint is $\mathcal{I}_i$, and the length of link $i$ is $L_i$. There is no friction at the joints.

(b) *Direct drive.* Assume each joint is directly driven by a DC motor with no gearing. Each motor comes with specifications of the mass $\mathfrak{m}_i^{\text{stator}}$ and inertia $\mathcal{I}_i^{\text{stator}}$ of the stator and the mass $\mathfrak{m}_i^{\text{rotor}}$ and inertia $\mathcal{I}_i^{\text{rotor}}$ of the rotor (the spinning portion). For the motor at joint $i$, the stator is attached to link $i-1$ and the rotor is attached to link $i$. The links are thin uniform-density rods of mass $\mathfrak{m}_i$ and length $L_i$.

In terms of the quantities given above, for each link $i \in \{1, 2\}$ give equations for the total inertia $\mathcal{I}_i$ about the joint, the mass $\mathfrak{m}_i$, and the distance $r_i$ from the joint to the center of mass. Think about how to assign the mass and inertia of the motors to the different links.

(c) *Geared robot.* Assume that motor $i$ has gearing with gear ratio $G_i$ and that the gearing itself is massless. As in part (b) above, for each link $i \in \{1, 2\}$, give equations for the total inertia $\mathcal{I}_i$ about the joint, mass $\mathfrak{m}_i$, and distance $r_i$ from the joint to the center of mass.

(d) *Simulation and control.* As in Exercise 11.7, write a simulator with (at least) four functions: a dynamics function, a numerical integrator, a trajectory generator, and a controller. Assume that there is zero friction at the joints, $g = 9.81$ m/s$^2$ in the direction indicated, $L_i = 1$ m, $r_i = 0.5$ m, $\mathfrak{m}_1 = 3$ kg, $\mathfrak{m}_2 = 2$ kg, $\mathcal{I}_1 = 2$ kg m$^2$, and $\mathcal{I}_2 = 1$ kg m$^2$. Program a PID controller, find gains that give a good response, and plot the joint angles as a function of time for a reference trajectory which is constant at rest at $(\theta_1, \theta_2) = (-\pi/2, 0)$

for $t < 1$ s and constant at $(\theta_1, \theta_2) = (0, -\pi/2)$ for $t \geq 1$ s. The initial state of the robot is at rest with $(\theta_1, \theta_2) = (-\pi/2, 0)$.

(e) *Torque limits.* Real motors have limits on the available torque. While these limits are generally velocity dependent, here we assume that each motor's torque limit is independent of velocity, $\tau_i \leq |\tau_i^{\max}|$. Assume that $\tau_1^{\max} = 100$ N m and $\tau_2^{\max} = 20$ N m. The control law may request greater torque but the actual torque is saturated at these values. Rerun the PID control simulation in (d) and plot the torques as well as the position as a function of time.

(f) *Friction.* Add a viscous friction coefficient of $b_i = 1$ N m s/rad to each joint and rerun the PID control simulation in (e).

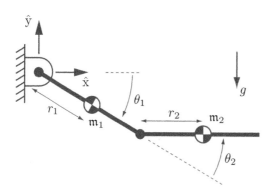

**Figure 11.25** A two-link robot arm. The length of link $i$ is $L_i$ and its inertia about the joint is $\mathcal{I}_i$. The acceleration due to gravity is $g = 9.81$ m/s$^2$.

**Exercise 11.10**   For the two-joint robot of Exercise 11.9, write a more sophisticated trajectory generator function. The trajectory generator should take the following as input:

- the desired initial position, velocity, and acceleration of each joint;
- the desired final position, velocity, and acceleration; and
- the total time of motion $T$.

A call of the form

```
[qd,qdotd,qdotdotd] = trajectory(time)
```

returns the desired position, velocity, and acceleration of each joint at time `time`. The trajectory generator should provide a trajectory that is a smooth function of time.

As an example, each joint could follow a fifth-order polynomial trajectory of the form

$$\theta_d(t) = a_0 + a_1 t + a_2 t^2 + a_3 t^3 + a_4 t^4 + a_5 t^5. \tag{11.67}$$

Given the desired positions, velocities, and accelerations of the joints at times $t = 0$ and $t = T$, you can uniquely solve for the six coefficients $a_0, \ldots, a_5$ by evaluating Equation (11.67) and its first and second derivatives at $t = 0$ and $t = T$.

Tune a PID controller to track a fifth-order polynomial trajectory moving from rest at $(\theta_1, \theta_2) = (-\pi/2, 0)$ to rest at $(\theta_1, \theta_2) = (0, -\pi/2)$ in $T = 2$ s. Give the values of your gains and plot the reference positions of both joints and the actual positions of both joints. You are free to ignore torque limits and friction.

**Exercise 11.11** For the two-joint robot of Exercise 11.9 and fifth-order polynomial trajectory of Exercise 11.10, simulate a computed torque controller to stabilize the trajectory. The robot has no joint friction or torque limits. The modeled link masses should be 20% greater than their actual values to create error in the feedforward model. Give the PID gains and plot the reference and actual joint angles for the computed torque controller as well as for PID control only.

**Exercise 11.12** The Krasovskii–LaSalle invariance principle states the following. Consider a system $\dot{x} = f(x), x \in \mathbb{R}^n$ such that $f(0) = 0$ and any energy-like function $V(x)$ such that:

- $V(x) > 0$ for all $x \neq 0$;
- $V(x) \to \infty$ as $x \to \infty$;
- $V(0) = \dot{V}(0) = 0$; and
- $\dot{V}(x) \leq 0$ along all trajectories of the system.

Let $S$ be the largest set of $\mathbb{R}^n$ such that $\dot{V}(x) = 0$ and trajectories beginning in $S$ remain in $S$ for all time. Then, if $S$ contains only the origin, the origin is globally asymptotically stable – all trajectories converge to the origin.

Using the energy function $V(x)$ from Equation (11.40), show how the Krasovskii–LaSalle principle is violated for centralized multi-joint PD setpoint control with gravity compensation if $K_p = 0$ or $K_d = 0$. For a practical robot system, is it possible to use the Krasovskii–LaSalle invariance principle to demonstrate global asymptotic stability even if $K_d = 0$? Explain your answer.

**Exercise 11.13** The two-joint robot of Exercise 11.9 can be controlled in task space using the endpoint task coordinates $X = (x, y)$, as shown in Figure 11.25. The task-space velocity is $\mathcal{V} = \dot{X}$. Give the Jacobian $J(\theta)$ and the task-space dynamics model $\{\tilde{\Lambda}(\theta), \tilde{\eta}(\theta, \mathcal{V})\}$ in the computed torque-control law (11.47).

**Exercise 11.14** Choose appropriate space and end-effector reference frames $\{s\}$ and $\{b\}$ and express natural and artificial constraints, six each, that achieve the following tasks: (a) opening a cabinet door; (b) turning a screw that advances linearly a distance $p$ for every revolution; and (c) drawing a circle on a chalkboard with a piece of chalk.

**Exercise 11.15** Assume that the end-effector of the two-joint robot in Figure 11.25 is constrained to move on the line $x - y = 1$. The robot's link lengths

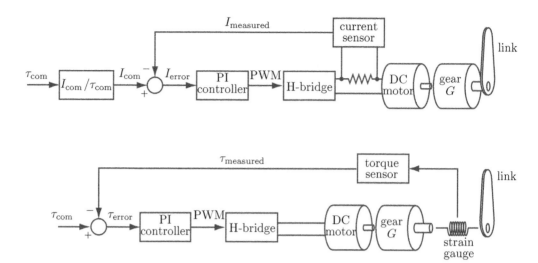

**Figure 11.26** Two methods for controlling the torque at a joint driven by a geared DC motor. (Upper) The current to the motor is measured by measuring the voltage across a small resistance in the current path. A PI controller works to make the actual current match better the requested current $I_{com}$. (Lower) The actual torque delivered to the link is measured by strain gauges.

are $L_1 = L_2 = 1$. Write the constraint as $A(\theta)\mathcal{V} = 0$, where $X = (x, y)$ and $\mathcal{V} = \dot{X}$.

**Exercise 11.16** Derive the constrained motion equations (11.59) and (11.60). Show all the steps.

**Exercise 11.17** We have been assuming that each actuator delivers the torque requested by the control law. In fact, there is typically an inner control loop running at each actuator, typically at a higher servo rate than the outer loop, to try to track the torque requested. Figure 11.26 shows two possibilities for a DC electric motor, where the torque $\tau$ delivered by the motor is proportional to the current $I$ through the motor, $\tau = k_t I$. The torque from the motor is amplified by the gearhead with gear ratio $G$.

In the upper control scheme the motor current is measured by a current sensor and compared with the desired current $I_{com}$; the error is passed through a PI controller which sets the duty cycle of a low-power pulse-width-modulation (PWM) digital signal and the PWM signal is sent to an H-bridge that generates the actual motor current. In the lower scheme, a strain gauge torque sensor is inserted between the output of the motor gearing and the link, and the measured torque is compared directly with the requested torque $\tau_{com}$. Since a strain gauge measures deflection, the element on which it is mounted must have a finite torsional stiffness. Series elastic actuators are designed to have particularly flexible torsional elements, so much so that encoders are used to measure the larger deflections. The torque is estimated from the encoder reading and the torsional spring constant.

(a) For the current sensing scheme, what multiplicative factor should go in the block labeled $I_{\text{com}}/\tau_{\text{com}}$? Even if the PI current controller does its job perfectly ($I_{\text{error}} = 0$) and the torque constant $k_t$ is perfectly known, what effect may contribute to error in the generated torque?

(b) For the strain gauge measurement method, explain the drawbacks, if any, of having a flexible element between the gearhead and the link.

**Exercise 11.18** Modify the `SimulateControl` function to allow initial state errors.

# 12 Grasping and Manipulation

Most of the book so far has been concerned with kinematics, dynamics, motion planning, and control of the robot itself. Only in Chapter 11, on the topics of force control and impedance control, did the robot finally begin interacting with an environment other than free space. Now the robot really becomes valuable – when it can perform useful work on objects in the environment.

In this chapter our focus moves outward from the robot itself to the interaction between the robot and its environment. The desired behavior of the robot hand or end-effector, whether motion control, force control, hybrid motion–force control, or impedance control, is assumed to be achieved perfectly using the methods discussed so far. Our focus now is on the contact interface between the robot and objects as well as on contacts among objects and between objects and constraints in the environment. In short, our focus is on *manipulation* rather than the *manipulator*. Examples of manipulation include grasping, pushing, rolling, throwing, catching, tapping, etc. To limit our scope, we will assume that the manipulator, objects, and obstacles in the environment are rigid.

To simulate, plan, and control robotic manipulation tasks, we need an understanding of (at least) three elements: contact kinematics; forces applied through contacts; and the dynamics of rigid bodies. In contact kinematics we study how rigid bodies can move relative to each other without penetration and classify these feasible motions according to whether the contacts are rolling, sliding, or separating. Contact force models address the normal and frictional forces that can be transmitted through rolling and sliding contacts. Finally, the actual motions of the bodies are those that simultaneously satisfy the kinematic constraints, contact force model, and rigid-body dynamics.

This chapter introduces contact kinematics (Section 12.1) and contact force modeling (Section 12.2) and applies these models to problems in robot grasping and other types of manipulation.

The following definitions from linear algebra will be useful in this chapter.

**Definition 12.1** Given a set of $j$ vectors $\mathcal{A} = a_1, \ldots, a_j \in \mathbb{R}^n$, we define the **linear span**, or the set of linear combinations, of the vectors to be

$$\text{span}(\mathcal{A}) = \left\{ \sum_{i=1}^{j} k_i a_i \mid k_i \in \mathbb{R} \right\},$$

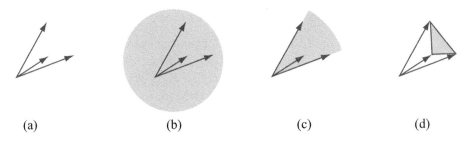

(a) (b) (c) (d)

**Figure 12.1** (a) Three vectors in $\mathbb{R}^2$, drawn as arrows from the origin. (b) The linear span of the vectors is the entire plane. (c) The positive linear span is the "cone" shaded gray. (d) The convex span is the polygon and its interior.

the **nonnegative linear combinations,** sometimes called the **positive** or **conical span,** to be

$$\text{pos}(\mathcal{A}) = \left\{ \sum_{i=1}^{j} k_i a_i \mid k_i \geq 0 \right\},$$

and the **convex span** to be

$$\text{conv}(\mathcal{A}) = \left\{ \sum_{i=1}^{j} k_i a_i \mid k_i \geq 0 \text{ and } \sum_i k_i = 1 \right\}.$$

Clearly $\text{conv}(\mathcal{A}) \subseteq \text{pos}(\mathcal{A}) \subseteq \text{span}(\mathcal{A})$ (see Figure 12.1). The following facts from linear algebra will also be useful.

1. The space $\mathbb{R}^n$ can be linearly spanned by $n$ vectors, but no fewer.
2. The space $\mathbb{R}^n$ can be positively spanned by $n + 1$ vectors, but no fewer.

The first fact is implicit in our use of $n$ coordinates to represent $n$-dimensional Euclidean space. Fact 2 follows from the fact that for any choice of $n$ vectors $\mathcal{A} = \{a_1, \ldots, a_n\}$ there exists a vector $c \in \mathbb{R}^n$ such that $a_i^{\mathrm{T}} c \leq 0$ for all $i$. In other words, no nonnegative combination of vectors in $\mathcal{A}$ can create a vector in the direction $c$. However, if we choose $a_1, \ldots, a_n$ to be orthogonal coordinate bases of $\mathbb{R}^n$ and then choose $a_{n+1} = -\sum_{i=1}^{n} a_i$, we see that this set of $n + 1$ vectors positively spans $\mathbb{R}^n$.

## 2.1    Contact Kinematics

Contact kinematics is the study of how two or more rigid bodies can move relative to each other while respecting the impenetrability constraint. It also classifies motion at a contact as either rolling or sliding. Let's start by looking at a single contact between two rigid bodies.

### 12.1.1    First-Order Analysis of a Single Contact

Consider two rigid bodies whose configurations are given by the local coordinate column vectors $q_1$ and $q_2$, respectively. Writing the composite configuration as $q = (q_1, q_2)$, we define a distance function $d(q)$ between the bodies that is positive when they are separated, zero when they are touching, and negative when they are in penetration. When $d(q) > 0$, there are no constraints on the motions of the bodies; each is free to move with six degrees of freedom. When the bodies are in contact $(d(q) = 0)$, we look at the time derivatives $\dot{d}$, $\ddot{d}$, etc., to determine whether the bodies stay in contact or break apart as they follow a particular trajectory $q(t)$. This can be determined by the following table of possibilities:

| $d$ | $\dot{d}$ | $\ddot{d}$ | $\cdots$ | |
|-----|-----------|------------|----------|--|
| $> 0$ | | | | no contact |
| $< 0$ | | | | infeasible (penetration) |
| $= 0$ | $> 0$ | | | in contact, but breaking free |
| $= 0$ | $< 0$ | | | infeasible (penetration) |
| $= 0$ | $= 0$ | $> 0$ | | in contact, but breaking free |
| $= 0$ | $= 0$ | $< 0$ | | infeasible (penetration) |
| etc. | | | | |

The contact is maintained only if all time derivatives are zero.

Now let's assume that the two bodies are initially in contact $(d = 0)$ at a single point. The first two time derivatives of $d$ are written

$$\dot{d} = \frac{\partial d}{\partial q} \dot{q}, \tag{12.1}$$

$$\ddot{d} = \dot{q}^{\mathrm{T}} \frac{\partial^2 d}{\partial q^2} \dot{q} + \frac{\partial d}{\partial q} \ddot{q}. \tag{12.2}$$

The terms $\partial d / \partial q$ and $\partial^2 d / \partial q^2$ carry information about the local contact geometry. The gradient vector $\partial d / \partial q$ corresponds to the separation direction in $q$ space associated with the **contact normal** (Figure 12.2). The matrix $\partial^2 d / \partial q^2$ encodes information about the relative curvature of the bodies at the contact point.

In this chapter we assume that only contact-normal information $\partial d / \partial q$ is available at contacts; other information about the local contact geometry, including the contact curvature $\partial^2 d / \partial q^2$ and higher derivatives, is unknown. With this assumption, we truncate our analysis at Equation (12.1) and assume that the bodies remain in contact if $\dot{d} = 0$. Since we are dealing only with the first-order contact derivative $\partial d / \partial q$, we refer to our analysis as a *first-order* analysis. In such a first-order analysis, the contacts in Figure 12.2 are treated identically since they have the same contact normal.

As indicated in the table above, a *second-order* analysis incorporating the contact curvature $\partial^2 d / \partial q^2$ may indicate that the contact is actually breaking or penetrating even when $d = \dot{d} = 0$. We will see examples of this, but a detailed analysis of second-order contact conditions is beyond the scope of this chapter.

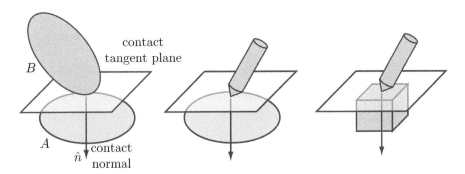

**Figure 12.2** (Left) The bodies $A$ and $B$ in single-point contact define a contact tangent plane and a contact normal vector $\hat{n}$ perpendicular to the tangent plane. By default, the positive direction of the normal is chosen into body $A$. Since contact curvature is not addressed in this chapter, the contact places the same restrictions on the motions of the rigid bodies in the middle and right panels.

### 12.1.2  Contact Types: Rolling, Sliding, and Breaking Free

Given two bodies in single-point contact, they undergo a **roll–slide motion** if the contact is maintained. The constraint that contact is maintained is a holonomic constraint, $d(q) = 0$. A necessary condition for maintaining contact is $\dot{d} = 0$.

Let's write the velocity constraint $\dot{d} = 0$ in a form, based on the contact normal, that does not require an explicit distance function (Figure 12.2). Let $\hat{n} \in \mathbb{R}^3$ be a unit vector aligned with the contact normal, expressed in a world frame. Let $p_A \in \mathbb{R}^3$ be the representation of the contact point on body $A$ in the world frame, and let $p_B \in \mathbb{R}^3$ be the representation of the contact point on body $B$. Although the contact-point vectors $p_A$ and $p_B$ are identical initially, the velocities $\dot{p}_A$ and $\dot{p}_B$ may be different. Thus the condition $\dot{d} = 0$ can be written

$$\hat{n}^{\mathrm{T}}(\dot{p}_A - \dot{p}_B) = 0. \tag{12.3}$$

Since the direction of the contact normal is defined as being into body $A$, the impenetrability constraint $\dot{d} \geq 0$ is written as

$$\hat{n}^{\mathrm{T}}(\dot{p}_A - \dot{p}_B) \geq 0. \tag{12.4}$$

Let us rewrite the constraint (12.4) in terms of the twists $\mathcal{V}_A = (\omega_A, v_A)$ and $\mathcal{V}_B = (\omega_B, v_B)$ of bodies $A$ and $B$ in a space frame.[1] Note that

$$\dot{p}_A = v_A + \omega_A \times p_A = v_A + [\omega_A]p_A,$$
$$\dot{p}_B = v_B + \omega_B \times p_B = v_B + [\omega_B]p_B.$$

We can define the wrench $\mathcal{F} = (m, f)$ corresponding to a unit force applied along the contact normal:

$$\mathcal{F} = (p_A \times \hat{n}, \hat{n}) = ([p_A]\hat{n}, \hat{n}).$$

---

[1]  All twists and wrenches are expressed in a space frame in this chapter.

It is not necessary to appeal to forces in a purely kinematic analysis of rigid bodies, but we will find it convenient to adopt this notation now in anticipation of the discussion of contact forces in Section 12.2.

With these expressions, the inequality constraint (12.4) can be written

$$\text{(impenetrability constraint)} \quad \mathcal{F}^{\mathrm{T}}(\mathcal{V}_A - \mathcal{V}_B) \geq 0 \qquad (12.5)$$

(see Exercise 12.1). If

$$\text{(active constraint)} \quad \mathcal{F}^{\mathrm{T}}(\mathcal{V}_A - \mathcal{V}_B) = 0 \qquad (12.6)$$

then, to first order, the constraint is active and the parts remain in contact.

In the case where $B$ is a stationary fixture, the impenetrability constraint (12.5) simplifies to

$$\mathcal{F}^{\mathrm{T}}\mathcal{V}_A \geq 0. \qquad (12.7)$$

If $\mathcal{F}^{\mathrm{T}}\mathcal{V}_A > 0$, $\mathcal{F}$ and $\mathcal{V}_A$ are said to be **repelling**. If $\mathcal{F}^{\mathrm{T}}\mathcal{V}_A = 0$, $\mathcal{F}$ and $\mathcal{V}_A$ are said to be **reciprocal** and the constraint is active.

Twists $\mathcal{V}_A$ and $\mathcal{V}_B$ satisfying (12.6) are called **first-order roll–slide motions** – the contact may be either sliding or rolling. **Roll–slide contacts** may be further separated into **rolling contacts** and **sliding contacts**. The contact is rolling if the bodies have no motion relative to each other at the contact:

$$\text{(rolling constraint)} \quad \dot{p}_A = v_A + [\omega_A]p_A = v_B + [\omega_B]p_B = \dot{p}_B. \qquad (12.8)$$

Note that "rolling" contacts include those where the two bodies remain stationary relative to each other, i.e., no relative rotation. Thus "sticking" is another term for these contacts.

If the twists satisfy Equation (12.6) but not the rolling equations of (12.8) then they are sliding.

We assign to a rolling contact the **contact label** R, to a sliding contact the label S, and to a contact that is breaking free (the impenetrability constraint (12.5) is satisfied but not the active constraint (12.6)) the label B.

The distinction between rolling and sliding contacts becomes especially important when we consider friction forces in Section 12.2.

**Example 12.2** Consider the contact shown in Figure 12.3. Bodies $A$ and $B$ are in contact at $p_A = p_B = [1 \ 2 \ 0]^{\mathrm{T}}$ with contact normal direction $\hat{n} = [0 \ 1 \ 0]^{\mathrm{T}}$. The impenetrability constraint (12.5) is

$$\mathcal{F}^{\mathrm{T}}(\mathcal{V}_A - \mathcal{V}_B) \geq 0,$$

which becomes

$$[([p_A]\hat{n})^{\mathrm{T}} \ \hat{n}^{\mathrm{T}}] \begin{bmatrix} \omega_A - \omega_B \\ v_A - v_B \end{bmatrix} \geq 0.$$

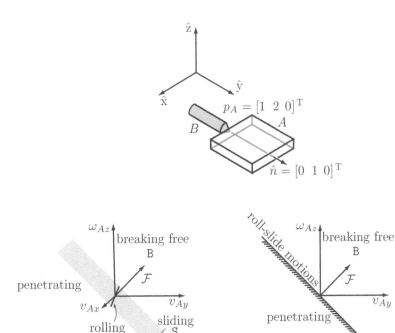

**Figure 12.3** Example 12.2. (Top) Body $B$ makes contact with $A$ at $p_A = p_B = [1\ 2\ 0]^{\mathrm{T}}$ with normal $\hat{n} = [0\ 1\ 0]^{\mathrm{T}}$. (Bottom left) The twists $\mathcal{V}_A$ and their corresponding contact labels for $B$ stationary and $A$ confined to a plane. The contact normal wrench $\mathcal{F}$ is $[m_x\ m_y\ m_z\ f_x\ f_y\ f_z]^{\mathrm{T}} = [0\ 0\ 1\ 0\ 1\ 0]^{\mathrm{T}}$. (Bottom right) Looking down the $-v_{Ax}$-axis.

Substituting values, we obtain

$$[0\ 0\ 1\ 0\ 1\ 0][\omega_{Ax} - \omega_{Bx}\ \omega_{Ay} - \omega_{By}\ \omega_{Az} - \omega_{Bz}\ v_{Ax} - v_{Bx}\ v_{Ay} - v_{By}\ v_{Az} - v_{Bz}]^{\mathrm{T}}$$
$$\geq 0$$

or

$$\omega_{Az} - \omega_{Bz} + v_{Ay} - v_{By} \geq 0;$$

and therefore roll–slide twists satisfy

$$\omega_{Az} - \omega_{Bz} + v_{Ay} - v_{By} = 0. \tag{12.9}$$

Equation (12.9) defines an 11-dimensional hyperplane in the 12-dimensional space of twists $(\mathcal{V}_A, \mathcal{V}_B)$.

The rolling constraint (12.8) is equivalent to

$$v_{Ax} - \omega_{Az} p_{Ay} + \omega_{Ay} p_{Az} = v_{Bx} - \omega_{Bz} p_{By} + \omega_{By} p_{Bz},$$
$$v_{Ay} + \omega_{Az} p_{Ax} - \omega_{Ax} p_{Az} = v_{By} + \omega_{Bz} p_{Bx} - \omega_{Bx} p_{Bz},$$
$$v_{Az} + \omega_{Ax} p_{Ay} - \omega_{Ay} p_{Ax} = v_{Bz} + \omega_{Bx} p_{By} - \omega_{By} p_{Bx};$$

substituting values for $p_A$ and $p_B$, we get

$$v_{Ax} - 2\omega_{Az} = v_{Bx} - 2\omega_{Bz}, \tag{12.10}$$

$$v_{Ay} + \omega_{Az} = v_{By} + \omega_{Bz}, \tag{12.11}$$

$$v_{Az} + 2\omega_{Ax} - \omega_{Ay} = v_{Bz} + 2\omega_{Bx} - \omega_{By}. \tag{12.12}$$

The constraint equations (12.10)–(12.12) define a nine-dimensional hyperplane subspace of the 11-dimensional hyperplane of roll–slide twists.

To visualize the constraints in a low-dimensional space, let's assume that $B$ is stationary ($\mathcal{V}_B = 0$) and $A$ is confined to the $z = 0$ plane, i.e., $\mathcal{V}_A = [\omega_{Ax}\ \omega_{Ay}\ \omega_{Az}\ v_{Ax}\ v_{Ay}\ v_{Az}]^{\mathrm{T}} = [0\ 0\ \omega_{Az}\ v_{Ax}\ v_{Ay}\ 0]^{\mathrm{T}}$. The wrench $\mathcal{F}$ is written $[m_z\ f_x\ f_y]^{\mathrm{T}} = [1\ 0\ 1]^{\mathrm{T}}$. The roll–slide constraint (12.9) reduces to

$$v_{Ay} + \omega_{Az} = 0,$$

while the rolling constraints simplify to

$$v_{Ax} - 2\omega_{Az} = 0,$$

$$v_{Ay} + \omega_{Az} = 0.$$

The single roll–slide constraint yields a plane in $(\omega_{Az}, v_{Ax}, v_{Ay})$ space, and the two rolling constraints yield a line in that plane. Because $\mathcal{V}_B = 0$, the constraint surfaces pass through the origin $\mathcal{V}_A = 0$. If $\mathcal{V}_B \neq 0$, this is no longer the case in general.

Figure 12.3 shows graphically that nonpenetrating twists $\mathcal{V}_A$ must have a nonnegative dot product with the constraint wrench $\mathcal{F}$ when $\mathcal{V}_B = 0$.

### 12.1.3    Multiple Contacts

Now suppose that a body $A$ is subject to $n$ contacts with $m$ other bodies, where $n \geq m$. The contacts are numbered $i = 1, \ldots, n$, and the other bodies are numbered $j = 1, \ldots, m$. Let $j(i) \in \{1, \ldots, m\}$ denote the number of the other body participating in contact $i$. Each contact $i$ constrains $\mathcal{V}_A$ to a half-space of its six-dimensional twist space that is bounded by a five-dimensional hyperplane of the form $\mathcal{F}^{\mathrm{T}}\mathcal{V}_A = \mathcal{F}^{\mathrm{T}}\mathcal{V}_{j(i)}$. Taking the union of the set of constraints from all the contacts, we get a **polyhedral convex set** (**polytope**[2] for short) $V$ of feasible twists in the $\mathcal{V}_A$ space, written as

$$V = \{\mathcal{V}_A \mid \mathcal{F}_i^{\mathrm{T}}(\mathcal{V}_A - \mathcal{V}_{j(i)}) \geq 0 \text{ for all } i\},$$

where $\mathcal{F}_i$ corresponds to the $i$th contact normal and $\mathcal{V}_{j(i)}$ is the twist of the other body at contact $i$. The constraint at contact $i$ is redundant if the half-space constraint contributed by contact $i$ does not change the feasible twist polytope $V$.

---

[2] We use the term "polytope" to refer generally to a convex set bounded by hyperplanes in an arbitrary vector space. The set need not be finite; it could be a cone with infinite volume. It could also be a point, or the null set if the constraints are incompatible with the rigid-body assumption.

In general, the feasible twist polytope for a body can consist of a six-dimensional interior (where no contact constraint is active), five-dimensional faces where one constraint is active, four-dimensional faces where two constraints are active, and so on, down to one-dimensional edges and zero-dimensional points. A twist $\mathcal{V}_A$ on a $k$-dimensional facet of the polytope indicates that $6 - k$ independent (non-redundant) contact constraints are active.

If all the bodies providing constraints are stationary, i.e., $\mathcal{V}_j = 0$ for all $j$, then each constraint hyperplane defined by (12.5) passes through the origin of $\mathcal{V}_A$ space. We call such a constraint **homogeneous**. The feasible twist set becomes a cone rooted at the origin, called a (homogeneous) **polyhedral convex cone**. Let $\mathcal{F}_i$ be the constraint wrench of stationary contact $i$. Then the feasible twist cone $V$ is

$$ V = \{\mathcal{V}_A \mid \mathcal{F}_i^{\mathrm{T}} \mathcal{V}_A \geq 0 \text{ for all } i\}. $$

If the $\mathcal{F}_i$ positively span the six-dimensional wrench space or, equivalently, the convex span of the $\mathcal{F}_i$ contains the origin in the interior then the feasible twist polytope $V$ reduces to a point at the origin, the stationary contacts completely constrain the motion of the body, and we have **form closure**, discussed in more detail in Section 12.1.7.

As mentioned in Section 12.1.2, each point contact $i$ can be given a label corresponding to the type of contact: B if the contact is breaking, R if the contact is rolling, and S if the contact is sliding, i.e., (12.6) is satisfied but (12.8) is not. The **contact mode** for the entire system can be written as the concatenation of the contact labels at the contacts. Since we have three distinct contact labels, a system of bodies with $n$ contacts can have a maximum of $3^n$ contact labels. Some of these contact modes may not be feasible, as their corresponding kinematic constraints may not be compatible.

**Example 12.3**   Figure 12.4 shows triangular fingers contacting a hexagonal body $A$. To more easily visualize the contact constraints the hexagon is restricted to translational motion in a plane only, so that its twist can be written $\mathcal{V}_A = (0, 0, 0, v_{Ax}, v_{Ay}, 0)$. In Figure 12.4(a) the single stationary finger creates a contact wrench $\mathcal{F}_1$ that can be drawn in $\mathcal{V}_A$ space. All feasible twists have a non-negative component in the direction of $\mathcal{F}_1$. Roll–slide twists satisfying $\mathcal{F}_1^{\mathrm{T}} \mathcal{V}_A = 0$ lie on the constraint line. Since no rotations are allowed, the only twist yielding a rolling contact is $\mathcal{V}_A = 0$. In Figure 12.4(b) the union of the constraints due to two stationary fingers creates a (polyhedral convex) cone of feasible twists. Figure 12.4(c) shows three fingers in contact, one of which is moving with twist $\mathcal{V}_3$. Because the moving finger has nonzero velocity, its constraint half-space is displaced from the origin by $\mathcal{V}_3$. The result is a closed polygon of feasible twists.

**Example 12.4**   Figure 12.5 shows the contact normals of three stationary contacts with a planar body $A$, not shown. The body moves in a plane, so $v_{Az} = \omega_{Ax} = \omega_{Ay} = 0$. In this example we do not distinguish between rolling and sliding motions, so the locations of the contacts along the normals are irrelevant. The

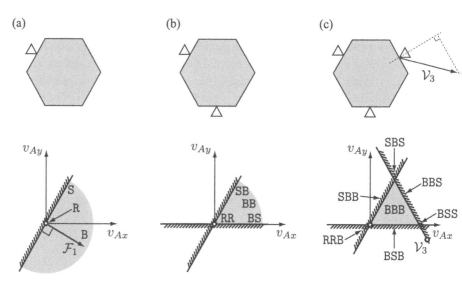

**Figure 12.4** Motion-controlled fingers contacting a hexagon that is constrained to translate in a plane only (Example 12.3). (a) A single stationary finger provides a single half-space constraint on the hexagon's twist $\mathcal{V}_A$. The feasible-motion half-space is shaded gray. The two-dimensional set of twists corresponding to breaking contact B, the one-dimensional set corresponding to sliding contact S, and the zero-dimensional set corresponding to rolling (fixed) contact R are shown. (b) The union of constraints from two stationary fingers creates a cone of feasible twists. This cone corresponds to four possible contact modes: RR, SB, BS, and BB. The contact label for the finger at upper left is given first. (c) Three fingers, one of which is moving with a linear velocity $\mathcal{V}_3$, create a closed polygon of feasible twists. There are seven possible contact modes corresponding to the feasible twists: a two-dimensional set where all contacts are breaking, three one-dimensional sets where one contact constraint is active, and three zero-dimensional sets where two contact constraints are active. Note that rolling contact at the moving finger is not feasible, since translation of the hexagon to "follow" the moving finger, as indicated by the ∘ at the lower right of the lower figure, would violate one of the impenetrability constraints. If the third finger were stationary, the only feasible motion of the hexagon would be zero velocity, with contact mode RRR.

three contact wrenches, written $(m_z, f_x, f_y)$, are $\mathcal{F}_1 = (-2, 0, 1)$, $\mathcal{F}_2 = (1, -1, 0)$, and $\mathcal{F}_3 = (1, 1, 0)$, yielding the motion constraints

$$v_{Ay} - 2\omega_{Az} \geq 0,$$
$$-v_{Ax} + \omega_{Az} \geq 0,$$
$$v_{Ax} + \omega_{Az} \geq 0.$$

These constraints describe a polyhedral convex cone of feasible twists rooted at the origin, as illustrated on the right in Figure 12.5.

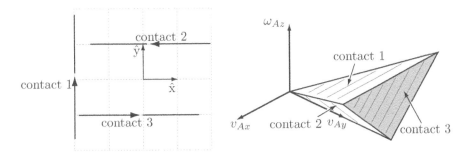

**Figure 12.5** Example 12.4. (Left) Arrows representing the lines of force corresponding to the contact normals of three stationary contacts on a planar body. If we are concerned only with feasible motions, and do not distinguish between rolling and sliding, contacts anywhere along the lines, with the contact normals shown, are equivalent. (Right) The three constraint half-spaces define a polyhedral convex cone of feasible twists. In the figure the cone is truncated at the plane $v_{Ay} = 2$. The outer faces of the cone are indicated by hatching on a white background, and the inner faces by hatching on a gray background. Twists in the interior of the cone correspond to all contacts breaking, while twists on the faces of the cone correspond to one active constraint and twists on one of the three edges of the cone correspond to two active constraints.

### 12.1.4   Collections of Bodies

The discussion above can be generalized to find the feasible twists of multiple bodies in contact. If bodies $i$ and $j$ make contact at a point $p$, where $\hat{n}$ points into body $i$ and $\mathcal{F} = ([p]\hat{n}, \hat{n})$ then their spatial twists $\mathcal{V}_i$ and $\mathcal{V}_j$ must satisfy the constraint

$$\mathcal{F}^{\mathrm{T}}(\mathcal{V}_i - \mathcal{V}_j) \geq 0 \tag{12.13}$$

to avoid penetration. This is a homogeneous half-space constraint in the composite $(\mathcal{V}_i, \mathcal{V}_j)$ twist space. In an assembly of $m$ bodies, each pairwise contact contributes another constraint in the $6m$-dimensional composite twist space ($3m$-dimensional for planar bodies) and the result is a polyhedral convex cone of kinematically feasible twists rooted at the origin of the composite twist space. The contact mode for the entire assembly is the concatenation of the contact labels at each contact in the assembly.

If there are bodies whose motion is controlled, e.g., robot fingers, the constraints on the motion of the remaining bodies are no longer homogeneous. As a result, the convex polyhedral set of feasible twists of the uncontrolled bodies, in their composite twist space, is no longer a cone rooted at the origin.

### 12.1.5   Other Types of Contact

We have been considering point contacts of the type shown in Figure 12.6(a), where at least one of the bodies in contact uniquely defines the contact normal. Figures 12.6(b)–(e) show other types of contact. The kinematic constraints pro-

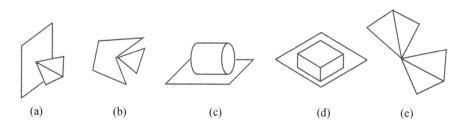

(a)            (b)            (c)            (d)            (e)

**Figure 12.6** (a) Vertex–face contact. (b) The contact between a convex vertex and a concave vertex can be treated as multiple point contacts, one at each face adjacent to the concave vertex. These faces define the contact normals. (c) A line contact can be treated as two point contacts at either end of the line. (d) A plane contact can be treated as point contacts at the corners of the convex hull of the contact area. (e) Convex vertex–vertex contact. This case is degenerate and so is not considered.

vided by the convex–concave vertex, line, and plane contacts of Figures 12.6(b)–(d) are, to first order, identical to those provided by finite collections of single-point contacts. The degenerate case in Figure 12.6(e) is ignored, as there is no unique definition of a contact normal.

The impenetrability constraint (12.5) derives from the fact that arbitrarily large contact forces can be applied in the normal direction to prevent penetration. In Section 12.2, we will see that tangential forces due to friction may also be applied, and these forces may prevent slipping between two bodies in contact. Normal and tangential contact forces are subject to constraints: the normal force must be pushing into a body, not pulling, and the maximum friction force is proportional to the normal force.

If we wish to apply a kinematic analysis that can approximate the effects of friction without explicitly modeling forces, we can define three purely kinematic models of point contacts: a **frictionless point contact**, a **point contact with friction**, and a **soft contact**, also called a soft-finger contact. A frictionless point contact enforces only the roll–slide constraint (12.5). A point contact with friction also enforces the rolling constraints (12.8), implicitly modeling friction forces sufficient to prevent slip at the contact. A soft contact enforces the rolling constraints (12.8) as well as one more constraint: the two bodies in contact may not spin relative to each other about the contact normal axis. This models deformation and the resulting friction moment resisting any spin due to the nonzero contact area between the two bodies. For planar problems, a point contact with friction and a soft contact are identical.

## 12.1.6 Planar Graphical Methods

Planar problems allow the possibility of using graphical methods to visualize the feasible motions for a single body, since the space of twists is three dimensional. An example planar twist cone is shown in Figure 12.5. Such a figure would be very difficult to draw for a system with more than three degrees of freedom.

A convenient way to represent a planar twist, $\mathcal{V} = (\omega_z, v_x, v_y)$, in $\{s\}$ is as a **center of rotation** (CoR) at $(-v_y/\omega_z, v_x/\omega_z)$ plus the angular velocity $\omega_z$. The CoR is the point in the (projective) plane that remains stationary under the motion, i.e., the point where the screw axis intersects the plane.[3] In the case where the speed of motion is immaterial, we may simply label the CoR with a '+', '−', or 0 sign representing the direction of rotation (Figure 12.7). The mapping from planar twists to CoRs is illustrated in Figure 12.8, which shows that the space of CoRs consists of a plane of '+' CoRs (counterclockwise), a plane of '−' CoRs (clockwise), and a circle of translation directions.

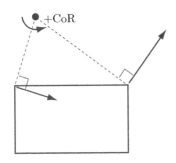

**Figure 12.7** Given the velocity of two points on a planar body, the lines normal to the velocities intersect at the CoR. The CoR shown is labeled '+' corresponding to the (counterclockwise) positive angular velocity of the body.

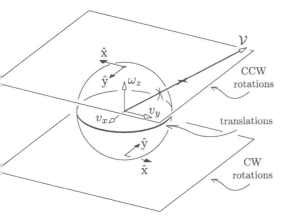

**Figure 12.8** Mapping a planar twist $\mathcal{V}$ to a CoR. The ray containing the vector $\mathcal{V}$ intersects the plane of '+' CoRs at $\omega_z = 1$, or the plane of '−' CoRs at $\omega_z = -1$, or the circle of translation directions.

Given two different twists $\mathcal{V}_1$ and $\mathcal{V}_2$ and their corresponding CoRs, the set of linear combinations of these twists, $k_1\mathcal{V}_1 + k_2\mathcal{V}_2$ where $k_1, k_2 \in \mathbb{R}$, corresponds to the line of CoRs passing through $\text{CoR}(\mathcal{V}_1)$ and $\text{CoR}(\mathcal{V}_2)$. Since $k_1$ and $k_2$ can have either sign, it follows that if either $\omega_{1z}$ or $\omega_{2z}$ is nonzero then the CoRs on this line can have either sign. If $\omega_{1z} = \omega_{2z} = 0$ then the set of linear combinations corresponds to the set of all translation directions.

A more interesting case is when $k_1, k_2 \geq 0$. Given two twists $\mathcal{V}_1$ and $\mathcal{V}_2$, the nonnegative linear combination of these two velocities is written

$$V = \text{pos}(\{\mathcal{V}_1, \mathcal{V}_2\}) = \{k_1\mathcal{V}_1 + k_2\mathcal{V}_2 \mid k_1, k_2 \geq 0\},$$

[3] Note that the case $\omega_z = 0$ must be treated with care, as it corresponds to a CoR at infinity.

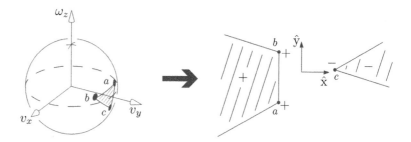

**Figure 12.9** The intersection of a twist cone with the unit twist sphere, and the representation of the cone as a set of CoRs (the two hatched regions join at infinity to form a single set).

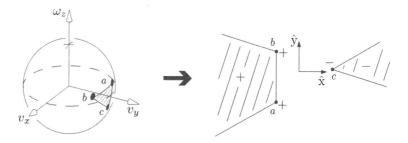

**Figure 12.10** (a) Positive linear combination of two CoRs labeled '+'. (b) Positive linear combination of a '+' CoR and a '−' CoR. (c) Positive linear combination of three '+' CoRs. (d) Positive linear combination of two '+' CoRs and a '−' CoR.

which is a planar twist cone rooted at the origin, with $\mathcal{V}_1$ and $\mathcal{V}_2$ defining the edges of the cone. If $\omega_{1z}$ and $\omega_{2z}$ have the same sign then the CoRs of their non-negative linear combinations $\mathrm{CoR}(\mathrm{pos}(\{\mathcal{V}_1, \mathcal{V}_2\}))$ all have that sign and lie on the line segment between the two CoRs. If $\mathrm{CoR}(\mathcal{V}_1)$ and $\mathrm{CoR}(\mathcal{V}_2)$ are labeled '+' and '−' respectively, then $\mathrm{CoR}(\mathrm{pos}(\{\mathcal{V}_1, \mathcal{V}_2\}))$ consists of the line containing the two CoRs, minus the segment between the CoRs. This set consists of a ray of CoRs labeled '+' attached to $\mathrm{CoR}(\mathcal{V}_1)$, a ray of CoRs labeled '−' attached to $\mathrm{CoR}(\mathcal{V}_2)$, and a point at infinity labeled 0, corresponding to translation. This collection of elements should be considered as a single line segment (though one passing through infinity), just like the first case mentioned above. Figures 12.9 and 12.10 show examples of CoR regions corresponding to positive linear combinations of planar twists.

The CoR representation of planar twists is particularly useful for representing the feasible motions of one movable body in contact with stationary bodies. Since the constraints are stationary, as noted in Section 12.1.3, the feasible twists form a polyhedral convex cone rooted at the origin. Such a cone can be represented uniquely by a set of CoRs with '+', '−', and 0 labels. A general twist polytope, as would be generated by moving constraints, cannot be uniquely represented by a set of CoRs with such labels.

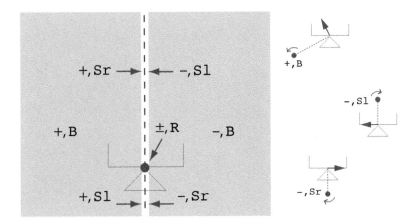

**Figure 12.11** The stationary triangle makes contact with a movable part. The CoRs to the left of the contact normal are labeled '+', to the right are labeled '−', and on the normal are labeled '±'. Also given are the contact labels for the CoRs. For points on the contact normal, the sign assigned to the Sl and Sr CoRs switches at the contact point. Three CORs and their associated labels are illustrated.

Given a contact between a stationary body and a movable body, we can plot the CoRs that do not violate the impenetrability constraint. Label all points on the contact normal '±', points to the left of the inward normal '+', and points to the right '−'. All points labeled '+' can serve as CoRs with positive angular velocity for the movable body, and all points labeled '−' can serve as CoRs with negative angular velocity, without violating the first-order contact constraint. We can further assign contact labels to each CoR corresponding to the first-order conditions for breaking contact B, sliding contact S, or rolling contact R. For planar sliding, we subdivide the label S into two subclasses: Sr, where the moving body slips to the right relative to the fixed constraint, and Sl, where the moving body slips to the left. Figure 12.11 illustrates the labeling.

If there is more than one contact, we simply take the union of the constraints and contact labels from the individual contacts. This unioning of the constraints implies that the feasible CoR region is convex, as is the homogeneous polyhedral twist cone.

**Example 12.5**   Figure 12.12(a) shows a planar body standing on a table while being contacted by a stationary robot finger. The finger defines an inequality constraint on the body's motion and the table defines two more. The cone of twists that do not violate the impenetrability constraints is represented by the CoRs that are consistently labeled for each contact (Figure 12.12(b)). Each feasible CoR is labeled with a contact mode that concatenates the labels for the individual contacts (Figure 12.12(c)).

Now look more closely at the CoR indicated by (+, SrBSr) in Figure 12.12(c). Is this motion really possible? It should be apparent that it is, in fact, *not* possible: the body would immediately penetrate the stationary finger. Our incorrect

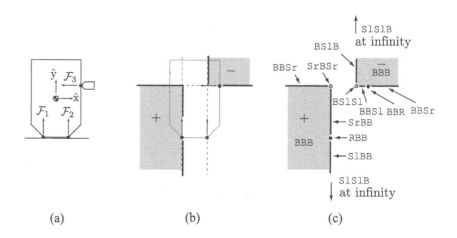

(a) (b) (c)

**Figure 12.12** Example 12.5. (a) A body resting on a table with two contact constraints provided by the table and a single contact constraint provided by the stationary finger. (b) The feasible twists represented as CoRs, shown in gray. Note that the lines that extend off to the left and to the bottom "wrap around" at infinity and come back in from the right and the top, respectively, so this CoR region should be interpreted as a single connected convex region. (c) The contact modes assigned to each feasible motion. The zero velocity contact mode is RRR.

conclusion that the motion was possible was due to the fact that our first-order analysis ignored the local contact curvature. A second-order analysis would show that this motion is indeed impossible. However, if the radius of curvature of the body at the contact were sufficiently small then the motion would be possible.

Thus a first-order roll–slide motion might be classified as penetrating or breaking by a second-order analysis. Similarly, if our second-order analysis indicates a roll–slide motion, a third- or higher-order analysis may indicate penetration or breaking free. In any case, if an $n$th-order analysis indicates that the contact is breaking or penetrating, then no analysis of order greater than $n$ will change the conclusion.

### 12.1.7　Form Closure

**Form closure** of a body is achieved if a set of stationary constraints prevents all motion of the body. If these constraints are provided by robot fingers, we call this a **form-closure grasp**. An example is shown in Figure 12.13.

#### 12.1.7.1　Number of Contacts Needed for First-Order Form Closure

Each stationary contact $i$ provides a half-space twist constraint of the form

$$\mathcal{F}_i^{\mathrm{T}} \mathcal{V} \geq 0.$$

Form closure holds if the only twist $\mathcal{V}$ satisfying the constraints is the zero twist. For $j$ contacts, this condition is equivalent to

$$\mathrm{pos}(\{\mathcal{F}_1, \ldots, \mathcal{F}_j\}) = \mathbb{R}^6$$

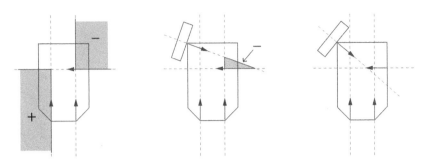

**Figure 12.13** (Left) The body from Figure 12.12, with three stationary point contacts and the body's feasible twist cone represented as a convex CoR region. (Middle) A fourth contact reduces the size of the feasible twist cone. (Right) By changing the angle of the fourth contact normal, no twist is feasible; the body is in form closure.

for bodies in three dimensions. Therefore, by fact 2 from the beginning of the chapter, at least $6 + 1 = 7$ contacts are needed for the first-order form closure of spatial bodies. For planar body, the condition is

$$\text{pos}(\{\mathcal{F}_1, \ldots, \mathcal{F}_j\}) = \mathbb{R}^3,$$

and $3 + 1 = 4$ contacts are needed for first-order form closure. These results are summarized in the following theorem.

**Theorem 12.6** *For a planar body, at least four point contacts are needed for first-order form closure. For a spatial body, at least seven point contacts are needed.*

Now consider the problem of grasping a circular disk in the plane. It should be clear that kinematically preventing motion of the disk is impossible regardless of the number of contacts; it will always be able to spin about its center. Such bodies are called **exceptional** – the positive span of the contact normal forces at all points on the body is not equal to $\mathbb{R}^n$, where $n = 3$ in the planar case and $n = 6$ in the spatial case. Examples of such bodies in three dimensions include surfaces of revolution, for example spheres and ellipsoids.

Figure 12.14 shows examples of planar grasps. The graphical methods of Section 12.1.6 indicate that the four contacts in Figure 12.14(a) immobilize the body. Our first-order analysis indicates that the bodies in Figures 12.14(b) and 12.14(c) can each rotate about their centers in the three-finger grasps, but in fact this is not possible for the body in Figure 12.14(b) – a second-order analysis would tell us that this body is actually immobilized. Finally, the first-order analysis tells us that the two-fingered grasps in Figures 12.14(d)–(f) are identical, but in fact the body in Figure 12.14(f) is immobilized by only two fingers owing to curvature effects.

To summarize, our first-order analysis always correctly labels breaking and penetrating motions but second- and higher-order effects may change first-order

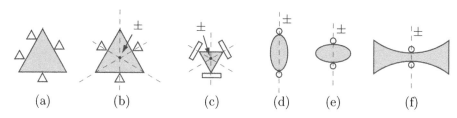

(a)                (b)                (c)            (d)        (e)            (f)

**Figure 12.14** (a) Four fingers yielding planar form closure. The first-order analysis treats (b) and (c) identically, saying that the triangle can rotate about its center in each case. A second-order analysis shows this is not possible for (b). The grasps in (d), (e), and (f) are identical by a first-order analysis, which says that rotation about any center on the vertical line is possible. This is true for (d), while for (e), rotation is possible about only some of these centers. No motion is possible in (f).

roll–slide motions to breaking or penetrating. If a body is in form closure by first-order analysis, it is in form closure for any analysis. If only roll-slide motions are feasible by first-order analysis, the body could be in form closure by a higher-order analysis; otherwise, the body is not in form closure by any analysis.

### 12.1.7.2    A Linear Programming Test for First-Order Form Closure

Let $F = [\mathcal{F}_1 \ \mathcal{F}_2 \ \cdots \ \mathcal{F}_j] \in \mathbb{R}^{n \times j}$ be a matrix whose columns are formed by the $j$ contact wrenches. For spatial bodies, $n = 6$ and for planar bodies, $n = 3$ with $\mathcal{F}_i = [m_{iz} \ f_{ix} \ f_{iy}]^\mathrm{T}$. The contacts yield form closure if there exists a vector of weights $k \in \mathbb{R}^j$, $k \geq 0$, such that $Fk + \mathcal{F}_{\text{ext}} = 0$ for all $\mathcal{F}_{\text{ext}} \in \mathbb{R}^n$.

Clearly the body is not in form closure if the rank of $F$ is not full ($\text{rank}(F) < n$). If $F$ is full rank, the form-closure condition is equivalent to the existence of strictly positive coefficients $k > 0$ such that $Fk = 0$. We can formulate this test as the following set of conditions, which is an example of a **linear program**:

$$\begin{aligned}
\text{find} \quad & k \\
\text{minimizing} \quad & 1^\mathrm{T}k \\
\text{such that} \quad & Fk = 0 \\
& k_i \geq 1, \ i = 1, \dots, j,
\end{aligned} \tag{12.14}$$

where 1 is a $j$-vector of ones. If $F$ is full rank and there exists a solution $k$ to (12.14), the body is in first-order form closure. Otherwise it is not. Note that the objective function $1^\mathrm{T}k$ is not necessary to answer the binary question, depending on the LP solver, but it is included to make sure the problem is well posed.

**Example 12.7**    The planar body in Figure 12.15 has a hole in the center. Two fingers each touch two different edges of the hole, creating four contact normals. The matrix $F = [\mathcal{F}_1 \ \mathcal{F}_2 \ \mathcal{F}_3 \ \mathcal{F}_4]$ is given by

$$F = \begin{bmatrix} 0 & 0 & -1 & 2 \\ -1 & 0 & 1 & 0 \\ 0 & -1 & 0 & 1 \end{bmatrix}.$$

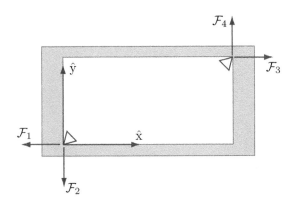

**Figure 12.15** Two fingers grasping the interior of a body.

The matrix $F$ is clearly rank 3. The linear program of (12.14) returns a solution with $k_1 = k_3 = 2$, $k_2 = k_4 = 1$, so with this grasp the body is in form closure. You could test this in MATLAB, for example, using the `linprog` function, which takes as arguments: the objective function, expressed as a vector of weights $f$ on the elements of $k$; a set of inequality constraints on $k$ of the form $Ak \leq b$ (used to encode $k_i \geq 1$); and a set of equality constraints of the form $A_{eq}k = b_{eq}$ (used to encode $Fk = 0$);

```
f = [1,1,1,1];
A = [[-1,0,0,0]; [0,-1,0,0]; [0,0,-1,0]; [0,0,0,-1]];
b = [-1,-1,-1,-1];
F = [[0,0,-1,2]; [-1,0,1,0]; [0,-1,0,1]];  % the F matrix
Aeq = F;
beq = [0,0,0];
k = linprog(f,A,b,Aeq,beq);
```

which yields the result

```
k =
      2.0000
      1.0000
      2.0000
      1.0000
```

If the right-hand finger were moved to the bottom right corner of the hole, the new $F$ matrix

$$F = \begin{bmatrix} 0 & 0 & 0 & -2 \\ -1 & 0 & 1 & 0 \\ 0 & -1 & 0 & -1 \end{bmatrix}$$

would still be full rank, but there would be no solution to the linear program. This grasp does not yield form closure: the body can slide downward on the page.

### 12.1.7.3 Measuring the Quality of a Form-Closure Grasp

Consider the two form-closure grasps shown in Figure 12.16. Which is a better grasp?

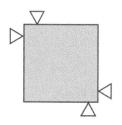 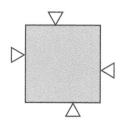

**Figure 12.16** Both grasps yield form closure, but which is better?

Answering this question requires a metric measuring the quality of a grasp. A **grasp metric** takes the set of contacts $\{\mathcal{F}_i\}$ and returns a single value $\mathrm{Qual}(\{\mathcal{F}_i\})$, where $\mathrm{Qual}(\{\mathcal{F}_i\}) < 0$ indicates that the grasp does not yield form closure, and larger positive values indicate better grasps.

There are many reasonable choices of grasp metric. As an example, suppose that, to avoid damaging the body, we require that the magnitude of the force at contact $i$ be less than or equal to $f_{i,\mathrm{max}} > 0$. Then the total set of contact wrenches that can be applied by the $j$ contacts is given by

$$CF = \left\{ \sum_{i=1}^{j} f_i \mathcal{F}_i \mid f_i \in [0, f_{i,\mathrm{max}}] \right\}. \tag{12.15}$$

See Figure 12.17 for an example in two dimensions. This shows the convex sets of wrenches that the contacts can apply to resist disturbance wrenches applied to the body. If the grasp yields form closure, the set includes the origin of the wrench space in its interior.

(a)
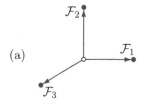 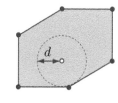

(b)
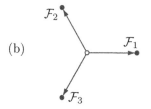 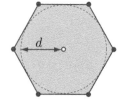

**Figure 12.17** (a) A set of three contact wrenches in a two-dimensional wrench space, and the radius $d$ of the largest ball of wrenches centered at the origin that fits inside the wrench polygon. (b) A different set of three wrenches yielding a larger inscribed ball.

Now the problem is to turn this polytope into a single number representing the quality of the grasp. Ideally this process would use some idea of the disturbance wrenches that the body can be expected to experience. A simpler choice is to set $\text{Qual}(\{\mathcal{F}_i\})$ to be the radius of the largest ball of wrenches, centered at the origin of the wrench space, that fits inside the convex polytope. In evaluating this radius, two caveats should be considered: (1) moments and forces have different units, so there is no obvious way to equate force and moment magnitudes, and (2) the moments due to contact forces depend on the location of the space-frame origin. To address (1), it is common to choose a characteristic length $r$ of the grasped body and convert contact moments $m$ to forces $m/r$. To address (2), the origin can be chosen somewhere near the geometric center of the body or at its center of mass.

Given the choice of the space frame and the characteristic length $r$, we simply calculate the signed distance from the origin of the wrench space to each hyperplane on the boundary of $CF$. The minimum of these distances is $\text{Qual}(\{\mathcal{F}_i\})$ (Figure 12.17).

Returning to our original example in Figure 12.16, we can see that if each finger is allowed to apply the same force then the grasp on the left may be considered better, as the contacts can resist greater moments about the center of the body.

#### 12.1.7.4 Choosing Contacts for Form Closure

Many methods have been suggested for choosing form-closure contacts for fixturing or grasping. One approach is to sample candidate grasp points on the surface of the body (four for planar bodies or seven for spatial) until a set is found yielding form closure. From there, the candidate grasp points may be incrementally repositioned according to gradient ascent, using the grasp metric, i.e., $\partial\,\text{Qual}(p)/\partial p$, where $p$ is the vector of all the coordinates of the contact locations.[4]

## 12.2 Contact Forces and Friction

### 12.2.1 Friction

A commonly used model of friction in robotic manipulation is **Coulomb friction**. This experimental law states that the tangential friction force magnitude $f_t$ is related to the normal force magnitude $f_n$ by $f_t \leq \mu f_n$, where $\mu$ is called the **friction coefficient**. If the contact is sliding, or currently rolling but with incipient slip (i.e., at the next instant the contacts are sliding), then $f_t = \mu f_n$ and the direction of the friction force is opposite to that of the sliding direction, i.e., friction dissipates energy. The friction force is independent of the speed of sliding.

Often two friction coefficients are defined, a static friction coefficient $\mu_s$ and

---

[4] The gradient vector $\partial\,\text{Qual}(p)/\partial p$ must be projected onto the tangent planes at the points of contact to keep the contact locations on the surface of the object.

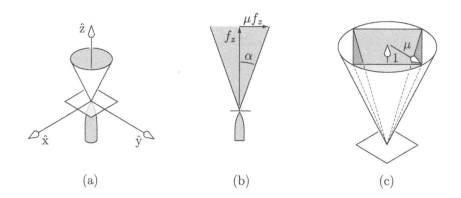

(a)         (b)         (c)

**Figure 12.18** (a) A friction cone illustrating all possible forces that can be transmitted through the contact. (b) A side view of the same friction cone showing the friction coefficient $\mu$ and the friction angle $\alpha = \tan^{-1}\mu$. (c) An inscribed polyhedral convex cone approximation to the circular friction cone.

a kinetic (or sliding) friction coefficient $\mu_k$, where $\mu_s \geq \mu_k$. This implies that a larger friction force is available to resist the initial motion but, once motion has begun, the resisting force is smaller. Many other friction models have been developed with different functional dependencies on factors such as the speed of sliding and the duration of static contact before sliding. All these are aggregate models of complex microscopic behavior. For simplicity, we will use the simplest Coulomb friction model with a single friction coefficient $\mu$. This model is reasonable for hard, dry, materials. The friction coefficient depends on the two materials in contact and typically ranges from 0.1 to 1.

For a contact normal in the $+\hat{z}$ direction, the set of forces that can be transmitted through the contact satisfies

$$\sqrt{f_x^2 + f_y^2} \leq \mu f_z, \qquad f_z \geq 0. \tag{12.16}$$

Figure 12.18(a) shows that this set of forces forms a **friction cone**. The set of forces that the finger can apply to the plane lies inside the cone shown. Figure 12.18(b) shows the same cone from a side view, illustrating the **friction angle** $\alpha = \tan^{-1}\mu$, which is the half-angle of the cone. If the contact is not sliding, the force may be anywhere inside the cone. If the finger slides to the right, the force it applies lies on the right-hand edge of the friction cone, with magnitude determined by the normal force. Correspondingly, the plane applies the opposing force to the finger, and the direction of the tangential (frictional) portion of this force opposes the sliding direction.

To allow linear formulations of contact mechanics problems, it is often convenient to represent the convex circular cone by a polyhedral convex cone. Figure 12.18(c) shows an inscribed four-sided pyramidal approximation of the friction cone, defined by the positive span of the $(f_x, f_y, f_z)$ cone edges $(\mu, 0, 1)$, $(-\mu, 0, 1)$, $(0, \mu, 1)$, and $(0, -\mu, 1)$. We can obtain a tighter approximation to the

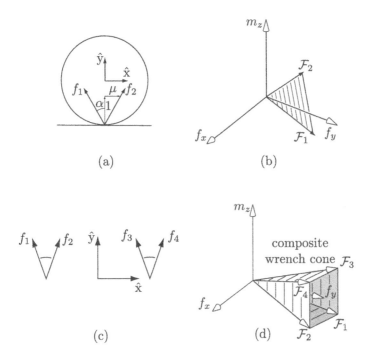

**Figure 12.19** (a) A planar friction cone with friction coefficient $\mu$ and corresponding friction angle $\alpha = \tan^{-1}\mu$. (b) The corresponding wrench cone. (c) Two friction cones. (d) The corresponding composite wrench cone.

circular cone by using more edges. An inscribed cone underestimates the friction forces available, while a circumscribed cone overestimates the friction forces. The choice of which to use depends on the application. For example, if we want to ensure that a robot hand can grasp an object, it is a good idea to underestimate the friction forces available.

For planar problems, no approximation is necessary – a friction cone is exactly represented by the positive span of the two edges of the cone, similarly to the side view illustrated in Figure 12.18(b).

Once we choose a coordinate frame, any contact force can be expressed as a wrench $\mathcal{F} = ([p]f, f)$, where $p$ is the contact location. This turns a friction cone into a wrench cone. A planar example is shown in Figure 12.19. The two edges of the planar friction cone give two rays in the wrench space, and the wrenches that can be transmitted to the body through the contact give the positive span of basis vectors along these edges. If $\mathcal{F}_1$ and $\mathcal{F}_2$ are basis vectors for these wrench cone edges, we write the wrench cone as $\mathcal{WC} = \mathrm{pos}(\{\mathcal{F}_1, \mathcal{F}_2\})$.

If multiple contacts act on a body, then the total set of wrenches that can be transmitted to the body through the contacts is the positive span of all the individual wrench cones $WC_i$,

$$WC = \text{pos}(\{WC_i\}) = \left\{ \sum_i k_i \mathcal{F}_i \mid \mathcal{F}_i \in WC_i, k_i \geq 0 \right\}.$$

This composite wrench cone is a convex cone rooted at the origin. An example of such a composite wrench cone is shown in Figure 12.19(d) for a planar object with the two friction cones shown in Figure 12.19(c). For planar problems, the composite wrench cone in the three-dimensional wrench space is polyhedral. For spatial problems, wrench cones in the six-dimensional wrench space are not polyhedral unless the individual friction cones are approximated by polyhedral cones, as in Figure 12.18(c).

If a contact or set of contacts acting on a body is ideally force-controlled, the wrench $\mathcal{F}_{\text{cont}}$ specified by the controller must lie within the composite wrench cone corresponding to those contacts. If there are other non-force-controlled contacts acting on the body, then the cone of possible wrenches on the body is equivalent to the wrench cone from the non-force-controlled contacts but translated to be rooted at $\mathcal{F}_{\text{cont}}$.

## 12.2.2    Planar Graphical Methods

### 12.2.2.1    Representing Wrenches

Any planar wrench $\mathcal{F} = (m_z, f_x, f_y)$ with a nonzero linear component can be represented as an arrow drawn in the plane, where the base of the arrow is at the point

$$(x, y) = \frac{1}{f_x^2 + f_y^2}(m_z f_y, -m_z f_x)$$

and the head of the arrow is at $(x + f_x, y + f_y)$. The moment is unchanged if we slide the arrow anywhere along its line, so any arrow of the same direction and length along this line represents the same wrench (Figure 12.20). If $f_x = f_y = 0$ and $m_z \neq 0$, the wrench is a pure moment, and we do not try to represent it graphically.

Two wrenches, represented as arrows, can be summed graphically by sliding the arrows along their lines until the bases of the arrows are coincident. The arrow corresponding to the sum of the two wrenches is obtained as shown in Figure 12.20. The approach can be applied sequentially to sum multiple wrenches represented as arrows.

### 12.2.2.2    Representing Wrench Cones

In the previous section each wrench had a specified magnitude. However, a rigid-body contact implies that the contact normal force can be arbitrarily large; the normal force achieves the magnitude needed to prevent two bodies from penetrating. Therefore it is useful to have a representation of all wrenches of the form $k\mathcal{F}$, where $k \geq 0$ and $\mathcal{F} \in \mathbb{R}^3$ is a basis vector.

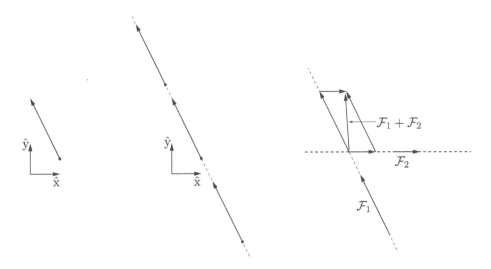

**Figure 12.20** (Left) The planar wrench $\mathcal{F} = (m_z, f_x, f_y) = (2.5, -1, 2)$ represented as an arrow in the $\hat{x}$–$\hat{y}$-plane. (Middle) The same wrench can be represented by an arrow anywhere along the line of action. (Right) Two wrenches are summed by sliding their arrows along their lines of action until the bases of the arrows are coincident, then doing a vector sum by the parallelogram construction.

One such representation is **moment labeling**. The arrow for the basis wrench $\mathcal{F}$ is drawn as described in Section 12.2.2.1. Then all points in the plane to the left of the line of the arrow are labeled '+', indicating that any positive scaling of $\mathcal{F}$ creates a positive moment $m_z$ about those points, and all points in the plane to the right of the arrow are labeled '−', indicating that any positive scaling of $\mathcal{F}$ creates a negative moment about those points. Points on the line are labeled '±'.

Generalizing, moment labels can represent any homogeneous convex planar wrench cone, much as a homogeneous convex planar twist cone can be represented as a convex CoR region. Given a collection of directed force lines corresponding to wrenches $k_i\mathcal{F}_i$ for all $k_i \geq 0$, the wrench cone $\text{pos}(\{\mathcal{F}_i\})$ can be represented by labeling each point in the plane with a '+' if each $\mathcal{F}_i$ makes a nonnegative moment about that point, with a '−' if each $\mathcal{F}_i$ makes a nonpositive moment about that point, with a '±' if each $\mathcal{F}_i$ makes zero moment about that point, or with a blank label if at least one wrench makes a positive moment and at least one wrench makes a negative moment about that point.

The idea is best illustrated by an example. In Figure 12.21(a), the basis wrench $\mathcal{F}_1$ is represented by labeling the points to the left of the force line with a '+' and points to the right of the line with a '−'. Points on the line are labeled '±'. In Figure 12.21(b), another basis wrench is added, which could represent the other edge of a planar friction cone. Only the points in the plane that are consistently labeled for both lines of force retain their labels; inconsistently labeled points lose their labels. Finally, a third basis wrench is added in Figure 12.21(c). The result is a single region labeled '+'. A nonnegative linear combination of the three basis wrenches can create any line of force in the plane that passes around this region in a counterclockwise sense. No other wrench can be created.

If an additional basis wrench were added passing clockwise around the region labeled '+' in Figure 12.21(c), then there would be no consistently labeled point in the plane; the positive linear span of the four wrenches would be the entire wrench space $\mathbb{R}^3$.

The moment-labeling representation is equivalent to a homogeneous convex wrench cone representation. The moment-labeling regions in each part, (a), (b) and (c), of Figure 12.21 are properly interpreted as a single convex region, much like the CoR regions of Section 12.1.6.

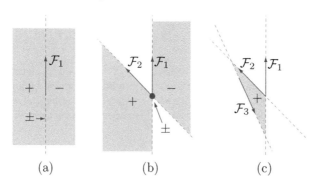

(a)          (b)          (c)

**Figure 12.21** (a) Representing a line of force by moment labels. (b) Representing the positive span of two lines of force by moment labels. (c) The positive span of three lines of force.

### 12.2.3    Force Closure

Consider a single movable object and a number of frictional contacts. We say the contacts result in **force closure** if the composite wrench cone contains the entire wrench space, so that any external wrench $\mathcal{F}_{\text{ext}}$ on the object can be balanced by contact forces.

We can derive a simple linear test for force closure which is exact for planar cases and approximate for spatial cases. Let $\mathcal{F}_i$, $i = 1, \dots, j$, be the wrenches corresponding to the edges of the friction cones for all the contacts. For planar problems, each friction cone contributes two edges and, for spatial problems, each friction cone contributes three or more edges, depending on the polyhedral approximation chosen (see Figure 12.18(c)). The columns of an $n \times j$ matrix $F$ are the $\mathcal{F}_i$, where $n = 3$ for planar problems and $n = 6$ for spatial problems. Now, the test for force closure is identical to that for form closure. The contacts yield force closure if

- rank $F = n$, and
- there exists a solution to the linear programming problem (12.14).

In the case of $\mu = 0$, each contact can provide forces only along the normal direction, and force closure is equivalent to first-order form closure.

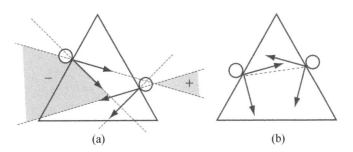

(a) (b)

**Figure 12.22** An equilateral triangle can be force-closure-grasped by two fingers on the edges of the triangle if $\mu \geq \tan 30° \approx 0.577$. (a) The grasp shown with $\mu = 0.25$ would not be in force closure, as indicated by the consistently labeled moment-labeling region. (b) The grasp shown is in force closure with $\mu = 1$; the dashed line indicates that the two contacts can "see" each other, i.e., their line of sight is inside both friction cones.

### 2.2.3.1 Number of Contacts Needed for Force Closure

For planar problems, four contact wrenches are sufficient to positively span the three-dimensional wrench space, which means that as few as two frictional contacts (with two friction cone edges each) are sufficient for force closure. Using moment labeling, we see that force closure is equivalent to having no consistent moment labels. For example, if the two contacts can "see" each other by a line of sight inside both friction cones, we have force closure (Figure 12.22(b)).

It is important to note that force closure simply means that the contact friction cones can generate any wrench. It does not necessarily mean that the body will not move in the presence of an external wrench. For the example of Figure 12.22(b), whether the triangle falls under gravity depends on the internal forces between the fingers. If the motors powering the fingers cannot provide sufficient forces, or if they are restricted to generate forces only in certain directions, the triangle may fall despite force closure.

Two frictional point contacts are insufficient to yield force closure for spatial bodies, as there is no way to generate a moment about the axis joining the two contacts. A force-closure grasp can be obtained with as few as three frictional contacts, however. A particularly simple and appealing result due to Li et al. (2003) reduces the force-closure analysis of spatial frictional grasps to a planar force-closure problem. Referring to Figure 12.23, suppose that a rigid body is constrained by three frictional point contacts. If the three contact points happen to be collinear then obviously any moment applied about this line cannot be resisted by the three contacts. We can therefore exclude this case and assume that the three contact points are not collinear. The three contacts then define a unique plane $S$ and, at each contact point, three possibilities arise (see Figure 12.24):

- the friction cone intersects $S$ in a planar cone;
- the friction cone intersects $S$ in a line;
- the friction cone intersects $S$ at a point.

The body is in force closure if and only if each friction cone intersects $S$ in a planar cone and $S$ is also in planar force closure.

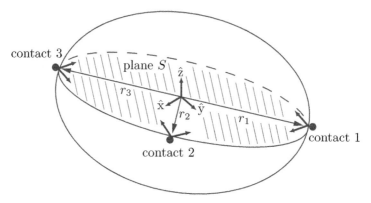

**Theorem 12.8** *Given a spatial rigid body restrained by three point contacts with friction, the body is in force closure if and only if the friction cone at each contact intersects the plane $S$ of the contacts in a cone and the plane $S$ is in planar force closure.*

*Proof* First, the necessity condition – if the spatial rigid body is in force closure then each friction cone intersects $S$ in a planar cone and $S$ is also in planar force closure – is easily verified: if the body is in spatial force closure then $S$ (which is a part of the body) must also be in planar force closure. Moreover, if even one friction cone intersects $S$ in a line or point then there will be external moments (e.g., about the line between the remaining two contact points) that cannot be resisted by the grasp.

To prove the sufficiency condition – if each friction cone intersects $S$ in a planar cone and $S$ is also in planar force closure then the spatial rigid body is in force closure – choose a fixed reference frame such that $S$ lies in the $\hat{x}$–$\hat{y}$-plane and let $r_i \in \mathbb{R}^3$ denote the vector from the fixed-frame origin to contact point $i$ (see Figure 12.23). Denoting the contact force at $i$ by $f_i \in \mathbb{R}^3$, the contact wrench $\mathcal{F}_i \in \mathbb{R}^6$ is then of the form

$$\mathcal{F}_i = \begin{bmatrix} m_i \\ f_i \end{bmatrix}, \tag{12.17}$$

where each $m_i = r_i \times f_i$, $i = 1, 2, 3$. Denote the arbitrary external wrench

**Figure 12.24** Three possibilities for the intersection between a friction cone and a plane.

$\mathcal{F}_{\text{ext}} \in \mathbb{R}^6$ by

$$\mathcal{F}_{\text{ext}} = \begin{bmatrix} m_{\text{ext}} \\ f_{\text{ext}} \end{bmatrix} \in \mathbb{R}^6. \tag{12.18}$$

Force closure then requires that there exist contact wrenches $\mathcal{F}_i$, $i = 1, 2, 3$, each lying inside its respective friction cone, such that, for any external disturbance wrench $\mathcal{F}_{\text{ext}}$, the following equality is satisfied:

$$\mathcal{F}_1 + \mathcal{F}_2 + \mathcal{F}_3 + \mathcal{F}_{\text{ext}} = 0 \tag{12.19}$$

or, equivalently,

$$f_1 + f_2 + f_3 + f_{\text{ext}} = 0, \tag{12.20}$$

$$(r_1 \times f_1) + (r_2 \times f_2) + (r_3 \times f_3) + m_{\text{ext}} = 0. \tag{12.21}$$

If each contact force and moment, as well as the external force and moment, is orthogonally decomposed into components lying on the plane spanned by $S$ (corresponding to the $\hat{x}$–$\hat{y}$-plane in our chosen reference frame) and its normal subspace $N$ (corresponding to the $\hat{z}$-axis in our chosen reference frame) then the previous force-closure equalities can be written as

$$f_{1S} + f_{2S} + f_{3S} = -f_{\text{ext},S}, \tag{12.22}$$

$$(r_1 \times f_{1S}) + (r_2 \times f_{2S}) + (r_3 \times f_{3S}) = -m_{\text{ext},S}, \tag{12.23}$$

$$f_{1N} + f_{2N} + f_{3N} = -f_{\text{ext},N}, \tag{12.24}$$

$$(r_1 \times f_{1N}) + (r_2 \times f_{2N}) + (r_3 \times f_{3N}) = -m_{\text{ext},N}. \tag{12.25}$$

In what follows we shall use $S$ to refer both to the slice of the rigid body corresponding to the $\hat{x}$–$\hat{y}$-plane and to the $\hat{x}$–$\hat{y}$-plane itself; we will always identify $N$ with the $\hat{z}$-axis.

Proceeding with the proof of sufficiency, we now show that if $S$ is in planar force closure then the body is in spatial force closure. In terms of Equations (12.24) and (12.25) we wish to show that, for any arbitrary forces $f_{\text{ext},S} \in S$, $f_{\text{ext},N} \in N$ and arbitrary moments $m_{\text{ext},S} \in S$, $m_{\text{ext},N} \in N$, there exist contact forces $f_{iS} \in S$, $f_{iN} \in N$, $i = 1, 2, 3$, that satisfy (12.24) and (12.25) such that, for each $i = 1, 2, 3$, the contact force $f_i = f_{iS} + f_{iN}$ lies in friction cone $i$.

First consider the force-closure equations (12.24) and (12.25) in the normal direction $N$. Given an arbitrary external force $f_{\text{ext},N} \in N$ and external moment $m_{\text{ext},S} \in S$, Equations (12.24) and (12.25) constitute a set of three linear equations in three unknowns. From our assumption that the three contact points are never collinear, these equations will always have a unique solution set $\{f_{1N}^*, f_{2N}^*, f_{3N}^*\}$ in $N$.

Since $S$ is assumed to be in planar force closure, for any arbitrary $f_{\text{ext},S} \in S$ and $m_{\text{ext},N} \in N$ there will exist planar contact forces $f_{iS} \in S$, $i = 1, 2, 3$, that lie inside their respective planar friction cones and also satisfy Equations (12.22) and (12.23). This solution set is not unique: one can always find a set of internal forces $\eta_i \in S$, $i = 1, 2, 3$, each lying inside its respective friction cone, satisfying

$$\eta_1 + \eta_2 + \eta_3 = 0, \tag{12.26}$$

$$(r_1 \times \eta_1) + (r_2 \times \eta_2) + (r_3 \times \eta_3) = 0. \tag{12.27}$$

(To see why such $\eta_i$ exist, recall that since $S$ is assumed to be in planar force closure, solutions to (12.22) and (12.23) must exist for $f_{\text{ext},S} = \mu_{\text{ext},N} = 0$; these solutions are precisely the internal forces $\eta_i$). Note that these two equations constitute three linear equality constraints involving six variables, so that there exists a three-dimensional linear subspace of solutions for $\{\eta_1, \eta_2, \eta_3\}$.

Now, if $\{f_{1S}, f_{2S}, f_{3S}\}$ satisfy (12.22) and (12.23) then so will $\{f_{1S} + \eta_1, f_{2S} + \eta_2, f_{3S} + \eta_3\}$. The internal forces $\{\eta_1, \eta_2, \eta_3\}$ can, in turn, be chosen to have sufficiently large magnitudes that the contact forces

$$f_1 = f_{1N}^* + f_{1S} + \eta_1, \tag{12.28}$$

$$f_2 = f_{2N}^* + f_{2S} + \eta_2, \tag{12.29}$$

$$f_3 = f_{3N}^* + f_{3S} + \eta_3 \tag{12.30}$$

all lie inside their respective friction cones. This completes the proof of the sufficiency condition.                                                   □

### 12.2.3.2   Measuring the Quality of a Force-Closure Grasp

Friction forces are not always repeatable. For example, try putting a coin on a book and tilting the book. The coin should begin to slide when the book is at an angle $\alpha = \tan^{-1} \mu$ with respect to the horizontal. If you do the experiment several times then you may find a range of measured values of $\mu$, owing to effects that are difficult to model. For that reason, when choosing between grasps it is reasonable to choose finger locations that minimize the friction coefficient needed to achieve force closure.

### 12.2.4   Duality of Force and Motion Freedoms

Our discussion of kinematic constraints and friction should have made it apparent that, for any point contact and contact label, the number of equality constraints on a body's motion caused by that contact is equal to the number of wrench freedoms it provides. For example, a breaking contact B provides zero equality constraints on the body's motion and also allows no contact force. A fixed contact R provides three motion constraints (the motion of a point on the body is specified) and three freedoms in the contact force: any wrench in the interior of the contact wrench cone is consistent with the contact mode. Finally, a slipping contact S provides one equality motion constraint (one equation on the body's motion must be satisfied to maintain the contact) and, for a given motion satisfying the constraint, the contact wrench has only one freedom: the magnitude of the contact wrench on the edge of the friction cone and opposing the slipping direction. In the planar case, the motion constraints and wrench freedoms for B, S, and R contacts are 0, 1, and 2, respectively.

## 2.3    Manipulation

So far we have studied the feasible twists and contact forces due to a set of contacts. We have also considered two types of manipulation: form-closure and force-closure grasping.

Manipulation consists of much more than just grasping, however. It includes almost anything where manipulators impose motions or forces with the purpose of achieving the motion or restraint of objects. Examples include carrying glasses on a tray without toppling them, pivoting a refrigerator about one of its feet, pushing a sofa along the floor, throwing and catching a ball, transporting parts on a vibratory conveyor, etc. Endowing a robot with methods of manipulation beyond grasp-and-carry allows it to manipulate several parts simultaneously, manipulate parts that are too large to be grasped or too heavy to be lifted, or even to send parts outside the workspace of the end-effector by throwing them.

To plan such manipulation tasks, we use the contact kinematic constraints of Section 12.1, the Coulomb friction law of Section 12.2, and the dynamics of rigid bodies. Restricting ourselves to a single rigid body and using the notation of Chapter 8, the body's dynamics are written as

$$\mathcal{F}_{ext} + \sum k_i \mathcal{F}_i = \mathcal{G}\dot{\mathcal{V}} - [\mathrm{ad}_{\mathcal{V}}]^{\mathrm{T}} \mathcal{G}\mathcal{V}, \qquad k_i \geq 0, \quad \mathcal{F}_i \in \mathcal{WC}_i, \qquad (12.31)$$

where $\mathcal{V}$ is the body's twist, $\mathcal{G}$ is its spatial inertia matrix, $\mathcal{F}_{ext}$ is the external wrench acting on the body due to gravity, etc., $\mathcal{WC}_i$ is the set of possible wrenches acting on the body due to contact $i$, and $\sum k_i \mathcal{F}_i$ is the wrench due to the contacts. All wrenches are written in the body's center-of-mass frame. Now, given a set of motion- or force-controlled contacts acting on the body, and the initial state of the system, one method for solving for the motion of the body is the following.

(a) Enumerate the set of possible contact modes considering the current state of the system (e.g., a contact that is currently sticking can transition to sliding or breaking). The contact modes consist of the contact labels R, S, and B at each contact.

(b) For each contact mode, determine whether there exists a contact wrench $\sum k_i \mathcal{F}_i$ that is consistent with the contact mode and Coulomb's law, and an acceleration $\dot{\mathcal{V}}$ consistent with the kinematic constraints of the contact mode, such that Equation (12.31) is satisfied. If so, this contact mode, contact wrench, and body acceleration comprises a consistent solution to the rigid-body dynamics.

This kind of "case analysis" may sound unusual; we are not simply solving a set of equations. It also leaves open the possibility that we could find more than one consistent solution, or perhaps no consistent solution. This is, in fact, the case: we can define problems with multiple solutions (**ambiguous** problems) and problems with no solutions (**inconsistent** problems). This state of affairs is a bit unsettling; surely there is exactly one solution to any real mechanics problem! But this is the price we pay for using the assumptions of perfectly rigid bodies and Coulomb friction. Despite the possibility of zero or multiple solutions, for many problems the method described above will yield a unique contact mode and motion.

Some of the manipulation tasks below are **quasistatic**, where the velocities and accelerations of the bodies are small enough that inertial forces may be ignored. Contact wrenches and external wrenches are always in force balance, and Equation (12.31) reduces to

$$\mathcal{F}_{\text{ext}} + \sum k_i \mathcal{F}_i = 0, \qquad k_i \geq 0, \quad \mathcal{F}_i \in \mathcal{WC}_i. \tag{12.32}$$

Below we illustrate the methods of this chapter with four examples.

**Example 12.9** (A block carried by two fingers)   Consider a planar block in gravity supported by two fingers, as in Figure 12.25(a). The friction coefficient between one finger and the block is $\mu = 1$, and the other contact is frictionless. Thus the cone of wrenches that can be applied by the fingers is $\text{pos}(\{\mathcal{F}_1, \mathcal{F}_2, \mathcal{F}_3\})$, as shown using moment labeling in Figure 12.25(b).

Our first question is whether the stationary fingers can keep the block at rest. To do so, the fingers must provide a wrench $\mathcal{F} = (m_z, f_x, f_y) = (0, 0, \text{mg})$ to balance the wrench $\mathcal{F}_{\text{ext}} = (0, 0, -\text{mg})$ due to gravity, where $g > 0$. As shown in Figure 12.25(b), this wrench is not in the composite cone of possible contact wrenches. Therefore the contact mode RR is not feasible, and the block will move relative to the fingers.

Now consider the case where the fingers each accelerate to the left at $2g$. In this case the contact mode RR requires that the block also accelerate to the left at $2g$. The wrench needed to cause this acceleration is $(0, -2\text{mg}, 0)$. Therefore the total wrench that the fingers must apply to the block is $(0, -2\text{mg}, 0) - \mathcal{F}_{\text{ext}} = (0, -2\text{mg}, \text{mg})$. As shown in Figures 12.25(c), (d), this wrench lies inside the composite wrench cone. Thus RR (the block stays stationary relative to the fingers) is a solution as the fingers accelerate to the left at $2g$.

This is called a **dynamic grasp** – inertial forces are used to keep the block pressed against the fingers while the fingers move. If we plan to manipulate the block using a dynamic grasp then we should make certain that no contact modes other than RR are feasible, for completeness.

Moment labels are convenient for understanding this problem graphically, but we can also solve it algebraically. The lower finger contacts the block at $(x, y) = (-3, -1)$ and the upper finger contacts the block at $(1, 1)$. This gives the basis contact wrenches

$$\mathcal{F}_1 = \frac{1}{\sqrt{2}}(-4, -1, 1),$$

$$\mathcal{F}_2 = \frac{1}{\sqrt{2}}(-2, 1, 1),$$

$$\mathcal{F}_3 = (1, -1, 0).$$

Let the fingers' acceleration in the $\hat{x}$-direction be written $a_x$. Then, under the

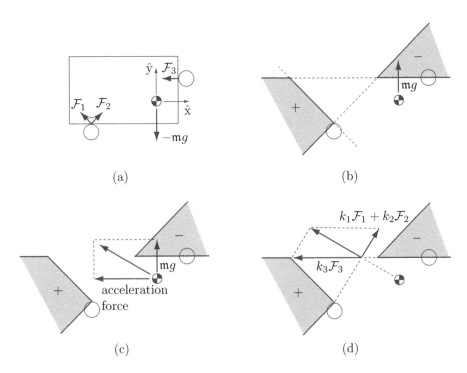

**Figure 12.25** (a) A planar block in gravity supported by two robot fingers, the lower with a friction cone with $\mu = 1$ and the upper with $\mu = 0$. (b) The composite wrench cone that can be applied by the fingers represented using moment labels. To balance the block against gravity, the fingers must apply the line of force shown. This line has a positive moment with respect to points labeled '$-$', and therefore it cannot be generated by the two fingers. (c) For the block to match the fingers' acceleration to the left, the contacts must apply the vector sum of the wrench to balance gravity and the wrench needed to accelerate the block to the left. This total wrench lies inside the composite wrench cone, as the line of force has a positive moment with respect to points labeled '$+$' and a negative moment with respect to points labeled '$-$'. (d) The total wrench applied by the fingers in (c) can be translated along the line of action without changing the wrench. This allows us to visualize easily the components $k_1\mathcal{F}_1 + k_2\mathcal{F}_2$ and $k_3\mathcal{F}_3$ provided by the fingers.

assumption that the block stays fixed with respect to the fingers (RR contact mode), Equation (12.31) can be written

$$k_1\mathcal{F}_1 + k_2\mathcal{F}_2 + k_3\mathcal{F}_3 + (0, 0, -mg) = (0, ma_x, 0). \qquad (12.33)$$

This yields three equations in the three unknowns, $k_1, k_2, k_3$. Solving, we get

$$k_1 = -\frac{1}{2\sqrt{2}}(a_x + g)\text{m}, \qquad k_2 = \frac{1}{2\sqrt{2}}(a_x + 5g)\text{m}, \qquad k_3 = -\frac{1}{2}(a_x - 3g)\text{m}.$$

For the $k_i$ to be nonnegative, we need $-5g \leq a_x \leq -g$. For $\hat{x}$-direction finger accelerations in this range, a dynamic grasp is a consistent solution.

**Example 12.10** (The meter-stick trick)  Try this experiment. Get a meter stick

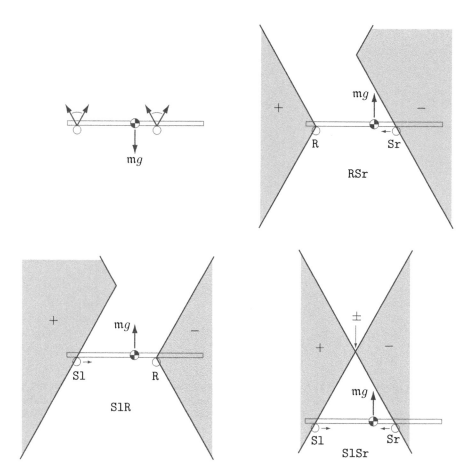

**Figure 12.26** Top left: Two frictional fingers supporting a meter stick in gravity. The other three panels show the moment labels for the RSr, S1R, and S1Sr contact modes. Only the S1R contact mode yields force balance.

(or any similar long smooth stick) and balance it horizontally on your two index fingers. Place your left finger near the 10 cm mark and your right finger near the 60 cm mark. The center of mass is closer to your right finger but still between your fingers, so that the stick is supported. Now, keeping your left finger stationary, slowly move your right finger towards your left until they touch. What happens to the stick?

If you didn't try the experiment, you might guess that your right finger passes under the center of mass of the stick, at which point the stick falls. If you did try the experiment, you saw something different. Let's see why.

Figure 12.26 shows the stick supported by two frictional fingers. Since all motions are slow, we use the quasistatic approximation that the stick's acceleration is zero and so the net contact wrench must balance the gravitational wrench. As the two fingers move together, the stick must slip on one or both fingers

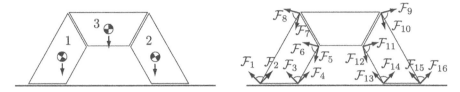

**Figure 12.27** (Left) An arch under gravity. (Right) The friction cones at the contacts of stone 1 and the contacts of stone 2.

to accommodate the fact that the fingers are getting closer to each other. Figure 12.26 shows the moment-labeling representation of the composite wrench cone for three different contact modes: RSr, where the stick remains stationary relative to the left finger and slips to the right relative to the right finger; SlR, where the stick slips to the left relative to the left finger and remains stationary relative to the right finger; and SlSr, where the stick slips on both fingers. It is clear from the figure that only the SlR contact mode can provide a wrench that balances the gravitational wrench. In other words, the right finger, which supports more of the stick's weight, remains fixed relative to the stick while the left finger slides under the stick. Since the right finger is moving to the left in the world frame, this means that the center of mass is moving to the left at the same speed. This continues until the center of mass is halfway between the fingers, at which point the stick transitions to the SlSr contact mode, and the center of mass stays centered between the fingers until they meet. The stick never falls.

Note that this analysis relies on the quasistatic assumption. It is easy to make the stick fall if you move your right finger quickly; the friction force at the right finger is not large enough to create the large stick acceleration needed to maintain a sticking contact. Also, in your experiment, you might notice that, when the center of mass is nearly centered, the stick does not actually achieve the idealized SlSr contact mode, but instead switches rapidly between the SlR and RSr contact modes. This occurs because the static friction coefficient is larger than the kinetic friction coefficient.

**Example 12.11** (Stability of an assembly) Consider the arch in Figure 12.27. Is it stable under gravity?

For a problem like this, graphical planar methods are difficult to use, since there are potentially multiple moving bodies. Instead we test algebraically for consistency of the contact mode with all contacts labeled R. The friction cones are shown in Figure 12.27. With these labelings of the friction cone edges, the arch can remain standing if there exist $k_i \geq 0$ for $i = 1, \ldots, 16$ satisfying the

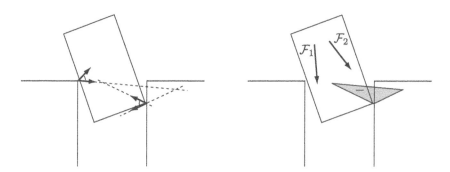

**Figure 12.28** (Left) A peg in two-point contact with a hole. (Right) The wrench $\mathcal{F}_1$ may cause the peg to jam, while the wrench $\mathcal{F}_2$ continues to push the peg into the hole.

following nine wrench-balance equations, three for each body:

$$\sum_{i=1}^{8} k_i \mathcal{F}_i + \mathcal{F}_{\text{ext1}} = 0,$$

$$\sum_{i=9}^{16} k_i \mathcal{F}_i + \mathcal{F}_{\text{ext2}} = 0,$$

$$-\sum_{i=5}^{12} k_i \mathcal{F}_i + \mathcal{F}_{\text{ext3}} = 0,$$

where $\mathcal{F}_{\text{ext}i}$ is the gravitational wrench on body $i$. The last set of equations comes from the fact that the wrenches that body 1 applies to body 3 are equal and opposite those that body 3 applies to body 1, and similarly for bodies 2 and 3.

This linear constraint satisfaction problem can be solved by a variety of methods, including linear programming.

**Example 12.12** (Peg insertion)   Figure 12.28 shows a force-controlled planar peg in two-point contact with a hole during insertion. Also shown are the contact friction cones acting on the peg and the corresponding composite wrench cone, illustrated using moment labels. If the force controller applies the wrench $\mathcal{F}_1$ to the peg, it may jam – the hole may generate contact forces that balance $\mathcal{F}_1$. Therefore the peg may get stuck in this position. If the force controller applies the wrench $\mathcal{F}_2$, however, the contacts cannot balance the wrench and insertion proceeds.

If the friction coefficients at the two contacts are large enough that the two friction cones "see" each other's base (Figure 12.22(b)), the peg is in force closure and the contacts may be able to resist any wrench (depending on the internal force between the two contacts). The peg is said to be wedged.

## 2.4 Summary

- Three ingredients are needed to solve rigid-body contact problems with friction: (1) the contact kinematics, which describes the feasible motions of rigid bodies in contact; (2) a contact force model, which describes the forces that can be transmitted through frictional contacts; and (3) rigid-body dynamics, as described in Chapter 8.

- Let two rigid bodies, $A$ and $B$, be in point contact at $p_A$ in a space frame. Let $\hat{n} \in \mathbb{R}^3$ be the unit contact normal, pointing into body $A$. Then the spatial contact wrench $\mathcal{F}$ associated with a unit force along the contact normal is $\mathcal{F} = [([p_A]\hat{n})^\mathrm{T}\ \hat{n}^\mathrm{T}]^\mathrm{T}$. The impenetrability constraint is

$$\mathcal{F}^\mathrm{T}(\mathcal{V}_A - \mathcal{V}_B) \geq 0,$$

  where $\mathcal{V}_A$ and $\mathcal{V}_B$ are the spatial twists of $A$ and $B$.

- A contact that is sticking or rolling is assigned the contact label R, a contact that is sliding is assigned the contact label S, and a contact that is breaking free is assigned the contact label B. For a body with multiple contacts, the contact mode is the concatenation of the labels of the individual contacts.

- A single rigid body subjected to multiple stationary point contacts has a homogeneous (rooted-at-the-origin) polyhedral convex cone of twists that satisfy all the impenetrability constraints.

- A homogeneous polyhedral convex cone of planar twists in $\mathbb{R}^3$ can be equivalently represented by a convex region of signed rotation centers in the plane.

- If a set of stationary contacts prevents a body from moving, purely by a kinematic analysis considering only the contact normals, the body is said to be in first-order form closure. The contact wrenches $\mathcal{F}_i$ for contacts $i = 1, \ldots, j$ positively span $\mathbb{R}^n$, where $n = 3$ for the planar case and $n = 6$ for the spatial case.

- At least four point contacts are required for first-order form closure of a planar body, and at least seven point contacts are required for first-order form closure of a spatial body.

- The Coulomb friction law states that the tangential frictional force magnitude $f_\mathrm{t}$ at a contact satisfies $f_\mathrm{t} \leq \mu f_\mathrm{n}$, where $\mu$ is the friction coefficient and $f_\mathrm{n}$ is the normal force. When the contact is sticking, the frictional force can be anything satisfying this constraint. When the contact is sliding, $f_\mathrm{t} = \mu f_\mathrm{n}$ and the direction of the friction force opposes the direction of sliding.

- Given a set of frictional contacts acting on a body, the wrenches that can be transmitted through these contacts is the positive span of the wrenches that can be transmitted through the individual contacts. These wrenches form a homogeneous convex cone. If the body is planar, or if the body is spatial but the contact friction cones are approximated by polyhedral cones, the wrench cone is also polyhedral.

- A homogeneous convex cone of planar wrenches in $\mathbb{R}^3$ can be represented as a convex region of moment labels in the plane.
- An body is in force closure if the homogeneous convex cone of contact wrenches from the stationary contacts is the entire wrench space ($\mathbb{R}^3$ or $\mathbb{R}^6$). If the contacts are frictionless, force closure is equivalent to first-order form closure.

## 12.5    Notes and References

The kinematics of contact draws heavily from concepts in linear algebra (see, for example, the texts by Strang (2009) and Meyer (2000)) and, more specifically, screw theory (Ball, 1900; Ohwovoriole and Roth, 1981; Bottema and Roth, 1990; Angeles, 2006; McCarthy, 1990). Graphical methods for the analysis of planar constraints were introduced by Reuleaux (1876), and Mason introduced the graphical construction of contact labels for planar kinematics and moment labels for the representation of homogeneous wrench cones (Mason, 1991, 2001). Polyhedral convex cones, and their application in representing feasible twist cones and contact wrench cones, are discussed in Mason (2001), Kao et al. (2016), Erdmann (1994), and Hirai and Asada (1993). The formalization of the friction law used in this chapter was given by Coulomb (1781). Surprising consequences of Coulomb friction are the problems of ambiguity and inconsistency (Lötstedt, 1981; Mason, 2001; Mason and Wang, 1988) and the fact that infinite friction does not necessarily prevent slipping at an active contact (Lynch and Mason, 1995).

Form closure and force closure are discussed in detail in Prattichizzo and Trinkle (2016). In particular, that reference uses the term "frictional form closure" to mean the same thing that "force closure" means in this chapter. According to Prattichizzo and Trinkle (2016), force closure additionally requires that the hand doing the grasping be sufficiently capable of controlling the internal "squeezing" forces. Similar distinctions are made in Bicchi (1995) and the reviews by Bicchi and Kumar (2000) and by Bicchi (2000). In this chapter we did not consider the details of the robot hand but adopted a definition of force closure based solely on the geometry and friction of the contacts.

The numbers of contacts needed for planar and spatial form closure were established by Reuleaux (1876) and Somoff (1900), respectively. Other foundational results in form- and force-closure grasping were developed in Lakshminarayana (1978), Mishra et al. (1987), and Markenscoff et al. (1990) and were reviewed in Bicchi (2000) and in Prattichizzo and Trinkle (2016), which also provides an overview of grasp-quality metrics. The fact that two friction cones "seeing" each other's base suffices for planar force closure was first reported in Nguyen (1988), and the result reviewed in this chapter on three-finger force-closure grasps in three dimensions appeared in Li et al. (2003). Grübler's formula was used to calculate the mobility of a grasped object using kinematic models of contact in (Mason and Salisbury, 1985).

Second-order models of contact constraints were introduced by Rimon and Burdick (1995, 1996, 1998a,b) and were used to show that curvature effects allow form closure by fewer contacts.

Jamming and wedging in robotic insertion were described in Simunovic (1975), Nevins and Whitney (1978), and Whitney (1982), and the notion of a dynamic grasp was first introduced in Mason and Lynch (1993).

An important class of methods for simulating systems of rigid bodies in frictional contact, not covered in this chapter, is based on solving linear and nonlinear complementarity problems (Stewart and Trinkle, 1996; Pang and Trinkle, 1996; Trinkle, 2003). These complementarity formulations directly encode the facts that: if a contact is breaking then no force is applied; if a contact is sticking then the force can be anywhere inside the friction cone; and if a contact is sliding then the force is on the edge of the friction cone.

General references on contact modeling and manipulation include chapters in the *Handbook of Robotics* (Kao et al., 2016; Prattichizzo and Trinkle, 2016), and the texts by Mason (2001) and Murray et al. (1994).

## 12.6    Exercises

**Exercise 12.1**    Prove that the impenetrability constraint (12.4) is equivalent to the constraint (12.7).

**Exercise 12.2**    Representing planar twists as centers of rotation.

(a) Consider the two planar twists $\mathcal{V}_1 = (\omega_{z1}, v_{x1}, v_{y1}) = (1, 2, 0)$ and $\mathcal{V}_2 = (\omega_{z2}, v_{x2}, v_{y2}) = (1, 0, -1)$. Draw the corresponding CoRs in a planar coordinate frame, and illustrate $\text{pos}(\{\mathcal{V}_1, \mathcal{V}_2\})$ as CoRs.
(b) Draw the positive span of $\mathcal{V}_1 = (\omega_{z1}, v_{x1}, v_{y1}) = (1, 2, 0)$ and $\mathcal{V}_2 = (\omega_{z2}, v_{x2}, v_{y2}) = (-1, 0, -1)$ as CoRs.

**Exercise 12.3**    A rigid body is contacted at $p = (1, 2, 3)$ with a contact normal into the body $\hat{n} = (0, 1, 0)$. Write the constraint on the body's twist $\mathcal{V}$ due to this contact.

**Exercise 12.4**    A space frame {s} is defined at a contact between a stationary constraint and an object. The contact normal, into the object, is along the $\hat{z}$-axis of the {s} frame.

(a) Write down the constraint on the object's twist $\mathcal{V}$ if the contact is a frictionless point contact.
(b) Write down the constraints on $\mathcal{V}$ if the contact is a point contact with friction.
(c) Write down the constraints on $\mathcal{V}$ if the contact is a soft contact.

**Exercise 12.5**   Figure 12.29 shows five stationary "fingers" contacting an object. The object is in first-order form closure and therefore force closure. If we take away one finger, the object may still be in form closure. For which subsets of four fingers is the object still in form closure? Prove your answers using graphical methods.

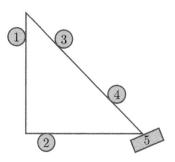

**Figure 12.29** A triangle in contact with five stationary fingers, yielding first-order form closure and therefore force closure. Analyze the contact when one or more fingers are removed. The hypotenuse of the triangle is 45° from the vertical on the page, and contact normal 5 is 22.5° from the vertical.

**Exercise 12.6**   Draw the set of feasible twists as CoRs when the triangle of Figure 12.29 is contacted only by finger 1. Label the feasible CoRs with their contact labels.

**Exercise 12.7**   Draw the set of feasible twists as CoRs when the triangle of Figure 12.29 is contacted only by fingers 1 and 2. Label the feasible CoRs with their contact labels.

**Exercise 12.8**   Draw the set of feasible twists as CoRs when the triangle of Figure 12.29 is contacted only by fingers 2 and 3. Label the feasible CoRs with their contact labels.

**Exercise 12.9**   Draw the set of feasible twists as CoRs when the triangle of Figure 12.29 is contacted only by fingers 1 and 5. Label the feasible CoRs with their contact labels.

**Exercise 12.10**   Draw the set of feasible twists as CoRs when the triangle of Figure 12.29 is contacted only by fingers 1, 2, and 3.

**Exercise 12.11**   Draw the set of feasible twists as CoRs when the triangle of Figure 12.29 is contacted only by fingers 1, 2, and 4.

**Exercise 12.12**   Draw the set of feasible twists as CoRs when the triangle of Figure 12.29 is contacted only by fingers 1, 3, and 5.

**Exercise 12.13**   Refer again to the triangle of Figure 12.29.

(a) Draw the wrench cone from contact 5, assuming a friction angle $\alpha = 22.5°$ (a friction coefficient $\mu = 0.41$), using moment labeling.

(b) Add contact 2 to the moment-labeling drawing. The friction coefficient at contact 2 is $\mu = 1$.

**Exercise 12.14** Refer again to the triangle of Figure 12.29. Draw the moment-labeling region corresponding to contact 1 with $\mu = 1$ and contact 4 with $\mu = 0$.

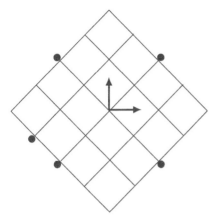

**Figure 12.30** A $4 \times 4$ planar square restrained by five frictionless point contacts.

**Exercise 12.15** The planar grasp of Figure 12.30 consists of five frictionless point contacts. The square's size is $4 \times 4$.

(a) Show that this grasp does not yield force closure.
(b) The grasp of part (a) can be modified to yield force closure by adding one frictionless point contact. Draw all the possible locations for this contact.

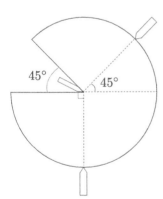

**Figure 12.31** A planar disk restrained by three frictionless point contacts.

**Exercise 12.16** Assume the contacts shown in Figure 12.31 are frictionless point contacts. Determine whether the grasp yields force closure. If it does not, how many additional frictionless point contacts are needed to construct a force closure grasp?

**Exercise 12.17** Consider the L-shaped planar object of Figure 12.32.

(a) Suppose that both contacts are point contacts with friction coefficient $\mu = 1$. Determine whether this grasp yields force closure.

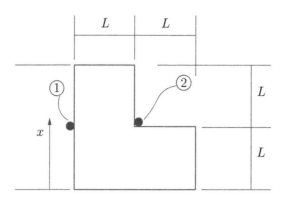

**Figure 12.32** An L-shaped planar object restrained by two point contacts with friction.

(b) Now suppose that point contact 1 has friction coefficient $\mu = 1$, while point contact 2 is frictionless. Determine whether this grasp yields force closure.

(c) The vertical position of contact 1 is allowed to vary; denote its height by $x$. Find all positions $x$ such that the grasp is force closure with $\mu = 1$ for contact 1 and $\mu = 0$ for contact 2.

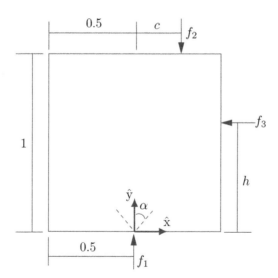

**Figure 12.33** A square restrained by three point contacts.

**Exercise 12.18**　A square is restrained by three point contacts as shown in Figure 12.33: $f_1$ is a point contact with friction coefficient $\mu$, while $f_2$ and $f_3$ are frictionless point contacts. If $c = \frac{1}{4}$ and $h = \frac{1}{2}$, find the range of values of $\mu$ such that grasp yields force closure.

**Exercise 12.19**　(a) For the planar grasp of Figure 12.34(a), assume contact $C$ is frictionless, while the friction coefficient at contacts $A$ and $B$ is $\mu = 1$. Determine whether this grasp yields force closure.

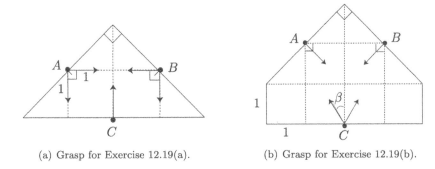

(a) Grasp for Exercise 12.19(a).          (b) Grasp for Exercise 12.19(b).

**Figure 12.34** Planar grasps.

(b) For the planar grasp of Figure 12.34(b), assume contacts $A$ and $B$ are frictionless, while contact $C$ has a friction cone of half-angle $\beta$. Find the range of values of $\beta$ for which this grasp yields force closure.

**Exercise 12.20**  Find a formula for the minimum friction coefficient, as a function of $n$, needed for a two-fingered planar force-closure grasp of an $n$-sided regular polygon, where $n$ is odd. Assume that the fingers can make contact only with the edges, not the vertices. If the fingers could also contact the vertices, how does your answer change? You can assume that the fingers are circular.

**Exercise 12.21**  Consider a table at rest, supported by four legs in frictional contact with the floor. The normal forces provided by each leg are not unique; there is an infinite set of solutions to the normal forces yielding a force balance with gravity. What is the dimension of the space of normal-force solutions? (Since there are four legs, the space of normal forces is four dimensional, and the space of solutions must be a subspace of this four-dimensional space.) What is the dimension of the space of contact-force solutions if we include tangential frictional forces?

**Exercise 12.22**  A thin rod in gravity is supported from below by a single stationary contact with friction, shown in Figure 12.35. One more frictionless contact can be placed anywhere else on the top or the bottom of the rod. Indicate all the places where this contact can be put so that the gravitational force is balanced. Use moment labeling to justify your answer. Prove the same using algebraic force balance, and comment on how the magnitude of the normal forces depends on the location of the second contact.

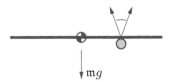

**Figure 12.35** A zero-thickness rod supported by a single contact.

**Exercise 12.23**    A frictionless finger begins pushing a box over a table (Figure 12.36). There is friction between the box and the table, as indicated in the figure. There are three possible contact modes between the box and the table: either the box slides to the right flat against the table, or it tips over at the right lower corner, or it tips over that corner while the corner also slides to the right. Which actually occurs? Assume a quasistatic force balance and answer the following questions.

(a) For each of the three contact modes, draw the moment-labeling regions corresponding to the table's friction cone edges active in that contact mode.
(b) For each moment-labeling drawing, determine whether the pushing force plus the gravitational force can be quasistatically balanced by the support forces. From this, determine which contact mode actually occurs.
(c) Graphically show a different support-friction cone for which the contact mode is different from your solution above.

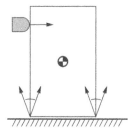

**Figure 12.36**  A frictionless finger pushes a box to the right. Gravity acts downward. Does the box slide flat against the table, does it tip over the lower right corner, or does it slide and tip over that corner?

**Exercise 12.24**    In Figure 12.37 body 1, of mass $m_1$ with center of mass at $(x_1, y_1)$, leans on body 2, of mass $m_2$ with center of mass at $(x_2, y_2)$. Both are supported by a horizontal plane, and gravity acts downward. The friction coefficient at all four contacts (at $(0,0)$, at $(x_L, y)$, at $(x_L, 0)$, and at $(x_R, 0)$) is $\mu > 0$. We want to know whether it is possible for the assembly to stay standing by some choice of contact forces within the friction cones. Write down the six equations of force balance for the two bodies in terms of the gravitational forces and the contact forces, and express the conditions that must be satisfied for this assembly to stay standing. How many equations and unknowns are there?

**Exercise 12.25**    Write a program that accepts a set of contacts acting on a planar body and determines whether the body is in first-order form closure.

**Exercise 12.26**    Write a program that accepts a set of contacts acting on a spatial body and determines whether the body is in first-order form closure.

**Exercise 12.27**    Write a program that accepts a friction coefficient and a set of contacts acting on a planar body and determines whether the body is in force closure.

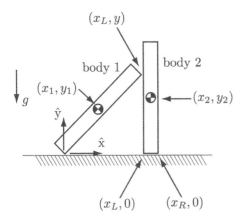

**Figure 12.37** One body leaning on another (Exercise 12.24).

**Exercise 12.28** Write a program that accepts a friction coefficient and a set of contacts acting on a spatial body and determines whether the body is in force closure. Use a polyhedral approximation to the friction cone at each contact point that underestimates the friction cone and that has four facets.

**Exercise 12.29** Write a program to simulate the quasistatic meter-stick trick of Example 12.10. The program takes as input: the initial $x$-position of the left finger, the right finger, and the stick's center of mass; the constant speed $\dot{x}$ of the right finger (toward the left finger); and the static and kinetic friction coefficients, where $\mu_{\mathrm{s}} \geq \mu_{\mathrm{k}}$. The program should continue the simulation until the two fingers touch or until the stick falls. It should plot the position of the left finger (which is constant), the right finger, and the center of mass as a function of time. Include an example where $\mu_{\mathrm{s}} = \mu_{\mathrm{k}}$, an example where $\mu_{\mathrm{s}}$ is only slightly larger than $\mu_{\mathrm{k}}$, and an example where $\mu_{\mathrm{s}}$ is much larger than $\mu_{\mathrm{k}}$.

**Exercise 12.30** Write a program that determines whether a given assembly of planar bodies can remain standing in gravity. Gravity $g$ acts in the $-\hat{y}$-direction. The assembly is described by $m$ bodies, $n$ contacts, and the friction coefficient $\mu$, all entered by the user. Each of the $m$ bodies is described by its mass $\mathrm{m}_i$ and the $(x_i, y_i)$ location of its center of mass. Each contact is described by the index $i$ of each of the two bodies involved in the contact and the unit normal direction (defined as into the first body). If the contact has only one body involved, the second body is assumed to be stationary (e.g., ground). The program should look for a set of coefficients $k_j \geq 0$ multiplying the friction-cone edges at the contacts (if there are $n$ contacts then there are $2n$ friction-cone edges and coefficients) such that each of the $m$ bodies is in force balance, considering gravity. Except in degenerate cases, if there are more force-balance equations ($3m$) than unknowns ($2n$) then there is no solution. In the usual case, where $2n > 3m$, there is a family of solutions, meaning that the force at each contact cannot be known with certainty.

One approach is to have your program generate an appropriate linear program and use the programming language's built-in linear-programming solver.

**Exercise 12.31**    This is a generalization of the previous exercise. Now, instead of simply deciding whether the assembly stays standing for a stationary base, the base moves according to a trajectory specified by the user, and the program determines whether the assembly can stay together during the trajectory (i.e., whether sticking contact at all contacts allows each body to follow the specified trajectory). The three-dimensional trajectory of the base can be specified as a polynomial in $(x(t), y(t), \theta(t))$, for a base reference frame defined at a particular position. For this problem, you also need to specify the scalar moment of inertia about the center of mass for each body in the assembly. You may find it convenient to express the motion and forces (gravitational, contact, inertial) in the frame of each body and solve for the dynamics in the body frames. Your program should check for stability (all contact normal forces are nonnegative while satisfying the dynamics) at finely spaced discrete points along the trajectory. It should return a binary result: the assembly can be maintained at all points along the trajectory, or not.

Wheeled Mobile Robots

A kinematic model of a mobile robot governs how wheel speeds map to robot velocities, while a dynamic model governs how wheel torques map to robot accelerations. In this chapter, we ignore the dynamics and focus on the kinematics. We also assume that the robots roll on hard, flat, horizontal ground without skidding (i.e., tanks and skid-steered vehicles are excluded). The mobile robot is assumed to have a single rigid-body chassis (not articulated like a tractor-trailer) with a configuration $T_{sb} \in SE(2)$ representing a chassis-fixed frame {b} relative to a fixed space frame {s} in the horizontal plane. We represent $T_{sb}$ by the three coordinates $q = (\phi, x, y)$. We also usually represent the velocity of the chassis as the time derivative of the coordinates, $\dot{q} = (\dot{\phi}, \dot{x}, \dot{y})$. Occasionally it will be convenient to refer to the chassis' planar twist $\mathcal{V}_b = (\omega_{bz}, v_{bx}, v_{by})$ expressed in {b}, where

$$
\mathcal{V}_b = \begin{bmatrix} \omega_{bz} \\ v_{bx} \\ v_{by} \end{bmatrix} = \begin{bmatrix} 1 & 0 & 0 \\ 0 & \cos\phi & \sin\phi \\ 0 & -\sin\phi & \cos\phi \end{bmatrix} \begin{bmatrix} \dot{\phi} \\ \dot{x} \\ \dot{y} \end{bmatrix}, \tag{13.1}
$$

$$
\dot{q} = \begin{bmatrix} \dot{\phi} \\ \dot{x} \\ \dot{y} \end{bmatrix} = \begin{bmatrix} 1 & 0 & 0 \\ 0 & \cos\phi & -\sin\phi \\ 0 & \sin\phi & \cos\phi \end{bmatrix} \begin{bmatrix} \omega_{bz} \\ v_{bx} \\ v_{by} \end{bmatrix}. \tag{13.2}
$$

This chapter covers kinematic modeling, motion planning, and feedback control for wheeled mobile robots, and concludes with a brief introduction to mobile manipulation, which is the problem of controlling the end-effector motion of a robot arm mounted on a mobile platform.

## 3.1 Types of Wheeled Mobile Robots

Wheeled mobile robots may be classified in two major categories, **omnidirectional** and **nonholonomic**. Omnidirectional mobile robots have no equality constraints on the chassis velocity $\dot{q} = (\dot{\phi}, \dot{x}, \dot{y})$, while nonholonomic robots are subject to a single Pfaffian velocity constraint $A(q)\dot{q} = 0$ (see Section 2.4 for an introduction to Pfaffian constraints). For a car-like robot, this constraint prevents the car from moving directly sideways. Despite this velocity constraint, the car can reach any $(\phi, x, y)$ configuration in an obstacle-free plane. In other

**Figure 13.1** (Left) A typical wheel that rolls without sideways slip – here a unicycle wheel. (Middle) An omniwheel. (Right) A mecanum wheel. Omniwheel and mecanum wheel images from VEX Robotics, Inc., used with permission.

words, the velocity constraint cannot be integrated to an equivalent configuration constraint, and therefore it is a nonholonomic constraint.

Whether a wheeled mobile robot is omnidirectional or nonholonomic depends in part on the type of wheels it employs (Figure 13.1). Nonholonomic mobile robots employ conventional wheels, such as you might find on your car: the wheel rotates about an axle perpendicular to the plane of the wheel at the wheel's center, and optionally it can be steered by spinning the wheel about an axis perpendicular to the ground at the contact point. The wheel rolls without sideways slip, which is the source of the nonholonomic constraint on the robot's chassis.

Omnidirectional wheeled mobile robots typically employ either **omniwheels** or **mecanum wheels**.[1] An omniwheel is a typical wheel augmented with rollers on its outer circumference. These rollers spin freely about axes in the plane of the wheel and tangential to the wheel's outer circumference, and they allow sideways sliding while the wheel drives forward or backward without slip in that direction. Mecanum wheels are similar, except that the spin axes of the circumferential rollers are not in the plane of the wheel (see Figure 13.1). The sideways sliding allowed by omniwheels and mecanum wheels ensures that there are no velocity constraints on the robot's chassis.

Omniwheels and mecanum wheels are not steered, only driven forward or backward. Because of their small diameter rollers, omniwheels and mecanum wheels work best on hard, flat ground.

The issues in the modeling, motion planning, and control of wheeled mobile robots depend intimately on whether the robot is omnidirectional or nonholonomic, so we treat these two cases separately in the following sections.

---

[1] These types of wheels are often called "Swedish wheels," as they were invented by Bengt Ilon working at the Swedish company Mecanum AB. The usage of, and the differentiation between, the terms "omniwheel," "mecanum wheel," and "Swedish wheel" is not completely standard, but here we use one popular choice.

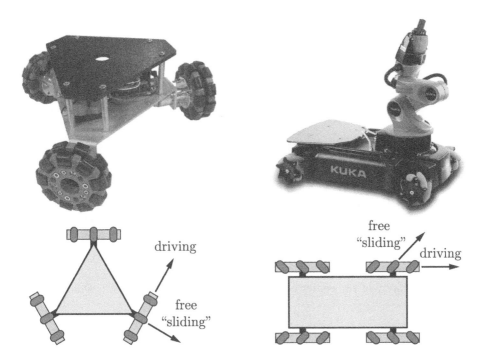

**Figure 13.2** (Left) A mobile robot with three omniwheels. Also shown for one omniwheel is the direction in which the wheel can freely slide due to the rollers, as well as the direction in which the wheel rolls without slipping when driven by the wheel motor. (The upper image is from www.superdroidrobots.com, used with permission.) (Right) The KUKA youBot mobile manipulator system, which uses four mecanum wheels for its mobile base. (The upper image is from KUKA Roboter GmbH, used with permission.)

## 3.2 Omnidirectional Wheeled Mobile Robots

### 3.2.1 Modeling

An omnidirectional mobile robot must have at least three wheels to achieve an arbitrary three-dimensional chassis velocity $\dot{q} = (\dot{\phi}, \dot{x}, \dot{y})$, since each wheel has only one motor (controlling its forward–backward velocity). Figure 13.2 shows two omnidirectional mobile robots, one with three omniwheels and one with four mecanum wheels. Also shown are the wheel motions obtained by driving the wheel motors as well as the free sliding motions allowed by the rollers.

Two important questions in kinematic modeling are the following.

(a) Given a desired chassis velocity $\dot{q}$, at what speeds must the wheels be driven?
(b) Given limits on the individual wheel driving speeds, what are the limits on the chassis velocity $\dot{q}$?

To answer these questions, we need to understand the wheel kinematics illustrated in Figure 13.3. In a frame $\hat{x}_w$–$\hat{y}_w$ at the center of the wheel, the linear

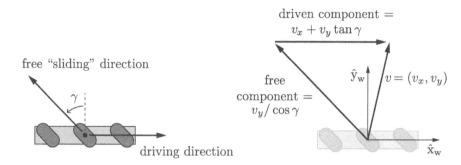

**Figure 13.3** (Left) The driving direction and the direction in which the rollers allow the wheel to slide freely. For an omniwheel $\gamma = 0$ and, for a mecanum wheel, typically $\gamma = \pm45°$. (Right) The driven and free sliding speeds for the wheel velocity $v = (v_x, v_y)$ expressed in the wheel frame $\hat{\mathrm{x}}_\mathrm{w}$–$\hat{\mathrm{y}}_\mathrm{w}$, where the $\hat{\mathrm{x}}_\mathrm{w}$-axis is aligned with the forward driving direction.

velocity of the center of the wheel is written $v = (v_x, v_y)$, which satisfies

$$\begin{bmatrix} v_x \\ v_y \end{bmatrix} = v_{\mathrm{drive}} \begin{bmatrix} 1 \\ 0 \end{bmatrix} + v_{\mathrm{slide}} \begin{bmatrix} -\sin\gamma \\ \cos\gamma \end{bmatrix}, \qquad (13.3)$$

where $\gamma$ denotes the angle at which free "sliding" occurs (allowed by the passive rollers on the circumference of the wheel), $v_{\mathrm{drive}}$ is the driving speed, and $v_{\mathrm{slide}}$ is the sliding speed. For an omniwheel $\gamma = 0$ and, for a mecanum wheel, typically $\gamma = \pm45°$. Solving Equation (13.3), we get

$$v_{\mathrm{drive}} = v_x + v_y \tan\gamma,$$
$$v_{\mathrm{slide}} = v_y / \cos\gamma.$$

Letting $r$ be the radius of the wheel and $u$ be the driving angular speed of the wheel,

$$u = \frac{v_{\mathrm{drive}}}{r} = \frac{1}{r}(v_x + v_y \tan\gamma). \qquad (13.4)$$

To derive the full transformation from the chassis velocity $\dot{q} = (\dot\phi, \dot x, \dot y)$ to the driving angular speed $u_i$ for wheel $i$, refer to the notation illustrated in Figure 13.4. The chassis frame {b} is at $q = (\phi, x, y)$ in the fixed space frame {s}. The center of the wheel and its driving direction are given by $(\beta_i, x_i, y_i)$ expressed in {b}, the wheel's radius is $r_i$, and the wheel's sliding direction is given by $\gamma_i$. Then $u_i$ is related to $\dot{q}$ by

$$u_i = h_i(\phi)\dot{q} =$$

$$\begin{bmatrix} \dfrac{1}{r_i} & \dfrac{\tan\gamma_i}{r_i} \end{bmatrix} \begin{bmatrix} \cos\beta_i & \sin\beta_i \\ -\sin\beta_i & \cos\beta_i \end{bmatrix} \begin{bmatrix} -y_i & 1 & 0 \\ x_i & 0 & 1 \end{bmatrix} \begin{bmatrix} 1 & 0 & 0 \\ 0 & \cos\phi & \sin\phi \\ 0 & -\sin\phi & \cos\phi \end{bmatrix} \begin{bmatrix} \dot\phi \\ \dot x \\ \dot y \end{bmatrix}.$$
$$(13.5)$$

Reading from right to left: the first transformation expresses $\dot{q}$ as $\mathcal{V}_b$; the second transformation produces the linear velocity at the wheel in {b}; the third transformation expresses this linear velocity in the wheel frame $\hat{x}_w$–$\hat{y}_w$; and the final transformation calculates the driving angular velocity using Equation (13.4).

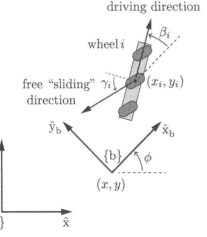

driving direction

wheel $i$

free "sliding" $\gamma_i$
direction

$\hat{y}_b$        $\hat{x}_b$

{b}        $\phi$

$(x, y)$

$\hat{x}$

**Figure 13.4** The fixed space frame {s}, a chassis frame {b} at $(\phi, x, y)$ in {s}, and wheel $i$ at $(x_i, y_i)$ with driving direction $\beta_i$, both expressed in {b}. The sliding direction of wheel $i$ is defined by $\gamma_i$.

Evaluating Equation (13.5) for $h_i(\phi)$, we get

$$h_i(\phi) = \frac{1}{r_i \cos \gamma_i} \left[ \begin{array}{c} x_i \sin(\beta_i + \gamma_i) - y_i \cos(\beta_i + \gamma_i) \\ \cos(\beta_i + \gamma_i + \phi) \\ \sin(\beta_i + \gamma_i + \phi) \end{array} \right]^{\text{T}}. \tag{13.6}$$

For an omnidirectional robot with $m \geq 3$ wheels, the matrix $H(\phi) \in \mathbb{R}^{m \times 3}$ mapping a desired chassis velocity $\dot{q} \in \mathbb{R}^3$ to the vector of wheel driving speeds $u \in \mathbb{R}^m$ is constructed by stacking the $m$ rows $h_i(\phi)$:

$$u = H(\phi)\dot{q} = \left[ \begin{array}{c} h_1(\phi) \\ h_2(\phi) \\ \vdots \\ h_m(\phi) \end{array} \right] \left[ \begin{array}{c} \dot{\phi} \\ \dot{x} \\ \dot{y} \end{array} \right]. \tag{13.7}$$

We can also express the relationship between $u$ and the body twist $\mathcal{V}_b$. This mapping does not depend on the chassis orientation $\phi$:

$$u = H(0)\mathcal{V}_b = \left[ \begin{array}{c} h_1(0) \\ h_2(0) \\ \vdots \\ h_m(0) \end{array} \right] \left[ \begin{array}{c} \omega_{bz} \\ v_{bx} \\ v_{by} \end{array} \right]. \tag{13.8}$$

The wheel positions and headings $(\beta_i, x_i, y_i)$ in {b}, and their free sliding directions $\gamma_i$, must be chosen so that $H(0)$ is rank 3. For example, if we constructed a mobile robot of omniwheels whose driving directions and sliding directions were all aligned, the rank of $H(0)$ would be 2, and there would be no way to controllably generate translational motion in the sliding direction.

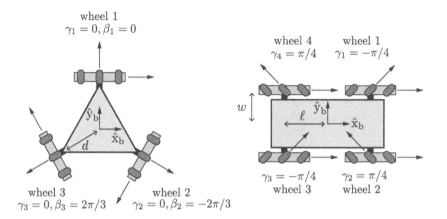

**Figure 13.5** Kinematic models for mobile robots with three omniwheels and four mecanum wheels. The radius of all wheels is $r$ and the driving direction for each of the mecanum wheels is $\beta_i = 0$.

In the case $m > 3$, as for the four-wheeled youBot of Figure 13.2, choosing $u$ such that Equation (13.8) is not satisfied for any $\mathcal{V}_b \in \mathbb{R}^3$ implies that the wheels must skid in their driving directions.

Using the notation in Figure 13.5, the kinematic model of the mobile robot with three omniwheels is

$$u = \begin{bmatrix} u_1 \\ u_2 \\ u_3 \end{bmatrix} = H(0)\mathcal{V}_b = \frac{1}{r} \begin{bmatrix} -d & 1 & 0 \\ -d & -1/2 & -\sin(\pi/3) \\ -d & -1/2 & \sin(\pi/3) \end{bmatrix} \begin{bmatrix} \omega_{bz} \\ v_{bx} \\ v_{by} \end{bmatrix} \qquad (13.9)$$

and the kinematic model of the mobile robot with four mecanum wheels is

$$u = \begin{bmatrix} u_1 \\ u_2 \\ u_3 \\ u_4 \end{bmatrix} = H(0)\mathcal{V}_b = \frac{1}{r} \begin{bmatrix} -\ell - w & 1 & -1 \\ \ell + w & 1 & 1 \\ \ell + w & 1 & -1 \\ -\ell - w & 1 & 1 \end{bmatrix} \begin{bmatrix} \omega_{bz} \\ v_{bx} \\ v_{by} \end{bmatrix}. \qquad (13.10)$$

For the mecanum robot, to move in the direction $+\hat{x}_b$, all wheels drive forward at the same speed; to move in the direction $+\hat{y}_b$, wheels 1 and 3 drive backward and wheels 2 and 4 drive forward at the same speed; and to rotate in the counterclockwise direction, wheels 1 and 4 drive backward and wheels 2 and 3 drive forward at the same speed. Note that the robot chassis is capable of the same speeds in the forward and sideways directions.

If the driving angular velocity of wheel $i$ is subject to the bound $|u_i| \leq u_{i,\max}$, i.e.,

$$-u_{i,\max} \leq u_i = h_i(0)\mathcal{V}_b \leq u_{i,\max},$$

then two parallel constraint planes defined by $-u_{i,\max} = h_i(0)\mathcal{V}_b$ and $u_{i,\max} = h_i(0)\mathcal{V}_b$ are generated in the three-dimensional space of body twists. Any $\mathcal{V}_b$ between these two planes does not violate the maximum driving speed of wheel

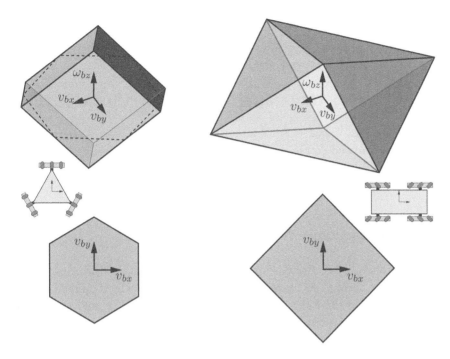

**Figure 13.6** (Top row) Regions of feasible body twists $V$ for the three-wheeled (left) and four-wheeled (right) robots of Figure 13.5. Also shown for the three-wheeled robot is the intersection (dashed line) with the $\omega_{bz} = 0$-plane. (Bottom row) The bounds in the $\omega_{bz} = 0$-plane (translational motions only).

$i$, while any $\mathcal{V}_b$ outside this slice is too fast for wheel $i$. The normal direction to the constraint planes is $h_i^{\mathrm{T}}(0)$, and the points on the planes closest to the origin are $-u_{i,\max} h_i^{\mathrm{T}}(0)/\|h_i(0)\|^2$ and $u_{i,\max} h_i^{\mathrm{T}}(0)/\|h_i(0)\|^2$.

If the robot has $m$ wheels then the region of feasible body twists $V$ is bounded by the $m$ pairs of parallel constraint planes. The region $V$ is therefore a convex three-dimensional polyhedron. The polyhedron has $2m$ faces and the origin (corresponding to zero twist) is in the center. Visualizations of the six-sided and eight-sided regions $V$ for the three-wheeled and four-wheeled models in Figure 13.5 are shown in Figure 13.6.

## 13.2.2 Motion Planning

Since omnidirectional mobile robots are free to move in any direction, any of the trajectory planning methods for kinematic systems in Chapter 9, and most of the motion planning methods of Chapter 10, can be adapted.

### 13.2.3    Feedback Control

Given a desired trajectory $q_d(t)$, we can adopt the feedforward plus PI feedback controller (11.15) to track the trajectory:

$$\dot{q}(t) = \dot{q}_d(t) + K_p(q_d(t) - q(t)) + K_i \int_0^t (q_d(\mathrm{t}) - q(\mathrm{t}))\, dt, \qquad (13.11)$$

where $K_p = k_p I \in \mathbb{R}^{3\times3}$ and $K_i = k_i I \in \mathbb{R}^{3\times3}$ have positive values along the diagonal and $q(t)$ is an estimate of the actual configuration derived from sensors. Then $\dot{q}(t)$ can be converted to the commanded wheel driving velocities $u(t)$ using Equation (13.7).

## 13.3    Nonholonomic Wheeled Mobile Robots

In Section 2.4, the $k$ Pfaffian velocity constraints acting on a system with configuration $q \in \mathbb{R}^n$ were written as $A(q)\dot{q} = 0$, where $A(q) \in \mathbb{R}^{k\times n}$. Instead of specifying the $k$ directions in which velocities are not allowed, we can write the allowable velocities of a kinematic system as a linear combination of $n - k$ velocity directions. This representation is equivalent, and it has the advantage that the coefficients of the linear combinations are precisely the controls available to us. We will see this representation in the kinematic models below.

The title of this section implies that the velocity constraints are not integrable to equivalent configuration constraints. We will establish this formally in Section 13.3.2.

### 13.3.1    Modeling

#### 13.3.1.1    The Unicycle

The simplest wheeled mobile robot is a single upright rolling wheel, or unicycle. Let $r$ be the radius of the wheel. We write the configuration of the wheel as $q = (\phi, x, y, \theta)$, where $(x, y)$ is the contact point, $\phi$ is the heading direction, and $\theta$ is the rolling angle of the wheel (Figure 13.7). The configuration of the "chassis" (e.g., the seat of the unicycle) is $(\phi, x, y)$. The kinematic equations of motion are

$$\dot{q} = \begin{bmatrix} \dot{\phi} \\ \dot{x} \\ \dot{y} \\ \dot{\theta} \end{bmatrix} = \begin{bmatrix} 0 & 1 \\ r\cos\phi & 0 \\ r\sin\phi & 0 \\ 1 & 0 \end{bmatrix} \begin{bmatrix} u_1 \\ u_2 \end{bmatrix} = G(q)u = g_1(q)u_1 + g_2(q)u_2. \quad (13.12)$$

The control inputs are $u = (u_1, u_2)$, with $u_1$ the wheel's driving speed and $u_2$ the heading direction turning speed. The controls are subject to the constraints $-u_{1,\max} \le u_1 \le u_{1,\max}$ and $-u_{2,\max} \le u_2 \le u_{2,\max}$.

The vector-valued functions $g_i(q) \in \mathbb{R}^4$ are the columns of the matrix $G(q)$, and they are called the **tangent vector fields** (also called the **control vector**

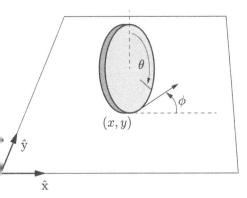

**Figure 13.7** A wheel rolling on a plane without slipping.

**fields** or simply the **velocity vector fields**) over $q$ associated with the controls $u_i = 1$. Evaluated at a specific configuration $q$, $g_i(q)$ is a **tangent vector** (or velocity vector) of the tangent vector field.

An example of a vector field on $\mathbb{R}^2$ is illustrated in Figure 13.8.

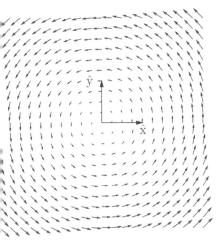

**Figure 13.8** The vector field $(\dot{x}, \dot{y}) = (-y, x)$.

All our kinematic models of nonholonomic mobile robots will have the form $\dot{q} = G(q)u$, as in Equation (13.12). Three things to notice about these models are: (1) there is no drift – zero controls mean zero velocity; (2) the vector fields $g_i(q)$ are generally functions of the configuration $q$; and (3) $\dot{q}$ is linear in the controls.

Since we are not usually concerned with the rolling angle of the wheel, we can drop the fourth row from (13.12) to get the simplified control system

$$\dot{q} = \begin{bmatrix} \dot{\phi} \\ \dot{x} \\ \dot{y} \end{bmatrix} = \begin{bmatrix} 0 & 1 \\ r\cos\phi & 0 \\ r\sin\phi & 0 \end{bmatrix} \begin{bmatrix} u_1 \\ u_2 \end{bmatrix}. \tag{13.13}$$

### 13.3.1.2    The Differential-Drive Robot

The **differential-drive robot**, or **diff-drive**, is perhaps the simplest wheeled mobile robot architecture. A diff-drive robot consists of two independently driven wheels of radius $r$ that rotate about the same axis, as well as one or more caster wheels, ball casters, or low-friction sliders that keep the robot horizontal. Let the distance between the driven wheels be $2d$ and choose the $(x, y)$ reference point halfway between the wheels (Figure 13.9). Writing the configuration as $q = (\phi, x, y, \theta_L, \theta_R)$, where $\theta_L$ and $\theta_R$ are the rolling angles of the left and right wheels, respectively, the kinematic equations are

$$
\dot{q} = \begin{bmatrix} \dot{\phi} \\ \dot{x} \\ \dot{y} \\ \dot{\theta}_L \\ \dot{\theta}_R \end{bmatrix} = \begin{bmatrix} -r/2d & r/2d \\ \frac{r}{2}\cos\phi & \frac{r}{2}\cos\phi \\ \frac{r}{2}\sin\phi & \frac{r}{2}\sin\phi \\ 1 & 0 \\ 0 & 1 \end{bmatrix} \begin{bmatrix} u_L \\ u_R \end{bmatrix}, \tag{13.14}
$$

where $u_L$ is the angular speed of the left wheel and $u_R$ that of the right. A positive angular speed of each wheel corresponds to forward motion at that wheel. The control value at each wheel is taken from the interval $[-u_{max}, u_{max}]$.

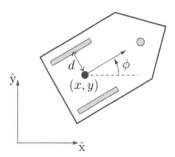

**Figure 13.9** A diff-drive robot consisting of two typical wheels and one ball caster wheel, shaded gray.

Since we are not usually concerned with the rolling angles of the two wheels, we can drop the last two rows to get the simplified control system

$$
\dot{q} = \begin{bmatrix} \dot{\phi} \\ \dot{x} \\ \dot{y} \end{bmatrix} = \begin{bmatrix} -r/2d & r/2d \\ \frac{r}{2}\cos\phi & \frac{r}{2}\cos\phi \\ \frac{r}{2}\sin\phi & \frac{r}{2}\sin\phi \end{bmatrix} \begin{bmatrix} u_L \\ u_R \end{bmatrix}. \tag{13.15}
$$

Two advantages of a diff-drive robot are its simplicity (typically the motor is attached directly to the axle of each wheel) and high maneuverability (the robot can spin in place by rotating the wheels in opposite directions). Casters are often not appropriate for outdoor use, however.

### 13.3.1.3    The Car-Like Robot

The most familiar wheeled vehicle is a car, with two steered front wheels and two fixed-heading rear wheels. To prevent slipping of the front wheels, they are steered using **Ackermann steering**, as illustrated in Figure 13.10. The center

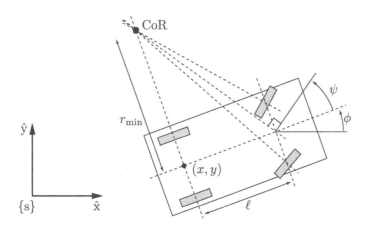

**Figure 13.10** The two front wheels of a car are steered at different angles using Ackermann steering in order that all wheels roll without slipping (i.e., the wheel heading direction is perpendicular to the line connecting the wheel to the CoR). The car is shown executing a turn at its minimum turning radius $r_{\min}$.

of rotation of the car's chassis lies on the line passing through the rear wheels at the intersection with the perpendicular bisectors of the front wheels.

To define the configuration of the car, we ignore the rolling angles of the four wheels and write $q = (\phi, x, y, \psi)$, where $(x, y)$ is the location of the midpoint between the rear wheels, $\phi$ is the car's heading direction, and $\psi$ is the steering angle of the car, defined at a virtual wheel at the midpoint between the front wheels. The controls are the forward speed $v$ of the car at its reference point and the angular speed $w$ of the steering angle. The car's kinematics are

$$
\dot{q} = \begin{bmatrix} \dot{\phi} \\ \dot{x} \\ \dot{y} \\ \dot{\psi} \end{bmatrix} = \begin{bmatrix} (\tan\psi)/\ell & 0 \\ \cos\phi & 0 \\ \sin\phi & 0 \\ 0 & 1 \end{bmatrix} \begin{bmatrix} v \\ w \end{bmatrix},
\tag{13.16}
$$

where $\ell$ is the wheelbase between the front and rear wheels. The control $v$ is limited to a closed interval $[v_{\min}, v_{\max}]$ where $v_{\min} < 0 < v_{\max}$, the steering rate is limited to $[-w_{\max}, w_{\max}]$ with $w_{\max} > 0$, and the steering angle $\psi$ is limited to $[-\psi_{\max}, \psi_{\max}]$ with $\psi_{\max} > 0$.

The kinematics (13.16) can be simplified if the steering control is actually just the steering angle $\psi$ and not its rate $w$. This assumption is justified if the steering rate limit $w_{\max}$ is high enough that the steering angle can be changed nearly instantaneously by a lower-level controller. In this case, $\psi$ is eliminated as a state variable, and the car's configuration is simply $q = (\phi, x, y)$. We use the control inputs $(v, \omega)$, where $v$ is still the car's forward speed and $\omega$ is now its rate of rotation. These can be converted to the controls $(v, \psi)$ by the relations

$$
v = v, \qquad \psi = \tan^{-1}\left(\frac{\ell\omega}{v}\right).
\tag{13.17}
$$

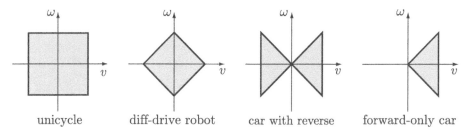

**Figure 13.11** The $(v, \omega)$ control sets for the simplified unicycle, diff-drive robot, and car kinematics. For the car with a reverse gear, control set illustrates that it is incapable of turning in place. The angle of the sloped lines in its bowtie control set is determined by its minimum turning radius. If a car has no reverse gear, only the right-hand half of the bowtie is available.

The constraints on the controls $(v, \omega)$ due to the constraints on $(v, \psi)$ take a somewhat complicated form, as we will shortly see.

The simplified car kinematics can now be written

$$\dot{q} = \begin{bmatrix} \dot{\phi} \\ \dot{x} \\ \dot{y} \end{bmatrix} = G(q)u = \begin{bmatrix} 0 & 1 \\ \cos\phi & 0 \\ \sin\phi & 0 \end{bmatrix} \begin{bmatrix} v \\ \omega \end{bmatrix}. \qquad (13.18)$$

The nonholonomic constraint implied by (13.18) can be derived using one of the equations from (13.18),

$$\dot{x} = v\cos\phi,$$
$$\dot{y} = v\sin\phi,$$

to solve for $v$, then substituting the result into the other equation to get

$$A(q)\dot{q} = [0 \ \sin\phi \ -\cos\phi]\dot{q} = \dot{x}\sin\phi - \dot{y}\cos\phi = 0.$$

### 13.3.1.4 Canonical Simplified Model for Nonholonomic Mobile Robots

The kinematics (13.18) gives a canonical simplified model for nonholonomic mobile robots. Using control transformations such as (13.17), the simplified unicycle kinematics (13.13) and the simplified differential-drive kinematics (13.15) can also be expressed in this form. The control transformation for the simplified unicycle kinematics (13.13) is

$$u_1 = \frac{v}{r}, \qquad u_2 = \omega$$

and the transformation for the simplified diff-drive kinematics (13.15) is

$$u_L = \frac{v - \omega d}{r}, \qquad u_R = \frac{v + \omega d}{r}.$$

With these input transformations, the only difference between the simplified unicycle, diff-drive robot, and car kinematics is the control limits on $(v, \omega)$. These are illustrated in Figure 13.11.

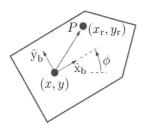

**Figure 13.12** The point $P$ is located at $(x_r, y_r)$ in the chassis-fixed frame {b}.

We can use the two control inputs $(v, \omega)$ in the canonical model (13.18) to directly control the two components of the linear velocity of a reference point $P$ fixed to the robot chassis. This is useful when a sensor is located at $P$, for example. Let $(x_P, y_P)$ be the coordinates of $P$ in the world frame, and $(x_r, y_r)$ be its (constant) coordinates in the chassis frame {b} (Figure 13.12). To find the controls $(v, \omega)$ needed to achieve a desired world-frame motion $(\dot{x}_P, \dot{y}_P)$, we first write

$$\begin{bmatrix} x_P \\ y_P \end{bmatrix} = \begin{bmatrix} x \\ y \end{bmatrix} + \begin{bmatrix} \cos\phi & -\sin\phi \\ \sin\phi & \cos\phi \end{bmatrix} \begin{bmatrix} x_r \\ y_r \end{bmatrix}. \tag{13.19}$$

Differentiating, we obtain

$$\begin{bmatrix} \dot{x}_P \\ \dot{y}_P \end{bmatrix} = \begin{bmatrix} \dot{x} \\ \dot{y} \end{bmatrix} + \dot{\phi} \begin{bmatrix} -\sin\phi & -\cos\phi \\ \cos\phi & -\sin\phi \end{bmatrix} \begin{bmatrix} x_r \\ y_r \end{bmatrix}. \tag{13.20}$$

Substituting $\omega$ for $\dot{\phi}$ and $(v\cos\phi, v\sin\phi)$ for $(\dot{x}, \dot{y})$ and solving, we get

$$\begin{bmatrix} v \\ \omega \end{bmatrix} = \frac{1}{x_r} \begin{bmatrix} x_r\cos\phi - y_r\sin\phi & x_r\sin\phi + y_r\cos\phi \\ -\sin\phi & \cos\phi \end{bmatrix} \begin{bmatrix} \dot{x}_P \\ \dot{y}_P \end{bmatrix}. \tag{13.21}$$

This equation may be read as $[v\ \omega]^T = J^{-1}(q)[\dot{x}_P\ \dot{y}_P]^T$, where $J(q)$ is the Jacobian relating $(v, \omega)$ to the world-frame motion of $P$. Note that the Jacobian $J(q)$ is singular when $P$ is chosen on the line $x_r = 0$. Points on this line, such as the midway point between the wheels of a diff-drive robot or between the rear wheels of a car, can only move in the heading direction of the vehicle.

### 3.3.2 Controllability

The feedback control for an omnidirectional robot is simple, as there is a set of wheel driving speeds for any desired chassis velocity $\dot{q}$ (Equation (13.7)). In fact, if the goal of the feedback controller is simply to stabilize the robot to the origin $q = (0, 0, 0)$, rather than trajectory tracking as in the control law (13.11), we could use the even simpler feedback controller

$$\dot{q}(t) = -Kq(t) \tag{13.22}$$

for any positive-definite $K$. The feedback gain matrix $-K$ acts like a spring to pull $q$ to the origin, and Equation (13.7) is used to transform $\dot{q}(t)$ to $u(t)$. The same type of "linear spring" controller could be used to stabilize the point $P$ on the canonical nonholonomic robot (Figure 13.12) to $(x_P, y_P) = (0, 0)$ since, by Equation (13.21), any desired $(\dot{x}_P, \dot{y}_P)$ can be achieved by the controls $(v, \omega)$.[2]

In short, the kinematics of the omnidirectional robot, as well as the kinematics of the point $P$ for the nonholonomic robot, can be rewritten in the single-integrator form

$$\dot{x} = \nu, \tag{13.23}$$

where $x$ is the configuration we are trying to control and $\nu$ is a "virtual control" that is actually implemented using the transformations in Equation (13.7) for an omnidirectional robot or Equation (13.21) for the control of $P$ by a nonholonomic robot. Equation (13.23) is a simple example of the more general class of linear control systems

$$\dot{x} = Ax + B\nu, \tag{13.24}$$

which are known to be **linearly controllable** if the **Kalman rank condition** is satisfied:

$$\text{rank}\,[B \; AB \; A^2B \; \cdots \; A^{n-1}B] = \dim(x) = n,$$

where $x \in \mathbb{R}^n, \nu \in \mathbb{R}^m, A \in \mathbb{R}^{n \times n}$, and $B \in \mathbb{R}^{n \times m}$. In Equation (13.23), $A = 0$ and $B$ is the identity matrix, trivially satisfying the rank condition for linear controllability since $m = n$. Linear controllability implies the existence of the simple linear control law

$$\nu = -Kx,$$

as in Equation (13.22), to stabilize the origin.

There is no linear controller that can stabilize the full chassis configuration to $q = 0$ for a nonholonomic robot, however; the nonholonomic robot is not linearly controllable. In fact, there is no controller that is a continuous function of $q$ which can stabilize $q = 0$. This fact is embedded in the following well-known result, which we state without proof.

**Theorem 13.1**   *A system $\dot{q} = G(q)u$ with rank $G(0) < \dim(q)$ cannot be stabilized to $q = 0$ by a continuous time-invariant feedback control law.*

This theorem applies to our canonical nonholonomic robot model, since the rank of $G(q)$ is 2 everywhere (there are only two control vector fields), while the chassis configuration is three dimensional.

For nonlinear systems of the form $\dot{q} = G(q)u$, there are other notions of controllability. We consider a few of these next and show that, even though the canonical nonholonomic robot is not linearly controllable, it still satisfies other important notions of controllability. In particular, the velocity constraint does not integrate to a configuration constraint – the set of reachable configurations is not reduced because of the velocity constraint.

---

[2] For the moment we ignore the different constraints on $(v, \omega)$ for the unicycle, diff-drive robot, and car-like robot, as they do not change the main result.

### .3.2.1 Definitions of Controllability

Our definitions of nonlinear controllability rely on the notion of the time- and space-limited reachable sets of the nonholonomic robot from a configuration $q$.

**Definition 13.2** Given a time $T > 0$ and a neighborhood[3] $W$ of an initial configuration $q$, the **reachable set** of configurations from $q$ at time $T$ by feasible trajectories remaining inside $W$ is written $\mathcal{R}^W(q, T)$. We further define the union of reachable sets at times $t \in [0, T]$:

$$\mathcal{R}^W(q, \leq T) = \bigcup_{0 \leq t \leq T} \mathcal{R}^W(q, t).$$

We now provide some standard definitions of nonlinear controllability.

**Definition 13.3** A robot is **controllable** from $q$ if, for any $q_{\text{goal}}$, there exists a control trajectory $u(t)$ that drives the robot from $q$ to $q_{\text{goal}}$ in finite time $T$. The robot is **small-time locally accessible** (STLA) from $q$ if, for any time $T > 0$ and any neighborhood $W$, the reachable set $\mathcal{R}^W(q, \leq T)$ is a full-dimensional subset of the configuration space. The robot is **small-time locally controllable** (STLC) from $q$ if, for any time $T > 0$ and any neighborhood $W$, the reachable set $\mathcal{R}^W(q, \leq T)$ is a neighborhood of $q$.

Small-time local accessibility and small-time local controllability are illustrated in Figure 13.13 for a two-dimensional configuration space. Clearly STLC at $q$ is a stronger condition than STLA at $q$. If a system is STLC at all $q$, then it is controllable from any $q$ by the patching together of paths in neighborhoods from $q$ to $q_{\text{goal}}$.

STLA    STLC

**Figure 13.13** Illustrations of small-time local accessibility and small-time local controllability in a two-dimensional space. The shaded regions are the sets reachable without leaving the neighborhood $W$.

For all the examples in this chapter, if a controllability property holds for any $q$ then it holds for all $q$, since the maneuverability of the robot does not change with its configuration.

Consider the examples of a car and of a forward-only car with no reverse gear. A forward-only car is STLA, as we will shortly see, but it is not STLC: if it is confined to a tight space (a small neighborhood $W$), it cannot reach configurations directly behind its initial configuration. A car with a reverse gear is STLC, however. Both cars are controllable in an obstacle-free plane, because even a forward-only car can drive anywhere.

If there are obstacles in the plane, there may be some free-space configurations

---

[3] A neighborhood $W$ of a configuration $q$ is any full-dimensional subset of configuration space containing $q$ in its interior. For example, the set of configurations in a ball of radius $r > 0$ centered at $q$ (i.e., all $q_b$ satisfying $\|q_b - q\| < r$) is a neighborhood of $q$.

that the forward-only car cannot reach but that the STLC car can reach. (Consider an obstacle directly in front of the car, for example.) If the obstacles are all defined as closed subsets of the plane containing their boundaries, the STLC car can reach any configuration in its connected component of the free space, despite its velocity constraint.

It is worth thinking about this last statement for a moment. All free configurations have collision-free neighborhoods, since the free space is defined as open and the obstacles are defined as closed (containing their boundaries). Therefore it is always possible to maneuver in any direction from any free configuration. If your car is shorter than the available parking space, you can parallel park into it, even if it takes a long time!

If any controllability property holds (controllability, STLA, or STLC) then the reachable configuration space is full dimensional, and therefore any velocity constraints on the system are nonholonomic.

### 13.3.2.2  Controllability Tests

Consider a driftless linear-in-the-control (**control-affine**) system

$$\dot{q} = G(q)u = \sum_{i=1}^{m} g_i(q)u_i, \qquad q \in \mathbb{R}^n, \qquad u \in \mathcal{U} \subset \mathbb{R}^m, \quad m < n, \qquad (13.25)$$

generalizing the canonical nonholonomic model where $n = 3$ and $m = 2$. The set of feasible controls is $\mathcal{U} \subset \mathbb{R}^m$. For example, the control sets $\mathcal{U}$ for the unicycle, diff-drive, car-like, and forward-only car-like robots were shown in Figure 13.11. In this chapter we consider two types of control sets $\mathcal{U}$: those whose positive linear span is $\mathbb{R}^m$, i.e., $\mathrm{pos}(\mathcal{U}) = \mathbb{R}^m$, such as the control sets for the unicycle, diff-drive robot, and car in Figure 13.11, and those whose positive linear span does not cover $\mathbb{R}^m$ but whose linear span does, i.e., $\mathrm{span}(\mathcal{U}) = \mathbb{R}^m$, such as the control set for the forward-only car in Figure 13.11.

The local controllability properties (STLA or STLC) of (13.25) depend on the noncommutativity of motions along the vector fields $g_i$. Let $F_\epsilon^{g_i}(q)$ be the configuration reached by following the vector field $g_i$ for time $\epsilon$ starting from $q$. Then two vector fields $g_i(q)$ and $g_j(q)$ commute if $F_\epsilon^{g_j}(F_\epsilon^{g_i}(q)) = F_\epsilon^{g_i}(F_\epsilon^{g_j}(q))$, i.e., the order of following the vector fields does not matter. If they do not commute, i.e., $F_\epsilon^{g_j}(F_\epsilon^{g_i}(q)) - F_\epsilon^{g_i}(F_\epsilon^{g_j}(q)) \neq 0$ then the order of application of the vector fields affects the final configuration. In addition, defining the noncommutativity as

$$\Delta q = F_\epsilon^{g_j}(F_\epsilon^{g_i}(q)) - F_\epsilon^{g_i}(F_\epsilon^{g_j}(q)) \qquad \text{for small } \epsilon,$$

if $\Delta q$ is in a direction that cannot be achieved directly by any other vector field $g_k$ then switching between $g_i$ and $g_j$ can create motion in a direction not present in the original set of vector fields. A familiar example is parallel parking a car: there is no vector field corresponding to direct sideways translation but, by alternating forward and backward motion along two different vector fields, it is possible to create a net motion to the side.

To calculate $q(2\epsilon) = F_\epsilon^{g_j}(F_\epsilon^{g_i}(q(0)))$ for small $\epsilon$ approximately, we use a Taylor expansion and truncate the expansion at $O(\epsilon^3)$. We start by following $g_i$ for a time $\epsilon$ and use the fact that $\dot{q} = g_i(q)$ and $\ddot{q} = (\partial g_i/\partial q)\dot{q} = (\partial g_i/\partial q)g_i(q)$:

$$q(\epsilon) = q(0) + \epsilon\dot{q}(0) + \frac{1}{2}\epsilon^2\ddot{q}(0) + O(\epsilon^3)$$

$$= q(0) + \epsilon g_i(q(0)) + \frac{1}{2}\epsilon^2\frac{\partial g_i}{\partial q}g_i(q(0)) + O(\epsilon^3).$$

Now, after following $g_j$ for a time $\epsilon$:

$$q(2\epsilon) = q(\epsilon) + \epsilon g_j(q(\epsilon)) + \frac{1}{2}\epsilon^2\frac{\partial g_j}{\partial q}g_j(q(\epsilon)) + O(\epsilon^3)$$

$$= q(0) + \epsilon g_i(q(0)) + \frac{1}{2}\epsilon^2\frac{\partial g_i}{\partial q}g_i(q(0))$$

$$+ \epsilon g_j\big(q(0) + \epsilon g_i(q(0))\big) + \frac{1}{2}\epsilon^2\frac{\partial g_j}{\partial q}g_j(q(0)) + O(\epsilon^3)$$

$$= q(0) + \epsilon g_i(q(0)) + \frac{1}{2}\epsilon^2\frac{\partial g_i}{\partial q}g_i(q(0))$$

$$+ \epsilon g_j(q(0)) + \epsilon^2\frac{\partial g_j}{\partial q}g_i(q(0)) + \frac{1}{2}\epsilon^2\frac{\partial g_j}{\partial q}g_j(q(0)) + O(\epsilon^3). \quad (13.26)$$

Note the presence of $\epsilon^2(\partial g_j/\partial q)g_i$, the only term that depends on the order of the vector fields. Using the expression (13.26), we can calculate the noncommutativity:

$$\Delta q = F_\epsilon^{g_j}(F_\epsilon^{g_i}(q)) - F_\epsilon^{g_i}(F_\epsilon^{g_j}(q)) = \epsilon^2\left(\frac{\partial g_j}{\partial q}g_i - \frac{\partial g_i}{\partial q}g_j\right)(q(0)) + O(\epsilon^3). \quad (13.27)$$

In addition to measuring the noncommutativity, $\Delta q$ is also equal to the net motion (to order $\epsilon^2$) obtained by following $g_i$ for time $\epsilon$, then $g_j$ for time $\epsilon$, then $-g_i$ for time $\epsilon$, and then $-g_j$ for time $\epsilon$.

The term $(\partial g_j/\partial q)g_i - (\partial g_i/\partial q)g_j$ in Equation (13.27) is important enough for us to give it its own name:

**Definition 13.4** The **Lie bracket** of the vector fields $g_i(q)$ and $g_j(q)$ is

$$[g_i, g_j](q) = \left(\frac{\partial g_j}{\partial q}g_i - \frac{\partial g_i}{\partial q}g_j\right)(q). \quad (13.28)$$

This Lie bracket is the same as that for twists, introduced in Section 8.2.2. The only difference is that the Lie bracket in Section 8.2.2 was thought of as the noncommutativity of two twists $\mathcal{V}_i, \mathcal{V}_j$ defined at a given instant, rather than of two velocity vector fields defined over all configurations $q$. The Lie bracket from Section 8.2.2 would be identical to the expression in Equation (13.28) if the constant twists were represented as vector fields $g_i(q), g_j(q)$ in local coordinates $q$. See Exercise 13.20, for example.

The Lie bracket of two vector fields $g_i(q)$ and $g_j(q)$ should itself be thought of as a vector field $[g_i, g_j](q)$, where approximate motion along the Lie bracket vector field can be obtained by switching between the original two vector fields.

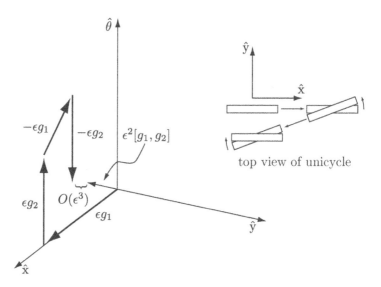

**Figure 13.14** The Lie bracket, $[g_1, g_2](q)$, of the forward–backward vector field $g_1(q)$ and the spin-in-place vector field $g_2(q)$ is a sideways vector field.

As we saw in our Taylor expansion, motion along the Lie bracket vector field is slow relative to the motions along the original vector fields; for small times $\epsilon$, motion of order $\epsilon$ can be obtained in the directions of the original vector fields, while motion in the Lie bracket direction is only of order $\epsilon^2$. This agrees with our common experience that moving a car sideways by parallel parking motions is slow relative to forward and backward or turning motions, as discussed in the next example.

**Example 13.5**    Consider the canonical nonholonomic robot with vector fields $g_1(q) = (0, \cos\phi, \sin\phi)$ and $g_2(q) = (1, 0, 0)$. Writing $g_1(q)$ and $g_2(q)$ as column vectors, the Lie bracket vector field $g_3(q) = [g_1, g_2](q)$ is given by

$$g_3(q) = [g_1, g_2](q) = \left(\frac{\partial g_2}{\partial q} g_1 - \frac{\partial g_1}{\partial q} g_2\right)(q)$$

$$= \begin{bmatrix} 0 & 0 & 0 \\ 0 & 0 & 0 \\ 0 & 0 & 0 \end{bmatrix} \begin{bmatrix} 0 \\ \cos\phi \\ \sin\phi \end{bmatrix} - \begin{bmatrix} 0 & 0 & 0 \\ -\sin\phi & 0 & 0 \\ \cos\phi & 0 & 0 \end{bmatrix} \begin{bmatrix} 1 \\ 0 \\ 0 \end{bmatrix}$$

$$= \begin{bmatrix} 0 \\ \sin\phi \\ -\cos\phi \end{bmatrix}.$$

The Lie bracket direction is a sideways "parallel parking" motion, as illustrated in Figure 13.14. The net motion obtained by following $g_1$ for $\epsilon$, $g_2$ for $\epsilon$, $-g_1$ for $\epsilon$, and $-g_2$ for $\epsilon$ is a motion of order $\epsilon^2$ in this Lie bracket direction, plus a term of order $\epsilon^3$.

From the result of Example 13.5, no matter how small the maneuvering space is for a car with a reverse gear, it can generate sideways motion. Thus we have shown that the Pfaffian velocity constraint implicit in the kinematics $\dot{q} = G(q)u$ for the canonical nonholonomic mobile robot is not integrable to a configuration constraint.

A Lie bracket $[g_i, g_j]$ is called a **Lie product** of degree 2, because the original vector fields appear twice in the bracket. For the canonical nonholonomic model, it is only necessary to consider the degree-2 Lie product to show that there are no configuration constraints. To test whether there are configuration constraints for more general systems of the form (13.25), it may be necessary to consider nested Lie brackets, such as $[g_i, [g_j, g_k]]$ or $[g_i, [g_i, [g_i, g_j]]]$, which are Lie products of degree 3 and 4, respectively. Just as it is possible to generate motions in Lie bracket directions by switching between the original vector fields, it is possible to generate motion in Lie product directions of degree greater than 2. Generating motions in these directions is even slower than for degree-2 Lie products.

The **Lie algebra** of a set of vector fields is defined by all Lie products of all degrees, including Lie products of degree 1 (the original vector fields themselves):

**Definition 13.6**   The **Lie algebra** of a set of vector fields $\mathcal{G} = \{g_1, \ldots, g_m\}$, written $\overline{\text{Lie}}(\mathcal{G})$, is the linear span of all Lie products of degree $1, \ldots, \infty$ of the vector fields $\mathcal{G}$.

For example, for $\mathcal{G} = \{g_1, g_2\}$, $\overline{\text{Lie}}(\mathcal{G})$ is given by the linear combinations of the following Lie products:

degree 1:  $g_1, g_2$
degree 2:  $[g_1, g_2]$
degree 3:  $[g_1, [g_1, g_2]]; [g_2, [g_1, g_2]]$
degree 4:  $[g_1, [g_1, [g_1, g_2]]]; [g_1, [g_2, [g_1, g_2]]]; [g_2, [g_1, [g_1, g_2]]]; [g_2, [g_2, [g_1, g_2]]]$
$\phantom{degree 4:} \vdots \phantom{xxx} \vdots$

Since Lie products obey the following identities,

- $[g_i, g_i] = 0$,
- $[g_i, g_j] = -[g_j, g_i]$,
- $[g_i, [g_j, g_k]] + [g_k, [g_i, g_j]] + [g_j, [g_k, g_i]] = 0$ (the Jacobi identity),

not all bracket combinations need to be considered at each degree level.

In practice there will be a finite degree $k$ beyond which higher-degree Lie products yield no more information about the Lie algebra. This happens, for example, when the dimension of the Lie products generated so far is $n$ at all $q$, i.e., $\dim(\overline{\text{Lie}}(\mathcal{G})(q)) = n$ for all $q$; no further Lie brackets can yield new motion directions, as all motion directions have already been obtained. If the dimension of the Lie products generated so far is less than $n$, however, in general there is no way to know when to stop trying higher-degree Lie products.[4]

---

[4] When the system (13.25) is known to be *regular*, however, if there is a degree $k$ that yields

With all this as background, we are finally ready to state our main theorem on controllability.

**Theorem 13.7**    *The control system (13.25), with $\mathcal{G} = \{g_1(q), \ldots, g_m(q)\}$, is small-time locally accessible from $q$ if $\dim(\overline{\mathrm{Lie}}(\mathcal{G})(q)) = \dim(q) = n$ and $\mathrm{span}(\mathcal{U}) = \mathbb{R}^m$. If additionally $\mathrm{pos}(\mathcal{U}) = \mathbb{R}^m$ then the system is small-time locally controllable from $q$.*

We omit a formal proof, but intuitively we can argue as follows. If the Lie algebra is full rank then the vector fields (followed both forward and backward) locally permit motion in any direction. If $\mathrm{pos}(\mathcal{U}) = \mathbb{R}^m$ (as for a car with a reverse gear) then it is possible to directly follow all vector fields forward or backward, or to switch between feasible controls in order to follow any vector field forward and backward arbitrarily closely, and therefore the Lie algebra rank condition implies STLC. If the controls satisfy only $\mathrm{span}(\mathcal{U}) = \mathbb{R}^m$ (like a forward-only car), then some vector fields may be followed only forward or backward. Nevertheless, the Lie algebra rank condition ensures that there are no equality constraints on the reachable set, so the system is STLA.

For any system of the form (13.25), the question whether the velocity constraints are integrable is finally answered by Theorem 13.7. If the system is STLA at any $q$, the constraints are not integrable.

Let's apply Theorem 13.7 to a few examples.

**Example 13.8** (Controllability of the canonical nonholonomic mobile robot)    In Example 13.5 we computed the Lie bracket $g_3 = [g_1, g_2] = (0, \sin\phi, -\cos\phi)$ for the canonical nonholonomic robot. Putting the column vectors $g_1(q)$, $g_2(q)$, and $g_3(q)$ side by side to form a matrix and calculating its determinant, we find

$$\det[\, g_1(q) \ \ g_2(q) \ \ g_3(q) \,] = \det \begin{bmatrix} 0 & 1 & 0 \\ \cos\phi & 0 & \sin\phi \\ \sin\phi & 0 & -\cos\phi \end{bmatrix} = \cos^2\phi + \sin^2\phi = 1,$$

i.e., the three vector fields are linearly independent at all $q$, and therefore the dimension of the Lie algebra is 3 at all $q$. By Theorem 13.7 and the control sets illustrated in Figure 13.11, the unicycle, diff-drive, and car with a reverse gear are STLC at all $q$, while the forward-only car is only STLA at all $q$. Each of the unicycle, diff-drive, car, and forward-only car is controllable in an obstacle-free plane.

**Example 13.9** (Controllability of the full configuration of the unicycle)    We already know from the previous example that the unicycle is STLC on its $(\phi, x, y)$ subspace; what if we include the rolling angle $\theta$ in the description of the configuration? According to Equation (13.12), for $q = (\phi, x, y, \theta)$, the two vector fields are $g_1(q) = (0, r\cos\phi, r\sin\phi, 1)$ and $g_2(q) = (1, 0, 0, 0)$. Calculating the degree-2

no new motion directions not included at lower degrees then there is no need to look at higher-degree Lie products.

and degree-3 Lie brackets

$$g_3(q) = [g_1, g_2](q) = (0, r \sin \phi, -r \cos \phi, 0),$$
$$g_4(q) = [g_2, g_3](q) = (0, r \cos \phi, r \sin \phi, 0),$$

we see that these directions correspond to sideways translation and to forward–backward motion without a change in the wheel rolling angle $\theta$, respectively. These directions are clearly linearly independent of $g_1(q)$ and $g_2(q)$, but we can confirm this by again writing the $g_i(q)$ as column vectors and evaluating

$$\det[\, g_1(q) \;\; g_2(q) \;\; g_3(q) \;\; g_4(q) \,] = -r^2,$$

i.e., $\dim(\overline{\text{Lie}}(\mathcal{G})(q)) = 4$ for all $q$. Since $\text{pos}(\mathcal{U}) = \mathbb{R}^2$ for the unicycle, by Figure 13.11, the unicycle is STLC at all points in its four-dimensional configuration space.

You can come to this same conclusion by constructing a short "parallel parking" type maneuver which results in a net change in the rolling angle $\theta$ with zero net change in the other configuration variables.

**Example 13.10** (Controllability of the full configuration of the diff-drive)   The full configuration of the diff-drive is $q = (\phi, x, y, \theta_\text{L}, \theta_\text{R})$, including the angles of both wheels. The two control vector fields are given in Equation (13.14). Taking the Lie brackets of these vector fields, we find that we can never create more than four linearly independent vector fields, i.e.,

$$\dim(\overline{\text{Lie}}(\mathcal{G})(q)) = 4$$

at all $q$. This is because there is a fixed relationship between the two wheel angles $(\theta_\text{L}, \theta_\text{R})$ and the angle of the robot chassis $\phi$. Therefore the three velocity constraints $(\dim(q) = 5, \dim(u) = 2)$ implicit in the kinematics (13.14) can be viewed as two nonholonomic constraints and one holonomic constraint. In the full five-dimensional configuration space, the diff-drive is nowhere STLA.

Since usually we worry only about the configuration of the chassis, this negative result is not of much concern.

### 13.3.3   Motion Planning

#### 13.3.3.1   Obstacle-Free Plane

It is easy to find feasible motions between any two chassis configurations $q_0$ and $q_\text{goal}$ in an obstacle-free plane for any of the four nonholonomic robot models (a unicycle, a diff-drive, a car with a reverse gear, and a forward-only car). The problem gets more interesting when we try to optimize an objective function. Below, we consider shortest paths for the forward-only car, shortest paths for the car with a reverse gear, and fastest paths for the diff-drive. The solutions to these problems depend on optimal control theory, and the proofs can be found in the original references (see Section 13.7).

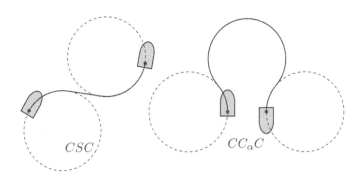

**Figure 13.15** The two classes of shortest paths for a forward-only car. The $CSC$ path could be written $RSL$, and the $CC_\alpha C$ path could be written $LR_\alpha L$.

**Shortest Paths for the Forward-Only Car**

The shortest-path problem involves finding a path from $q_0$ to $q_{\text{goal}}$ that minimizes the length of the path that is followed by the robot's reference point. This is not an interesting question for the unicycle or the diff-drive; a shortest path for each of them comprises a rotation to point toward the goal position $(x_{\text{goal}}, y_{\text{goal}})$, a translation, and then a rotation to the goal orientation. The total path length is $\sqrt{(x_0 - x_{\text{goal}})^2 + (y_0 - y_{\text{goal}})^2}$.

The problem is more interesting for the forward-only car, sometimes called the **Dubins car** in honor of the mathematician who first studied the structure of the shortest planar curves with bounded curvature between two oriented points.

**Theorem 13.11**   *For a forward-only car with the control set shown in Figure 13.11, the shortest paths consist only of arcs at the minimum turning radius and straight-line segments. Denoting a circular arc segment as $C$ and a straight-line segment as $S$, the shortest path between any two configurations follows either (a) the sequence $CSC$ or (b) the sequence $CC_\alpha C$, where $C_\alpha$ indicates a circular arc of angle $\alpha > \pi$. Any of the $C$ or $S$ segments can be of length zero.*

The two optimal path classes for a forward-only car are illustrated in Figure 13.15. We can calculate the shortest path by enumerating the possible $CSC$ and $CC_\alpha C$ paths. First, construct two minimum-turning-radius circles for the vehicle at both $q_0$ and $q_{\text{goal}}$ and then solve for (a) the points where lines (with the correct heading direction) are tangent to one of the circles at $q_0$ and one of the circles at $q_{\text{goal}}$, and (b) the points where a minimum-turning-radius circle (with the correct heading direction) is tangent to one of the circles at $q_0$ and one of the circles at $q_{\text{goal}}$. The solutions to (a) correspond to $CSC$ paths and the solutions to (b) correspond to $CC_\alpha C$ paths. The shortest of all the solutions is the optimal path. The shortest path may not be unique.

If we break the $C$ segments into two categories, $L$ (when the steering wheel is pegged to the left) and $R$ (when the steering wheel is pegged to the right), we

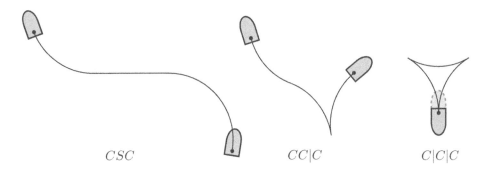

**Figure 13.16** Three of the nine classes of shortest paths for a car.

see that there are four types of $CSC$ paths ($LSL$, $LSR$, $RSL$, and $RSR$) and two types of $CC_\alpha C$ paths ($RL_\alpha R$ and $LR_\alpha L$).

**Shortest Paths for the Car with a Reverse Gear**

The shortest paths for a car with a reverse gear, sometimes called the **Reeds–Shepp car** in honor of the mathematicians who first studied the problem, again use only straight-line segments and minimum-turning-radius arcs. Using the notation $C$ for a minimum-turning-radius arc, $C_a$ for an arc of angle $a$, $S$ for a straight-line segment, and $|$ for a cusp (a reversal of the linear velocity), Theorem 13.12 enumerates the possible shortest path sequences.

**Theorem 13.12**  *For a car with a reverse gear with the control set shown in Figure 13.11, the shortest path between any two configurations is in one of the following nine classes:*

$$C|C|C \qquad CC|C \qquad C|CC \qquad CC_a|C_aC \quad C|C_aC_a|C$$
$$C|C_{\pi/2}SC \quad CSC_{\pi/2}|C \quad C|C_{\pi/2}SC_{\pi/2}|C \quad CSC$$

*Any of the $C$ or $S$ segments can be of length zero.*

Three of the nine shortest path classes are illustrated in Figure 13.16. Again, the actual shortest path may be found by enumerating the finite set of possible solutions in the path classes in Theorem 13.12. The shortest path may not be unique.

If we break the $C$ segments into four categories, $L^+$, $L^-$, $R^+$, and $R^-$, where $L$ and $R$ mean that the steering wheel is turned all the way to the left or right and the superscripts '+' and '−' indicate the gear shift (forward or reverse), then the nine path classes of Theorem 13.12 can be expressed as $(6 \times 4) + (3 \times 8) = 48$ different types:

| six path classes, each with four path types: | $C|C|C$, $CC|C$, $C|CC$, $CC_a|C_aC$, $C|C_aC_a|C$, $C|C_{\pi/2}SC_{\pi/2}|C$ |

three path classes, each with eight path types:    $C|C_{\pi/2}SC$, $CSC_{\pi/2}|C$, $CSC$

| Motion segments | Number of types | Motion sequences |
|---|---|---|
| 1 | 4 | $F$, $B$, $R$, $L$ |
| 2 | 8 | $FR$, $FL$, $BR$, $BL$, $RF$, $RB$, $LF$, $LB$ |
| 3 | 16 | $FRB$, $FLB$, $FR_\pi B$, $FL_\pi B$,<br>$BRF$, $BLF$, $BR_\pi F$, $BL_\pi F$,<br>$RFR$, $RFL$, $RBR$, $RBL$,<br>$LFR$, $LFL$, $LBR$, $LBL$ |
| 4 | 8 | $FRBL$, $FLBR$, $BRFL$, $BLFR$,<br>$RFLB$, $RBLF$, $LFRB$, $LBRF$ |
| 5 | 4 | $FRBLF$, $FLBRF$, $BRFLB$, $BLFRB$ |

**Table 13.1** The 40 time-optimal trajectory types for the diff-drive. The notation $R_\pi$ and $L_\pi$ indicate spins of angle $\pi$.

The four types for six path classes are determined by the four different initial motion directions, $L^+$, $L^-$, $R^+$, and $R^-$. The eight types for three path classes are determined by the four initial motion directions and whether the turn is to the left or the right after the straight-line segment. There are only four types in the $C|C_{\pi/2}SC_{\pi/2}|C$ class because the turn after the $S$ segment is always opposite to the turn before the $S$ segment.

If it takes zero time to reverse the linear velocity, a shortest path is also a minimum-time path for the control set for the car with a reverse gear illustrated in Figure 13.11, where the only controls $(v, \omega)$ ever used are the two controls $(\pm v_{\max}, 0)$, an $S$ segment, or the four controls $(\pm v_{\max}, \pm \omega_{\max})$, a $C$ segment.

### Minimum-Time Motions for the Diff-Drive

For a diff-drive robot with the diamond-shaped control set in Figure 13.11, any minimum-time motion consists of only translational motions and spins in place.

**Theorem 13.13**  *For a diff-drive robot with the control set illustrated in Figure 13.11, minimum-time motions consist of forward and backward translations ($F$ and $B$) at maximum speed $\pm v_{max}$ and spins in place ($R$ and $L$ for right turns and left turns) at maximum angular speed $\pm \omega_{max}$. There are 40 types of time-optimal motions, which are categorized in Table 13.1 by the number of motion segments. The notations $R_\pi$ and $L_\pi$ indicate spins of angle $\pi$.*

Note that Table 13.1 includes both $FR_\pi B$ and $FL_\pi B$, which are equivalent, as well as $BR_\pi F$ and $BL_\pi F$. Each trajectory type is time optimal for some pair $\{q_0, q_{\text{goal}}\}$, and the time-optimal trajectory may not be unique. Notably absent are three-segment sequences where the first and last motions are translations in the same direction (i.e., $FRF$, $FLF$, $BRB$, and $BLB$).

While any reconfiguration of the diff-drive can be achieved by spinning, translating, and spinning, in some cases other three-segment sequences have a shorter travel time. For example consider a diff-drive with $v_{\max} = \omega_{\max} = 1$, $q_0 = 0$, and $q_{\text{goal}} = (-7\pi/8, 1.924, 0.383)$, as shown in Figure 13.17. The time needed

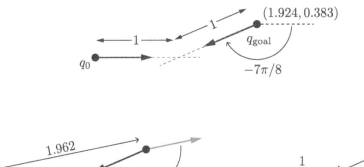

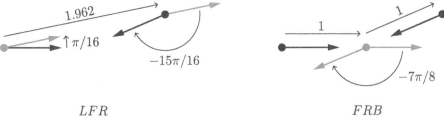

$LFR$                                    $FRB$

**Figure 13.17** (Top) A motion planning problem specified as a motion from $q_0 = (0, 0, 0)$ to $q_{\text{goal}} = (-7\pi/8, 1.924, 0.383)$. (Bottom left) A non-optimal $LFR$ solution taking time 5.103. (Bottom right) The time-optimal $FRB$ solution, through a "via point," taking time 4.749.

for a spin of angle $\alpha$ is $|\alpha|/\omega_{\text{max}} = |\alpha|$ and the time for a translation of $d$ is $|d|/v_{\text{max}} = |d|$. Therefore, the time needed for the $LFR$ sequence is

$$\frac{\pi}{16} + 1.962 + \frac{15\pi}{16} = 5.103,$$

while the time needed for the $FRB$ sequence through a "via point" is

$$1 + \frac{7\pi}{8} + 1 = 4.749.$$

### 13.3.3.2 With Obstacles

If there are obstacles in the plane, the grid-based motion planning methods of Section 10.4.2 can be applied to the unicycle, diff-drive, car with a reverse gear, or forward-only car using discretized versions of the control sets in Figure 13.11. See, for example, the discretizations in Figure 10.14, which use the extremal controls from Figure 13.11. Using extremal controls takes advantage of our observation that shortest paths for reverse-gear cars and the diff-drive consist of minimum-turning-radius turns and straight-line segments. Also, because the C-space is only three dimensional, the grid size should be manageable for reasonable resolutions along each dimension.

We can also apply the sampling methods of Section 10.5. For RRTs, we can again use a discretization of the control set, as mentioned above or, for both PRMs and RRTs, a local planner that attempts to connect two configurations could use the shortest paths from Theorems 13.11, 13.12, or 13.13.

Another option for a reverse-gear car is to use any efficient obstacle-avoiding

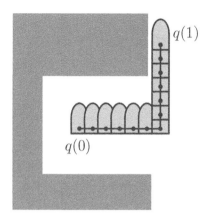 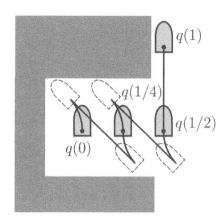

**Figure 13.18** (Left) The original path from $q(0)$ to $q(1)$ found by a motion planner that does not respect the reverse-gear car's motion constraints. (Right) The path found by the recursive subdivision has via points at $q(1/4)$ and $q(1/2)$.

path planner, even if it ignores the motion constraints of the vehicle. Since such a car is STLC and since the free configuration space is defined to be open (obstacles are closed, containing their boundaries), the car can follow the path found by the planner arbitrarily closely. To follow the path closely, however, the motion may have to be slow – imagine using parallel parking to travel a kilometer down the road.

Alternatively, an initial constraint-free path can be quickly transformed into a fast, feasible, path that respects the car's motion constraints. To do this, represent the initial path as $q(s), s \in [0, 1]$. Then try to connect $q(0)$ to $q(1)$ using a shortest path from Theorem 13.12. If this path is in collision, then divide the original path in half and try to connect $q(0)$ to $q(1/2)$ and $q(1/2)$ to $q(1)$ using shortest paths. If either of these paths are in collision, divide that path, and so on. Because the car is STLC and the initial path lies in open free space, the process will eventually terminate; the new path consists of a sequence of subpaths from Theorem 13.12. The process is illustrated in Figure 13.18.

### 13.3.4    Feedback Control

We can consider three types of feedback control problems for the canonical nonholonomic mobile robot (13.18) with controls $(v, \omega)$:

(a) **Stabilization of a configuration.** Given a desired configuration $q_d$, drive the error $q_d - q(t)$ to zero as time goes to infinity. As we saw in Theorem 13.1, no time-invariant feedback law that is continuous in the state variables can stabilize a configuration for a nonholonomic mobile robot. There do exist time-varying and discontinuous feedback laws that accomplish the task, but we do not consider this problem further here.

(b) **Trajectory tracking.** Given a desired trajectory $q_d(t)$, drive the error $q_d(t) - q(t)$ to zero as time goes to infinity.

(c) **Path tracking.** Given a path $q(s)$, follow the geometric path without regard to the time of the motion. This provides more control freedom than the trajectory tracking problem; essentially, we can choose the speed of the reference configuration along the path so as to help reduce the tracking error, in addition to choosing $(v, \omega)$.

Path tracking and trajectory tracking are "easier" than stabilizing a configuration, in the sense that there exist continuous time-invariant feedback laws to stabilize the desired motions. In this section we consider the problem of trajectory tracking.

Assume that the reference trajectory is specified as $q_d(t) = (\phi_d(t), x_d(t), y_d(t))$ for $t \in [0, T]$, with a corresponding nominal control $(v_d(t), \omega_d(t)) \in \text{int}(\mathcal{U})$ for $t \in [0, T]$. The requirement that the nominal control be in the interior of the feasible control set $\mathcal{U}$ ensures that some control effort is "left over" to correct small errors. This implies that the reference trajectory is neither a shortest path nor a time-optimal trajectory, since optimal motions saturate the controls. The reference trajectory could be planned using not-quite-extremal controls.

A simple first controller idea is to choose a reference point $P$ on the chassis of the robot (but not on the axis of the two driving wheels), as in Figure 13.12. The desired trajectory $q_d(t)$ is then represented by the desired trajectory of the reference point $(x_{Pd}(t), y_{Pd}(t))$. To track this reference point trajectory, we can use a proportional feedback controller

$$\begin{bmatrix} \dot{x}_P \\ \dot{y}_P \end{bmatrix} = \begin{bmatrix} k_p(x_{Pd} - x_P) \\ k_p(y_{Pd} - y_P) \end{bmatrix}, \tag{13.29}$$

where $k_p > 0$. This simple linear control law is guaranteed to pull the actual position $p$ along with the moving desired position. The velocity $(\dot{x}_P, \dot{y}_P)$ calculated by the control law (13.29) is converted to $(v, \omega)$ by Equation (13.21).

The idea is that, as long as the reference point is moving, over time the entire robot chassis will line up with the desired orientation of the chassis. The problem is that the controller may choose the opposite orientation of what is intended; there is nothing in the control law to prevent this. Figure 13.19 shows two simulations, one where the control law (13.29) produces the desired chassis motion and one where the control law causes an unintended reversal in the sign of the driving velocity $v$. In both simulations the controller succeeds in causing the reference point to track the desired motion.

To fix this, let us explicitly incorporate chassis angle error in the control law. The fixed space frame is {s}, the chassis frame {b} is at the point between the two wheels of the diff-drive (or the two rear wheels for a reverse-gear car) with the forward driving direction along the $\hat{x}_b$-axis, and the frame corresponding to

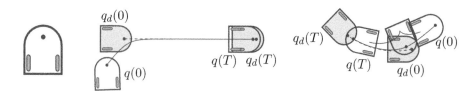

**Figure 13.19** (Left) A nonholonomic mobile robot with a reference point. (Middle) A scenario where the linear control law (13.29) tracking a desired reference point trajectory yields the desired trajectory tracking behavior for the entire chassis. (Right) A scenario where the point-tracking control law causes an unintended cusp in the robot motion. The reference point converges to the desired path but the robot's orientation is opposite to the intended orientation.

$q_d(t)$ is $\{d\}$. We define the error coordinates

$$q_e = \begin{bmatrix} \phi_e \\ x_e \\ y_e \end{bmatrix} = \begin{bmatrix} 1 & 0 & 0 \\ 0 & \cos\phi_d & \sin\phi_d \\ 0 & -\sin\phi_d & \cos\phi_d \end{bmatrix} \begin{bmatrix} \phi - \phi_d \\ x - x_d \\ y - y_d \end{bmatrix}, \tag{13.30}$$

as illustrated in Figure 13.20. The vector $(x_e, y_e)$ is the $\{s\}$-coordinate error vector $(x - x_d, y - y_d)$ expressed in the reference frame $\{d\}$.

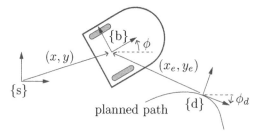

**Figure 13.20** The space frame $\{s\}$, the robot frame $\{b\}$, and the desired configuration $\{d\}$ driving forward along the planned path. The heading-direction error is $\phi_e = \phi - \phi_d$.

Consider the nonlinear feedforward plus feedback control law

$$\begin{bmatrix} v \\ \omega \end{bmatrix} = \begin{bmatrix} (v_d - k_1|v_d|(x_e + y_e\tan\phi_e))/\cos\phi_e \\ \omega_d - (k_2 v_d y_e + k_3|v_d|\tan\phi_e)\cos^2\phi_e \end{bmatrix}, \tag{13.31}$$

where $k_1, k_2, k_3 > 0$. Note two things about this control law: (1) if the error is zero, the control is simply the nominal control $(v_d, \omega_d)$; and (2) the controls grow without bound as $\phi_e$ approaches $\pi/2$ or $-\pi/2$. In practice, we assume that $|\phi_e|$ is less than $\pi/2$ during trajectory tracking.

In the controller for $v$, the second term, $-k_1|v_d|x_e/\cos\phi_e$, attempts to reduce $x_e$ by driving the robot so as to catch up with or slow down to the reference frame. The third term, $-k_1|v_d|y_e\tan\phi_e/\cos\phi_e$, attempts to reduce $y_e$ using the component of the forward or backward velocity that impacts $y_e$.

In the controller for the turning velocity $\omega$, the second term, $-k_2 v_d y_e\cos^2\phi_e$, attempts to reduce $y_e$ in the future by turning the heading direction of the robot toward the reference-frame origin. The third term, $-k_3|v_d|\tan\phi_e\cos^2\phi_e$, attempts to reduce the heading error $\phi_e$.

A simulation of the control law (13.31) is shown in Figure 13.21.

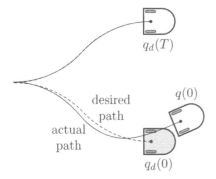

**Figure 13.21** A mobile robot implementing the nonlinear control law (13.31).

The control law requires $v_d \neq 0$, so it is not appropriate for stabilizing "spin-in-place" motions for a diff-drive. The proof of the stability of the control law requires methods beyond the scope of this book. In practice, the gains should be chosen large enough to provide significant corrective action but not so large that the controls chatter at the boundary of the feasible control set $\mathcal{U}$.

## 3.4 Odometry

Odometry is the process of estimating the chassis configuration $q$ from the wheel motions, essentially integrating the effect of the wheel velocities. Since wheel-rotation sensing is available on all mobile robots, odometry is cheap and convenient. Estimation errors tend to accumulate over time, though, due to unexpected slipping and skidding of the wheels and to numerical integration error. Therefore, it is common to supplement odometry with other position sensors, such as GPS, the visual recognition of landmarks, ultrasonic beacons, laser or ultrasonic range sensing, etc. Those sensing modalities have their own measurement uncertainty but errors do not accumulate over time. As a result, odometry generally gives superior results on short time scales, but odometric estimates should either (1) be periodically corrected by other sensing modalities or, preferably, (2) integrated with other sensing modalities in an estimation framework based on a Kalman filter, particle filter, or similar.

In this section we focus on odometry. We assume that each wheel of an omnidirectional robot, and each rear wheel of a diff-drive or car, has an encoder that senses how far the wheel has rotated in its driving direction. If the wheels are driven by stepper motors then we know the driving rotation of each wheel from the steps we have commanded to it.

The goal is to estimate the new chassis configuration $q_{k+1}$ as a function of the previous chassis configuration $q_k$, given the change in wheel angles from the instant $k$ to the instant $k + 1$.

Let $\Delta\theta_i$ be the change in wheel $i$'s driving angle since the wheel angle was last queried a time $\Delta t$ ago. Since we know only the net change in the wheel driving angle, not the time history of how the wheel angle evolved during the time interval, the simplest assumption is that the wheel's angular velocity was constant during the time interval, $\dot{\theta}_i = \Delta\theta_i/\Delta t$. The choice of units used to measure the time interval is not relevant (since we will eventually integrate the chassis body twist $\mathcal{V}_b$ over the same time interval), so we set $\Delta t = 1$, i.e., $\dot{\theta}_i = \Delta\theta$.

For omnidirectional mobile robots, the vector of wheel speeds $\dot{\theta}$, and therefore $\Delta\theta$, is related to the body twist $\mathcal{V}_b = (\omega_{bz}, v_{bx}, v_{by})$ of the chassis by Equation (13.8):

$$\Delta\theta = H(0)\mathcal{V}_b,$$

where $H(0)$ for the three-omniwheel robot is given by Equation (13.9) and for the four-mecanum-wheel robot is given by Equation (13.10). Therefore, the body twist $\mathcal{V}_b$ corresponding to $\Delta\theta$ is

$$\mathcal{V}_b = H^\dagger(0)\Delta\theta = F\Delta\theta,$$

where $F = H^\dagger(0)$ is the pseudoinverse of $H(0)$. For the three-omniwheel robot,

$$\mathcal{V}_b = F\Delta\theta = r \begin{bmatrix} -1/(3d) & -1/(3d) & -1/(3d) \\ 2/3 & -1/3 & -1/3 \\ 0 & -1/(2\sin(\pi/3)) & 1/(2\sin(\pi/3)) \end{bmatrix} \Delta\theta \quad (13.32)$$

and for the four-mecanum-wheel robot,

$$\mathcal{V}_b = F\Delta\theta = \frac{r}{4} \begin{bmatrix} -1/(\ell+w) & 1/(\ell+w) & 1/(\ell+w) & -1/(\ell+w) \\ 1 & 1 & 1 & 1 \\ -1 & 1 & -1 & 1 \end{bmatrix} \Delta\theta.$$
$$(13.33)$$

The relationship $\mathcal{V}_b = F\dot{\theta} = F\Delta\theta$ also holds for the diff-drive robot and the car (Figure 13.22), where $\Delta\theta = (\Delta\theta_L, \Delta\theta_R)$ (the increments for the left and right wheels) and

$$\mathcal{V}_b = F\Delta\theta = r \begin{bmatrix} -1/(2d) & 1/(2d) \\ 1/2 & 1/2 \\ 0 & 0 \end{bmatrix} \begin{bmatrix} \Delta\theta_L \\ \Delta\theta_R \end{bmatrix}. \quad (13.34)$$

Since the wheel speeds are assumed constant during the time interval, so is the body twist $\mathcal{V}_b$. Calling $\mathcal{V}_{b6}$ the six-dimensional version of the planar twist $\mathcal{V}_b$ (i.e.,

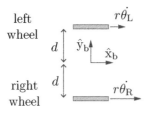

**Figure 13.22** The left and right wheels of a diff-drive or the left and right rear wheels of a car.

$\mathcal{V}_{b6} = (0, 0, \omega_{bz}, v_{bx}, v_{by}, 0))$, $\mathcal{V}_{b6}$ can be integrated to generate the displacement created by the wheel-angle increment vector $\Delta\theta$:

$$T_{bb'} = e^{[\mathcal{V}_{b6}]}.$$

From $T_{bb'} \in SE(3)$, which expresses the new chassis frame {b′} relative to the initial frame {b}, we can extract the change in coordinates relative to the body frame {b}, $\Delta q_b = (\Delta\phi_b, \Delta x_b, \Delta y_b)$, in terms of $(\omega_{bz}, v_{bx}, v_{by})$:

$$\text{if } \omega_{bz} = 0, \ \Delta q_b = \begin{bmatrix} \Delta\phi_b \\ \Delta x_b \\ \Delta y_b \end{bmatrix} = \begin{bmatrix} 0 \\ v_{bx} \\ v_{by} \end{bmatrix}; \tag{13.35}$$

$$\text{if } \omega_{bz} \neq 0, \ \Delta q_b = \begin{bmatrix} \Delta\phi_b \\ \Delta x_b \\ \Delta y_b \end{bmatrix} = \begin{bmatrix} \omega_{bz} \\ (v_{bx}\sin\omega_{bz} + v_{by}(\cos\omega_{bz} - 1))/\omega_{bz} \\ (v_{by}\sin\omega_{bz} + v_{bx}(1 - \cos\omega_{bz}))/\omega_{bz} \end{bmatrix}.$$

Transforming $\Delta q_b$ in {b} to $\Delta q$ in the fixed frame {s} using the chassis angle $\phi_k$,

$$\Delta q = \begin{bmatrix} 1 & 0 & 0 \\ 0 & \cos\phi_k & -\sin\phi_k \\ 0 & \sin\phi_k & \cos\phi_k \end{bmatrix} \Delta q_b, \tag{13.36}$$

the updated odometry estimate of the chassis configuration is finally

$$q_{k+1} = q_k + \Delta q.$$

In summary, $\Delta q$ is calculated using Equations (13.35) and (13.36) as a function of $\mathcal{V}_b$ and the previous chassis angle $\phi_k$, and Equation (13.32), (13.33), or (13.34) is used to calculate $\mathcal{V}_b$ as a function of the wheel-angle changes $\Delta\theta$ for the three-omniwheel robot, the four-mecanum-wheel robot, or a nonholonomic robot (the diff-drive or the car), respectively.

## 13.5 Mobile Manipulation

For a robot arm mounted on a mobile base, **mobile manipulation** describes the coordination of the motion of the base and the robot joints to achieve a desired motion at the end-effector. Typically the motion of the arm can be controlled more precisely than the motion of the base, so the most popular type of mobile manipulation involves driving the base, parking it, letting the arm perform the precise motion task, then driving away.

In some cases, however, it is advantageous, or even necessary, for the end-effector motion to be achieved by a combination of motion of the base and motion of the arm. Defining the fixed space frame {s}, the chassis frame {b}, a frame at the base of the arm {0}, and an end-effector frame {e}, the configuration of {e} in {s} is

$$X(q, \theta) = T_{se}(q, \theta) = T_{sb}(q)\, T_{b0}\, T_{0e}(\theta) \in SE(3),$$

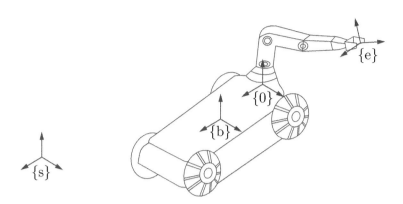

**Figure 13.23** The space frame {s} and the frames {b}, {0}, and {e} attached to the mobile manipulator.

where $\theta \in \mathbb{R}^n$ is the set of arm joint positions for the $n$-joint robot, $T_{0e}(\theta)$ is the forward kinematics of the arm, $T_{b0}$ is the fixed offset of {0} from {b}, $q = (\phi, x, y)$ is the planar configuration of the mobile base, and

$$
T_{sb}(q) =
\begin{bmatrix}
\cos\phi & -\sin\phi & 0 & x \\
\sin\phi & \cos\phi & 0 & y \\
0 & 0 & 1 & z \\
0 & 0 & 0 & 1
\end{bmatrix},
$$

where $z$ is a constant indicating the height of the {b} frame above the floor. See Figure 13.23.

Let $X(t)$ be the path of the end-effector as a function of time. Then $[\mathcal{V}_e(t)] = X^{-1}(t)\dot{X}(t)$ is the $se(3)$ representation of the end-effector twist expressed in {e}. Further, let the vector of wheel velocities, whether the robot is omnidirectional or nonholonomic, be written $u \in \mathbb{R}^m$. For kinematic control of the end-effector frame using the wheel and joint velocities, we need the Jacobian $J_e(\theta) \in \mathbb{R}^{6 \times (m+n)}$ satisfying

$$
\mathcal{V}_e = J_e(\theta)
\begin{bmatrix}
u \\
\dot{\theta}
\end{bmatrix}
= [J_{\text{base}}(\theta) \ J_{\text{arm}}(\theta)]
\begin{bmatrix}
u \\
\dot{\theta}
\end{bmatrix}.
$$

Note that the Jacobian $J_e(\theta)$ does not depend on $q$: the end-effector velocity expressed in {e} is independent of the configuration of the mobile base. Also, we can partition $J_e(\theta)$ into $J_{\text{base}}(\theta) \in \mathbb{R}^{6 \times m}$ and $J_{\text{arm}}(\theta) \in \mathbb{R}^{6 \times n}$. The term $J_{\text{base}}(\theta)u$ expresses the contribution of the wheel velocities $u$ to the end-effector's velocity, and the term $J_{\text{arm}}(\theta)\dot{\theta}$ expresses the contribution of the joint velocities to the end-effector's velocity.

In Chapter 5 we developed a method to derive $J_{\text{arm}}(\theta)$, which is called the body Jacobian $J_b(\theta)$ in that chapter. All that remains is to find $J_{\text{base}}(\theta)$. As we saw in Section 13.4, for any type of mobile base there exists an $F$ satisfying

$$
\mathcal{V}_b = Fu.
$$

To create a six-dimensional twist $\mathcal{V}_{b6}$ corresponding to the planar twist $\mathcal{V}_b$, we can define the $6 \times m$ matrix

$$F_6 = \begin{bmatrix} 0_m \\ 0_m \\ F \\ 0_m \end{bmatrix},$$

where two rows of $m$ zeros are stacked above $F$ and one row is situated below it. Now we have

$$\mathcal{V}_{b6} = F_6 u.$$

This chassis twist can be expressed in the end-effector frame as

$$[\text{Ad}_{T_{eb}(\theta)}]\mathcal{V}_{b6} = [\text{Ad}_{T_{0e}^{-1}(\theta)T_{b0}^{-1}}]\mathcal{V}_{b6} = [\text{Ad}_{T_{0e}^{-1}(\theta)T_{b0}^{-1}}]F_6 u = J_{\text{base}}(\theta)u.$$

Therefore

$$J_{\text{base}}(\theta) = [\text{Ad}_{T_{0e}^{-1}(\theta)T_{b0}^{-1}}]F_6.$$

Now that we have the complete Jacobian $J_e(\theta) = [\, J_{\text{base}}(\theta) \;\; J_{\text{arm}}(\theta) \,]$, we can perform numerical inverse kinematics (Section 6.2) or implement kinematic feedback control laws to track a desired end-effector trajectory. For example, given a desired end-effector trajectory $X_d(t)$, we can choose the kinematic task-space feedforward plus feedback control law (11.16),

$$\mathcal{V}(t) = [\text{Ad}_{X^{-1}X_d}]\mathcal{V}_d(t) + K_p X_{\text{err}}(t) + K_i \int_0^t X_{\text{err}}(t) \, dt, \qquad (13.37)$$

where $[\mathcal{V}_d(t)] = X_d^{-1}(t)\dot{X}_d(t)$, the transform $[\text{Ad}_{X^{-1}X_d}]$ changes the frame of representation of the feedforward twist $\mathcal{V}_d$ from the frame at $X_d$ to the actual end-effector frame at $X$, and $[X_{\text{err}}] = \log(X^{-1}X_d)$. The commanded end-effector-frame twist $\mathcal{V}(t)$ is implemented as

$$\begin{bmatrix} u \\ \dot{\theta} \end{bmatrix} = J_e^\dagger(\theta)\mathcal{V}.$$

As discussed in Section 6.3, it is possible to use a weighted pseudoinverse to penalize certain wheel or joint velocities.

An example is shown in Figure 13.24. The mobile base is a diff-drive and the arm moves in the plane with only one revolute joint. The desired motion of the end-effector $X_d(t), t \in [0,1]$, is parametrized by $\alpha = -\pi t$, $x_d(t) = -3\cos(\pi t)$, and $y_d(t) = 3\sin(\pi t)$, where $\alpha$ indicates the planar angle from the $\hat{x}_s$-axis to the $\hat{x}_e$-axis (see Figure 13.24). The performance in Figure 13.24 demonstrates a bit of overshoot, indicating that the diagonal gain matrix $K_i = k_i I$ should use somewhat lower gains. Alternatively, the gain matrix $K_p = k_p I$ could use larger gains, provided that there are no practical problems with these larger gains (see the discussion in Section 11.3.1.2).

Note that, for an arbitrary $X_d(t)$ to be feasible for the mobile manipulator, the Jacobian $J_e(\theta)$ should be full rank everywhere; see Exercise 13.30.

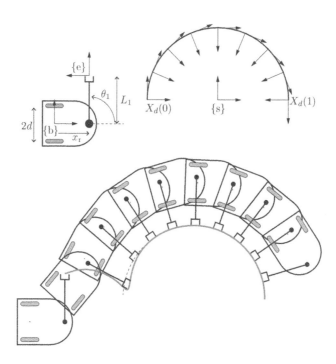

**Figure 13.24** A diff-drive with a 1R planar arm with end-effector frame {e}. (Top) The initial configuration of the robot and the desired end-effector trajectory $X_d(t)$. (Bottom) Trajectory tracking using the control law (13.37). The end-effector shoots past the desired path before settling into accurate trajectory tracking.

## 13.6    Summary

- The chassis configuration of a wheeled mobile robot moving in the plane is $q = (\phi, x, y)$. The velocity can be represented either as $\dot{q}$ or as the planar twist $\mathcal{V}_b = (\omega_{bz}, v_{bx}, v_{by})$ expressed in the chassis-fixed frame {b}, where

$$
\mathcal{V}_b = \begin{bmatrix} \omega_{bz} \\ v_{bx} \\ v_{by} \end{bmatrix} = \begin{bmatrix} 1 & 0 & 0 \\ 0 & \cos\phi & \sin\phi \\ 0 & -\sin\phi & \cos\phi \end{bmatrix} \begin{bmatrix} \dot{\phi} \\ \dot{x} \\ \dot{y} \end{bmatrix}.
$$

- The chassis of a nonholonomic mobile robot is subject to a single nonintegrable Pfaffian velocity constraint $A(q)\dot{q} = [0 \ \sin\phi \ -\cos\phi]\dot{q} = \dot{x}\sin\phi - \dot{y}\cos\phi = 0$. An omnidirectional robot, employing omniwheels or mecanum wheels, has no such constraint.
- For a properly constructed omnidirectional robot with $m \geq 3$ wheels, there exists a rank 3 matrix $H(\phi) \in \mathbb{R}^{m \times 3}$ that maps the chassis velocity $\dot{q}$ to

the wheel driving velocities $u$:

$$u = H(\phi)\dot{q}.$$

In terms of the body twist $\mathcal{V}_b$,

$$u = H(0)\mathcal{V}_b.$$

The driving speed limits of each wheel place two parallel planar constraints on the feasible body twists, creating a polyhedron $V$ of feasible body twists.

- Motion planning and feedback control for omnidirectional robots is simplified by the fact that there are no chassis velocity equality constraints.
- Nonholonomic mobile robots are described as driftless linear-in-the-control systems

$$\dot{q} = G(q)u, \qquad u \in \mathcal{U} \subset \mathbb{R}^m,$$

where $G(q) \in \mathbb{R}^{n \times m}, n > m$. The $m$ columns $g_i(q)$ of $G(q)$ are called the control vector fields.

- The canonical simplified nonholonomic mobile robot model is

$$\dot{q} = \begin{bmatrix} \dot{\phi} \\ \dot{x} \\ \dot{y} \end{bmatrix} = G(q)u = \begin{bmatrix} 0 & 1 \\ \cos\phi & 0 \\ \sin\phi & 0 \end{bmatrix} \begin{bmatrix} v \\ \omega \end{bmatrix}.$$

The control sets $\mathcal{U}$ differ for the unicycle, diff-drive, reverse-gear car, and forward-only car.

- A control system is small-time locally accessible (STLA) from $q$ if, for any time $T > 0$ and any neighborhood $W$, the reachable set in time less than $T$ without leaving $W$ is a full-dimensional subset of the configuration space. A control system is small-time locally controllable (STLC) from $q$ if, for any time $T > 0$ and any neighborhood $W$, the reachable set in time less than $T$ without leaving $W$ is a neighborhood of $q$. If the system is STLC from a given $q$, it can maneuver locally in any direction.

- The Lie bracket of two vector fields $g_1$ and $g_2$ is the vector field

$$[g_1, g_2] = \left( \frac{\partial g_2}{\partial q} g_1 - \frac{\partial g_1}{\partial q} g_2 \right).$$

- A Lie product of degree $k$ is a Lie bracket term where the original vector fields appear a total of $k$ times. A Lie product of degree 1 is just one of the original vector fields.
- The Lie algebra of a set of vector fields $\mathcal{G} = \{g_1, \ldots, g_m\}$, written $\overline{\text{Lie}}(\mathcal{G})$, is the linear span of all Lie products of degree $1, \ldots, \infty$ of the vector fields $\mathcal{G}$.
- A driftless control-affine system is small-time locally accessible from $q$ if $\dim(\overline{\text{Lie}}(\mathcal{G})(q)) = \dim(q) = n$ and $\text{span}(\mathcal{U}) = \mathbb{R}^m$. If additionally $\text{pos}(\mathcal{U}) = \mathbb{R}^m$ then the system is small-time locally controllable from $q$.

- For a forward-only car in an obstacle-free plane, shortest paths always follow a turn at the tightest turning radius ($C$) or straight-line motions ($S$). There are two classes of shortest paths: $CSC$ and $CC_\alpha C$, where $C_\alpha$ is a turn of angle $|\alpha| > \pi$. Any $C$ or $S$ segment can be of length zero.
- For a car with a reverse gear, shortest paths always consist of a sequence of straight-line segments or turns at the tightest turning radius. Shortest paths always belong to one of nine classes.
- For the diff-drive, minimum-time motions always consist of turn-in-place motions and straight-line motions.
- For the canonical nonholonomic robot, there is no time-invariant control law which is continuous in the configuration and which will stabilize the origin configuration. Continuous time-invariant control laws exist that will stabilize a trajectory, however.
- Odometry is the process of estimating the chassis configuration on the basis of how far the robot's wheels have rotated in their driving direction, assuming no skidding in the driving direction and, for typical wheels (not omniwheels or mecanum wheels), no slip in the orthogonal direction.
- For a mobile manipulator with $m$ wheels and $n$ joints in the robot arm, the end-effector twist $\mathcal{V}_e$ in the end-effector frame $\{e\}$ is written

$$\mathcal{V}_e = J_e(\theta) \begin{bmatrix} u \\ \dot\theta \end{bmatrix} = [J_{\text{base}}(\theta)\ J_{\text{arm}}(\theta)] \begin{bmatrix} u \\ \dot\theta \end{bmatrix}.$$

The $6 \times m$ Jacobian $J_{\text{base}}(\theta)$ maps the wheel velocities $u$ to a velocity at the end-effector, and the $6 \times n$ Jacobian $J_{\text{arm}}(\theta)$ is the body Jacobian derived in Chapter 5. The Jacobian $J_{\text{base}}(\theta)$ is given by

$$J_{\text{base}}(\theta) = [\text{Ad}_{T_{0e}^{-1}(\theta)T_{b0}^{-1}}]F_6$$

where $F_6$ is the transformation from the wheel velocities to the chassis twist, $\mathcal{V}_{b6} = Fu$.

## 13.7    Notes and References

Excellent references on the modeling, motion planning, and control of mobile robots include the books by de Wit et al. (2012) and Laumond (1998), chapters in Siciliano et al. (2009) and in the *Handbook of Robotics* (Morin and Samson, 2008; Samson et al., 2016), and the chapter (Oriolo, 2015) in the *Encyclopedia of Systems and Control*.

General references on nonholonomic systems, underactuated systems, and notions of nonlinear controllability include Bloch (2003), Bullo and Lewis (2004), Choset et al. (2005), Isidori (1995), Jurdjevic (1997), Murray et al. (1994), Nijmeijer and van der Schaft (1990), and Sastry (1999), a chapter in the *Control Handbook* chapter (Lynch et al., 2011), and the chapter (Lynch, 2015) in the

*Encyclopedia of Systems and Control.* Theorem 13.1 is a strengthening of a result originally reported in Brockett (1983a). Theorem 13.7 is an application of Chow's theorem (Chow, 1939) considering different possible control sets. A more general condition describing the conditions under which Chow's theorem can be used to determine local controllability was given by Sussmann (1987).

The original results for the shortest paths for a forward-only car and for a car with a reverse gear were given by Dubins (1957) and Reeds and Shepp (1990), respectively. These results were extended and applied to motion planning problems in Boissonnat et al. (1994) and Souères and Laumond (1996) and independently derived using principles of differential geometry in Sussmann and Tang (1991). The minimum-time motions for a diff-drive were derived in Balkcom and Mason (2002). The motion planner for a car-like mobile robot based on replacing segments of an arbitrary path with the shortest feasible paths for the car, described in Section 13.3.3.2, was given in Laumond et al. (1994).

The nonlinear control law (13.31) for tracking a reference trajectory for a nonholonomic mobile robot was taken from Morin and Samson (2008) and Samson et al. (2016).

## 13.8 Exercises

**Exercise 13.1** In the omnidirectional mobile robot kinematic modeling of Section 13.2.1, we derived the relationship between wheel velocities and chassis velocity in what seemed to be an unusual way. First, we specified the chassis velocity, then we calculated how the wheels must be driving (and sliding). At first glance, this approach does not seem to make sense causally; we should specify the velocities of the wheels, then calculate the chassis velocity. Explain mathematically why this modeling approach makes sense, and under what condition the method cannot be used.

**Exercise 13.2** According to the kinematic modeling of Section 13.2.1, each wheel of an omnidirectional robot adds two more velocity constraints on the chassis twist $\mathcal{V}_b$. This might seem counterintuitive, since more wheels means more motors and we might think that having more motors should result in more motion capability, not more constraints. Explain clearly why having extra wheels implies extra velocity constraints, in our kinematic modeling, and which assumptions in the kinematic modeling may be unrealistic.

**Exercise 13.3** For the three-omniwheel robot of Figure 13.5, is it possible to drive the wheels so that they skid? (In other words, so that the wheels slip in the driving direction.) If so, give an example set of wheel velocities.

**Exercise 13.4** For the four-mecanum-wheel robot of Figure 13.5, is it possible to drive the wheels so that they skid? (In other words, the wheels slip in the driving direction.) If so, give an example set of wheel velocities.

**Exercise 13.5**   Replace the wheels of the four-mecanum-wheel robot of Figure 13.5 by wheels with $\gamma = \pm 60°$. Derive the matrix $H(0)$ in the relationship $u = H(0)\mathcal{V}_b$. Is it rank 3? If necessary, you can assume values for $\ell$ and $w$.

**Exercise 13.6**   Consider the three-omniwheel robot of Figure 13.5. If we replace the omniwheels by mecanum wheels with $\gamma = 45°$, is it still a properly constructed omnidirectional mobile robot? In other words, in the relationship $u = H(0)\mathcal{V}_b$, is $H(0)$ rank 3?

**Exercise 13.7**   Consider a mobile robot with three mecanum wheels for which $\gamma = \pm 45°$ at the points of an equilateral triangle. The chassis frame {b} is at the center of the triangle. The driving directions of all three wheels are the same (e.g., along the body $\hat{x}_b$-axis) and the free sliding directions are $\gamma = 45°$ for two wheels and $\gamma = -45°$ for the other wheel. Is this a properly constructed omnidirectional mobile robot? In other words, in the relationship $u = H(0)\mathcal{V}_b$, is $H(0)$ rank 3?

**Exercise 13.8**   Using your favorite graphics software (e.g., MATLAB), plot the two planes bounding the set of feasible body twists $\mathcal{V}_b$ for wheel 2 of the three-omniwheel robot of Figure 13.5.

**Exercise 13.9**   Using your favorite graphics software (e.g., MATLAB), plot the two planes bounding the set of feasible body twists $\mathcal{V}_b$ for wheel 1 of the four-mecanum-wheel robot of Figure 13.5.

**Exercise 13.10**   Consider a four-omniwheel mobile robot with wheels at the points of a square. The chassis frame {b} is at the center of the square, and the driving direction of each wheel is in the direction 90° counterclockwise from the vector from the the origin of {b} to the wheel. You may assume that the sides of the squares have length 2. Find the matrix $H(0)$. Is it rank 3?

**Exercise 13.11**   Implement a collision-free grid-based planner for an omnidirectional robot. You may assume that the robot has a circular chassis, so for collision-detection purposes, you need to consider only the $(x, y)$ location of the robot. The obstacles are circles with random center locations and random radii. You can use Dijkstra's algorithm or $A^*$ to find a shortest path that avoids obstacles.

**Exercise 13.12**   Implement an RRT planner for an omnidirectional robot. As above, you may assume a circular chassis and circular obstacles.

**Exercise 13.13**   Implement a feedforward plus proportional feedback controller to track a desired trajectory for an omnidirectional mobile robot. Test it on the desired trajectory $(\phi_d(t), x_d(t), y_d(t)) = (t, 0, t)$ for $t \in [0, \pi]$. The initial configuration of the robot is $q(0) = (-\pi/4, 0.5, -0.5)$. Plot the configuration error as a function of time. You can also show an animation of the robot converging to the trajectory.

**Exercise 13.14** Write down the Pfaffian constraints $A(q)\dot{q} = 0$ corresponding to the unicycle model in Equation (13.12).

**Exercise 13.15** Write down the Pfaffian constraints corresponding to the diff-drive model in Equation (13.14).

**Exercise 13.16** Write down the Pfaffian constraints corresponding to the car-like model in Equation (13.16).

**Exercise 13.17** Give examples of two systems that are STLA but not STLC. The systems should not be wheeled mobile robots.

**Exercise 13.18** Continue the Taylor expansion in Equation (13.26) to find the net motion (to order $\epsilon^2$) obtained by following $g_i$ for time $\epsilon$, then $g_j$ for time $\epsilon$, then $-g_i$ for time $\epsilon$, then $-g_j$ for time $\epsilon$. Show that it is equivalent to the expression (13.27).

**Exercise 13.19** Write down the canonical nonholonomic mobile robot model (13.18) in the chassis-fixed form

$$\mathcal{V}_b = B \begin{bmatrix} v \\ \omega \end{bmatrix},$$

where $B$ is a $3 \times 2$ matrix whose columns correspond to the chassis twist associated with the controls $v$ and $\omega$.

**Exercise 13.20** Throughout this book, we have been using the configuration spaces $SE(3)$ and $SO(3)$, and their planar subsets $SE(2)$ and $SO(2)$. These are known as matrix Lie groups. Body and spatial twists are represented in matrix form as elements of $se(3)$ (or $se(2)$ in the plane) and body and spatial angular velocities are represented in matrix form as elements of $so(3)$ (or $so(2)$ in the plane). The spaces $se(3)$ and $so(3)$ correspond to all possible $\dot{T}$ and $\dot{R}$ when $T$ or $R$ is the identity matrix. Since these spaces correspond to all possible velocities, each is called the Lie algebra of its respective matrix Lie group. Let us call $G$ a matrix Lie group and $\mathfrak{g}$ its Lie algebra, and let $X$ be an element of $G$, and $A$ and $B$ be elements of $\mathfrak{g}$. In other words, $A$ and $B$ can be thought of as possible values of $\dot{X}$ when $X = I$.

Any "velocity" $A$ in $\mathfrak{g}$ can be "translated" to a velocity $\dot{X}$ at any $X \in G$ by pre-multiplying or post-multiplying by $X$, i.e., $\dot{X} = XA$ or $\dot{X} = AX$. If we choose $\dot{X} = XA$, i.e., $A = X^{-1}\dot{X}$ then we can think of $A$ as a "body velocity" (e.g., the matrix form of a body twist if $G = SE(3)$) and if we choose $\dot{X} = AX$, i.e., $A = \dot{X}X^{-1}$, we can think of $A$ as a "spatial velocity" (e.g., the matrix form of a spatial twist if $G = SE(3)$). In this way, $A$ can be extended to an entire vector field over $G$. If the extension is obtained by multiplying by $X$ on the left then the vector field is called left-invariant (constant in the body frame), and if the extension is by multiplying by $X$ on the right then the vector field is called right-invariant (constant in the space frame). Velocities that are constant in the body frame, like the vector fields for the canonical nonholonomic mobile robot, correspond to left-invariant vector fields.

Just as we can define a Lie bracket of two vector fields, as in Equation (13.28), we can define the Lie bracket of $A, B \in \mathfrak{g}$ as

$$[A, B] = AB - BA, \tag{13.38}$$

as described in Section 8.2.2 for $\mathfrak{g} = se(3)$. Confirm that this formula describes the same Lie bracket vector field as Equation (13.28) for the canonical non-holonomic vector fields $g_1(q) = (0, \cos \phi, \sin \phi)$ and $g_2(q) = (1, 0, 0)$. To do this, first express the two vector fields as $A_1, A_2 \in se(2)$, considered to be the generators of the left-invariant vector fields $g_1(q)$ and $g_2(q)$ (since the vector fields correspond to constant velocities in the chassis frame). Then take the Lie bracket $A_3 = [A_1, A_2]$ and extend $A_3$ to a left-invariant vector field defined at all $X \in SE(2)$. Show that this is the same result obtained using the formula (13.28).

The Lie bracket formula in Equation (13.28) is general for any vector fields expressed as a function of coordinates $q$, while the formula (13.38) is particularly for left- and right-invariant vector fields defined by elements of the Lie algebra of a matrix Lie group.

**Exercise 13.21**   Using your favorite symbolic math software (e.g., Mathematica), write or experiment with software that symbolically calculates the Lie bracket of two vector fields. Show that it correctly calculates Lie brackets for vector fields of any dimension.

**Exercise 13.22**   For the full five-dimensional diff-drive model described by Equation (13.14), calculate Lie products that generate two motion directions not present in the original two vector fields. Write down the holonomic constraint corresponding to the direction in which it is not possible to generate motions.

**Exercise 13.23**   Implement a collision-free grid-based motion planner for a car-like robot among obstacles using techniques from Section 10.4.2. Decide how to specify the obstacles.

**Exercise 13.24**   Implement a collision-free RRT-based motion planner for a car-like robot among obstacles using techniques from Section 10.5. Decide how to specify obstacles.

**Exercise 13.25**   Implement the reference point trajectory-tracking control law (13.29) for a diff-drive robot. Show in a simulation that it succeeds in tracking a desired trajectory for the point.

**Exercise 13.26**   Implement the nonlinear feedforward plus feedback control law (13.31). Demonstrate its performance in tracking a reference trajectory with different sets of control gains, including one set that yields "good" performance.

**Exercise 13.27**   Write a program that accepts a time history of wheel encoder values for the two rear wheels of a car and estimates the chassis configuration as a function of time using odometry. Prove that it yields correct results for a chassis motion that involves rotations and translations.

**Exercise 13.28**   Write a program that accepts a time history of wheel encoder values for the three wheels of a three-omniwheel robot and estimates the chassis configuration as a function of time using odometry. Prove that it yields correct results for a chassis motion that involves rotations and translations.

**Exercise 13.29**   Write a program that accepts a time history of wheel encoder values for the four wheels of a four-mecanum-wheel robot and estimates the chassis configuration as a function of time using odometry. Prove that it yields correct results for a chassis motion that involves rotations and translations.

**Exercise 13.30**   Consider the mobile manipulator in Figure 13.24. Write down the Jacobian $J_e(\theta)$ as a $3 \times 3$ matrix function of $d$, $x_r$, $L_1$, and $\theta_1$. Is the Jacobian rank 3 for all choices of $d$, $x_r$, $L_1$, and $\theta_1$? If not, under what conditions is it not full rank?

**Exercise 13.31**   Write a simulation for a mobile manipulation controller similar to that demonstrated in Figure 13.24. For this simulation you need to include a simulation of odometry to keep track of the mobile base's configuration. Demonstrate your controller on the same example trajectory and initial conditions shown in Figure 13.24 for good and bad choices of control gains.

**Exercise 13.32**   Wheel-based odometry can be supplemented by odometry based on an inertial measurement unit (IMU). A typical IMU includes a three-axis gyro, for sensing angular velocities of the chassis, and a three-axis accelerometer, for sensing linear accelerations of the chassis. From a known initial state of the mobile robot (e.g., at rest at a known position), the sensor data from the IMU can be integrated over time to yield a position estimate of the robot. Since the data are numerically integrated once in the case of angular velocities, and twice in the case of linear accelerations, the estimate will drift from the actual value over time, just as the wheel-based odometry estimate will.

In a paragraph describe operating conditions for the mobile robot, including properties of the wheels' interactions with the ground, where IMU-based odometry might be expected to yield better configuration estimates, and conditions where wheel-based odometry might be expected to yield better estimates. In another paragraph describe how the two methods can be used simultaneously, to improve the performance beyond that of either method alone. You should feel free to do an internet search and comment on specific data-fusion tools or filtering techniques that might be useful.

**Exercise 13.33**   The KUKA youBot (Figure 13.25) is a mobile manipulator consisting of a 5R arm mounted on an omnidirectional mobile base with four mecanum wheels. The chassis frame {b} is centered between the four wheels at a

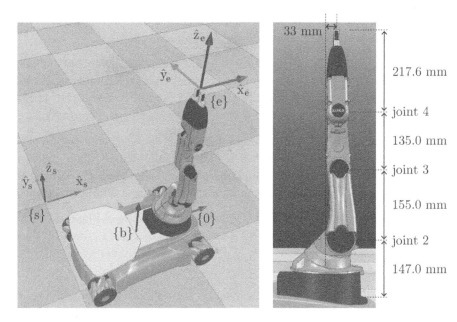

**Figure 13.25** (Left) The KUKA youBot mobile manipulator and the fixed space frame {s}, the chassis frame {b}, the arm base frame {0}, and the end-effector frame {e}. The arm is at its zero configuration. (Right) A close-up of the arm at its zero configuration. Joint axes 1 and 5 (not shown) point upward and joint axes 2, 3, and 4 are out of the page.

height $z = 0.0963$ m above the floor, and the configuration of the chassis relative to a fixed space frame {s} is

$$T_{sb}(q) = \begin{bmatrix} \cos\phi & -\sin\phi & 0 & x \\ \sin\phi & \cos\phi & 0 & y \\ 0 & 0 & 1 & 0.0963 \\ 0 & 0 & 0 & 1 \end{bmatrix},$$

where $q = (\phi, x, y)$. The kinematics of the four-wheeled mobile base is described in Figure 13.5 and the surrounding text, where the front–back distance between the wheels is $2\ell = 0.47$ m, the side-to-side distance between the wheels is $2w = 0.3$ m, and the radius of each wheel is $r = 0.0475$ m.

The fixed offset from the chassis frame {b} to the base frame of the arm {0} is

$$T_{b0} = \begin{bmatrix} 1 & 0 & 0 & 0.1662 \\ 0 & 1 & 0 & 0 \\ 0 & 0 & 1 & 0.0026 \\ 0 & 0 & 0 & 1 \end{bmatrix},$$

i.e., the arm base frame {0} is aligned with the chassis frame {b} and is displaced by 166.2 mm in $\hat{x}_b$ and 2.6 mm in $\hat{z}_b$. The end-effector frame {e} at the zero configuration of the arm (as shown in Figure 13.25) relative to the base frame $\{0\}$ is

$$M_{0e} = \begin{bmatrix} 1 & 0 & 0 & 0.0330 \\ 0 & 1 & 0 & 0 \\ 0 & 0 & 1 & 0.6546 \\ 0 & 0 & 0 & 1 \end{bmatrix}.$$

You can ignore the arm joint limits in this exercise.

(a) Examining the right-hand side of Figure 13.25 – and keeping in mind that (i) joint axes 1 and 5 point up on the page and joint axes 2, 3, and 4 point out of the page, and (ii) positive rotation about an axis is by the right-hand rule – either confirm that the screw axes in the end-effector frame $\mathcal{B}_i$ are as shown in the following table:

| $i$ | $\omega_i$ | $v_i$ |
|---|---|---|
| 1 | $(0,0,1)$ | $(0,0.0330,0)$ |
| 2 | $(0,-1,0)$ | $(-0.5076,0,0)$ |
| 3 | $(0,-1,0)$ | $(-0.3526,0,0)$ |
| 4 | $(0,-1,0)$ | $(-0.2176,0,0)$ |
| 5 | $(0,0,1)$ | $(0,0,0)$ |

or provide the correct $\mathcal{B}_i$.

(b) The robot arm has only five joints, so it is incapable of generating an arbitrary end-effector twist $\mathcal{V}_e \in \mathbb{R}^6$ when the mobile base is parked. If we are able to move the mobile base and the arm joints simultaneously, are there configurations $\theta$ of the arm at which arbitrary twists are not possible? If so, indicate these configurations. Also explain why the configuration $q = (\phi, x, y)$ of the mobile base is irrelevant to this question.

(c) Use numerical inverse kinematics to find a chassis and arm configuration $(q, \theta)$ that places the end-effector at

$$X(q, \theta) = \begin{bmatrix} 1 & 0 & 0 & 0 \\ 0 & 0 & 1 & 1.0 \\ 0 & -1 & 0 & 0.4 \\ 0 & 0 & 0 & 1 \end{bmatrix}.$$

You can try $q_0 = (\phi_0, x_0, y_0) = (0,0,0)$ and $\theta_0 = (0,0,-\pi/2,0,0)$ as your initial guess.

(d) You will write a robot simulator to test the kinematic task-space feedforward plus feedback control law (13.37) tracking the end-effector trajectory defined by the path

$$X_d(s) = \begin{bmatrix} \sin(s\pi/2) & 0 & \cos(s\pi/2) & s \\ 0 & 1 & 0 & 0 \\ -\cos(s\pi/2) & 0 & \sin(s\pi/2) & 0.491 \\ 0 & 0 & 0 & 1 \end{bmatrix}, \quad s \in [0,1],$$

and the time scaling

$$s(t) = \frac{3}{25}t^2 - \frac{2}{125}t^3, \qquad t \in [0,5].$$

In other words, the total time of the motion is 5 s.

Your program should take the trajectory and initial configuration of the robot as input, in addition to the control gains and any other parameters you see fit. In this exercise, due to initial error, the initial configuration of the robot is not on the path: $q_0 = (\phi_0, x_0, y_0) = (-\pi/8, -0.5, 0.5)$ and $\theta_0 = (0, -\pi/4, \pi/4, -\pi/2, 0)$.

Your main program loop should run 100 times per simulated second, i.e., each time step is $\Delta t = 0.01$ s, for a total of 500 time steps for the simulation. Each time through the loop, your program should:

- Calculate the desired configuration $X_d$ and twist $\mathcal{V}_d$ at the current time.
- Calculate the current configuration error $X_{\mathrm{err}} = (\omega_{\mathrm{err}}, v_{\mathrm{err}})$ and save $X_{\mathrm{err}}$ in an array for later plotting.
- Evaluate the control law (13.37) to find the commanded wheel speeds and joint velocities $(u, \dot{\theta})$.
- Step the simulation of the robot's motion forward in time by $\Delta t$ to find the new configuration of the robot. You may use a simple first-order Euler integration for the arm: the new joint angle is just the old angle plus the commanded joint velocity multiplied by $\Delta t$. To calculate the new configuration of the chassis, use odometry from Section 13.4, remembering that the change in wheel angles during one simulation step is $u\Delta t$.

You are encouraged to test the feedforward portion of your controller first. A starting configuration on the path is approximately $q_0 = (\phi_0, x_0, y_0) = (0, -0.526, 0)$ and $\theta_0 = (0, -\pi/4, \pi/4, -\pi/2, 0)$. Once you have confirmed that your feedforward controller works as expected, add a nonzero proportional gain to get good performance from the configuration with initial error. Finally, you can add a nonzero integral gain to see transient effects such as overshoot.

After the simulation completes, plot the six components of $X_{\mathrm{err}}$ as a function of time. If possible, choose gains $K_p$ and $K_i$ such that it is possible to see typical features of a PI velocity controller: a little bit of overshoot and oscillation and eventually nearly zero error. You should choose control gains such that the 2% settling time is one or two seconds, so the transient response is clearly visible. If you have the visualization tools available, create a movie of the robot's motion corresponding to your plots.

(e) While retaining stability, choose a set of different control gains that gives a visibly different behavior of the robot. Provide the plots and movie and comment on why the different behavior agrees, or does not agree, with what you know about PI velocity control.

**Exercise 13.34** One type of wheeled mobile robot, not considered in this chapter, has three or more conventional wheels which are all individually steerable. Steerable conventional wheels allow the robot chassis to follow arbitrary paths without relying on the passive sideways rolling of mecanum wheels or omni-wheels.

In this exercise you will model a mobile robot with four steerable wheels. Assume that each wheel has two actuators, one to steer it and one to drive it. The wheel locations relative to the chassis frame {b} mimic the case of the four-wheeled robot in Figure 13.5: they are located at the four points $(\pm\ell, \pm w)$ in {b}. The steering angle $\theta_i$ of wheel $i$ is zero when it rolls in the $+\hat{x}_b$-direction for a positive driving speed $u_i > 0$, and a positive rotation of the steering angle is defined as counterclockwise on the page. The linear speed at wheel $i$ is $ru_i$, where $r$ is the radius of the wheel.

(a) Given a desired chassis twist $\mathcal{V}_b$, derive equations for the four wheel steering angles $\theta_i$ and the four wheel driving speeds $u_i$. (Note that the pair $(\theta_i, u_i)$ yields the same linear motion at wheel $i$ as $(-\theta_i, -u_i)$.)

(b) The "controls" for wheel $i$ are the steering angle $\theta_i$ and the driving speed $u_i$. In practice, however, there are bounds on how quickly the wheel steering angle can be changed. Comment on the implications for the modeling, path planning, and control of a steerable-wheel mobile robot.

# Appendix A Summary of Useful Formulas

## Chapter 2

- dof = (sum of freedoms of bodies) − (number of independent configuration constraints)
- Grübler's formula is an expression of the above formula for mechanisms with $N$ links (including ground) and $J$ joints, where joint $i$ has $f_i$ degrees of freedom and $m = 3$ for planar mechanisms or $m = 6$ for spatial mechanisms:

$$\text{dof} = m(N - 1 - J) + \sum_{i=1}^{J} f_i.$$

- Pfaffian velocity constraints take the form $A(\theta)\dot{\theta} = 0$.

## Chapter 3

| Rotations | Rigid-Body Motions |
|---|---|
| $R \in SO(3) : 3 \times 3$ matrices | $T \in SE(3) : 4 \times 4$ matrices |
| $R^{\mathrm{T}}R = I,\ \det R = 1.$ | $T = \begin{bmatrix} R & p \\ 0 & 1 \end{bmatrix},$ where $R \in SO(3), p \in \mathbb{R}^3.$ |
| $R^{-1} = R^{\mathrm{T}}.$ | $T^{-1} = \begin{bmatrix} R^{\mathrm{T}} & -R^{\mathrm{T}}p \\ 0 & 1 \end{bmatrix}.$ |
| Change of coordinate frame: $R_{ab}R_{bc} = R_{ac},\ \ R_{ab}p_b = p_a.$ | Change of coordinate frame: $T_{ab}T_{bc} = T_{ac},\ \ T_{ab}p_b = p_a.$ |
| Rotating a frame {b}: $R = \text{Rot}(\hat{\omega}, \theta).$ $R_{sb'} = RR_{sb}$: rotate by $\theta$ about $\hat{\omega}_s = \hat{\omega}$ $R_{sb''} = R_{sb}R$: rotate by $\theta$ about $\hat{\omega}_b = \hat{\omega}.$ | Displacing a frame {b}: $T = \begin{bmatrix} \text{Rot}(\hat{\omega}, \theta) & p \\ 0 & 1 \end{bmatrix}.$ $T_{sb'} = TT_{sb}$: rotate by $\theta$ about $\hat{\omega}_s = \hat{\omega}$ (moves {b} origin), translate $p$ in {s}. $T_{sb''} = T_{sb}T$: translate $p$ in {b}, rotate by $\theta$ about $\hat{\omega}$ in new body frame. |
| continued... | |

| Rotations | Rigid-Body Motions |
|---|---|
| Unit rotation axis is $\hat{\omega} \in \mathbb{R}^3$, where $\|\hat{\omega}\| = 1$. | "Unit" screw axis is $\mathcal{S} = \begin{bmatrix} \omega \\ v \end{bmatrix} \in \mathbb{R}^6$, where either (i) $\|\omega\| = 1$ or (ii) $\omega = 0$ and $\|v\| = 1$.<br><br>For a screw axis $\{q, \hat{s}, h\}$ with finite $h$, $\mathcal{S} = \begin{bmatrix} \omega \\ v \end{bmatrix} = \begin{bmatrix} \hat{s} \\ -\hat{s} \times q + h\hat{s} \end{bmatrix}.$ |
| Angular velocity is $\omega = \hat{\omega}\dot{\theta}$. | Twist is $\mathcal{V} = \mathcal{S}\dot{\theta}$. |
| For any 3-vector, e.g., $\omega \in \mathbb{R}^3$,<br><br>$[\omega] = \begin{bmatrix} 0 & -\omega_3 & \omega_2 \\ \omega_3 & 0 & -\omega_1 \\ -\omega_2 & \omega_1 & 0 \end{bmatrix} \in so(3)$<br><br>Identities for $\omega, x \in \mathbb{R}^3, R \in SO(3)$:<br>$[\omega] = -[\omega]^{\mathrm{T}}, \quad [\omega]x = -[x]\omega,$<br>$[\omega][x] = ([x][\omega])^{\mathrm{T}}, \quad R[\omega]R^{\mathrm{T}} = [R\omega].$ | For $\mathcal{V} = \begin{bmatrix} \omega \\ v \end{bmatrix} \in \mathbb{R}^6$,<br><br>$[\mathcal{V}] = \begin{bmatrix} [\omega] & v \\ 0 & 0 \end{bmatrix} \in se(3)$<br><br>(the pair $(\omega, v)$ can be a twist $\mathcal{V}$ or a "unit" screw axis $\mathcal{S}$, depending on the context). |
| $\dot{R}R^{-1} = [\omega_s], \quad R^{-1}\dot{R} = [\omega_b].$ | $\dot{T}T^{-1} = [\mathcal{V}_s], \quad T^{-1}\dot{T} = [\mathcal{V}_b]$ .<br>$[\mathrm{Ad}_T] = \begin{bmatrix} R & 0 \\ [p]R & R \end{bmatrix} \in \mathbb{R}^{6\times6}$<br>Identities: $[\mathrm{Ad}_T]^{-1} = [\mathrm{Ad}_{T^{-1}}],$<br>$[\mathrm{Ad}_{T_1}][\mathrm{Ad}_{T_2}] = [\mathrm{Ad}_{T_1 T_2}].$ |
| Change of coordinate frame:<br>$\hat{\omega}_a = R_{ab}\hat{\omega}_b, \quad \omega_a = R_{ab}\omega_b.$ | Change of coordinate frame:<br>$\mathcal{S}_a = [\mathrm{Ad}_{T_{ab}}]\mathcal{S}_b, \quad \mathcal{V}_a = [\mathrm{Ad}_{T_{ab}}]\mathcal{V}_b.$ |
| Exp. coords. for $R \in SO(3)$: $\hat{\omega}\theta \in \mathbb{R}^3$:<br>$\exp : [\hat{\omega}]\theta \in so(3) \rightarrow R \in SO(3),$<br>$R = \mathrm{Rot}(\hat{\omega}, \theta) = e^{[\hat{\omega}]\theta}$<br>$= I + \sin\theta[\hat{\omega}] + (1 - \cos\theta)[\hat{\omega}]^2$<br><br>$\log : R \in SO(3) \rightarrow [\hat{\omega}]\theta \in so(3).$<br>Algorithm in §3.2.3.3 | Exp. coords. for $T \in SE(3)$: $\mathcal{S}\theta \in \mathbb{R}^6$:<br>$\exp : [\mathcal{S}]\theta \in se(3) \rightarrow T \in SE(3),$<br>$T = e^{[\mathcal{S}]\theta} = \begin{bmatrix} e^{[\omega]\theta} & * \\ 0 & 1 \end{bmatrix},$<br>where $* =$ is given by<br>$(I\theta + (1 - \cos\theta)[\omega] + (\theta - \sin\theta)[\omega]^2)v.$<br>$\log : T \in SE(3) \rightarrow [\mathcal{S}]\theta \in se(3).$<br>Algorithm in §3.3.3.2 |
| Moment change of coord. frame:<br>$m_a = R_{ab}m_b.$ | Wrench change of coord. frame:<br>$\mathcal{F}_a = (m_a, f_a) = [\mathrm{Ad}_{T_{ba}}]^{\mathrm{T}}\mathcal{F}_b.$ |

## Chapter 4

- The product of exponentials formula for a serial chain manipulator is

$$\text{for a space frame,} \quad T = e^{[\mathcal{S}_1]\theta_1} \cdots e^{[\mathcal{S}_n]\theta_n} M,$$

$$\text{for a body frame,} \quad T = M e^{[\mathcal{B}_1]\theta_1} \cdots e^{[\mathcal{B}_n]\theta_n},$$

where $M$ is the frame of the end-effector in the space frame when the manipulator is at its home position, $\mathcal{S}_i$ is the spatial twist when joint $i$ rotates (or translates) at unit speed while all other joints are at their zero position, and $\mathcal{B}_i$ is the body twist of the end-effector frame when joint $i$ moves at unit speed and all other joints are at their zero positions.

## Chapter 5

- For a manipulator end-effector configuration written in coordinates $x$, the forward kinematics is $x = f(\theta)$, and the differential kinematics is given by $\dot{x} = (\partial f/\partial\theta)\dot{\theta} = J(\theta)\dot{\theta}$, where $J(\theta)$ is the manipulator Jacobian.
- Written using twists, the relation is $\mathcal{V}_* = J_*(\theta)\dot{\theta}$, where $*$ is either $s$ (for a space Jacobian) or $b$ (for abody Jacobian). The columns $J_{si}$, $i = 2, \ldots, n$, of the space Jacobian are

$$J_{si}(\theta) = [\mathrm{Ad}_{e^{[\mathcal{S}_1]\theta_1} \cdots e^{[\mathcal{S}_{i-1}]\theta_{i-1}}}]\mathcal{S}_i,$$

  with $J_{s1} = \mathcal{S}_1$, and the columns $J_{bi}$, $i = 1, \ldots, n-1$, of the body Jacobian are

$$J_{bi}(\theta) = [\mathrm{Ad}_{e^{-[\mathcal{B}_n]\theta_n} \cdots e^{-[\mathcal{B}_{i+1}]\theta_{i+1}}}]\mathcal{B}_i,$$

  with $J_{bn} = \mathcal{B}_n$. The spatial twist caused by joint-$i$ motion is altered only by the configurations of joints inboard from joint $i$ (i.e., between the joint and the space frame), while the body twist caused by joint $i$ is altered only by the configurations of joints outboard from joint $i$ (i.e., between the joint and the body frame).

  The two Jacobians are related by

$$J_b(\theta) = [\mathrm{Ad}_{T_{bs}(\theta)}]J_s(\theta), \qquad J_s(\theta) = [\mathrm{Ad}_{T_{sb}(\theta)}]J_b(\theta).$$

- The generalized forces $\tau$ at the joints are related to the wrenches expressed in the space frame or end-effector body frame by

$$\tau = J_*^{\mathrm{T}}(\theta)\mathcal{F}_*,$$

  where $*$ is $s$ (space frame) or $b$ (body frame).
- The manipulability ellipsoid is defined by

$$\mathcal{V}^{\mathrm{T}}(JJ^{\mathrm{T}})^{-1}\mathcal{V} = 1,$$

  where $\mathcal{V}$ may be a set of task-space coordinate velocities $\dot{q}$, a spatial or body twist, or the angular or linear components of a twist, and $J$ is the appropriate Jacobian satisfying $\mathcal{V} = J(\theta)\dot{\theta}$. The principal axes of the manipulability ellipsoid are aligned with the eigenvectors of $JJ^{\mathrm{T}}$, and the semi-axis lengths are the square roots of the corresponding eigenvalues.
- The force ellipsoid is defined by

$$\mathcal{F}^{\mathrm{T}}JJ^{\mathrm{T}}\mathcal{F} = 1,$$

  where $J$ is a Jacobian (possibly given in terms of a minimum set of task-space coordinates or in terms of the spatial or body wrench) and $\mathcal{F}$ is an end-effector force or wrench satisfying $\tau = J^{\mathrm{T}}\mathcal{F}$. The principal axes of the manipulability ellipsoid are aligned with the eigenvectors of $(JJ^{\mathrm{T}})^{-1}$, and the semi-axis lengths are the square roots of the corresponding eigenvalues.

## Chapter 6

- The law of cosines states that $c^2 = a^2 + b^2 - 2ab\cos\gamma$, where $a$, $b$, and $c$ are the lengths of the sides of a triangle and $\gamma$ is the interior angle opposite side $c$. This formula is often useful to solve inverse kinematics problems.
- Numerical methods are used to solve the inverse kinematics for systems for which closed-form solutions do not exist. A Newton–Raphson method using the Jacobian pseudoinverse $J^\dagger(\theta)$ is outlined below.
  - (a) **Initialization**: Given $T_{sd}$ and an initial guess $\theta^0 \in \mathbb{R}^n$. Set $i = 0$.
  - (b) Set $[\mathcal{V}_b] = \log\left(T_{sb}^{-1}(\theta^i)T_{sd}\right)$. While $\|\omega_b\| > \epsilon_\omega$ or $\|v_b\| > \epsilon_v$ for small $\epsilon_\omega, \epsilon_v$:
    - Set $\theta^{i+1} = \theta^i + J_b^\dagger(\theta^i)\mathcal{V}_b$.
    - Increment $i$.

  If $J$ is square and full rank, then $J^\dagger = J^{-1}$. If $J \in \mathbb{R}^{m \times n}$ is full rank (rank $m$ for $n > m$ or rank $n$ for $n < m$), that is, the robot is not at a singularity, the pseudoinverse can be calculated as follows:

  $$J^\dagger = J^{\mathrm{T}}(JJ^{\mathrm{T}})^{-1} \quad \text{if } n > m \quad \text{(called a right inverse since } JJ^\dagger = I\text{)}$$
  $$J^\dagger = (J^{\mathrm{T}}J)^{-1}J^{\mathrm{T}} \quad \text{if } n < m \quad \text{(called a left inverse since } J^\dagger J = I\text{)}.$$

## Chapter 8

- The Lagrangian is the kinetic minus the potential energy: $\mathcal{L}(\theta, \dot{\theta}) = \mathcal{K}(\theta, \dot{\theta}) - \mathcal{P}(\theta)$.
- The Euler–Lagrange equations are

  $$\tau = \frac{d}{dt}\frac{\partial \mathcal{L}}{\partial \dot{\theta}} - \frac{\partial \mathcal{L}}{\partial \theta}.$$

- The equations of motion of a robot can be written in the following equivalent forms:

  $$\begin{aligned}
  \tau &= M(\theta)\ddot{\theta} + h(\theta, \dot{\theta}) \\
  &= M(\theta)\ddot{\theta} + c(\theta, \dot{\theta}) + g(\theta) \\
  &= M(\theta)\ddot{\theta} + \dot{\theta}^{\mathrm{T}}\Gamma(\theta)\dot{\theta} + g(\theta) \\
  &= M(\theta)\ddot{\theta} + C(\theta, \dot{\theta})\dot{\theta} + g(\theta),
  \end{aligned}$$

  where $M(\theta)$ is the $n \times n$ symmetric positive-definite mass matrix, $h(\theta, \dot{\theta})$ is the sum of the generalized forces due to gravity and to quadratic velocity terms, $c(\theta, \dot{\theta})$ are quadratic velocity forces, $g(\theta)$ are gravitational forces, $\Gamma(\theta)$ is an $n \times n \times n$ matrix of Christoffel symbols of the first kind obtained from partial derivatives of $M(\theta)$ with respect to $\theta$, and $C(\theta, \dot{\theta})$ is the $n \times n$ Coriolis matrix whose $(i, j)$th entry is given by

  $$c_{ij}(\theta, \dot{\theta}) = \sum_{k=1}^{n} \Gamma_{ijk}(\theta)\dot{\theta}_k.$$

  If the end-effector of the robot is applying a wrench $\mathcal{F}_{\text{tip}}$ to the environment, the term $J^{\mathrm{T}}(\theta)\mathcal{F}_{\mathrm{tip}}$ should be added to the right-hand side of the robot's dynamic equations.

- The symmetric positive-definite rotational inertia matrix of a rigid body is

$$\mathcal{I}_b = \begin{bmatrix} \mathcal{I}_{xx} & \mathcal{I}_{xy} & \mathcal{I}_{xz} \\ \mathcal{I}_{xy} & \mathcal{I}_{yy} & \mathcal{I}_{yz} \\ \mathcal{I}_{xz} & \mathcal{I}_{yz} & \mathcal{I}_{zz} \end{bmatrix},$$

where

$$\begin{aligned}
\mathcal{I}_{xx} &= \int_{\mathcal{B}}(y^2 + z^2)\rho(x,y,z)dV, & \mathcal{I}_{yy} &= \int_{\mathcal{B}}(x^2 + z^2)\rho(x,y,z)dV, \\
\mathcal{I}_{zz} &= \int_{\mathcal{B}}(x^2 + y^2)\rho(x,y,z)dV, & \mathcal{I}_{xy} &= -\int_{\mathcal{B}}xy\rho(x,y,z)dV, \\
\mathcal{I}_{xz} &= -\int_{\mathcal{B}}xz\rho(x,y,z)dV, & \mathcal{I}_{yz} &= -\int_{\mathcal{B}}yz\rho(x,y,z)dV,
\end{aligned}$$

$\mathcal{B}$ is the body, $dV$ is a differential volume element, and $\rho(x,y,z)$ is the density function.

- If $\mathcal{I}_b$ is defined in a frame {b} at the center of mass then $\mathcal{I}_q$, the inertia in a frame {q} aligned with {b}, but displaced from the origin of {b} by $q \in \mathbb{R}^3$ in {b} coordinates, is

$$\mathcal{I}_q = \mathcal{I}_b + \mathfrak{m}(q^{\mathrm{T}}qI - qq^{\mathrm{T}})$$

by Steiner's theorem.

- The spatial inertia matrix $\mathcal{G}_b$ expressed in a frame {b} at the center of mass is defined as the $6 \times 6$ matrix

$$\mathcal{G}_b = \begin{bmatrix} \mathcal{I}_b & 0 \\ 0 & \mathfrak{m}I \end{bmatrix}.$$

In a frame {a} at a configuration $T_{ba}$ relative to {b}, the spatial inertia matrix is

$$\mathcal{G}_a = [\mathrm{Ad}_{T_{ba}}]^{\mathrm{T}}\mathcal{G}_b[\mathrm{Ad}_{T_{ba}}].$$

- The Lie bracket of two twists $\mathcal{V}_1$ and $\mathcal{V}_2$ is

$$\mathrm{ad}_{\mathcal{V}_1}(\mathcal{V}_2) = [\mathrm{ad}_{\mathcal{V}_1}]\mathcal{V}_2,$$

where

$$[\mathrm{ad}_{\mathcal{V}}] = \begin{bmatrix} [\omega] & 0 \\ [v] & [\omega] \end{bmatrix} \in \mathbb{R}^{6\times 6}.$$

- The twist–wrench formulation of the rigid-body dynamics of a single rigid body is

$$\mathcal{F}_b = \mathcal{G}_b\dot{\mathcal{V}}_b - [\mathrm{ad}_{\mathcal{V}_b}]^{\mathrm{T}}\mathcal{G}_b\mathcal{V}_b.$$

The equations have the same form if $\mathcal{F}$, $\mathcal{V}$, and $\mathcal{G}$ are each expressed in the same frame, regardless of the frame.

- The kinetic energy of a rigid body is $\frac{1}{2}\mathcal{V}_b^{\mathrm{T}}\mathcal{G}_b\mathcal{V}_b$, and the kinetic energy of an open-chain robot is $\frac{1}{2}\dot{\theta}^{\mathrm{T}}M(\theta)\dot{\theta}$.

- The forward–backward Newton–Euler inverse dynamics algorithm is the following:

  **Initialization:** Attach a frame $\{0\}$ to the base, frames $\{1\}$ to $\{n\}$ to the centers of mass of links $\{1\}$ to $\{n\}$, and a frame $\{n+1\}$ at the end-effector, fixed in the frame $\{n\}$. Define $M_{i,i-1}$ to be the configuration of $\{i-1\}$ in $\{i\}$ when $\theta_i = 0$. Let $\mathcal{A}_i$ be the screw axis of joint $i$ expressed in $\{i\}$, and $\mathcal{G}_i$ be the $6 \times 6$ spatial inertia matrix of link $i$. Define $\mathcal{V}_0$ to be the twist of the base frame $\{0\}$ expressed in base-frame coordinates. (This quantity is typically zero.) Let $\mathfrak{g} \in \mathbb{R}^3$ be the gravity vector expressed in base-frame-$\{0\}$ coordinates, and define $\dot{\mathcal{V}}_0 = (0, -\mathfrak{g})$. (Gravity is treated as an acceleration of the base in the opposite direction.) Define $\mathcal{F}_{n+1} = \mathcal{F}_{\text{tip}} = (m_{\text{tip}}, f_{\text{tip}})$ to be the wrench applied to the environment by the end-effector expressed in the end-effector frame $\{n+1\}$.

  **Forward iterations:** Given $\theta, \dot{\theta}, \ddot{\theta}$, for $i = 1$ to $n$ do

  $$T_{i,i-1} = e^{-[\mathcal{A}_i]\theta_i} M_{i,i-1},$$
  $$\mathcal{V}_i = \mathrm{Ad}_{T_{i,i-1}}(\mathcal{V}_{i-1}) + \mathcal{A}_i \dot{\theta}_i,$$
  $$\dot{\mathcal{V}}_i = \mathrm{Ad}_{T_{i,i-1}}(\dot{\mathcal{V}}_{i-1}) + \mathrm{ad}_{\mathcal{V}_i}(\mathcal{A}_i)\dot{\theta}_i + \mathcal{A}_i \ddot{\theta}_i.$$

  **Backward iterations:** For $i = n$ to $1$ do

  $$\mathcal{F}_i = \mathrm{Ad}_{T_{i+1,i}}^{\mathrm{T}}(\mathcal{F}_{i+1}) + \mathcal{G}_i \dot{\mathcal{V}}_i - \mathrm{ad}_{\mathcal{V}_i}^{\mathrm{T}}(\mathcal{G}_i \mathcal{V}_i),$$
  $$\tau_i = \mathcal{F}_i^{\mathrm{T}} \mathcal{A}_i.$$

- Let $J_{ib}(\theta)$ be the Jacobian relating $\dot{\theta}$ to the body twist $\mathcal{V}_i$ in link $i$'s center-of-mass frame $\{i\}$. Then the mass matrix $M(\theta)$ of the manipulator can be expressed as

  $$M(\theta) = \sum_{i=1}^{n} J_{ib}^{\mathrm{T}}(\theta) \mathcal{G}_i J_{ib}(\theta).$$

- The robot's dynamics $M(\theta)\ddot{\theta} + h(\theta, \dot{\theta})$ can be expressed in the task space as

  $$\mathcal{F} = \Lambda(\theta)\dot{\mathcal{V}} + \eta(\theta, \mathcal{V}),$$

  where $\mathcal{F}$ is the wrench applied to the end-effector, $\mathcal{V}$ is the twist of the end-effector, and $\mathcal{F}$, $\mathcal{V}$, and the Jacobian $J(\theta)$ are all defined in the same frame. The task-space mass matrix $\Lambda(\theta)$ and gravity and quadratic velocity forces $\eta(\theta, \mathcal{V})$ are

  $$\Lambda(\theta) = J^{-\mathrm{T}} M(\theta) J^{-1},$$
  $$\eta(\theta, \mathcal{V}) = J^{-\mathrm{T}} h(\theta, J^{-1}\mathcal{V}) - \Lambda(\theta) \dot{J} J^{-1} \mathcal{V},$$

  where $J^{-\mathrm{T}} = (J^{-1})^{\mathrm{T}}$.

- Define two $n \times n$ projection matrices of rank $n - k$

  $$P(\theta) = I - A^{\mathrm{T}}(AM^{-1}A^{\mathrm{T}})^{-1}AM^{-1}$$
  $$P_{\ddot{\theta}}(\theta) = M^{-1}PM = I - M^{-1}A^{\mathrm{T}}(AM^{-1}A^{\mathrm{T}})^{-1}A$$

corresponding to the $k$ Pfaffian constraints acting on the robot, $A(\theta)\dot{\theta} = 0$, $A \in \mathbb{R}^{k \times n}$. Then the $n + k$ constrained equations of motion

$$\tau = M(\theta)\ddot{\theta} + h(\theta, \dot{\theta}) + A^{\mathrm{T}}(\theta)\lambda,$$
$$A(\theta)\dot{\theta} = 0$$

can be reduced to the following equivalent forms by eliminating the Lagrange multipliers $\lambda$:

$$P\tau = P(M\ddot{\theta} + h),$$
$$P_{\ddot{\theta}}\ddot{\theta} = P_{\ddot{\theta}}M^{-1}(\tau - h).$$

The matrix $P$ projects away joint force–torque components that act on the constraints without doing work on the robot, and the matrix $P_{\ddot{\theta}}$ projects away acceleration components that do not satisfy the constraints.

## Chapter 9

- A straight-line path in joint space is given by $\theta(s) = \theta_{\mathrm{start}} + s(\theta_{\mathrm{end}} - \theta_{\mathrm{start}})$ as $s$ goes from 0 to 1.
- A constant-screw-axis motion of the end-effector from $X_{\mathrm{start}} \in SE(3)$ to $X_{\mathrm{end}}$ is $X(s) = X_{\mathrm{start}} \exp(\log(X_{\mathrm{start}}^{-1}X_{\mathrm{end}})s)$ as $s$ goes from 0 to 1.
- The path-constrained dynamics of a robot can be written

$$m(s)\ddot{s} + c(s)\dot{s}^2 + g(s) = \tau \in \mathbb{R}^n$$

as $s$ goes from 0 to 1.

## Chapter 13

- The Lie bracket of two vector fields $g_1$ and $g_2$ is the vector field

$$[g_1, g_2] = \left( \frac{\partial g_2}{\partial q} g_1 - \frac{\partial g_1}{\partial q} g_2 \right).$$

# Appendix B  Other Representations of Rotations

## .1    Euler Angles

As we established earlier, the orientation of a rigid body can be parametrized by three independent coordinates. For example, consider a rigid body with a body frame {b} attached to it, initially aligned with the space frame {s}. Now rotate the body by $\alpha$ about the body $\hat{z}_b$-axis, then by $\beta$ about the body $\hat{y}_b$-axis, and finally by $\gamma$ about the body $\hat{x}_b$-axis. Then $(\alpha, \beta, \gamma)$ are the **ZYX Euler angles** representing the final orientation of the body (see Figure B.1). If the successive rotations are made with respect to the body frame, the result corresponds to the final rotation matrix

$$R(\alpha, \beta, \gamma) = I \, \text{Rot}(\hat{z}, \alpha) \, \text{Rot}(\hat{y}, \beta) \, \text{Rot}(\hat{x}, \gamma),$$

where

$$\text{Rot}(\hat{z}, \alpha) = \begin{bmatrix} \cos\alpha & -\sin\alpha & 0 \\ \sin\alpha & \cos\alpha & 0 \\ 0 & 0 & 1 \end{bmatrix}, \qquad \text{Rot}(\hat{y}, \beta) = \begin{bmatrix} \cos\beta & 0 & \sin\beta \\ 0 & 1 & 0 \\ -\sin\beta & 0 & \cos\beta \end{bmatrix},$$

$$\text{Rot}(\hat{x}, \gamma) = \begin{bmatrix} 1 & 0 & 0 \\ 0 & \cos\gamma & -\sin\gamma \\ 0 & \sin\gamma & \cos\gamma \end{bmatrix}.$$

Writing out the entries explicitly, we get

$$R(\alpha, \beta, \gamma) = \begin{bmatrix} c_\alpha c_\beta & c_\alpha s_\beta s_\gamma - s_\alpha c_\gamma & c_\alpha s_\beta c_\gamma + s_\alpha s_\gamma \\ s_\alpha c_\beta & s_\alpha s_\beta s_\gamma + c_\alpha c_\gamma & s_\alpha s_\beta c_\gamma - c_\alpha s_\gamma \\ -s_\beta & c_\beta s_\gamma & c_\beta c_\gamma \end{bmatrix}, \qquad (\text{B.1})$$

where $s_\alpha$ is shorthand for $\sin\alpha$, $c_\alpha$ for $\cos\alpha$, etc.

We now ask the following question: given an arbitrary rotation matrix $R$, does there exist $(\alpha, \beta, \gamma)$ satisfying Equation (B.1)? In other words, can the ZYX Euler angles represent all orientations? The answer is yes, and we prove this fact constructively as follows. Let $r_{ij}$ be the $(i, j)$th element of $R$. Then, from Equation (B.1), we know that $r_{11}^2 + r_{21}^2 = \cos^2\beta$; as long as $\cos\beta \neq 0$, or equivalently $\beta \neq \pm 90°$, we have two possible solutions for $\beta$:

$$\beta = \text{atan2}\left(-r_{31}, \sqrt{r_{11}^2 + r_{21}^2}\right)$$

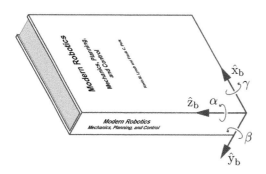

**Figure B.1** To understand the ZYX Euler angles, use the corner of a box or a book as the body frame. The ZYX Euler angles correspond to successive rotations of the body about the $\hat{z}_b$-axis by $\alpha$, the $\hat{y}_b$-axis by $\beta$, and the $\hat{x}_b$-axis by $\gamma$.

and

$$\beta = \text{atan2}\left(-r_{31}, -\sqrt{r_{11}^2 + r_{21}^2}\right).$$

(The atan2 two-argument arctangent is described at the beginning of Chapter 6.) In the first case $\beta$ lies in the range $[-90°, 90°]$, while in the second case it lies in the range $[90°, 270°]$. Assuming that the $\beta$ obtained above is not $\pm 90°$, $\alpha$ and $\gamma$ can then be determined from the following relations:

$$\alpha = \text{atan2}(r_{21}, r_{11}),$$

$$\gamma = \text{atan2}(r_{32}, r_{33}).$$

In the event that $\beta = \pm 90°$, there exists a one-parameter family of solutions for $\alpha$ and $\gamma$. This is most easily seen from Figure B.3. If $\beta = 90°$ then $\alpha$ and $\gamma$ represent rotations (in the opposite direction) about the same vertical axis. Hence, if $(\alpha, \beta, \gamma) = (\bar{\alpha}, 90°, \bar{\gamma})$ is a solution for a given rotation $R$ then any triple $(\bar{\alpha}', 90°, \bar{\gamma}')$, where $\bar{\alpha}' - \bar{\gamma}' = \bar{\alpha} - \bar{\gamma}$, is also a solution.

### B.1.1    Algorithm for Computing the ZYX Euler Angles

Given $R \in SO(3)$, we wish to find angles $\alpha, \gamma \in (-\pi, \pi]$ and $\beta \in [-\pi/2, \pi/2)$ that satisfy

$$R = \begin{bmatrix} c_\alpha c_\beta & c_\alpha s_\beta s_\gamma - s_\alpha c_\gamma & c_\alpha s_\beta c_\gamma + s_\alpha s_\gamma \\ s_\alpha c_\beta & s_\alpha s_\beta s_\gamma + c_\alpha c_\gamma & s_\alpha s_\beta c_\gamma - c_\alpha s_\gamma \\ -s_\beta & c_\beta s_\gamma & c_\beta c_\gamma \end{bmatrix}. \tag{B.2}$$

Denote by $r_{ij}$ the $(i, j)$th entry of $R$.

(a) If $r_{31} \neq \pm 1$, set

$$\beta = \text{atan2}\left(-r_{31}, \sqrt{r_{11}^2 + r_{21}^2}\right), \tag{B.3}$$

$$\alpha = \text{atan2}(r_{21}, r_{11}), \tag{B.4}$$

$$\gamma = \text{atan2}(r_{32}, r_{33}), \tag{B.5}$$

where the square root is taken to be positive.

(b) If $r_{31} = -1$ then $\beta = \pi/2$, and a one-parameter family of solutions for $\alpha$ and $\gamma$ exists. One possible solution is $\alpha = 0$ and $\gamma = \text{atan2}(r_{12}, r_{22})$.

(c) If $r_{31} = 1$ then $\beta = -\pi/2$, and a one-parameter family of solutions for $\alpha$ and $\gamma$ exists. One possible solution is $\alpha = 0$ and $\gamma = -\text{atan2}(r_{12}, r_{22})$.

## .1.2 Other Euler Angle Representations

The ZYX Euler angles can be visualized using the wrist mechanism shown in Figure B.2. The ZYX Euler angles $(\alpha, \beta, \gamma)$ refer to the angle of rotation about the three joint axes of this mechanism. In the figure the wrist mechanism is shown in its zero position, i.e., when all three joints are set to zero.

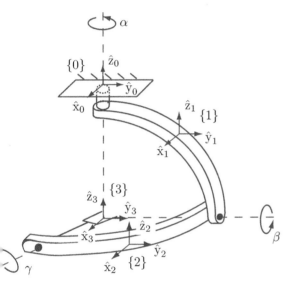

**Figure B.2** Wrist mechanism illustrating the ZYX Euler angles.

Four reference frames are defined as follows: frame $\{0\}$ is the fixed frame, while frames $\{1\}$, $\{2\}$, and $\{3\}$ are attached to the three links of the wrist mechanism as shown. When the wrist is in the zero position, all four reference frames have the same orientation. At the joint angles $(\alpha, \beta, \gamma)$, frame $\{1\}$ relative to $\{0\}$ is $R_{01}(\alpha) = \text{Rot}(\hat{z}, \alpha)$, and similarly $R_{12}(\beta) = \text{Rot}(\hat{y}, \beta)$ and $R_{23}(\gamma) = \text{Rot}(\hat{x}, \gamma)$. Therefore $R_{03}(\alpha, \beta, \gamma) = \text{Rot}(\hat{z}, \alpha)\,\text{Rot}(\hat{y}, \beta)\,\text{Rot}(\hat{x}, \gamma)$ as in Equation (B.1).

It should be evident that the choice of zero position for $\beta$ is, in some sense, arbitrary. That is, we could just as easily have defined the home position of the wrist mechanism to be as in Figure B.3; this would then lead to another three-parameter representation $(\alpha, \beta, \gamma)$ for $SO(3)$. In fact, Figure B.3 illustrates the **ZYZ Euler angles**. The resulting rotation matrix can be obtained via the following sequence of rotations, equivalent to rotating the body in Figure B.1

first about the body's $\hat{z}_b$-axis, then about the $\hat{y}_b$-axis, then about the $\hat{z}_b$-axis:

$$R(\alpha, \beta, \gamma) = \text{Rot}(\hat{z}, \alpha)\text{Rot}(\hat{y}, \beta)\text{Rot}(\hat{z}, \gamma)$$

$$= \begin{bmatrix} c_\alpha & -s_\alpha & 0 \\ s_\alpha & c_\alpha & 0 \\ 0 & 0 & 1 \end{bmatrix} \begin{bmatrix} c_\beta & 0 & s_\beta \\ 0 & 1 & 0 \\ -s_\beta & 0 & c_\beta \end{bmatrix} \begin{bmatrix} c_\gamma & -s_\gamma & 0 \\ s_\gamma & c_\gamma & 0 \\ 0 & 0 & 1 \end{bmatrix}$$

$$= \begin{bmatrix} c_\alpha c_\beta c_\gamma - s_\alpha s_\gamma & -c_\alpha c_\beta s_\gamma - s_\alpha c_\gamma & c_\alpha s_\beta \\ s_\alpha c_\beta c_\gamma + c_\alpha s_\gamma & -s_\alpha c_\beta s_\gamma + c_\alpha c_\gamma & s_\alpha s_\beta \\ -s_\beta c_\gamma & s_\beta s_\gamma & c_\beta \end{bmatrix}. \tag{B.6}$$

Just as before, we can show that for every rotation $R \in SO(3)$, there exists a triple $(\alpha, \beta, \gamma)$ that satisfies $R = R(\alpha, \beta, \gamma)$ for $R(\alpha, \beta, \gamma)$ as given in Equation (B.6). (Of course, the resulting formulas will differ from those for the ZYX Euler angles.)

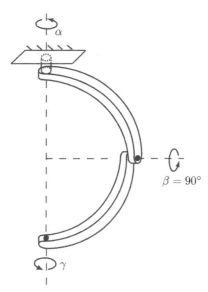

$\beta = 90°$

**Figure B.3** Configuration corresponding to $\beta = 90°$ for ZYX Euler angles.

From the wrist mechanism interpretation of the ZYX and ZYZ Euler angles, it should be evident that, for Euler-angle parametrizations of $SO(3)$, what really matters is that rotation axis 1 is orthogonal to rotation axis 2, and that rotation axis 2 is orthogonal to rotation axis 3 (axes 1 and 3 need not necessarily be orthogonal to each other). Specifically, any sequence of rotations of the form

$$\text{Rot}(\text{axis } 1, \alpha)\text{Rot}(\text{axis } 2, \beta)\text{Rot}(\text{axis } 3, \gamma), \tag{B.7}$$

where axis 1 is orthogonal to axis 2, and axis 2 is orthogonal to axis 3, can serve as a valid three-parameter representation for $SO(3)$. The angle of rotation for the first and third rotations ranges in value over a $2\pi$ interval, while that of the second rotation ranges in value over an interval of length $\pi$.

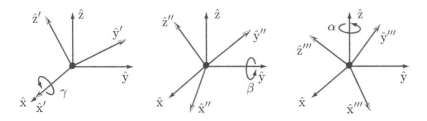

**Figure B.4** Illustration of XYZ roll–pitch–yaw angles.

## .2 Roll–Pitch–Yaw Angles

While Euler angles refer to the angles in a sequence of rotations in a body-fixed frame, the **roll–pitch–yaw angles** refer to the angles in a sequence of rotations about axes fixed in the space frame. Referring to Figure B.4, given a frame in the identity configuration (that is, $R = I$), we first rotate this frame by an angle $\gamma$ about the $\hat{x}$-axis of the fixed frame, then by an angle $\beta$ about the $\hat{y}$-axis of the fixed frame, and finally by an angle $\alpha$ about the $\hat{z}$-axis of the fixed frame.

Since the three rotations are in the fixed frame, the final orientation is

$$R(\alpha, \beta, \gamma) = \text{Rot}(\hat{z}, \alpha)\text{Rot}(\hat{y}, \beta)\text{Rot}(\hat{x}, \gamma)I$$

$$= \begin{bmatrix} c_\alpha & -s_\alpha & 0 \\ s_\alpha & c_\alpha & 0 \\ 0 & 0 & 1 \end{bmatrix} \begin{bmatrix} c_\beta & 0 & s_\beta \\ 0 & 1 & 0 \\ -s_\beta & 0 & c_\beta \end{bmatrix} \begin{bmatrix} 1 & 0 & 0 \\ 0 & c_\gamma & -s_\gamma \\ 0 & s_\gamma & c_\gamma \end{bmatrix} I$$

$$= \begin{bmatrix} c_\alpha c_\beta & c_\alpha s_\beta s_\gamma - s_\alpha c_\gamma & c_\alpha s_\beta c_\gamma + s_\alpha s_\gamma \\ s_\alpha c_\beta & s_\alpha s_\beta s_\gamma + c_\alpha c_\gamma & s_\alpha s_\beta c_\gamma - c_\alpha s_\gamma \\ -s_\beta & c_\beta s_\gamma & c_\beta c_\gamma \end{bmatrix}. \tag{B.8}$$

This product of three rotations is exactly the same as that for the ZYX Euler angles given in (B.2). We see that the same product of three rotations admits two different physical interpretations: as a sequence of rotations with respect to the body frame (ZYX Euler angles) or, reversing the order of the rotations, as a sequence of rotations with respect to the fixed frame (the XYZ roll–pitch–yaw angles).

The terms roll, pitch, and yaw are often used to describe the rotational motion of a ship or aircraft. In the case of a typical fixed-wing aircraft, for example, suppose a body frame is attached such that the $\hat{x}$-axis is in the direction of forward motion, the $\hat{z}$-axis is the vertical axis pointing downward toward ground (assuming the aircraft is flying level with respect to ground), and the $\hat{y}$-axis extends in the direction of the wing. The roll, pitch, and yaw motions are then defined according to the XYZ roll–pitch–yaw angles $(\alpha, \beta, \gamma)$ of Equation (B.8).

## B.3     Unit Quaternions

One disadvantage of the exponential coordinates on $SO(3)$ is that, because of the division by $\sin\theta$ in the logarithm formula, the logarithm can be numerically sensitive to small rotation angles $\theta$. The necessary singularity of the three-parameter representation occurs at $R = I$. The **unit quaternions** are an alternative representation of rotations that alleviates this singularity, but at the cost of having a fourth variable in the representation. We now illustrate the definition and use of these coordinates.

Let $R \in SO(3)$ have the exponential coordinate representation $\hat{\omega}\theta$, i.e., $R = e^{[\hat{\omega}]\theta}$, where as usual $\|\hat{\omega}\| = 1$ and $\theta \in [0, \pi]$. The unit quaternion representation of $R$ is constructed as follows. Define $q \in \mathbb{R}^4$ according to

$$q = \begin{bmatrix} q_0 \\ q_1 \\ q_2 \\ q_3 \end{bmatrix} = \begin{bmatrix} \cos(\theta/2) \\ \hat{\omega}\sin(\theta/2) \end{bmatrix} \in \mathbb{R}^4. \tag{B.9}$$

As defined, $q$ clearly satisfies $\|q\| = 1$. Geometrically, $q$ is a point lying on the three-dimensional unit sphere in $\mathbb{R}^4$, and for this reason the unit quaternions are also identified with the 3-sphere, denoted $S^3$. Naturally, among the four coordinates of $q$, only three can be chosen independently. Recalling that $1 + 2\cos\theta = \text{tr } R$, and using the cosine double-angle formula $\cos 2\phi = 2\cos^2\phi - 1$, the elements of $q$ can be obtained directly from the entries of $R$ as follows:

$$q_0 = \frac{1}{2}\sqrt{1 + r_{11} + r_{22} + r_{33}}, \tag{B.10}$$

$$\begin{bmatrix} q_1 \\ q_2 \\ q_3 \end{bmatrix} = \frac{1}{4q_0}\begin{bmatrix} r_{32} - r_{23} \\ r_{13} - r_{31} \\ r_{21} - 2_{12} \end{bmatrix}. \tag{B.11}$$

Going the other way, given a unit quaternion $(q_0, q_1, q_2, q_3)$ the corresponding rotation matrix $R$ is obtained as a rotation about the unit axis, in the direction of $(q_1, q_2, q_3)$, by an angle $2\cos^{-1} q_0$. Explicitly,

$$R = \begin{bmatrix} q_0^2 + q_1^2 - q_2^2 - q_3^2 & 2(q_1q_2 - q_0q_3) & 2(q_0q_2 + q_1q_3) \\ 2(q_0q_3 + q_1q_2) & q_0^2 - q_1^2 + q_2^2 - q_3^2 & 2(q_2q_3 - q_0q_1) \\ 2(q_1q_3 - q_0q_2) & 2(q_0q_1 + q_2q_3) & q_0^2 - q_1^2 - q_2^2 + q_3^2 \end{bmatrix}. \tag{B.12}$$

From the above explicit formula it should be apparent that both $q \in S^3$ and its antipodal point $-q \in S^3$ produce the same rotation matrix $R$: for every rotation matrix there exists two unit-quaternion representations that are antipodal to each other.

The final property of the unit quaternions concerns the product of two rotations. Let $R_q, R_p \in SO(3)$ denote two rotation matrices, with unit-quaternion representations $\pm q, \pm p \in S^3$, respectively. The unit-quaternion representation for the product $R_q R_p$ can then be obtained by first arranging the elements of $q$ and $p$ in the form of the following $2 \times 2$ complex matrices:

$$
Q = \begin{bmatrix} q_0 + iq_1 & q_2 + ip_3 \\ -q_2 + iq_3 & q_0 - iq_1 \end{bmatrix}, \qquad P = \begin{bmatrix} p_0 + ip_1 & p_2 + ip_3 \\ -p_2 + ip_3 & p_0 - ip_1 \end{bmatrix}, \qquad \text{(B.13)}
$$

where $i$ denotes the imaginary unit. Now take the product $N = QP$, where the entries of $N$ are given by

$$
N = \begin{bmatrix} n_0 + in_1 & n_2 + in_3 \\ -n_2 + in_3 & n_0 - in_1 \end{bmatrix}. \qquad \text{(B.14)}
$$

The unit quaternion for the product $R_q R_p$ is then $\pm(n_0, n_1, n_2, n_3)$, obtained from the entries of $N$:

$$
\begin{bmatrix} n_0 \\ n_1 \\ n_2 \\ n_3 \end{bmatrix} = \begin{bmatrix} q_0 p_0 - q_1 p_1 - q_2 p_2 - q_3 p_3 \\ q_0 p_1 + p_0 q_1 + q_2 p_3 - q_3 p_2 \\ q_0 p_2 + p_0 q_2 - q_1 p_3 + q_3 p_1 \\ q_0 p_3 + p_0 q_3 + q_1 p_2 - q_2 p_1 \end{bmatrix}. \qquad \text{(B.15)}
$$

## .4     Cayley–Rodrigues Parameters

The Cayley–Rodrigues parameters form another set of widely used local coordinates for $SO(3)$. These parameters can be obtained from the exponential representation on $SO(3)$ as follows: given $R = e^{[\hat{\omega}]\theta}$ for some unit vector $\hat{\omega}$ and angle $\theta$, the Cayley–Rodrigues parameters $r \in \mathbb{R}^3$ are obtained by setting

$$
r = \hat{\omega} \tan \frac{\theta}{2}. \qquad \text{(B.16)}
$$

Referring again to the radius-$\pi$ solid-ball picture of $SO(3)$ (Figure 3.13), the above parametrization has the effect of infinitely "stretching" the radius of this ball via the tangent half-angle function. These parameters can be derived from a general formula attributed to Cayley that is also valid for rotation matrices of arbitrary dimension: if $R \in SO(3)$ such that $\operatorname{tr} R \neq -1$ then $(I - R)(I + R)^{-1}$ is skew symmetric. Denoting this skew-symmetric matrix by $[r]$, it is known that $R$ and $[r]$ are related as follows:

$$
R = (I - [r])(I + [r])^{-1}, \qquad \text{(B.17)}
$$
$$
[r] = (I - R)(I + R)^{-1}. \qquad \text{(B.18)}
$$

The above two formulas establish a one-to-one correspondence between $so(3)$ and those elements of $SO(3)$ with trace not equal to $-1$. In the event that tr $R = -1$, the following alternative formulas can be used to relate $SO(3)$ (this time excluding those with unit trace) and $so(3)$ in a one-to-one fashion:

$$R = -(I - [r])(I + [r])^{-1}, \tag{B.19}$$

$$[r] = (I + R)(I - R)^{-1} \tag{B.20}$$

Furthermore, Equation (B.18) can be explicitly computed as

$$R = \frac{(1 - r^{\mathrm{T}} r)I + 2rr^{\mathrm{T}} + 2[r]}{1 + r^{\mathrm{T}} r} \tag{B.21}$$

with its inverse mapping given by

$$[r] = \frac{R - R^{\mathrm{T}}}{1 + \operatorname{tr} R}. \tag{B.22}$$

(This formula is valid when $\operatorname{tr} R \neq -1$). The vector $r = 0$ therefore corresponds to the identity matrix, and $-r$ represents the inverse of the rotation corresponding to $r$.

The following two identities also follow from the above formulas:

$$1 + \operatorname{tr} R = \frac{4}{1 + r^{\mathrm{T}} r}, \tag{B.23}$$

$$R - R^{\mathrm{T}} = \frac{4[r]}{1 + r^{\mathrm{T}} r}. \tag{B.24}$$

An attractive feature of the Cayley–Rodrigues parameters is the particularly simple form for the composition of two rotation matrices. If $r_1$ and $r_2$ denote the Cayley–Rodrigues parameters for two rotations $R_1$ and $R_2$, respectively, then the Cayley–Rodrigues parameters for $R_3 = R_1 R_2$, denoted $r_3$, are given by

$$r_3 = \frac{r_1 + r_2 + (r_1 \times r_2)}{1 - r_1^{\mathrm{T}} r_2} \tag{B.25}$$

In the event that $r_1^{\mathrm{T}} r_2 = 1$, or equivalently $\operatorname{tr}(R_1 R_2) = -1$, the following alternative composition formula can be used. Define

$$s = \frac{r}{\sqrt{1 + r^{\mathrm{T}} r}} \tag{B.26}$$

so that the rotation corresponding to $r$ can be written

$$R = I + 2\sqrt{1 - s^{\mathrm{T}} s}\,[s] + 2[s]^2. \tag{B.27}$$

The direction of $s$ coincides with that of $r$, and $\|s\| = \sin(\theta/2)$. The composition law now becomes

$$s_3 = s_1 \sqrt{1 - s_2^{\mathrm{T}} s_2} + s_2 \sqrt{1 - s_1^{\mathrm{T}} s_1} + (s_1 \times s_2) \tag{B.28}$$

Angular velocities and accelerations also admit a simple form in terms of the

Cayley–Rodrigues parameters. If $r(t)$ denotes the Cayley–Rodrigues representation of the orientation trajectory $R(t)$ then, in vector form,

$$\omega_s = \frac{2}{1 + \|r\|^2}(r \times \dot{r} + \dot{r}), \tag{B.29}$$

$$\omega_b = \frac{2}{1 + \|r\|^2}(-r \times \dot{r} + \dot{r}). \tag{B.30}$$

The angular acceleration with respect to the fixed and body frames can now be obtained by time-differentiating the above expressions:

$$\dot{\omega}_s = \frac{2}{1 + \|r\|^2}(r \times \ddot{r} + \ddot{r} - r^{\mathrm{T}}\dot{r}\,\omega_s), \tag{B.31}$$

$$\dot{\omega}_b = \frac{2}{1 + \|r\|^2}(-r \times \ddot{r} + \ddot{r} - r^{\mathrm{T}}\dot{r}\,\omega_b). \tag{B.32}$$

# Appendix C  Denavit–Hartenberg Parameters

The basic idea underlying the Denavit–Hartenberg approach to forward kinematics is to attach reference frames to each link of the open chain and then to derive the forward kinematics from the knowledge of the relative displacements between adjacent link frames. Assume that a fixed reference frame has been established and that a reference frame (the end-effector frame) has been attached to some point on the last link of the open chain. For a chain consisting of $n$ one-degree-of-freedom joints, the links are numbered sequentially from 0 to $n$: the ground link is labeled 0, and the end-effector frame is attached to link $n$. Reference frames attached to the links are also correspondingly labeled from $\{0\}$ (the fixed frame) to $\{n\}$ (the end-effector frame). The joint variable corresponding to the $i$th joint is denoted $\theta_i$. The forward kinematics of the $n$-link open chain can then be expressed as

$$T_{0n}(\theta_1,\ldots,\theta_n) = T_{01}(\theta_1)T_{12}(\theta_2)\cdots T_{n-1,n}(\theta_n), \tag{C.1}$$

where $T_{i,i-1} \in SE(3)$ denotes the relative displacement between link frames $\{i-1\}$ and $\{i\}$. Depending on how the link reference frames have been chosen, each $T_{i-1,i}$ can be obtained in a straightforward fashion.

## C.1    Assigning Link Frames

Rather than attaching reference frames to each link in an arbitrary fashion, in the Denavit–Hartenberg convention a set of rules for assigning link frames is observed. Figure C.1 illustrates the frame-assignment convention for two adjacent revolute joints $i-1$ and $i$ that are connected by link $i-1$.

The first rule is that the $\hat{z}_i$-axis coincides with joint axis $i$ and the $\hat{z}_{i-1}$-axis coincides with joint axis $i-1$. The direction of positive rotation about each link's $\hat{z}$-axis is determined by the right-hand rule.

Once the $\hat{z}$-axis-direction has been assigned, the next rule determines the origin of the link reference frame. First, find the line segment that orthogonally intersects both the joint axes $\hat{z}_{i-1}$ and $\hat{z}_i$. For now let us assume that this line segment is unique; the case where it is not unique (i.e., when the two joint axes are parallel), or fails to exist (i.e., when the two joint axes intersect), is addressed later. Connecting joint axes $i-1$ and $i$ by a mutually perpendicular line, the

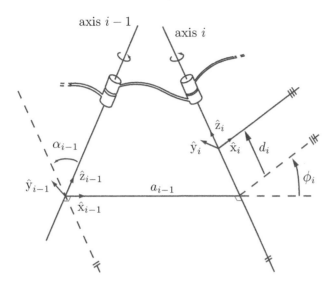

**Figure C.1** Illustration of the Denavit–Hartenberg parameters.

origin of frame $\{i-1\}$ is then located at the point where this line intersects joint axis $i-1$.

Determining the remaining $\hat{x}$- and $\hat{y}$-axes of each link reference frame is now straightforward: the $\hat{x}$-axis is chosen to be in the direction of the mutually perpendicular line pointing from the $(i-1)$-axis to the $i$-axis. The $\hat{y}$-axis is then uniquely determined from the cross product $\hat{x} \times \hat{y} = \hat{z}$. Figure C.1 depicts the link frames $\{i\}$ and $\{i-1\}$ chosen according to this convention.

Having assigned reference frames in this fashion for links $i$ and $i-1$, we now define four parameters that exactly specify $T_{i-1,i}$:

- The length of the mutually perpendicular line, denoted by the scalar $a_{i-1}$, is called the **link length** of link $i-1$. Despite its name, this link length does not necessarily correspond to the actual length of the physical link.
- The **link twist** $\alpha_{i-1}$ is the angle from $\hat{z}_{i-1}$ to $\hat{z}_i$, measured about $\hat{x}_{i-1}$.
- The **link offset** $d_i$ is the distance from the intersection of $\hat{x}_{i-1}$ and $\hat{z}_i$ to the origin of the link-$i$ frame (the positive direction is defined to be along the $\hat{z}_i$-axis).
- The **joint angle** $\phi_i$ is the angle from $\hat{x}_{i-1}$ to $\hat{x}_i$, measured about the $\hat{z}_i$-axis.

These parameters constitute the Denavit–Hartenberg parameters (D–H parameters). For an open chain with $n$ one-degree-of-freedom joints, the $4n$ D–H parameters are sufficient to completely describe the forward kinematics. In the case of an open chain with all joints revolute, the link lengths $a_{i-1}$, twists $\alpha_{i-1}$, and offset parameters $d_i$ are all constant, while the joint angle parameters $\phi_i$ act as the joint variables.

We now consider the cases where the mutually perpendicular line is undefined

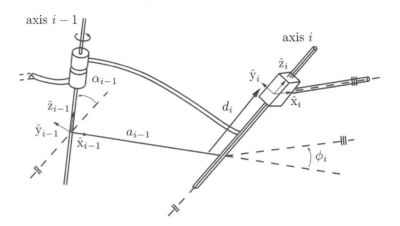

**Figure C.2** Link frame assignment convention for prismatic joints. Joint $i - 1$ is a revolute joint, while joint $i$ is a prismatic joint.

or fails to be unique, or where some of the joints are prismatic; finally, we consider how to choose the ground and end-effector frames.

### When Adjacent Revolute Joint Axes Intersect

If two adjacent revolute joint axes intersect each other then a mutually perpendicular line between the joint axes fails to exist. In this case the link length is set to zero, and we choose $\hat{x}_{i-1}$ to be perpendicular to the plane spanned by $\hat{z}_{i-1}$ and $\hat{z}_i$. There are two possibilities, both of which are acceptable: one leads to a positive value of the twist angle $\alpha_{i-1}$ while the other leads to a negative value.

### When Adjacent Revolute Joint Axes Are Parallel

The second special case occurs when two adjacent revolute joint axes are parallel. In this case there exist many possibilities for a mutually perpendicular line, all of which are valid (more precisely, a one-dimensional family of mutual perpendicular lines is said to exist). A useful guide is to try to choose the mutually perpendicular line that is the most physically intuitive and that results in as many zero parameters as possible.

### Prismatic Joints

For prismatic joints, the $\hat{z}$-direction of the link reference frame is chosen to be along the positive direction of translation. This convention is consistent with that for revolute joints, in which the $\hat{z}$-axis indicates the positive axis of rotation. With this choice the link offset $d_i$ is the joint variable and the joint angle $\phi_i$ is constant (see Figure C.2). The procedure for choosing the link-frame origin, as well as the remaining $\hat{x}$- and $\hat{y}$-axes, remains the same as for revolute joints.

### Assigning the Ground and End-Effector Frames

Our frame-assignment procedure described thus far does not specify how to choose the ground and final link frames. Here, as before, a useful guideline is

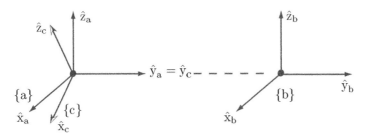

**Figure C.3** An example of three frames {a}, {b}, and {c}, for which the transformations $T_{ab}$ and $T_{ac}$ cannot be described by any set of D–H parameters.

to choose initial and final frames that are the most physically intuitive and that simplify as many D–H parameters as possible. This usually implies that the ground frame is chosen to coincide with the link-1 frame in its zero (rest) position; in the event that the joint is revolute this choice forces $a_0 = \alpha_0 = d_1 = 0$, while for a prismatic joint we have $a_0 = \alpha_0 = \phi_1 = 0$. The end-effector frame is attached to some reference point on the end-effector, usually at a location that makes the description of the task intuitive and natural and also simplifies as many of the D–H parameters as possible (e.g., their values become zero).

It is important to realize that arbitrary choices of the ground and end-effector frames may not always be possible, since there may not exist a valid set of D–H parameters to describe the relative transformation. We elaborate on this point below.

## .2 Why Four Parameters are Sufficient

In our earlier study of spatial displacements we argued that a minimum of six independent parameters were required to describe the relative displacement between two frames in space, three for the orientation and three for the position. On the basis of this result, it would seem that, for an $n$-link arm, a total of $6n$ parameters would be required to completely describe the forward kinematics (each $T_{i-1,i}$ in the above equation would require six parameters). Surprisingly, in the D–H parameter representation only four parameters are required for each transformation $T_{i-1,i}$. Although this may at first appear to contradict our earlier results, the reduction in the number of parameters is accomplished by the carefully stipulated rules for assigning link reference frames. If the link reference frames are assigned in arbitrary fashion, then more parameters are required.

Consider, for example, the link frames shown in Figure C.3. The transformation from frame {a} to frame {b} is a pure translation along the $\hat{y}$-axis of frame {a}. If one were to try to express the transformation $T_{ab}$ in terms of the D–H parameters $(\alpha, a, d, \theta)$ as prescribed above, it should become apparent that no such set of parameter values exists. Similarly, the transformation $T_{ac}$ also does not admit a description in terms of D–H parameters, as only rotations about the $\hat{x}$- and $\hat{z}$-axes are permissible. Under our D–H convention, only rotations and translations along the $\hat{x}$- and $\hat{z}$-axes are allowed, and no combination of such motions can achieve the transformation shown in Figure C.3.

Given that the D–H convention uses exactly four parameters to describe the transformation between link frames, one might naturally wonder whether the number of parameters can be reduced even further, by an even more clever set of link-frame assignment rules. Denavit and Hartenberg showed that this is not possible and that four is the minimum number of parameters (Denavit and Hartenberg, 1955).

We end this section with a reminder that there are alternative conventions for assigning link frames. Whereas we chose the $\hat{z}$-axis to coincide with the joint axis, some authors choose the $\hat{x}$-axis and reserve the $\hat{z}$-axis to be the direction of the mutually perpendicular line. To avoid ambiguities in the interpretation of the D–H parameters, it is essential to include a concise description of the link frames together with the parameter values.

## C.3     Manipulator Forward Kinematics

Once all the transformations $T_{i-1,i}$ between adjacent link frames are known in terms of their D–H parameters, the forward kinematics is obtained by sequentially multiplying these link transformations. Each link frame transformation is of the form

$$T_{i-1,i} = \text{Rot}(\hat{x}, \alpha_{i-1})\text{Trans}(\hat{x}, a_{i-1})\text{Trans}(\hat{z}, d_i)\text{Rot}(\hat{z}, \phi_i)$$
$$= \begin{bmatrix} \cos\phi_i & -\sin\phi_i & 0 & a_{i-1} \\ \sin\phi_i \cos\alpha_{i-1} & \cos\phi_i \cos\alpha_{i-1} & -\sin\alpha_{i-1} & -d_i \sin\alpha_{i-1} \\ \sin\phi_i \sin\alpha_{i-1} & \cos\phi_i \sin\alpha_{i-1} & \cos\alpha_{i-1} & d_i \cos\alpha_{i-1} \\ 0 & 0 & 0 & 1 \end{bmatrix},$$

where

$$
\text{Rot}(\hat{x}, \alpha_{i-1}) =
\begin{bmatrix}
1 & 0 & 0 & 0 \\
0 & \cos\alpha_{i-1} & -\sin\alpha_{i-1} & 0 \\
0 & \sin\alpha_{i-1} & \cos\alpha_{i-1} & 0 \\
0 & 0 & 0 & 1
\end{bmatrix},
\tag{C.2}
$$

$$
\text{Trans}(\hat{x}, a_{i-1}) =
\begin{bmatrix}
1 & 0 & 0 & a_{i-1} \\
0 & 1 & 0 & 0 \\
0 & 0 & 1 & 0 \\
0 & 0 & 0 & 1
\end{bmatrix},
\tag{C.3}
$$

$$
\text{Trans}(\hat{z}, d_i) =
\begin{bmatrix}
1 & 0 & 0 & 0 \\
0 & 1 & 0 & 0 \\
0 & 0 & 1 & d_i \\
0 & 0 & 0 & 1
\end{bmatrix},
\tag{C.4}
$$

$$
\text{Rot}(\hat{z}, \phi_i) =
\begin{bmatrix}
\cos\phi_i & -\sin\phi_i & 0 & 0 \\
\sin\phi_i & \cos\phi_i & 0 & 0 \\
0 & 0 & 1 & 0 \\
0 & 0 & 0 & 1
\end{bmatrix}.
\tag{C.5}
$$

A useful way to visualize $T_{i,i-1}$ is that it transports frame $\{i-1\}$ to frame $\{i\}$ via the following sequence of four transformations:

(a) A rotation of frame $\{i-1\}$ about its $\hat{x}$-axis by an angle $\alpha_{i-1}$.
(b) A translation of this new frame along its $\hat{x}$-axis by a distance $a_{i-1}$.
(c) A translation of the new frame formed by (b) along its $\hat{z}$-axis by a distance $d_i$.
(d) A rotation of the new frame formed by (c) about its $\hat{z}$-axis by an angle $\phi_i$.

Note that switching the order of the first and second steps will not change the final form of $T_{i-1,i}$. Similarly, the order of the third and fourth steps can also be switched without affecting $T_{i-1,i}$.

## .4 Examples

We now derive the D–H parameters for some common spatial open-chain structures.

**Example C.1** (A 3R spatial open chain)  Consider the 3R spatial open chain of Figure 4.3, shown in its zero position (i.e., with all its joint variables set to zero). The assigned link reference frames are shown in the figure, and the corresponding D–H parameters are listed in the following table:

| $i$ | $\alpha_{i-1}$ | $a_{i-1}$ | $d_i$ | $\phi_i$ |
|---|---|---|---|---|
| 1 | 0 | 0 | 0 | $\theta_1$ |
| 2 | 90° | $L_1$ | 0 | $\theta_2 - 90°$ |
| 3 | $-90°$ | $L_2$ | 0 | $\theta_3$ |

Note that frames $\{1\}$ and $\{2\}$ are uniquely specified from our frame assignment convention, but that we have some latitude in choosing frames $\{0\}$ and $\{3\}$. Here we choose the ground frame $\{0\}$ to coincide with frame $\{1\}$ (resulting in $\alpha_0 = a_0 = d_1 = 0$) and frame $\{3\}$ to be such that $\hat{x}_3 = \hat{x}_2$ (resulting in no offset to the joint angle $\theta_3$).

**Example C.2** (A spatial RRRP open chain)  The next example we consider is the four-dof RRRP spatial open chain of Figure C.4, here shown in its zero position. The link frame assignments are as shown, and the corresponding D–H parameters are listed in the figure.

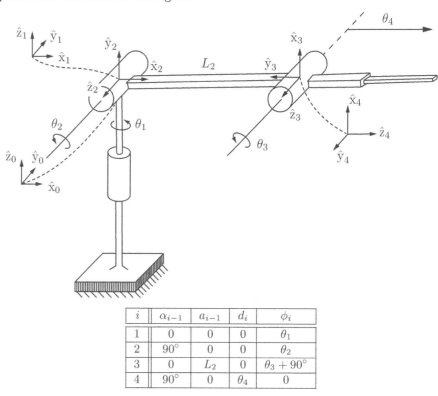

| $i$ | $\alpha_{i-1}$ | $a_{i-1}$ | $d_i$ | $\phi_i$ |
|---|---|---|---|---|
| 1 | 0 | 0 | 0 | $\theta_1$ |
| 2 | 90° | 0 | 0 | $\theta_2$ |
| 3 | 0 | $L_2$ | 0 | $\theta_3 + 90°$ |
| 4 | 90° | 0 | $\theta_4$ | 0 |

**Figure C.4** An RRRP spatial open chain.

The four joint variables are $(\theta_1, \theta_2, \theta_3, \theta_4)$, where $\theta_4$ is the displacement of the prismatic joint. As in the previous example, the ground frame $\{0\}$ and final link frame $\{4\}$ have been chosen to make as many of the D–H parameters zero as possible.

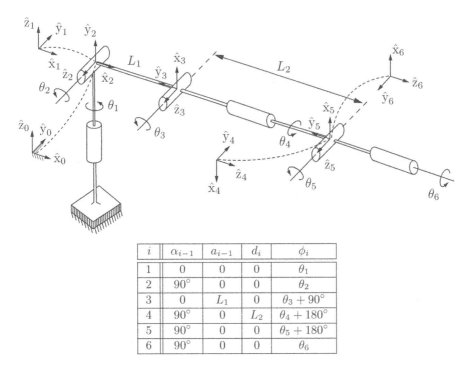

| $i$ | $\alpha_{i-1}$ | $a_{i-1}$ | $d_i$ | $\phi_i$ |
|-----|----------------|-----------|-------|----------|
| 1 | 0 | 0 | 0 | $\theta_1$ |
| 2 | 90° | 0 | 0 | $\theta_2$ |
| 3 | 0 | $L_1$ | 0 | $\theta_3 + 90°$ |
| 4 | 90° | 0 | $L_2$ | $\theta_4 + 180°$ |
| 5 | 90° | 0 | 0 | $\theta_5 + 180°$ |
| 6 | 90° | 0 | 0 | $\theta_6$ |

**Figure C.5** A 6R spatial open chain.

**Example C.3** (A spatial 6R open chain)   The final example we consider is the widely used 6R robot arm (Figure C.5). This open chain has six rotational joints: the first three joints function as a Cartesian positioning device, while the last three joints act as a ZYZ Euler angle-type wrist. The link frames are shown in the figure, and the corresponding D–H parameters are listed in the table accompanying the figure.

## .5 Relation Between the PoE and D–H Representations

The product of exponentials formula can be derived directly from the D–H parameter-based representation of the forward kinematics. As before, denote the relative displacement between adjacent link frames by

$$T_{i-1,i} = \mathrm{Rot}(\hat{x}, \alpha_{i-1})\mathrm{Trans}(\hat{x}, a_{i-1})\mathrm{Trans}(\hat{z}, d_i)\mathrm{Rot}(\hat{z}, \phi_i).$$

If joint $i$ is revolute, the first three matrices can be regarded as constant and $\phi_i$ becomes the revolute joint variable. Define $\theta_i = \phi_i$ and

$$M_i = \mathrm{Rot}(\hat{x}, \alpha_{i-1})\mathrm{Trans}(\hat{x}, a_{i-1})\mathrm{Trans}(\hat{z}, d_i), \tag{C.6}$$

and write $\text{Rot}(\hat{z}, \theta_i)$ as the following matrix exponential:

$$\text{Rot}(\hat{z}, \theta_i) = e^{[\mathcal{A}_i]\theta_i}, \qquad [\mathcal{A}_i] = \begin{bmatrix} 0 & -1 & 0 & 0 \\ 1 & 0 & 0 & 0 \\ 0 & 0 & 0 & 0 \\ 0 & 0 & 0 & 0 \end{bmatrix}. \tag{C.7}$$

With the above definitions we can write $T_{i-1,i} = M_i e^{[\mathcal{A}_i]\theta_i}$.

If joint $i$ is prismatic then $d_i$ becomes the joint variable, $\phi_i$ is a constant parameter, and the order of $\text{Trans}(\hat{z}, d_i)$ and $\text{Rot}(\hat{z}, \phi_i)$ in $T_{i-1,i}$ can be reversed (recall that reversing translations and rotations taken along the same axis still results in the same motion). In this case we can still write $T_{i-1,i} = M_i e^{[\mathcal{A}_i]\theta_i}$, where $\theta_i = d_i$ and

$$M_i = \text{Rot}(\hat{x}, \alpha_{i-1})\text{Trans}(\hat{x}, a_{i-1})\text{Rot}(\hat{z}, \phi_i), \tag{C.8}$$

$$[\mathcal{A}_i] = \begin{bmatrix} 0 & 0 & 0 & 0 \\ 0 & 0 & 0 & 0 \\ 0 & 0 & 0 & 1 \\ 0 & 0 & 0 & 0 \end{bmatrix}. \tag{C.9}$$

From the above, for an $n$-link open chain containing both revolute and prismatic joints, the forward kinematics can be written as

$$T_{0,n} = M_1 e^{[\mathcal{A}_1]\theta_1} M_2 e^{[\mathcal{A}_2]\theta_2} \cdots M_n e^{[\mathcal{A}_n]\theta_n} \tag{C.10}$$

where $\theta_i$ denotes joint variable $i$, and $[\mathcal{A}_i]$ is either of the form (C.7) or of the form (C.9), depending on whether joint $i$ is revolute or prismatic.

We now make use of the matrix identity $Me^P M^{-1} = e^{MPM^{-1}}$, which holds for any nonsingular $M \in \mathbb{R}^{n \times n}$ and arbitrary $P \in \mathbb{R}^{n \times n}$. This identity can also be rearranged as $Me^P = e^{MPM^{-1}}M$. Beginning from the left of Equation (C.10), if we repeatedly apply the identity, after $n$ iterations we obtain the product of exponentials formula as originally derived:

$$\begin{aligned} T_{0n} &= e^{M_1[\mathcal{A}_1]M_1^{-1}\theta_1}(M_1M_2)e^{[\mathcal{A}_2]\theta_2} \cdots e^{[\mathcal{A}_n]\theta_n} \\ &= e^{M_1[\mathcal{A}_1]M_1^{-1}\theta_1} e^{(M_1M_2)[\mathcal{A}_2](M_1M_2)^{-1}\theta_2}(M_1M_2M_3)e^{[\mathcal{A}_3]\theta_3} \cdots e^{[\mathcal{A}_n]\theta_n} \\ &= e^{[\mathcal{S}_1]\theta_1} \cdots e^{[\mathcal{S}_n]\theta_n} M, \end{aligned} \tag{C.11}$$

where

$$[\mathcal{S}_i] = (M_1 \cdots M_i)[\mathcal{A}_i](M_1 \cdots M_i)^{-1}, \qquad i = 1, \ldots, n, \tag{C.12}$$

$$M = M_1 M_2 \cdots M_n. \tag{C.13}$$

We now re-examine the physical meaning of the $\mathcal{S}_i$ by recalling how a screw twist transforms under a change of reference frames. If $\mathcal{S}_a$ represents the screw twist for a given screw motion with respect to frame $\{a\}$, and $\mathcal{S}_b$ represents the screw twist for the same physical screw motion but this time with respect to frame $\{b\}$, then recall that $\mathcal{S}_a$ and $\mathcal{S}_b$ are related by

$$[\mathcal{S}_b] = T_{ba}[\mathcal{S}_a]T_{ba}^{-1} \tag{C.14}$$

or, using the adjoint notation $\mathrm{Ad}_{T_{ba}}$,

$$\mathcal{S}_b = \mathrm{Ad}_{T_{ba}}(\mathcal{S}_a). \tag{C.15}$$

Seen from the perspective of this transformation rule, Equation (C.12) suggests that $\mathcal{A}_i$ is the screw twist for joint axis $i$ as seen from link frame $\{i\}$, while $\mathcal{S}_i$ is the screw twist for joint axis $i$ as seen from the fixed frame $\{0\}$.

## .6  A Final Comparison

We now summarize the relative advantages and disadvantages of the PoE formula as compared with the D–H representation. Recall that the D–H parameters constitute a minimal parameter set, i.e., only four parameters are needed to describe the transformation between adjacent link frames. However, it is necessary to assign link frames in a way such that valid D–H parameters exist; they cannot be chosen arbitrarily. The same applies when choosing the base and end-effector frames. Moreover, there is more than one convention for assigning link frames; in some conventions the link frame is attached so that the joint axis is aligned in the $\hat{x}$ rather than the $\hat{z}$-direction as we have done. Further note that for revolute joints, the joint variable is taken to be $\theta$ whereas for prismatic joints the joint variable is $d$.

Another disadvantage of the D–H parameters is that they can become ill-conditioned. For example, when adjacent joint axes are nearly parallel, the common normal between the joint axes can vary wildly with small changes in the axes' orientation. This ill-conditioned behavior of the D–H parameters makes their accurate measurement and identification difficult, since robots typically have manufacturing and other errors so that, e.g., a collection of joint axes may indeed deviate from being exactly parallel or from intersecting at a single common point.

The requirements for identifying the D–H parameters of a robot can be contrasted with those for the PoE formula. Once a zero position for the robot has been specified, and a base frame and end-effector frame established (recall that, unlike the case with D–H parameters, the base and end-effector frames can be chosen arbitrarily with no restrictions), then the product of exponentials formula is completely defined. No link reference frames are necessary and no additional bookkeeping is needed to distinguish between revolute and prismatic joints. The interpretation of the parameters in the PoE formula, as the screws representing the joint axes, is natural and intuitive. Moreover the columns of the Jacobian can also be interpreted as the (configuration-dependent) screws of the joint axes.

The only disadvantage – that the PoE representation of the joint axes uses more parameters than the D–H representation – is more than offset by the many advantages. In short, there is little practical or other reason to use the D–H parameters in modeling the forward kinematics of open chains.

# Appendix D  Optimization and Lagrange Multipliers

Suppose that $x^* \in \mathbb{R}$ is a local minimum of a twice-differentiable objective function $f(x)$, $f : \mathbb{R} \to \mathbb{R}$, in the sense that for all $x$ near $x^*$, we have $f(x) \geq f(x^*)$. We can then expect that the slope of $f(x)$ at $x^*$ is zero, i.e.,

$$\frac{\partial f}{\partial x}(x^*) = 0,$$

and also that

$$\frac{\partial^2 f}{\partial x^2}(x^*) \geq 0.$$

If $f$ is multi-dimensional, i.e., $f : \mathbb{R}^n \to \mathbb{R}$, and all partial derivatives of $f$ exist up to second-order, then a necessary condition for $x^* \in \mathbb{R}^n$ to be a local minimum is that its gradient be zero:

$$\nabla f(x^*) = \left[ \begin{array}{ccc} \dfrac{\partial f}{\partial x_1}(x^*) & \cdots & \dfrac{\partial f}{\partial x_n}(x^*) \end{array} \right]^{\mathrm{T}} = 0.$$

For example, consider the linear equation $Ax = b$, where $A \in \mathbb{R}^{m \times n}$ and $b \in \mathbb{R}^m$ $(m > n)$ are given. Because there are more constraints $(m)$ than variables $(n)$, in general a solution to $Ax = b$ will not exist. Suppose we seek the $x$ that best approximates a solution, in the sense of satisfying

$$\min_{x \in \mathbb{R}^n} f(x) = \frac{1}{2}\|Ax - b\|^2 = \frac{1}{2}(Ax - b)^{\mathrm{T}}(Ax - b) = \frac{1}{2}x^{\mathrm{T}}A^{\mathrm{T}}Ax - 2b^{\mathrm{T}}Ax + b^{\mathrm{T}}b.$$

The first-order necessary condition is given by

$$A^{\mathrm{T}}Ax - A^{\mathrm{T}}b = 0. \tag{D.1}$$

If $\operatorname{rank} A = n$ then $A^{\mathrm{T}}A \in \mathbb{R}^{n \times n}$ is invertible, and the solution to (D.1) is

$$x^* = (A^{\mathrm{T}}A)^{-1}A^{\mathrm{T}}b.$$

Now suppose that we wish to find, among all $x \in \mathbb{R}^n$ that satisfy $g(x) = 0$ for some differentiable $g : \mathbb{R}^n \to \mathbb{R}^m$ (typically $m \leq n$ to ensure that there exists an infinity of solutions to $g(x) = 0$), the $x^*$ that minimizes the objective function $f(x)$. Suppose that $x^*$ is a local minimum of $f$ that is also a regular point of the surface parametrized implicitly by $g(x) = 0$, i.e., $x^*$ satisfies $g(x^*) = 0$ and

$$\operatorname{rank} \frac{\partial g}{\partial x}(x^*) = m.$$

Then, from the fundamental theorem of linear algebra, it can be shown that there exists some $\lambda^* \in \mathbb{R}^m$ (called the **Lagrange multiplier**) that satisfies

$$\nabla f(x^*) + \frac{\partial g}{\partial x}^{\mathrm{T}}(x^*)\lambda^* = 0 \tag{D.2}$$

Equation (D.2) together with $g(x^*) = 0$ constitute the first-order necessary conditions for $x^*$ to be a feasible local mininum of $f(x)$. Note that these two equations represent $n + m$ equations in the $n + m$ unknowns $x$ and $\lambda$.

As an example, consider the quadratic objective function $f(x)$ such that

$$\min_{x \in \mathbb{R}^n} f(x) = \frac{1}{2}x^{\mathrm{T}}Qx + c^{\mathrm{T}}x,$$

subject to the linear constraint $Ax = b$, where $Q \in \mathbb{R}^n$ is symmetric positive-definite (that is, $x^{\mathrm{T}}Qx > 0$ for all $x \in \mathbb{R}^n$) and the matrix $A \in \mathbb{R}^{m \times n}$, $m \leq n$, is of maximal rank $m$. The first-order necessary conditions for this equality-constrained optimization problem are

$$Qx + A^{\mathrm{T}}\lambda = -c,$$
$$Ax = b.$$

Since $A$ is of maximal rank and $Q$ is invertible, the solutions to the first-order necessary conditions can be obtained, after some manipulation, as

$$x = Gb + (I - GA)Q^{-1}c,$$
$$\lambda = Bb + BAQ^{-1}c,$$

where $G \in \mathbb{R}^{n \times m}$ and $B \in \mathbb{R}^{m \times m}$ are defined as

$$G = Q^{-1}A^{\mathrm{T}}B, \qquad B = (AQ^{-1}A^{\mathrm{T}})^{-1}.$$

# Bibliography

Angeles, J. 2006. *Fundamentals of Robotic Mechanical Systems: Theory, Methods, and Algorithms*. Springer.

Ansari, A. R., and Murphey, T. D. 2016. Sequential action control: closed-form optimal control for nonlinear and nonsmooth systems. *IEEE Transactions on Robotics*, **32**(5), 1196–1214.

Åström, K. J., and Murray, R. M. 2008. *Feedback Systems: An Introduction for Scientists and Engineers*. Princeton University Press.

Balkcom, D. J., and Mason, M. T. 2002. Time optimal trajectories for differential drive vehicles. *International Journal of Robotics Research*, **21**(3), 199–217.

Ball, R. S. 1900. *A Treatise on the Theory of Screws* (Reprinted 1998). Cambridge University Press.

Barraquand, J., and Latombe, J.-C. 1993. Nonholonomic multibody mobile robots: controllability and motion planning in the presence of obstacles. *Algorithmica*, **10**, 121–155.

Barraquand, J., Langlois, B., and Latombe, J.-C. 1992. Numerical potential field techniques for robot path planning. *IEEE Transactions on Systems, Man, and Cybernetics*, **22**(2), 224–241.

Bejczy, A. K. 1974. Robot arm dynamics and control. Technical Memorandum 33-669. Jet Propulsion Lab.

Bellman, R., and Dreyfus, S. 1962. *Applied Dynamic Programming*. Princeton University Press.

Bicchi, A. 1995. On the closure properties of robotic grasping. *International Journal of Robotics Research*, **14**(4), 319–334.

Bicchi, A. 2000. Hands for dexterous manipulation and robust grasping: a difficult road toward simplicity. *IEEE Transactions on Robotics and Automation*, **16**(6), 652–662.

Bicchi, A., and Kumar, V. 2000. Robotic grasping and contact: a review. In: *Proc. IEEE International Conference on Robotics and Automation*.

Bloch, A. M. 2003. *Nonholonomic Mechanics and Control*. Springer.

Bobrow, J. E., Dubowsky, S., and Gibson, J. S. 1985. Time-optimal control of robotic manipulators along specified paths. *International Journal of Robotics Research*, **4**(3), 3–17.

Boissonnat, J.-D., Cérézo, A., and Leblond, J. 1994. Shortest paths of bounded curvature in the plane. *Journal of Intelligent Robotic Systems*, **11**, 5–20.

Boothby, W. M. 2002. *An Introduction to Differentiable Manifolds and Riemannian Geometry*. Academic Press.

Bottema, O., and Roth, B. 1990. *Theoretical Kinematics*. Dover Publications.

Brockett, R. W. 1983a. Asymptotic stability and feedback stabilization. In: *Differential Geometric Control Theory*, Brockett, R. W., Millman, R. S., and Sussmann, H. J. (eds.). Birkhauser.

Brockett, R. W. 1983b. Robotic manipulators and the product of exponentials formula. In: *Proc. International Symposium on the Mathematical Theory of Networks and Systems.*

Bullo, F., and Lewis, A. D. 2004. *Geometric Control of Mechanical Systems.* Springer.

Bullo, F., and Murray, R. M. 1999. Tracking for fully actuated mechanical systems: a geometric framework. *Automatica,* **35,** 17–34.

Canny, J. 1988. *The Complexity of Robot Motion Planning.* MIT Press.

Canny, J., Reif, J., Donald, B., and Xavier, P. 1988. On the complexity of kinodynamic planning. Pages 306–316 of: *Proc. IEEE Symposium on the Foundations of Computer Science.*

Ceccarelli, M. 2000. Screw axis defined by Giulio Mozzi in 1763 and early studies on helicoidal motion. *Mechanism and Machine Theory,* **35,** 761–770.

Chiaverini, S., Oriolo, G., and Maciejewski, A. A. 2016. Redundant Robots. Pages 221–242 of: *Handbook of Robotics, Second Edition,* Siciliano, B., and Khatib, O. (eds.). Springer.

Choset, H., Lynch, K. M., Hutchinson, S., Kantor, G., Burgard, W., Kavraki, L. E., and Thrun, S. 2005. *Principles of Robot Motion: Theory, Algorithms, and Implementations.* MIT Press.

Chow, W.-L. 1939. Über Systeme von linearen partiellen Differentialgleichungen erster Ordnung. *Math. Ann.,* **117,** 98–105.

Chung, W. K., Fu, L.-C., and Kröger, T. 2016. Motion control. Pages 163–194 of: *Handbook of Robotics, Second Edition,* Siciliano, B., and Khatib, O. (eds.). Springer.

Corke, P. 2017. *Robotics, Vision and Control: Fundamental Algorithms in MATLAB, Second Edition.* Springer.

Coulomb, C. A. 1781. Théorie des machines simples en ayant égard au frottement de leurs parties et à la roideur des cordages. *Mémoires des mathématique et de physique présentés à l'Académie des Sciences.*

Craig, J. 2004. *Introduction to Robotics: Mechanics and Control, Third edition.* Prentice-Hall.

de Wit, C. C., Siciliano, B., and Bastin, G. (eds.). 2012. *Theory of Robot Control.* Springer.

Denavit, J., and Hartenberg, R. S. 1955. A kinematic notation for lower-pair mechanisms based on matrices. *ASME Journal of Applied Mechanics,* **23,** 215–221.

Diftler, M. A., Mehling, J. S., Abdallah, M. E., Radford, N. A., Bridgwater, L. B., Sanders, A. M., Askew, R. S., Linn, D. M., Yamokoski, J. D., Permenter, F. A., Hargrave, B. K., Platt, R., Savely, R. T., and Ambrose, R. O. 2011. Robonaut 2 – the first humanoid robot in space. In: *Proc. IEEE International Conference on Robotics and Automation.*

di Gregorio, R., and Parenti-Castelli, V. 2002. Mobility analysis of the 3-UPU parallel mechanism assembled for a pure translational motion. *ASME Journal of Mechanical Design,* **124**(2), 259–264.

Dijkstra, E. W. 1959. A note on two problems in connexion with graphs. *Numerische Mathematik,* **1,** 269–271.

do Carmo, M. 1976. *Differential Geometry of Curves and Surfaces.* Prentice-Hall.

Donald, B. R., and Xavier, P. 1995a. Provably good approximation algorithms for optimal kinodynamic planning for Cartesian robots and open chain manipulators. *Algorithmica,* **4**(6), 480–530.

Donald, B. R., and Xavier, P. 1995b. Provably good approximation algorithms for optimal kinodynamic planning: robots with decoupled dynamics bounds. *Algorithmica,* **4**(6), 443–479.

Dubins, L. E. 1957. On curves of minimal length with a constraint on average curvature and with prescribed initial and terminal positions and tangents. *American Journal of Mathematics*, **79**, 497–516.

Duffy, J. 1990. The fallacy of modern hybrid control theory that is based on "orthogonal complements" of twist and wrench spaces. *Journal of Robotic Systems*, **7**(2), 139–144.

Erdman, A. G., and Sandor, G. N. 1996. *Advanced Mechanism Design: Analysis and Synthesis Volumes I and II*. Prentice-Hall.

Erdmann, M. A. 1994. On a representation of friction in configuration space. *International Journal of Robotics Research*, **13**(3), 240–271.

Faverjon, B. 1984. Obstacle avoidance using an octree in the configuration space of a manipulator. Pages 504–512 of: *IEEE International Conference on Robotics and Automation*.

Featherstone, R. 1983. The calculation of robot dynamics using articulated-body inertias. *International Journal of Robotics Research*, **2**(1), 13–30.

Featherstone, R. 2008. *Rigid Body Dynamics Algorithms*. Springer.

Featherstone, R., and Orin, D. 2016. Dynamics. Pages 37–66 of: *Handbook of Robotics, Second Edition*, Siciliano, B., and Khatib, O. (eds.). Springer.

Franklin, G. F., Powell, J. D., and Emami-Naeini, A. 2014. *Feedback Control of Dynamic Systems, Seventh Edition*. Pearson.

Gilbert, E. G., Johnson, D. W., and Keerthi, S. S. 1988. A fast procedure for computing the distance between complex objects in three-dimensional space. *IEEE Journal of Robotics and Automation*, **4**(2), 193–203.

Greenwood, D. T. 2006. *Advanced Dynamics*. Cambridge University Press.

Han, C., Kim, J., and Park, F. C. 2002. Kinematic sensitivity analysis of the 3-UPU parallel mechanism. *Mechanism and Machine Theory*, **37**(8), 787–798.

Hart, P. E., Nilsson, N. J., and Raphael, B. 1968. A formal basis for the heuristic determination of minimum cost paths. *IEEE Transactions on Systems Science and Cybernetics*, **4**(2), 100–107.

Herman, M. 1986. Fast, three-dimensional, collision-free motion planning. Pages 1056–1063 of: *IEEE International Conference on Robotics and Automation*.

Hirai, S., and Asada, H. 1993. Kinematics and statics of manipulation using the theory of polyhedral convex cones. *International Journal of Robotics Research*, **12**(5), 434–447.

Hogan, N. 1985a. Impedance control: an approach to manipulation: Part I – Theory. *ASME Journal of Dynamic Systems, Measurement, and Control*, **7**(Mar.), 1–7.

Hogan, N. 1985b. Impedance control: an approach to manipulation: Part II – Implementation. *ASME Journal of Dynamic Systems, Measurement, and Control*, **7**(Mar.), 8–16.

Hogan, N. 1985c. Impedance control: an approach to manipulation: Part III – Applications. *ASME Journal of Dyanmic Systems, Measurement, and Control*, **7**, 17–24.

Hollerbach, J. M. 1984. Dynamic scaling of manipulator trajectories. *ASME Journal of Dynamic Systems, Measurement, and Control*, **106**, 102–106.

Howard, S., Žefran, M., and Kumar, V. 1998. On the $6 \times 6$ Cartesian stiffness matrix for three-dimensional motions. *Mechanism and Machine Theory*, **33**(4), 389–408.

Hubbard, P. M. 1996. Approximating polyhedra with spheres for time-critical collision detection. *ACM Transactions on Graphics*, **15**(3), 179–210.

Husty, M. L. 1996. An algorithm for solving the direct kinematics of general Stewart–Gough platforms. *Mechanism and Machine Theory*, **31**(4), 365–380.

Isidori, A. 1995. *Nonlinear Control Systems*. Springer.

Jurdjevic, V. 1997. *Geometric Control Theory*. Cambridge University Press.

Kambhampati, S., and Davis, L. S. 1986. Multiresolution path planning for mobile robots. *IEEE Journal of Robotics and Automation*, **2**(3), 135–145.

Kao, I., Lynch, K. M., and Burdick, J. W. 2016. Contact modeling and manipulation. Pages 931–954 of: *Handbook of Robotics, Second Edition*, Siciliano, B., and Khatib, O. (eds.). Springer.

Karaman, S., and Frazzoli, E. 2010. Incremental sampling-based algorithms for optimal motion planning. In: *Proc. of Robotics: Science and Systems*.

Karaman, S., and Frazzoli, E. 2011. Sampling-based algorithms for optimal motion planning. *International Journal of Robotics Research*, **30**(7), 846–894.

Kavraki, L., Švestka, P., Latombe, J.-C., and Overmars, M. 1996. Probabilistic roadmaps for fast path planning in high dimensional configuration spaces. *IEEE Transactions on Robotics and Automation*, **12**, 566–580.

Kavraki, L. E., and LaValle, S. M. 2016. Motion Planning. Pages 139–161 of: *Handbook of Robotics, Second Edition*, Siciliano, B., and Khatib, O. (eds.). Springer.

Kelly, R. 1997. PD control with desired gravity compensation of robotic manipulators: a review. *International Journal of Robotics Research*, **16**(5), 660–672.

Khalil, H. K. 2014. *Nonlinear Control*. Pearson.

Khatib, O. 1986. Real-time obstacle avoidance for manipulators and mobile robots. *International Journal of Robotics Research*, **5**(1), 90–98.

Khatib, O. 1987. A unified approach to motion and force control of robot manipulators: the operational space formulation. *IEEE Journal of Robotics and Automation*, **3**(1), 43–53.

Klein, C. A., and Blaho, B. E. 1987. Dexterity measures for the design and control of kinematically redundant manipulators. *International Journal of Robotics Research*, **6**(2), 72–83.

Koditschek, D. E. 1991a. The control of natural motion in mechanical systems. *Journal of Dynamic Systems, Measurement, and Control*, **113**(Dec.), 547–551.

Koditschek, D. E. 1991b. Some applications of natural motion control. *Journal of Dynamic Systems, Measurement, and Control*, **113**(Dec.), 552–557.

Koditschek, D. E., and Rimon, E. 1990. Robot navigation functions on manifolds with boundary. *Advances in Applied Mathematics*, **11**, 412–442.

Lakshminarayana, K. *Mechanics of form closure*. ASME Rep. 78-DET-32, 1978.

Latombe, J.-C. 1991. *Robot Motion Planning*. Kluwer Academic Publishers.

Laumond, J.-P. (ed.). 1998. *Robot Motion Planning and Control*. Springer.

Laumond, J.-P., Jacobs, P. E., Taïx, M., and Murray, R. M. 1994. A motion planner for nonholonomic mobile robots. *IEEE Transactions on Robotics and Automation*, **10**(5), 577–593.

LaValle, S. M. 2006. *Planning Algorithms*. Cambridge University Press.

LaValle, S. M., and Kuffner, J. J. 1999. Randomized kinodynamic planning. In: *Proc. IEEE International Conference on Robotics and Automation*.

LaValle, S. M., and Kuffner, J. J. 2001a. Randomized kinodynamic planning. *International Journal of Robotics Research*, **20**(5), 378–400.

LaValle, S. M., and Kuffner, J. J. 2001b. Rapidly-exploring random trees: progress and prospects. In: *Algorithmic and Computational Robotics: New Directions*, Donald, B. R., Lynch, K. M., and Rus, D. (eds.). A. K. Peters.

Lee, H. Y., and Liang, C. G. 1988. A new vector theory for the analysis of spatial mechanisms. *Mechanism and Machine Theory*, **23**(3), 209–217.

Lee, S.-H., Kim, J., Park, F. C., Kim, M., and Bobrow, J. E. 2005. Newton-type algorithms for dynamics-based robot movement optimization. *IEEE Transactions on Robotics*, **21**(4), 657–667.

Li, J. W., Liu, H., and Cai, H. G. 2003. On computing three-finger force-closure grasps of 2-D and 3-D objects. *IEEE Transactions on Robotics and Automation*, **19**(1), 155–161.

Likhachev, M., Gordon, G., and Thrun, S. 2003. ARA*: Anytime A* with provable bounds on sub-optimality. In: *Advances in Neural Information Processing Systems (NIPS)*.

Liu, G., and Li, Z. 2002. A unified geometric approach to modeling and control of constrained mechanical systems. *IEEE Transactions on Robotics and Automation*, **18**(4), 574–587.

Lončarić, J. 1985. Geometrical Analysis of Compliant Mechanisms in Robotics. Ph.D. thesis, Division of Applied Sciences, Harvard University.

Lončarić, J. 1987. Normal forms of stiffness and compliance matrices. *IEEE Journal of Robotics and Automation*, **3**(6), 567–572.

Lötstedt, P. 1981. Coulomb friction in two-dimensional rigid body systems. *Zeitschrift für Angewandte Mathematik und Mechanik*, **61**, 605–615.

Lozano-Perez, T. 1980. Spatial planning: a configuration space approach. AI Memorandum 605, MIT Artificial Intelligence Laboratory.

Lozano-Perez, T. 2001. Automatic planning of manipulator transfer movements. *IEEE Transactions on Systems, Man, and Cybernetics*, **11**(10), 681–698.

Lozano-Pérez, T., and Wesley, M. A. 1979. An algorithm for planning collision-free paths among polyhedral obstacles. *Communications of the ACM*, **22**(10), 560–570.

Luenberger, D. G., and Ye, Y. 2008. *Linear and Nonlinear Programming*. Springer US.

Luh, J. Y. S., Walker, M. W., and Paul, R. P. C. 1980. Resolved-acceleration control of mechanical manipulators. *IEEE Transactions on Automatic Control*, **25**(3), 468–474.

Lynch, K. M. 2015. Underactuated robots. Pages 1503–1510 of: *Encyclopedia of Systems and Control*, Baillieul, J., and Samad, T. (eds.). Springer.

Lynch, K. M., and Mason, M. T. 1995. Pulling by pushing, slip with infinite friction, and perfectly rough surfaces. *International Journal of Robotics Research*, **14**(2), 174–183.

Lynch, K. M., and Mason, M. T. 1999. Dynamic nonprehensile manipulation: controllability, planning, and experiments. *International Journal of Robotics Research*, **18**(1), 64–92.

Lynch, K. M., Bloch, A. M., Drakunov, S. V., Reyhanoglu, M., and Zenkov, D. 2011. Control of nonholonomic and underactuated systems. In: *The Control Handbook*, Levine, W. (ed.). Taylor and Francis.

Manocha, D., and Canny, J. 1989. Real time inverse kinematics for general manipulators. Pages 383–389 of: *Proc. IEEE International Conference on Robotics and Automation*, vol. 1.

Markenscoff, X., Ni, L., and Papadimitriou, C. H. 1990. The geometry of grasping. *International Journal of Robotics Research*, **9**(1), 61–74.

Markiewicz, B. R. 1973. Analysis of the computed torque drive method and comparison with conventional position servo for a computer-controlled manipulator. Technical Memorandum 33-601. Jet Propulsion Laboratory.

Mason, M. T. 1981. Compliance and force control for computer controlled manipulators. *IEEE Transactions on Systems, Man, and Cybernetics*, **11**(June), 418–432.

Mason, M. T. 1991. Two graphical methods for planar contact problems. Pages 443–448 of: *Proc. IEEE/RSJ International Conference on Intelligent Robots and Systems*.

Mason, M. T. 2001. *Mechanics of Robotic Manipulation*. MIT Press.

Mason, M. T., and Lynch, K. M. 1993. Dynamic manipulation. Pages 152–159 of: *Proc. IEEE/RSJ International Conference on Intelligent Robots and Systems*.

Mason, M. T., and Salisbury, J. K. 1985. *Robot Hands and the Mechanics of Manipulation*. MIT Press.

Mason, M. T., and Wang, Y. 1988. On the inconsistency of rigid-body frictional planar mechanics. In: *Proc. IEEE International Conference on Robotics and Automation*.

McCarthy, J. M. 1990. *Introduction to Theoretical Kinematics*. MIT Press.

McCarthy, J. M., and Soh, G. S. 2011. *Geometric Design of Linkages*. Springer.

Mehling, J. S. 2015. Impedance Control Approaches for Series Elastic Actuators. Ph.D. thesis, Rice University.

Merlet, J.-P. 2006. *Parallel Robots*. Springer.

Merlet, J.-P., Gosselin, C., and Huang, Tian. 2016. Parallel Mechanisms. Pages 443–461 of: *Handbook of Robotics, Second Edition*, Siciliano, B., and Khatib, O. (eds.). Springer.

Meyer, C. D. 2000. *Matrix Analysis and Applied Linear Algebra*. SIAM.

Millman, R. S., and Parker, G. D. 1977. *Elements of Differential Geometry*. Prentice-Hall.

Mishra, B., Schwartz, J. T., and Sharir, M. 1987. On the existence and synthesis of multifinger positive grips. *Algorithmica*, **2**(4), 541–558.

Morin, P., and Samson, C. 2008. Motion control of wheeled mobile robots. Pages 799–826 of: *Handbook of Robotics, First Edition*, Siciliano, B., and Khatib, O. (eds.). Springer.

Murray, R., Li, Z., and Sastry, S. 1994. *A Mathematical Introduction to Robotic Manipulation*. CRC Press.

Nef, T., Guidali, M., and Riener, R. 2009. ARMin III – arm therapy exoskeleton with an ergonomic shoulder actuation. *Applied Bionics and Biomechanics*, **6**(2), 127–142.

Nevins, J. L., and Whitney, D. E. 1978. Computer-controlled assembly. *Scientific American*, **238**(2), 62–74.

Nguyen, V.-D. 1988. Constructing force-closure grasps. *International Journal of Robotics Research*, **7**(3).

Nijmeijer, H., and van der Schaft, A. J. 1990. *Nonlinear Dynamical Control Systems*. Springer.

Ohwovoriole, M. S., and Roth, B. 1981. An extension of screw theory. *Journal of Mechanical Design*, **103**(4), 725–735.

Oriolo, G. 2015. Wheeled robots. In: *Encyclopedia of Systems and Control*, Baillieul, J., and Samad, T. (eds.). Springer.

Paden, B. 1986. Kinematics and Control of Robot Manipulators. Ph.D. thesis, Department of Electrical Engineering and Computer Sciences, University of California, Berkeley.

Pang, J. S., and Trinkle, J. C. 1996. Complementarity formulations and existence of solutions of dynamic multi-rigid-body contact problems with Coulomb friction. *Mathematical Programming*, **73**(2), 199–226.

Park, F. C. 1991. Optimal Kinematic Design of Mechanisms. Ph.D. thesis, Division of Applied Sciences, Harvard University.

Park, F. C. 1994. Computational aspects of the product of exponentials formula for robot kinematics. *IEEE Transactions on Automatic Control*, **39**(3), 643–647.

Park, F. C., and Brockett, R. W. 1994. Kinematic dexterity of robotic mechanisms. *International Journal of Robotics Research*, **13**(1), 1–15.

Park, F. C., and Kang, I. G. 1999. Cubic spline algorithms for orientation interpolation. *International Journal of Numerical Methods in Engineering*, **46**, 46–54.

Park, F. C., and Kim, J. 1999. Singularity analysis of closed kinematic chains. *ASME Journal of Mechanical Design*, **121**(1), 32–38.

Park, F. C., Bobrow, J. E., and Ploen, S. R. 1995. A Lie group formulation of robot dynamics. *International Journal of Robotics Research*, **14**(6), 609–618.

Paul, R. C. 1972. Modeling trajectory calculation and servoing of a computer controlled arm. AI Memorandum 177. Stanford University Artificial Intelligence Laboratory.

Pfeiffer, F., and Johanni, R. 1987. A concept for manipulator trajectory planning. *IEEE Journal of Robotics and Automation*, **RA-3**(2), 115–123.

Pham, Q.-C. 2014. A general, fast, and robust implementation of the time-optimal path parameterization algorithm. *IEEE Transactions on Robotics*, **30**(6), 1533–1540. Code available at `https://github.com/quangounet/TOPP`.

Pham, Q.-C., and Stasse, O. 2015. Time-optimal path parameterization for redundantly actuated robots: a numerical integration approach. *IEEE/ASME Transactions on Mechatronics*, **20**(6), 3257–3263.

Pratt, G. A., and Williamson, M. M. 1995. Series elastic actuators. In: *Proc. IEEE/RSJ International Conference on Intelligent Robots and Systems*.

Prattichizzo, D., and Trinkle, J. C. 2016. Grasping. Pages 955–988 of: *Handbook of Robotics, Second Edition*, Siciliano, B., and Khatib, O. (eds.). Springer.

Raghavan, M., and Roth, B. 1990. Kinematic analysis of the 6R manipulator of general geometry. In: *Proc. International Symposium on Robotics Research*.

Raibert, M. H., and Craig, J. J. 1981. Hybrid position/force control of manipulators. *ASME Journal of Dyanmic Systems, Measurement, and Control*, **102**, 126–133.

Raibert, M. H., and Horn, B. K. P. 1978. Manipulator control using the configuration space method. *Industrial Robot*, **5**(June), 69–73.

Reeds, J. A., and Shepp, L. A. 1990. Optimal paths for a car that goes both forwards and backwards. *Pacific Journal of Mathematics*, **145**(2), 367–393.

Reuleaux, F. 1876. *The Kinematics of Machinery*. MacMillan. Reprinted by Dover, 1963.

Rimon, E., and Burdick, J. 1996. On force and form closure for multiple finger grasps. Pages 1795–1800 of: *Proc. IEEE International Conference on Robotics and Automation*.

Rimon, E., and Burdick, J. W. 1995. A configuration space analysis of bodies in contact – II. 2nd-order mobility. *Mechanism and Machine Theory*, **30**(6), 913–928.

Rimon, E., and Burdick, J. W. 1998a. Mobility of bodies in contact – Part I. A 2nd-order mobility index for multiple-finger grasps. *IEEE Transactions on Robotics and Automation*, **14**(5), 696–708.

Rimon, E., and Burdick, J. W. 1998b. Mobility of bodies in contact – Part II. How forces are generated by curvature effects. *IEEE Transactions on Robotics and Automation*, **14**(5), 709–717.

Rimon, E., and Koditschek, D. E. 1991. The construction of analytic diffeomorphisms for exact robot navigation on star worlds. *Transactions of the American Mathematical Society*, **327**, 71–116.

Rimon, E., and Koditschek, D. E. 1992. Exact robot navigation using artificial potential functions. *IEEE Transactions on Robotics and Automation*, **8**(5), 501–518.

Robonaut 2. `http://robonaut.jsc.nasa.gov/R2`. Accessed November 2, 2016.

Rohmer, E., Singh, S. P. N., and Freese, M. 2013. V-REP: A versatile and scalable robot simulation framework. In: *Proc. IEEE/RSJ International Conference on Intelligent Robots and Systems*.

Russell, S., and Norvig, P. 2009. *Artificial Intelligence: a Modern Approach Third Edition*. Pearson.

Samet, H. 1984. The quadtree and related hierarchical data structures. *Computing Surveys*, **16**(2), 187–260.

Samson, C., Morin, P., and Lenain, R. 2016. Modeling and control of wheeled mobile robots. Pages 1235–1265 of: *Handbook of Robotics, Second Edition*, Siciliano, B., and Khatib, O. (eds.). Springer.

Sastry, S. S. 1999. *Nonlinear Systems: Analysis, Stability, and Control*. Springer.

Schwartz, J. T., and Sharir, M. 1983a. On the "piano movers'" problem. I. The case of a two-dimensional rigid polygonal body moving amidst polygonal barriers. *Communications on Pure and Applied Mathematics*, **36**(3), 345–398.

Schwartz, J. T., and Sharir, M. 1983b. On the "piano movers'" problem. II. General techniques for computing topological properties of real algebraic manifolds. *Advances in Applied Mathematics*, **4**(3), 298–351.

Schwartz, J. T., and Sharir, M. 1983c. On the piano movers' problem. III. Coordinating the motion of several independent bodies: the special case of circular bodies moving amidst polygonal barriers. *International Journal of Robotics Research*, **2**(3), 46–75.

Shamir, T., and Yomdin, Y. 1988. Repeatability of redundant manipulators: mathematical solution to the problem. *IEEE Transactions on Automatic Control*, **33**(11), 1004–1009.

Shiller, Z., and Dubowsky, S. 1985. On the optimal control of robotic manipulators with actuator and end-effector constraints. Pages 614–620 of: *Proc. IEEE International Conference on Robotics and Automation*.

Shiller, Z., and Dubowsky, S. 1988. Global time optimal motions of robotic manipulators in the presence of obstacles. Pages 370–375 of: *Proc. IEEE International Conference on Robotics and Automation*.

Shiller, Z., and Dubowsky, S. 1989. Robot path planning with obstacles, actuator, gripper, and payload constraints. *International Journal of Robotics Research*, **8**(6), 3–18.

Shiller, Z., and Dubowsky, S. 1991. On computing the global time-optimal motions of robotic manipulators in the Presence of Obstacles. *IEEE Transactions on Robotics and Automation*, **7**(6), 785–797.

Shiller, Z., and Lu, H.-H. 1992. Computation of path constrained time optimal motions with dynamic ssingularities. *ASME Journal of Dynamic Systems, Measurement, and Control*, **114**(Mar.), 34–40.

Shin, K. G., and McKay, N. D. 1985. Minimum-time control of robotic manipulators with geometric path constraints. *IEEE Transactions on Automatic Control*, **30**(6), 531–541.

Shuster, M. D. 1993. A survey of attitude representations. *Journal of the Astronautical Sciences*, **41**(4), 439–517.

Siciliano, B., and Khatib, O. 2016. *Handbook of Robotics, Second Edition*. Springer.

Siciliano, B., Sciavicco, L., Villani, L., and Oriolo, G. 2009. *Robotics: Modelling, Planning and Control*. Springer.

Simunovic, S. N. 1975. Force information in assembly processes. In: *Proc. Fifth International Symposium on Industrial Robots*.

Slotine, J.-J. E., and Yang, H. S. 1989. Improving the efficiency of time-optimal path-following algorithms. *IEEE Transactions on Robotics and Automation*, **5**(1), 118–124.

Somoff, P. 1900. Uber Gebiete von Schraubengeschwindigkeiten eines starren Korpers bie verschiedner Zahl von Stutzflachen. *Z. Math. Phys.*, **45**, 245–306.

Souères, P., and Laumond, J.-P. 1996. Shortest paths synthesis for a car-like robot. *IEEE Transactions on Automatic Control*, **41**(5), 672–688.

Spong, M. W. 2015. Robot motion control. Pages 1168–1176 of: *Encyclopedia of Systems and Control*, Baillieul, J., and Samad, T. (eds.). Springer.

Spong, M. W., Hutchinson, S., and Vidyasagar, M. 2005. *Robot Modeling and Control*. Wiley.

Stewart, D. E., and Trinkle, J. C. 1996. An implicit time-stepping scheme for rigid body dynamics with inelastic collisions and Coulomb friction. *International Journal for Numerical Methods in Engineering*, **39**(15), 2673–2691.

Strang, G. 2009. *Introduction to Linear Algebra, Fourth Edition*. Wellesley–Cambridge Press.

Şucan, I. A., and Chitta, S. MoveIt! Online at http://moveit.ros.org. Accessed November 2016

Şucan, I. A., Moll, M., and Kavraki, L. E. 2012. The Open Motion Planning Library. *IEEE Robotics & Automation Magazine*, **19**(4), 72–82. http://ompl.kavrakilab.org.

Sussmann, H. J. 1987. A general theorem on local controllability. *SIAM Journal on Control and Optimization*, **25**(1), 158–194.

Sussmann, H. J., and Tang, W. 1991. Shortest paths for the Reeds–Shepp car: a worked out example of the use of geometric techniques in nonlinear optimal control. Technical Report SYCON-91-10. Rutgers University.

Takegaki, M., and Arimoto, S. 1981. A new feedback method for dynamic control of manipulators. *ASME Journal of Dyanmic Systems, Measurement, and Control*, **112**, 119–125.

Trinkle, J. C. 2003. Formulation of multibody dynamics as complementarity problems. In: *proc. ASME International Design Engineering Technical Conferences*.

Tsiotras, P., Junkins, J. L., and Schaub, H. 1997. Higher-order Cayley transforms with applications to attitude representations. *AIAA Journal of Guidance, Control, and Dynamics*, **20**(3), 528–534.

Tzorakoleftherakis, E., Ansari, A., Wilson, A., Schultz, J., and Murphey, T. D. 2016. Model-based reactive control for hybrid and high-dimensional robotic systems. *IEEE Robotics and Automation Letters*, **1**(1), 431–438.

Uno, Y., Kawato, M., and Suzuki, R. 1989. Formation and control of optimal trajectory in human multijoint arm movement. *Biological Cybernetics*, **61**, 89–101.

Vanderbroght, B., Albu-Schaeffer, A., Bicchi, A., Burdet, E., Caldwell, D. G., Carloni, R., Catalano, M., Eiberger, O., Friedl, W., Ganesh, G., Garabini, M., Grebenstein, M., Grioli, G., Haddadin, S., Hoppner, H., Jafari, A., Laffranchi, M., Lefeber, D., Petit, F., Stramigioli, S., Tsagarakis, N., Damme, M. Van, Ham, R. Van, Visser, L. C., and Wolf, S. 2013. Variable impedance actuators: a review. *Robotics and Autonomous Systems*, **61**(12), 1601–1614.

Villani, L., and De Schutter, J. 2016. Force Control. Pages 195–219 of: *Handbook of Robotics, Second Edition*, Siciliano, B., and Khatib, O. (eds.). Springer.

Vukobratović, M., and Kirćanski, M. 1982. A method for optimal synthesis of manipulation robot trajectories. *ASME Journal of Dynamic Systems, Measurement, and Control*, **104**, 188–193.

Whitney, D. E. 1982. Quasi-static assembly of compliantly supported rigid parts. *ASME Journal of Dynamic Systems, Measurement, and Control*, **104**(Mar.), 65–77.

Whittaker, E. T. 1917. *A Treatise on the Analytical Dynamics of Particles and Rigid Bodies*. Cambridge University Press.

Witkin, A., and Kass, M. 1988. Spacetime constraints. *Computer Graphics*, **22**(4), 159–168.

Yoshikawa, T. 1985. Manipulability of robotic mechanisms. *International Journal of Robotics Research*, **4**(2), 3–9.

# Index